T0328612

Sucker-Rod Pumping Handbook

Sucker-Rod Pumping Handbook

Production Engineering Fundamentals and Long-Stroke Rod Pumping

Gabor Takacs, PhD
Petroleum Engineering Department,
University of Miskolc,
Hungary

AMSTERDAM • BOSTON • HEIDELBERG • LONDON
NEW YORK • OXFORD • PARIS • SAN DIEGO
SAN FRANCISCO • SINGAPORE • SYDNEY • TOKYO

Gulf Professional Publishing is an imprint of Elsevier

Gulf Professional Publishing is an imprint of Elsevier
225 Wyman Street, Waltham, MA 02451, USA
The Boulevard, Langford Lane, Kidlington, Oxford, OX5 1GB, UK

ISBN: 978-0-12-417204-3

Library of Congress Cataloging-in-Publication Data
A catalogue record for this book is available from the Library of Congress

British Library Cataloguing-in-Publication Data
A catalogue record for this book is available from the British Library

For information on all Gulf Professional Publishing publications
visit our website at http://store.elsevier.com/

Contents

Preface

This book retains the basic structure and approach of my "Modern Sucker-Rod Pumping" published by PennWell Books in 1993 but is a completely updated and enlarged version of the original. Sucker-rod pumping, although a 150-year-old technology, still continuously evolves; new equipment, new procedures, and new optimization schemes appear every day, of which practicing engineers must be aware. I wanted to help anyone interested in this technology by presenting the state of the art of modern-day sucker-rod pumping theories and practices. For this reason I diligently researched the literature for new developments throughout the years so that no significant contribution to our industry could escape my attention. Proof of this is the total number of references cited in the text, which amounts to almost 400.

Having spent most of my career in academia and regularly teaching industry short courses on artificial lift topics, I fully understand the importance of having proper textbooks. When I was preparing this volume, one of my objectives was to provide the trainee as well as the experienced engineer with a comprehensive textbook. The book follows the systematic way of teaching new material: after covering the fundamentals of production engineering the different components of the equipment and their features are introduced, then basic design and selection procedures are covered. Advanced topics like optimization and analysis are fully explained with worked examples. The book can be used at the undergraduate and graduate levels, as well as in industrial training of basic and advanced topics.

All phases of rod pumping are introduced by presenting the technical background first; then the most significant technical and theoretical developments are fully covered. This approach ensures that the book can serve as a complete reference to its topic. Since design of individual parts and of the total pumping system is covered as well, one can utilize the book for solving most of the design problems encountered in practice.

New technological developments in the last twenty years have significantly widened the application ranges of sucker-rod pumping, and today deeper wells with greater liquid rates can be produced than ever before. In addition to new materials and innovative equipment covered in the book, I chose to present long-stroke rod pumping in a separate chapter, never published before by anyone else. The two main types of technologies available today are introduced and their technical and operational features are described in detail.

April 2015
Gabor Takacs

CHAPTER

INTRODUCTION TO SUCKER-ROD PUMPING

1

CHAPTER OUTLINE

1.1 ARTIFICIAL LIFT METHODS

Usually, oil wells in the early stages of their lives flow naturally to the surface and are called flowing wells. Flowing production means that the pressure at the well bottom is sufficient to overcome the sum of pressure losses occurring along the flow path to the separator. When this criterion is not met, natural flow ends and the well **dies**. The two main reasons for a well's dying are:

- its flowing bottom hole pressure drops below the total pressure losses in the well; or
- pressure losses in the well become greater than the bottom hole pressure needed for moving the well stream to the surface.

The first case occurs due to the removal of fluids from the underground reservoir; the second case involves an increasing flow resistance in the well. This can be caused by:

- an increase in the density of the flowing fluid as a result of decreased gas production; or
- various mechanical problems like a small tubing size, downhole restrictions, etc.

Artificial lifting methods are used to produce fluids from wells that are already dead or to increase the production rate from flowing wells. The importance of artificial lifting is clearly seen from the total

Sucker-Rod Pumping Handbook. http://dx.doi.org/10.1016/B978-0-12-417204-3.00001-7

number of installations: according to one estimate there are approximately two million oil wells worldwide, of which about 50% are placed on some kind of artificial lift [1].

There are several lifting mechanisms available for the production engineer to choose from. One widely used group of artificial lift methods uses some kind of a **pump** set below the liquid level to increase the pressure of the well stream so as to overcome the pressure losses occurring along the flow path. Other lifting methods use **compressed gas**, injected from the surface into the well tubing to help the lifting of well fluids to the surface.

Although all artificial lift methods could be grouped based on the two basic mechanisms just discussed, their traditional classification is somewhat different, as discussed in the following.

1.1.1 GAS LIFTING

All versions of gas lifting use high-pressure gas (in most cases natural gas, but other gases like N_2 or CO_2 can also be used) injected in the well stream at some downhole point. In **continuous-flow** gas lift, a steady rate of gas is injected in the well tubing, aerating the liquid and thus reducing the pressure losses occurring along the flow path. Due to the reduction of flow resistance, the well's original bottom hole pressure becomes sufficient to move the gas/liquid mixture to the surface and the well starts to flow again. Therefore, continuous-flow gas lifting can be considered as the continuation of flowing production.

In **intermittent** gas lift, gas is injected periodically into the tubing string whenever a sufficient length of liquid has accumulated at the well bottom. A relatively high volume of gas injected below the liquid column pushes that column to the surface as a slug. Gas injection is then interrupted until a new liquid slug of the proper column length builds up again. Production of well liquids, therefore, is done by cycles. The **plunger-assisted** version of intermittent gas lift, a.k.a. **plunger lift**, uses a special free plunger traveling in the well tubing and inserted just below the accumulated liquid slug in order to separate the upward-moving liquid from the gas below it. These versions of gas lift physically displace the accumulated liquids from the well, a mechanism totally different from that of continuous-flow gas lifting.

1.1.2 PUMPING

Pumping involves the use of a **downhole pump** to increase the pressure in the well to overcome the sum of flowing pressure losses occurring along the flow path up to the surface. Pumping can be further classified using several different criteria, e.g., the operational principle of the pump used. However, the generally accepted classification is based on the way the downhole pump is driven and distinguishes between rod and rodless pumping.

Rod pumping methods utilize a string of metal rods connecting the downhole pump to the surface driving mechanism which, depending on the type of pump used, generates an oscillating or rotating movement. Historically, the first kinds of pumps to be applied in water and oil wells were of the positive-displacement type, requiring an alternating vertical movement to operate. The dominant and oldest type of rod pumping is **walking-beam pumping**, or simply called **sucker-rod pumping** (SRP). It uses a positive-displacement plunger pump and its most well-known surface feature is the pumping unit featuring a pivoted **walking beam**.

The need for producing deeper and deeper wells with increased liquid volumes necessitated the evolution of **long-stroke** sucker-rod pumping. Different units were developed with the common feature of using the same pumps and rod strings as in conventional sucker-rod pumping, but with substantially longer pump stroke lengths. The desired long strokes did not permit the use of a walking beam, and completely different surface driving mechanisms had to be invented. The basic types in this class are distinguished according to the type of surface drive used: pneumatic drive, hydraulic drive, or mechanical drive long-stroke pumping.

A newly emerged rod pumping system uses a **progressing cavity pump** that requires the rod string to be rotated for its operation. This pump, like the plunger pumps used in other types of rod pumping systems, also works on the principle of positive displacement, but does not contain any valves.

Rodless pumping methods, as the name implies, do not utilize a rod string to operate the downhole pump from the surface. Accordingly, other means (other than mechanical) are used to provide energy to the downhole pump, such as electric or hydraulic. A variety of pump types can be utilized in rodless pumping installations, including centrifugal, positive displacement, or hydraulic pumps.

The most important kind of rodless pumping is **electrical submersible pumping** (ESP), utilizing a multistage centrifugal pump driven by an electrical motor, both contained in a single package and submerged below the fluid level in the well. Power is supplied to the motor by an electric cable run from the surface. Such units are ideally suited to produce high liquid volumes.

The other lifting systems in the rodless category all employ a high-pressure power fluid that is pumped down the hole. **Hydraulic pumping** was the first method developed; such units have a positive-displacement pump driven by a hydraulic engine, contained in one downhole unit. The engine or motor provides an alternating movement necessary to operate the pump section. The hydraulic **turbine-driven** pumping unit consists of a multistage turbine and a multistage centrifugal pump section connected in series. The turbine is supplied with power fluid from the surface and drives the centrifugal pump at high rotational speeds, which lifts well fluids to the surface.

Jet pumping, although it is a hydraulically driven method of fluid lifting, completely differs from the rodless pumping principles discussed so far. Its downhole equipment contains a **nozzle** through which the power fluid pumped from the surface creates a high-velocity jet stream. The kinetic energy of this jet is converted into useful work by the jet pump, lifting the commingled stream of the power fluid and the well's produced liquids to the surface. The downhole unit of a jet pump installation is the only oil-well pumping equipment known today that contains **no moving parts**.

1.1.3 ARTIFICIAL LIFT POPULATIONS

There are no reliable estimates on the distribution of each artificial lift method in the different parts of the world. One generally accepted fact is, however, that sucker-rod pumping installations are the most numerous worldwide; in the US there were about 350,000 such installations in 2007 [1].

The charts in Fig. 1.1, available from the Artificial Lift Research and Development Council (ALRDC) web page [2], present estimates on the number of different installations and their share in the world's total oil production. As seen, there is no correlation for most of the artificial lift methods between the number of installations and the liquid volumes produced.

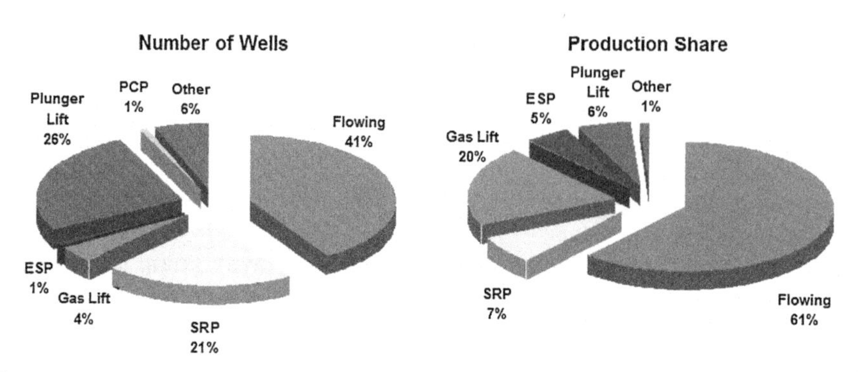

FIGURE 1.1

Estimates on the number of different oil well installations and their share in the world's total oil production.

1.2 COMPARISON OF LIFT METHODS

Although there are other types of artificial lifts in addition to those discussed so far, their importance is negligible. In any case, there is a multitude of choices available to an engineer for selecting the type of lift to be used. Although the use of many of those lifting mechanisms may be restricted or even ruled out by actual field conditions, such as well depth, production rates desired, fluid properties, etc., usually more than one lift system turns out to be technically feasible. It is then the production engineer's responsibility to select the type of lift that provides the **most profitable** way of producing the desired liquid volume from the given well(s). After a decision is made on the lifting method to be applied, a complete design of the installation for initial and future conditions should follow.

1.2.1 LIFTING CAPACITIES

In order to provide a rough comparison of the available artificial lifting methods, two figures are presented where approximate maximum **liquid production rates** of the different installations are given in the function of **lifting depth**. The lifting capacities listed are estimates based on latest information, including data from a major equipment manufacturer [3].

Figure 1.2 shows three lifting mechanisms capable of producing exceptionally high liquid rates: gas lifting, electrical submersible pumping, and jet pumping. As seen, gas lifting (continuous flow) allows the production of the greatest amounts of liquid from medium to high depths. In shallow wells like those used for water supply, however, electrical submersible pumping is capable to produce rates in excess of 60,000 bpd.

Figure 1.3, on the other hand, includes artificial lift methods of moderate liquid production capacity: hydraulic pumping, progressive cavity pumping (PCP), rod pumping, and plunger lifting. In most cases, lifting depth has a profound importance for the liquid volume lifted, and well rates rapidly decrease in deeper wells. It can be noted that sucker-rod and PCP pumps produce very similar rates from the lifting depth range of 3,000–6,000 ft; this fact, combined with the much lower investment and production costs of PCP installations, explains the great popularity of PCP pumps over rod-pumping applications in recent years.

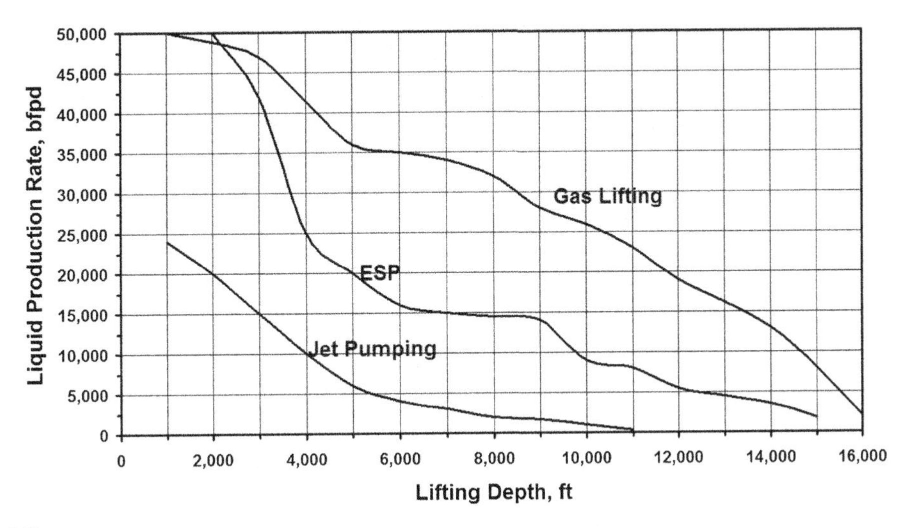

FIGURE 1.2

Maximum liquid capacities of high-capacity artificial lift installations.

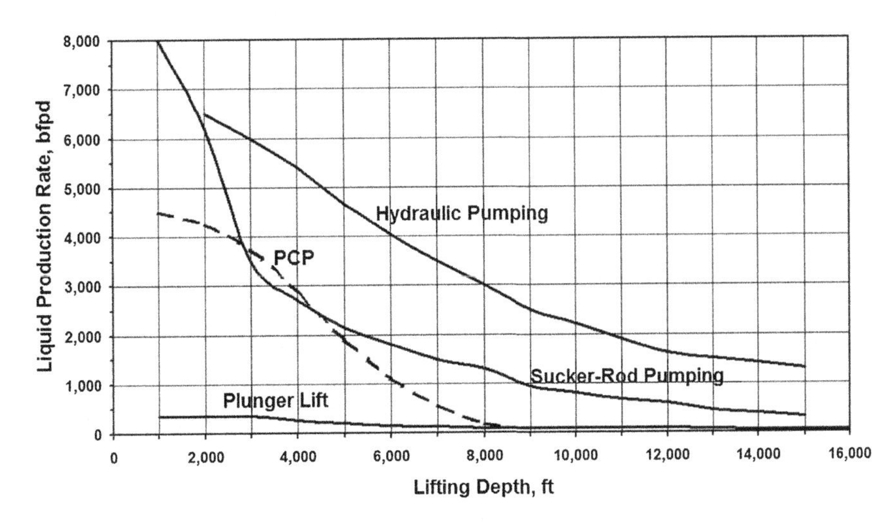

FIGURE 1.3

Maximum liquid capacities of moderate-capacity artificial lift installations.

1.2.2 SYSTEM EFFICIENCIES

The artificial lift methods available today offer very different energy efficiencies. The total efficiency of an artificial lift installation is found from the total energy required to operate the system and the hydraulic power spent on lifting the fluids to the surface. This efficiency is the product of the individual efficiencies of the system's components. The decisive part of the overall efficiency is due to the effectiveness of the lifting mechanism, e.g., the energy efficiency of the pump used, but power losses in the well and on the surface can also have a great impact on the final figure. The basic prerequisite for high total energy efficiency, therefore, is the application of a highly efficient lifting mechanism.

The most excellent device available for artificial lifting is the PCP pump, which can be more than 70% efficient in converting mechanical energy to hydraulic work. Since the use of PCP pumps in oil wells requires relatively simple surface and downhole installations with low levels of energy losses in system components, PCP systems are the most efficient among the artificial lift methods. No wonder that, wherever well conditions fall in their application ranges, the number of PCP pumping installations is growing very fast.

Next in the line, as shown in Fig. 1.4, are sucker-rod pumping and ESP installations, with maximum system efficiencies of about 60%. Although sucker-rod and ESP pumps alone can have quite high energy efficiencies, both lifting methods are plagued by high downhole losses in their power transmission system. In addition to these losses, free gas entering the pumps dramatically reduces their hydraulic output and, consequently, the overall system efficiency.

Hydraulic pumping installations utilizing positive displacement pumps usually have power efficiencies around 50%. Jet pumping and continuous-flow gas lifting are relatively low-efficiency artificial lift methods, with maximums around 30%. Intermittent gas lift has the lowest energy efficiency among the available lift methods.

1.2.3 FURTHER CONSIDERATIONS

The selection of the proper lifting method for a given well or field requires more than comparing the production capabilities and efficiencies of the possible systems. Fluid properties, field conditions, operating and investment cost estimates, and possible production problems are all to be considered before a final decision is arrived at. Table 1.1 (modified after [3]) provides help for a preliminary selection of possible candidates and for eliminating those methods not suitable for actual conditions.

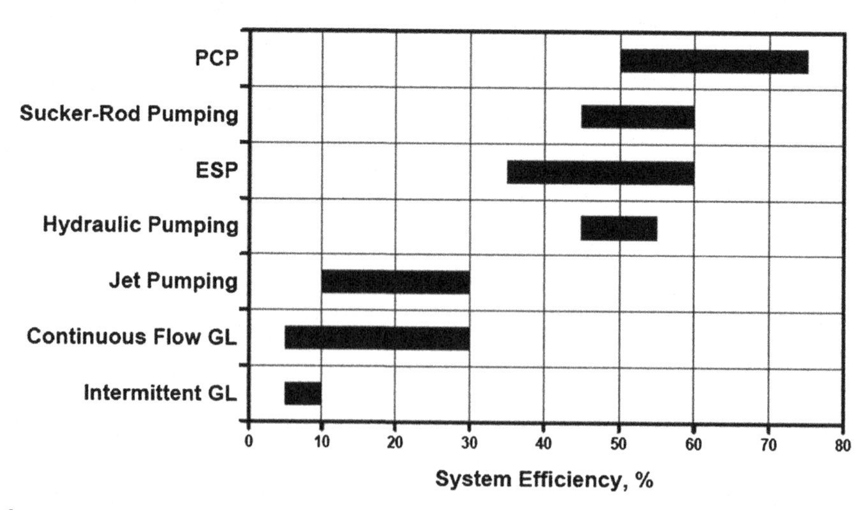

FIGURE 1.4

Approximate system efficiencies of different artificial lift methods.

Table 1.1 Main features of artificial lift installations

AI Method	SRP	Gas lifting	ESP	PCP	Hydraulic pumping	Jet pumping	Plunger lift
Max. operating depth, ft	16,000	18,000	15,000	12,000	17,000	15,000	19,000
Max. operating rate, bpd	6,000	50,000	60,000	6,000	8,000	20,000	400
Max. operating temp., F	550°	450°	400°	250°	550°	550°	550°
Gas handling	Fair to good	Excellent	Fair	Good	Fair	Good	Excellent
Corrosion handling	Good to excellent	Good to excellent	Good	Fair	Good	Excellent	Excellent
Solids handling	Fair to good	Good	Fair	Excellent	Fair	Good	Fair
Fluid gravity, API°	>8°	>15	>10°	<40°	>8°	>8°	>15°
Offshore application	Limited	Excellent	Excellent	Limited	Good	Excellent	N/A

1.3 MAIN FEATURES OF SUCKER-ROD PUMPING

1.3.1 SHORT HISTORY

During the early history of mankind different methods were used to deliver drinking water from rock formations close to the surface. These included drilling and pumping of shallow wells, probably first invented by the Chinese many centuries ago. Creating plunger pumps made of bamboo pipes and valves carved out of stone, they used wooden rods to operate the pump from the surface.

The history of artificial lifting of oil wells had begun shortly after the birth of the petroleum industry. At those times, instead of the rotary drilling technique predominantly used today, **cable tools** were used to drill the wells; this technology relied on a wooden walking beam that lifted and dropped the drilling bit hung on a rope or cable. In early times, the drilling rig was driven by a steam engine. When the well ceased to flow, it was quite simple to use the existing walking beam to operate a bottom-hole plunger pump. It became common practice to leave the cable-tool drilling rig on the well so that it could later be used for pumping. The sucker-rod pumping system was born, and its operational principles have not changed since [4].

Although today's sucker-rod pumping equipment does not rely on wooden materials and steam power, its basic parts are still the same as they were before; see Fig. 1.5 First of all, the **walking beam**, the symbol of this pumping method, is still used to convert the prime mover's rotary motion to the alternating motion needed to drive the pump. The second basic part is the **rod string**, which connects

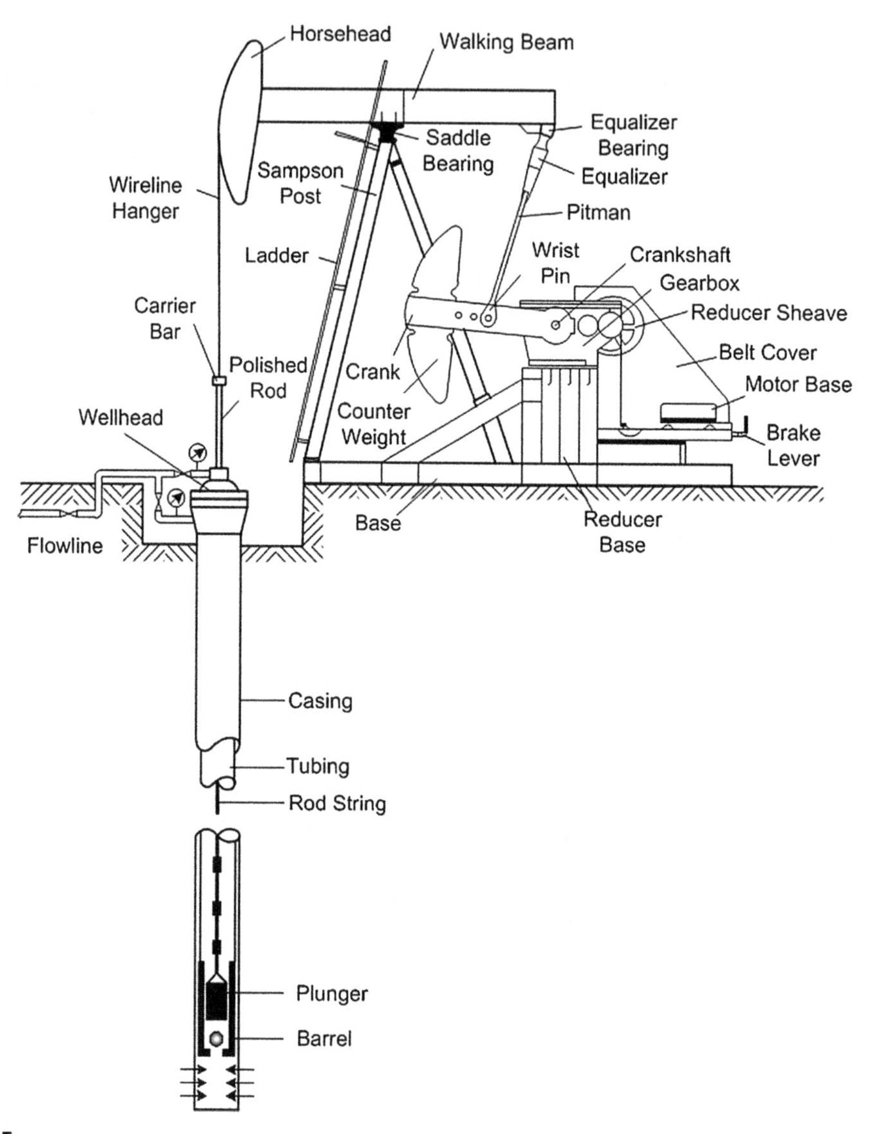

FIGURE 1.5

Basic components of the sucker-rod pumping system.

the surface pumping unit to the downhole pump. The third basic element is the **pump** itself, which, from the earliest times on, works on the positive displacement principle and consists of a stationary cylinder, a moving plunger, and two valves. This whole system has stood the test of time and is still a reliable alternative for the majority of artificial lift installations.

Thanks to its long history, it is no wonder that the sucker-rod pumping system is still a very popular means of artificial lifting all over the world. Table 1.2 lists some vital statistics, using data from [5–7],

Table 1.2 Oil well statistics for the United States [2,3] and the former Soviet Union [4]

Country	USA		USSR	
Year	**1978**	**1985**	**1963**	**1973**
Number of oil wells (thousands) =	528	679	38	57
Artificially Lifted				
(thousands) =	490	638	31	49
% of total =	93%	94%	81%	85%
Rod Pumped				
(thousands) =	419		27	40
% of total =	79.3%		72%	69%
% of artificially lifted =	85.4%		89%	81%

of the oil wells produced in two of the world's biggest producers: the United States and the former USSR. As seen from this table, the majority of the oil wells in both countries are on some kind of artificial lift. The percentage of these wells, as compared to the total number of wells, has increased with the years, both in the United States and in the former USSR. More than two-thirds of the total wells are on sucker-rod pumping. When compared to the total number of artificial lift installations, the sucker-rod pumping system has an even greater share, more than 80%.

The above numbers show the quantity of oil well installations only and do not inform about the total oil volume attained with the different production methods. The importance of sucker-rod pumping is much reduced if the total volume of liquids lifted is considered, because a great number of installations operate on low producers. In the United States alone, about 441,500 wells belonged to the stripper well category in the late 1980s [8], with production rates below 10 bpd, all produced by sucker-rod pumping. These facts notwithstanding, sucker-rod pumping is, and will remain, a major factor in artificial lifting of oil wells.

1.3.2 APPLICATIONS

For a long time in the history of oil production, sucker-rod pumping was the first and, many times, the only choice when artificial lift was needed. This is the reason why sucker-rod pumping is the oldest and most widely used type of oil well production all over the world. With the introduction of several other lift methods known today, however, the application ranges of sucker-rod pumping became more defined. Because of the depth limitations due to the limited strength of steel materials used in rod manufacturing, sucker-rod pumping is usually considered to lift moderate to low liquid production rates from shallow to medium well depths.

The production capacities of sucker-rod pumping installations range from very low to high production rates. Approximate maximum rates are shown in Fig. 1.6, as originally given by **Clegg** [9]. The values plotted represent the highest liquid rates attainable while using the largest conventional

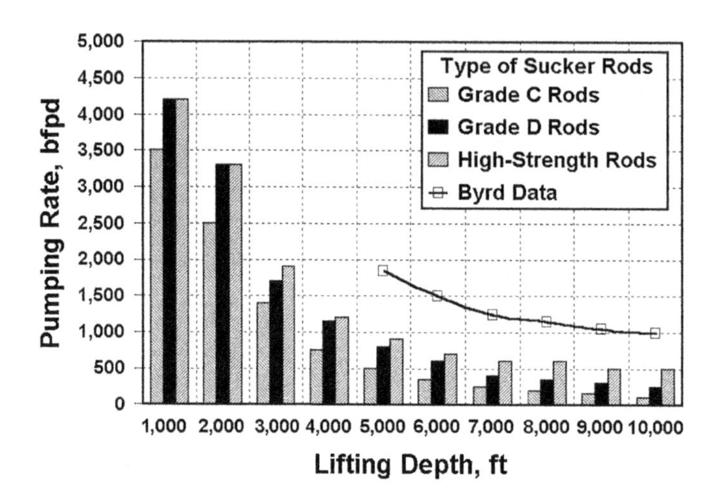

FIGURE 1.6

Maximum liquid production rates available with sucker-rod pumping after **Clegg** [9] and **Byrd** [10].

pumping unit (C-912D-365-168) and technologies of the late 1980s. The figure also contains data from **Byrd** [10], who gave theoretical maximum rates by considering other than conventional pumping units.

Observation of the production capacities proves that increased lifting depths cause a rapid drop in attainable production rates. At any particular depth, different volumes can be lifted depending on the strength of the rod material used. Stronger material grades allow greater tensile stresses in the string and thus permit higher liquid production rates. These facts lead to the conclusion that the main factors limiting liquid production from sucker-rod pumping are lifting depth and rod strength.

Developments in sucker-rod pumping technology in recent decades have outdated the pumping capacities given in Fig. 1.6. One recent study from a major manufacturer [11] reported on computer simulations to find maximum pumping rates attainable from sucker-rod-pumped installations. The study assumed the use of (a) high-strength rod materials, (b) pump sizes up to 5¾ in, (c) the biggest currently available pumping units of each geometrical arrangement, (d) anchored tubing strings, and (e) pumped-off conditions. The calculated liquid production rates are shown in Fig. 1.7 in the function of lifting depth for three commonly used pumping unit geometries. As seen, conventional pumping units can lift more than 8,000 bpd from 1,000 ft and all geometries can produce about 200 bpd from a depth of 15,000 ft.

The world's deepest sucker-rod pumping installation had its pump set at 16,850 ft [12]. It was used to remove liquids from a gas well with a correspondingly low liquid rate of 20 bfpd, and had a composite rod string made up of fiberglass and steel rods. Another pump was run to a depth of 14,500 ft [13,14] and initially produced an oil rate of 150 bpd. These and other deep installations clearly prove that the use of the latest developments in pumping technology (special-geometry pumping units, special high-strength rods or composite rod strings, ultrahigh-slip electric motors, etc.) can substantially increase the depth range and the production capacity of this well-proven artificial lift system.

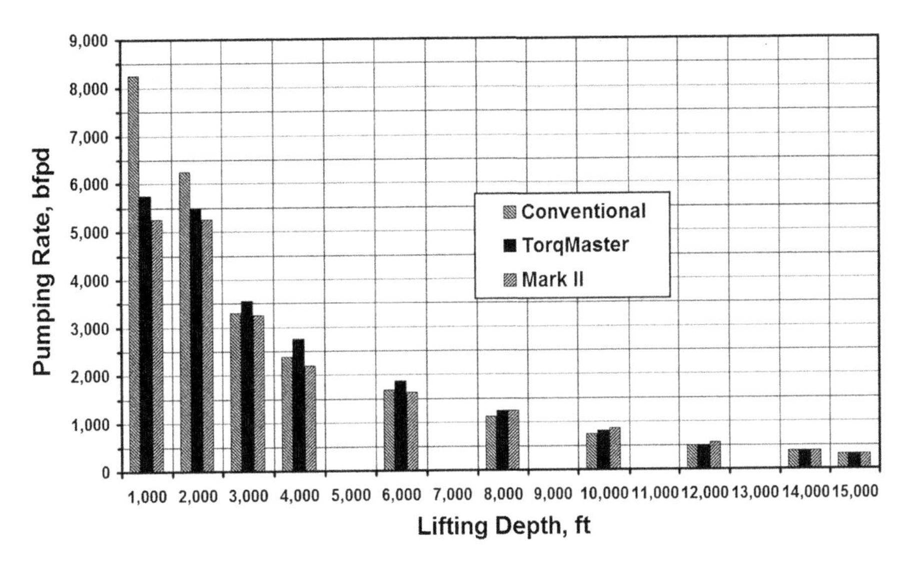

FIGURE 1.7

Calculated maximum pumping rates vs. pumping depth for different geometry pumping units after [11].

1.3.3 ADVANTAGES, LIMITATIONS

Sucker-rod pumping, as any other kind of artificial lifting, has several advantages when used to produce oil, but its limitations are also numerous, as detailed in the following.

Advantages:

- It is a **well-known** lifting method to field personnel everywhere, and is simple to operate and analyze.
- Proper installation design is relatively **simple** and can also be made in the field.
- Under average conditions, it can be used until the end of a well's productive life, up to **abandonment**.
- Pumping **capacity**, within limits, can easily be changed to accommodate changes in well inflow performance. Intermittent operation is also feasible using pump-off control devices.
- It can work with very low pump intake pressures and can thus achieve very high **draw-downs** (greater than 90% of reservoir pressure), so production of wells with low to extremely low bottom hole pressures is possible.
- **Corrosion** or scale treatments are relatively easy through the annulus.
- System components and replacement parts are standardized, interchangeable, and readily **available** worldwide.
- Pumping units and gearboxes have a high **salvage value** because they can be operational for several decades.

Limitations:

- **Free gas** present at pump intake drastically reduces liquid production and causes mechanical problems.

- Pumping depth is **limited**, mainly by the mechanical strength of the sucker-rod material.
- In **deviated** or crooked wells, friction of metal parts can lead to mechanical failures of the tubing and/or the rod string.
- The rod string must be protected against **corrosion** and mechanical damage; otherwise, material fatigue leads to early failures.
- Pumping of **sand** or abrasive-laden fluids reduces the life of most kinds of downhole pumps.
- The polished rod stuffing box, if not properly maintained, may present an environmental **hazard** by leaking well fluids to the atmosphere.
- A heavy **workover** unit is needed for servicing downhole equipment.
- The surface pumping unit requires a big space and is heavy and **obtrusive**.

REFERENCES

[1] Lea JF. Artificial lift selection. Chapter 10 in SPE Petroleum Engineering Handbook, vol. IV. Society of Petroleum Engineers; 2007.
[2] www.alrdc.com.
[3] Artificial lift systems brochure. Houston, Texas: Weatherford Co; 2007.
[4] History of petroleum engineering. New York: American Petroleum Institute; 1961.
[5] Rothrock Jr R. Maintenance, workover costs to top $3 billion. PEI; July 1978. 19–21.
[6] Moore SD. Well servicing expenditures, activity drop substantially. PEI; July 1986. 20-1,24,26.
[7] Grigorashtsenko GI. General features of the technical and technological developments in oil production. in Russian. Nef'tyanoe Khozyaystvo; July 1974. 28–33.
[8] Clegg JD. Artificial lift producing at high rates. Proc. 32nd Southwestern petroleum short course. 1985. p. 333–353.
[9] Clegg JD. High-rate artificial lift. JPT; March 1988. 277–82.
[10] Byrd JP. Pumping deep wells with a beam and sucker rod system. Paper SPE 6436 presented at the deep drilling and production symposium of the SPE, Amarillo, Texas. April 17-19, 1977.
[11] Ghareeb MM, Shedid SA, Ibrahim M. Simulation investigations for enhanced performance of beam pumping system for deep, high-volume Wells. Paper SPE 108284 presented at the international oil conference and exhibition held in Veracruz, Mexico. June 27-30, 2007.
[12] Henderson LJ. Deep sucker rod pumping for gas well unloading. Paper SPE 13199 presented at the 59th annual technical conference and exhibition of SPE, Houston, Texas. September 16-19, 1984.
[13] Wilson JW. Shell runs 14,500-ft sucker rod completion. PEI; Dec. 1982. 48–9.
[14] Gott CI. Successful rod pumping at 14,500 ft. Paper SPE 12198 presented at the 58th annual technical conference and exhibition of the SPE, San Francisco, California. October 5-8, 1983.

CHAPTER

A REVIEW OF PRODUCTION ENGINEERING FUNDAMENTALS

2

CHAPTER OUTLINE

Sucker-Rod Pumping Handbook. http://dx.doi.org/10.1016/B978-0-12-417204-3.00002-9

2.1 INTRODUCTION

Chapter 2 presents a concise treatment of the basic knowledge required from a practicing engineer working on sucker-rod pumping problems. Production engineering principles and practices related to the design and analysis of rod pumping installations are detailed in the necessary depth. The first section discusses the most significant properties of oilfield fluids oil, water, and natural gas. Since no artificial lift design can be made without the proper knowledge of the oil well's production capability, the section on inflow performance presents a thorough survey of the available methods for describing inflow to oil wells. A separate section deals with the single-phase and multiphase flow problems encountered in production engineering design. The chapter concludes with a review of the latest tool of the production engineer: the system analysis approach of evaluating well performance.

The calculation models presented in the chapter are supported by worked examples throughout the text. Later chapters rely on the concepts and procedures covered here, and a basic understanding of this chapter is advised.

2.2 PROPERTIES OF OILFIELD FLUIDS

2.2.1 INTRODUCTION

The fluids most often encountered in oil well production operations are hydrocarbons and water. These can be in either liquid or gaseous state, depending on prevailing conditions. As conditions like pressure and temperature change along the flow path (while fluids are moving from the well bottom to the surface), phase relations and physical parameters of the flowing fluids continuously change as well. Therefore, in order to determine operating conditions or to design production equipment it is essential to describe those fluid parameters that have an effect on the process of fluid lifting. In the following, calculation methods for the determination of physical properties of oil, water, and natural gas will be detailed.

Naturally, the most reliable approach to the calculation of accurate fluid properties is to use actual well data. Collection of such data usually involves pVT measurements and necessitates the use of delicate instrumentation. This is why, in the majority of cases, such data are not complete or are missing entirely. In these cases, one has to rely on previously published correlations. The correlations presented in Section 2.2 help the practicing engineer to solve the problems that insufficient information on fluid properties can cause. The reader is warned, however, that this is not a complete review of available methods, nor are the procedures given superior by any means to those not mentioned. It is intended only to give some theoretical background and practical methods for calculating fluid properties required for rod pumping calculations. As such, only black-oil hydrocarbon systems are investigated, as these are usually encountered in pumping wells.

Before a detailed treatment of the different correlations is presented, generally accepted definitions and relevant equations of the fluid physical properties are given.

2.2.2 BASIC THERMODYNAMIC PROPERTIES

Density, ρ

Density gives the ratio of the mass of the fluid to the volume occupied. It usually varies with pressure and temperature and can be calculated as:

$$\rho = \frac{m}{V} \tag{2.1}$$

Specific Gravity, γ

Specific gravity for liquids, γ_l, is the ratio of the liquid's density to the density of pure water, both taken at standard conditions:

$$\gamma_l = \frac{\rho_{\mathrm{lsc}}}{\rho_{\mathrm{wsc}}} \tag{2.2}$$

Specific gravity of gases is calculated by using air density at standard conditions:

$$\gamma_g = \frac{\rho_{\mathrm{gsc}}}{\rho_{\mathrm{asc}}} = \frac{\rho_{\mathrm{gsc}}}{0.0764} \tag{2.3}$$

Specific gravity, as seen above, is a dimensionless measure of liquid or gas density. It is widely used in correlating physical parameters and is by definition a constant for any given gas or liquid.

In the petroleum industry, gravity of liquids is often expressed in API gravity, with °API units. Specific gravity and API gravity are related by the formula:

$$\gamma_l = \frac{141.5}{131.5 + {}^\circ\mathrm{API}} \tag{2.4}$$

Viscosity, μ

Viscosity, more properly called dynamic viscosity, directly relates to the fluid's resistance to flow. This resistance is caused by friction generated between moving fluid molecules. Viscosity of a given fluid usually varies with pressure and temperature and has a great effect on the pressure drop of single-phase or multiphase flows. The customary unit of dynamic viscosity is the centipoise, cP.

Bubble Point Pressure, p_b

In multicomponent hydrocarbon fluid systems, changes of pressure and temperature result in phase changes. If we start from a liquid phase and decrease the pressure at a constant temperature, the first gas bubble comes out of solution at the bubble point pressure of the system. At higher pressures only liquid phase is present; at lower pressures a two-phase mixture exists. Bubble point pressure, therefore, is a very important parameter and can be used to determine prevailing phase conditions.

Solution Gas/Oil Ratio, R_s

Under elevated pressures, crude oil dissolves available gas and then can release it when pressure is decreased. To quantify the gas volume a crude can dissolve at given conditions, solution GOR (gas/oil ratio) is used. This parameter gives the volume at standard conditions of the dissolved gas in a crude oil. Oil volume being measured at atmospheric conditions in a stock tank in barrels (STB = stock tank barrel), the measurement unit of R_s is standard cubic foot (scf)/STB.

R_s is a function of system composition, pressure, and temperature and is defined as:

$$R_s = \frac{V_{gdissolved}}{V_{osc}} \tag{2.5}$$

Gas solubility in water can also be described with the above principles, but due to its lower magnitude it is usually neglected.

Volume Factor, *B*

The volume factor of a given fluid (gas, oil, or water) is utilized to calculate actual fluid volumes at any condition from volumes measured at standard conditions. It includes the effects of pressure and temperature, which have an impact on actual volume. In the case of liquids, the volume factor represents the effects of dissolved gases as well. The volume factor is sometimes called formation volume factor (FVF), which is a misnomer because it implies the specific conditions of the formation. The name FVF, therefore, should only be used to designate the value of the volume factor at reservoir conditions.

In general, volume factor is defined as follows:

$$B = \frac{V(p,T)}{V_{sc}} \tag{2.6}$$

where:

B = volume factor, bbl/STB, or cf/scf,
$V(p,T)$ = volume at pressure p and temperature T, bbl or cu ft,
V_{sc} = fluid volume at standard conditions, STB or scf.

By using Eq. (2.1) and substituting volume V into the above equation, an alternate definition of volume factor can be written:

$$B = \frac{\rho_{sc}}{\rho(p,T)} \tag{2.7}$$

where:

$\rho(p,T)$ = fluid density at given p,T,
ρ_{sc} = fluid density at standard conditions.

2.2.3 LIQUID PROPERTY CORRELATIONS

2.2.3.1 Water

Liquids are usually compared to pure water, of which the basic properties at standard conditions are:

Specific gravity	$\gamma_w = 1.00$
API gravity	10 °API
Density	$\rho_{wsc} = 62.4$ lb/cu ft = 350 lb/bbl
Hydrostatic gradient	0.433 psi/ft.
Dynamic viscosity	$\mu_w = 1$ cP.

In production engineering calculations gas solubility in formation water is usually neglected and water viscosity is assumed to be constant. This is why in the following no correlations will be given for R_s and μ_w of water.

Volume Factor of Water

The volume factor of water at p and T can be approximated by the use of the correlation of **Gould** [1]:

$$B_w = 1.0 + 1.21 \times 10^{-4} T_x + 10^{-6} T_x^2 - 3.33 \times 10^{-6} p \tag{2.8}$$

where:

B_w = volume factor of water, bbl/STB,
$T_x = T - 60$, °F
T = temperature, °F
p = pressure, psi.

EXAMPLE 2.1: CALCULATE THE PRESSURE EXERTED BY A STATIC COLUMN OF WATER OF HEIGHT 200 FT. AVERAGE TEMPERATURE AND AVERAGE PRESSURE IN THE COLUMN ARE $P = 50$ PSI AND $T = 120$ °F, RESPECTIVELY. USE A WATER SPECIFIC GRAVITY OF 1.12.

Solution

Water density at standard conditions from Eq. (2.2):

$$\rho_{lsc} = \gamma_w \rho_{wsc} = 1.12 \times 62.4 = 69.9 \text{ lb/cu ft}.$$

The volume factor of water at the given conditions is found from Eq. (2.8) with $T_x = 120\text{–}60 = 60$ °F:

$$B_w = 1.0 + 1.2 \times 10^{-4} \times 60 + 10^{-6} \times 60^2 - 3.33 \times 10^{-6} \times 50 = 1.01 \text{ bbl/STB}.$$

The density at p and T from Eq. (2.7):

$$\rho(p, T) = 69.9/1.01 = 69.2 \text{ lb/cu ft}.$$

The hydrostatic pressure at the bottom of the column is:

$$p = \rho h = 69.2 \times 200 = 13,840 \text{ lb/sq ft} = 13,840/144 = 96 \text{ psi}.$$

2.2.3.2 Crude oil

Bubble Point Pressure

For estimating bubble point pressure of crude oils, the use of the **Standing** correlation [2] is widely accepted:

$$p_b = 18 \left(\frac{R_b}{\gamma_g} \right)^{0.83} 10^y \tag{2.9}$$

where:

p_b = bubble point pressure, psi,
R_b = solution GOR at pressures above p_b, scf/STB,
γ_g = gas specific gravity, –,
$y = 0.00091\ T - 0.0125$ °API, and
T = temperature, °F.

EXAMPLE 2.2: DECIDE WHETHER FREE GAS IS PRESENT IN A WELL AT THE DEPTH OF THE SUCKER-ROD PUMP, WHERE $P = 700$ psi AND $T = 100\ ^{\circ}$F. THE PRODUCED OIL IS OF 30 °API GRAVITY AND HAS A TOTAL SOLUTION *GOR* OF $R_b = 120$ SCF/STB, GAS SPECIFIC GRAVITY BEING 0.75.

Solution

The bubble point pressure, using Eq. (2.9):

$$y = 0.00091 \times 100 - 0.01255 \times 30 = -0.285.$$

$$p_b = 18\ (120/0.75)^{0.83} \times 10^{-0.284} = 631\ \text{psi}.$$

Thus, only liquid phase exists, because the given pressure is above the calculated bubble point pressure.

Solution GOR

If, under some conditions, crude oil and liberated gas are present, they must coexist in equilibrium state. This means that any small decrease in pressure results in some gas going out of solution from the oil. The oil is, therefore, at its bubble point pressure at the prevailing temperature, and **Standing**'s bubble point pressure correlation can be applied. Substituting local pressure and temperature into Eq. (2.9), this can be solved for R_s:

$$R_s = \gamma_g \left(\frac{p}{18 \times 10^y}\right)^{1.205} \tag{2.10}$$

where:

R_s = solution gas/oil ratio, scf/STB,
γ_g = gas specific gravity, –,
p = pressure, psi,
$y = 0.00091\ T - 0.0125\ ^{\circ}\text{API}$, and
T = temperature, °F.

R_s can be estimated also by the use of the **Lasater** [3] or the **Vazquez–Beggs** [4] correlation.

The typical variation of solution gas/oil ratios with pressure and temperature is given in Fig. 2.1 for a given oil. As seen, the solution GOR increases with pressure but is constant above the bubble point where the oil is referred to as being undersaturated. This means that it can dissolve more gas, if available. GOR decreases with an increase of system temperature because gas comes out of solution easier at elevated temperatures.

Volume Factor

Several correlations are available for the calculation of oil volume factors [2,4]. Of these, **Standing**'s B_o correlation [2] is discussed here; it was developed on the same database as his bubble point pressure correlation. It is valid for pressures less than bubble point pressure, i.e., in the two-phase region.

$$B_o = 0.972 + 1.47 \times 10^{-4} F^{1.175} \tag{2.11}$$

where:

B_o = oil volume factor, bbl/STB,
$F = R_s\ (\gamma_g/\gamma_o)^{0.5} + 1.25\ T$,
γ_g, γ_o = gas and oil specific gravities, –,
R_s = solution gas/oil ratio, scf/STB,
T = temperature, °F.

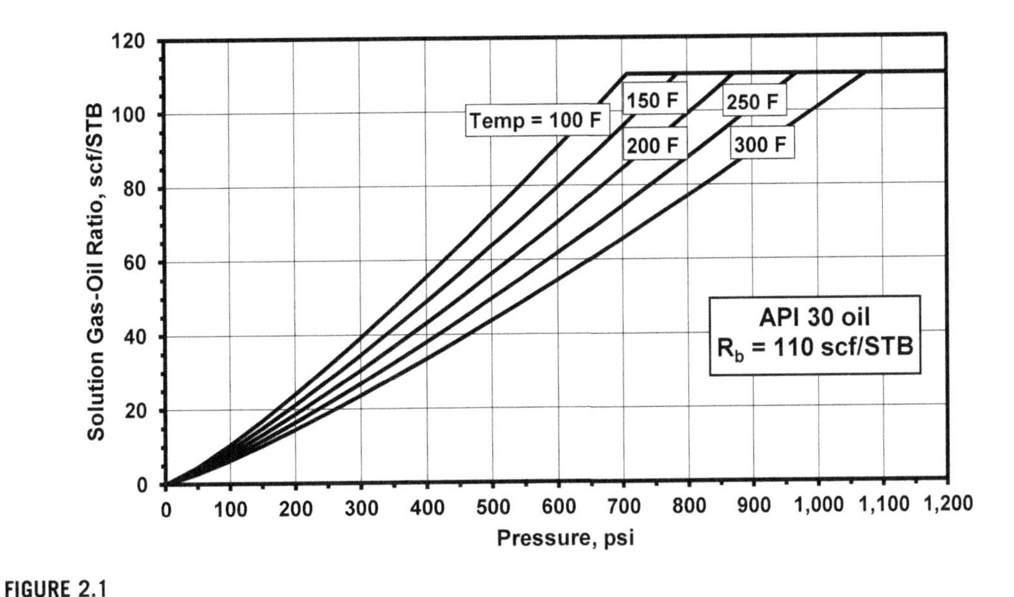

FIGURE 2.1

Variation of solution gas/oil ratio with pressure and temperature for an example case.

By definition, B_o is always greater than unity and is affected by expansion due to temperature, by the amount of gas dissolved in the oil, and by compression due to actual pressure. An example is provided in Fig. 2.2, where the variation of B_o with pressure and temperature is plotted for the same oil as in Fig. 2.1. At pressures above the bubble point pressure the oil is undersaturated and B_o is constant.

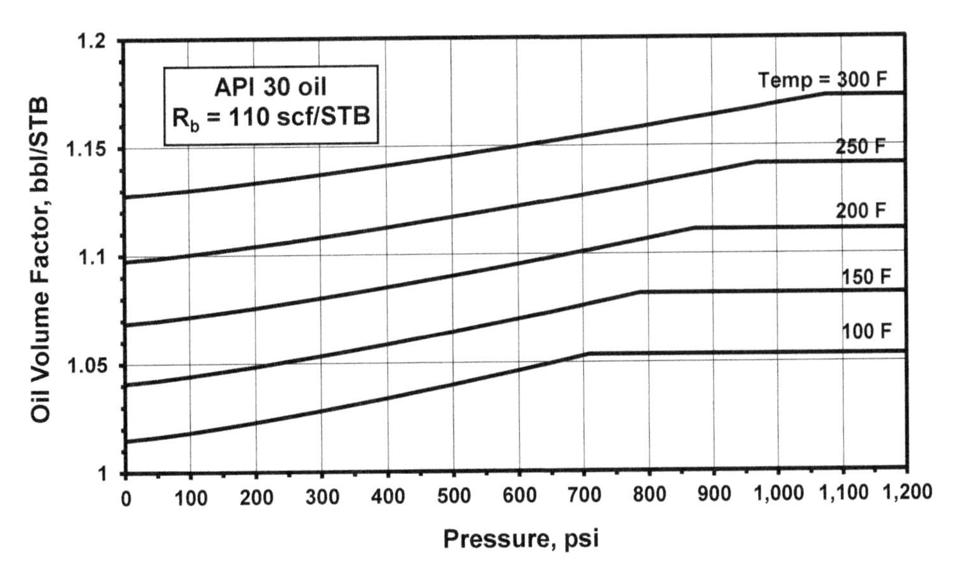

FIGURE 2.2

Variation of oil volume factor with pressure and temperature for an example case.

EXAMPLE 2.3: LIQUID DISPLACEMENT OF A SUCKER-ROD PUMP WAS CALCULATED TO BE $Q = 600$ BPD OIL AT DOWNHOLE CONDITIONS, WHERE $P = 800$ PSI AND $T = 120$ °F. FIND THE CRUDE OIL VOLUME THAT WILL BE MEASURED AT THE SURFACE IN THE STOCK TANK. OIL GRAVITY IS 30 °API, GAS SPECIFIC GRAVITY IS 0.6.

Solution

The oil volume in the stock tank can be found from the definition of oil volume factor, Eq. (2.6):

$$q_{osc} = q(p, T)/B_o$$

To find B_o, R_s has to be calculated first, from Eq. (2.10):

$$y = 0.00091 \times 120 - 0.0125 \times 30 = -0.266.$$

$$R_s = 0.6\left(800/18/10^{-0.266}\right)^{1.205} = 121.4 \text{ scf/STB}.$$

Before applying Standing's B_o correlation, oil specific gravity is calculated from API gravity, by Eq. (2.4):

$$\gamma_o = 141.5/(131.5 + 30) = 0.876.$$

With the above values, from Eq. (2.11):

$$F = 121.4(0.6/0.876)^{0.5} + 1.25 \times 120 = 250.5.$$

$$B_o = 0.972 + 1.47 \times 10^{-4} \times 250.5^{1.175} = 1.07 \text{ bbl/STB}.$$

Finally, oil volume at standard conditions:

$$q_{osc} = 600/1.07 = 561 \text{ STB/d}.$$

Oil volume has "shrunk" from 600 bpd at the well bottom to 561 STB/d in the stock tank. This volume change is the effect of gas going continuously out of solution in the well and in the tank, and is also the effect of volume changes due to variations in pressure and temperature.

Viscosity

The following applies to crude oils exhibiting Newtonian behavior, as non-Newtonian oils or emulsions cannot be characterized by viscosity alone.

Factors affecting the viscosity of Newtonian oils are composition, pressure, temperature, and the amount of dissolved gas. The viscosity of dead (gasless) oils at atmospheric pressure was correlated with API gravity and temperature by **Beal** [5], given in Fig. 2.3.

Dissolved gas volume has a very profound impact on the viscosity of live oil. A generally accepted method to account for this is the correlation developed by **Chew** and **Connally** [6]. They presented the following equation:

$$\mu_o = A\mu_{oD}^b \tag{2.12}$$

where:

μ_o = live oil viscosity, cP
μ_{oD} = viscosity of dead (gasless) oil, cP.

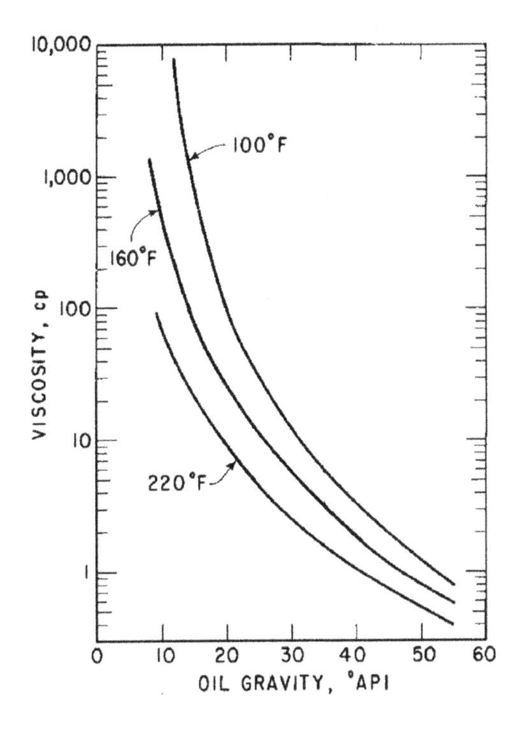

FIGURE 2.3

Viscosity of dead oil according to **Beal** [5].

The authors gave factors *A* and *b* on a graph, in the function of solution *GOR*. The two curves can be closely approximated by the equations given below:

$$A = 0.2 + 0.810^{-0.00081R_s} \tag{2.13}$$

$$b = 0.43 + 0.5710^{-0.00072R_s} \tag{2.14}$$

where:

R_s = solution gas/oil ratio, scf/STB.

2.2.4 PROPERTIES OF NATURAL GASES

2.2.4.1 Behavior of gases

Before discussing the behavior of natural gases, some considerations on ideal gases have to be detailed. An ideal gas has molecules that are negligibly small compared to the volume occupied by the gas. The kinetic theory of such gases states that the volume of a given ideal gas, its absolute pressure, its absolute temperature, and the total number of molecules are interrelated. This relationship is described by the ideal gas law. From this, any parameter can be calculated, provided the other three are known.

Natural gases are gas mixtures and contain mainly hydrocarbon gases with usually lower concentrations of other components. Due to their complex nature and composition they cannot be considered as ideal gases and cannot be described by the ideal gas law. Several different methods were devised to characterize the behavior of real gases. These are called equations of state and are usually empirical equations that try to describe the relationships between the so-called state parameters of gas: absolute pressure, absolute temperature, and volume. The most frequently used and simplest equation of state for real gases is the engineering equation of state:

$$pV = ZnRT_a \tag{2.15}$$

where:

p = absolute pressure, psia,
V = gas volume, cu ft,
Z = gas deviation factor, –,
n = number of moles,
R = 10.73, gas constant,
T_a = absolute temperature = T(°F) + 460, °R.

The engineering equation of state differs from the ideal gas law by the inclusion of the gas deviation factor only, which is sometimes referred to as compressibility or supercompressibility factor. It accounts for the deviation of the real gas volume from the volume of an ideal gas under the same conditions, and is defined as:

$$Z = \frac{V_{\text{actual}}}{V_{\text{ideal}}} \tag{2.16}$$

The problem of describing the behavior of natural gases was thus reduced to the proper determination of deviation factors.

The most common approach to the determination of deviation factors is based on the theorem of corresponding states. This principle states that real gas mixtures, like natural gases, behave similarly if their pseudo-reduced parameters are identical. The definition of pseudo-reduced pressure and temperature are:

$$p_{pr} = \frac{p}{p_{pc}} \tag{2.17}$$

$$T_{pr} = \frac{T}{T_{pc}} \tag{2.18}$$

where:

p_{pr}, T_{pr} = pseudo-reduced pressure and temperature,
p, T = pressure and temperature,
P_{pc}, T_{pc} = pseudocritical pressure and temperature.

The pseudocritical pressure and temperature of the gas are determined as weighted averages from the critical parameters of the individual components. Using this principle, deviation factors for common natural gas mixtures can be correlated as functions of their pseudo-reduced parameters. The most widely accepted Z-factor correlation was given by **Standing–Katz** [7] and is reproduced in Fig. 2.4.

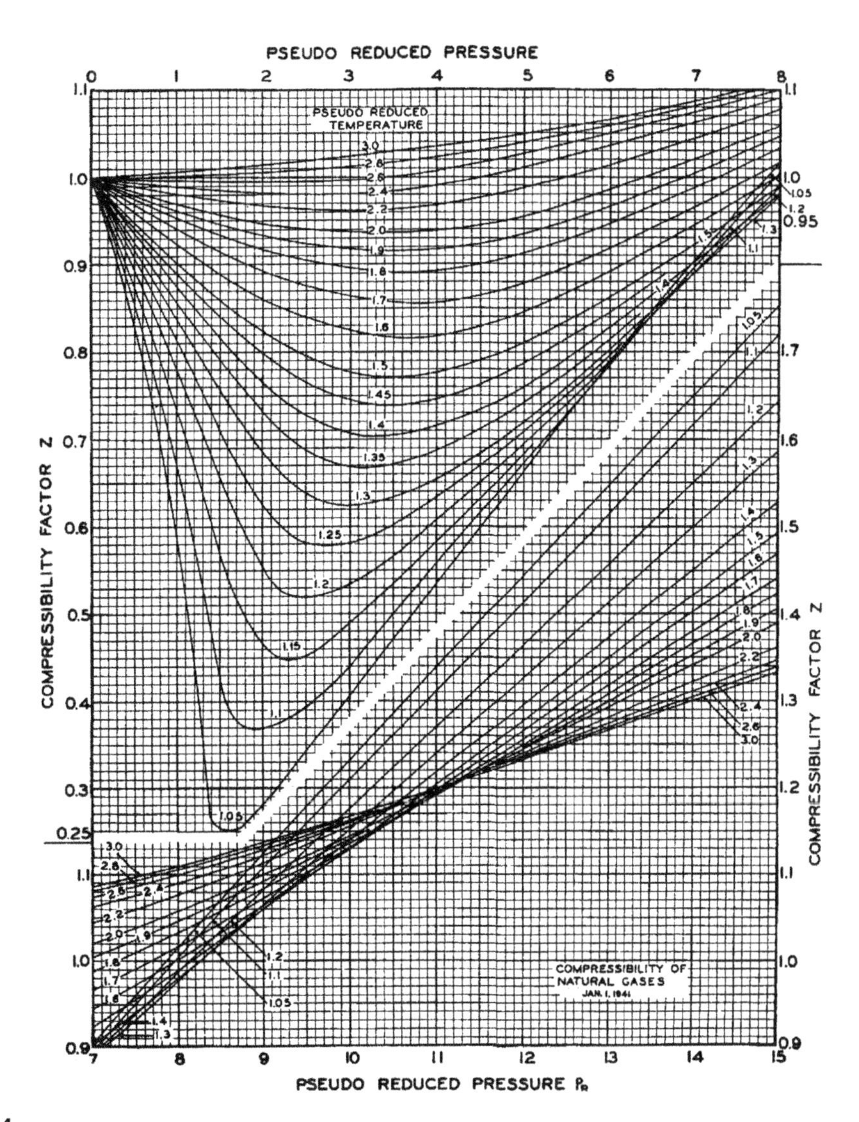

FIGURE 2.4

Deviation factor for natural gases after **Standing–Katz** [7].

EXAMPLE 2.4: FIND THE DEVIATION FACTOR FOR A NATURAL GAS AT $P = 1{,}200$ PSIA AND $T = 200\ °F$, IF THE PSEUDOCRITICAL PARAMETERS ARE $P_{PC} = 630$ PSIA AND $T_{PC} = 420\ °R$.

Solution

The pseudo-reduced parameters are:

$$p_{pr} = 1{,}200/630 = 1.9.$$

$$T_{pr} = (200 + 460)/420 = 1.57.$$

Using these values, from Fig. 2.4: $Z = 0.85$.

Gas Volume Factor

The engineering equation of state enables the direct calculation of volume factors for gases. Equation (2.15) can be written for a given number of moles in the following form:

$$\frac{pV}{ZT_a} = \left(\frac{pV}{ZT_a}\right)_{sc} \tag{2.19}$$

This equation can be solved for B_g, which is the ratio of actual volume to the volume at standard conditions:

$$B_g = \frac{V}{V_{sc}} = \frac{p_{sc} Z T_a}{p Z_{sc} T_{sc}} \tag{2.20}$$

Substituting into this the values $p_{sc} = 14.7$ psia, $T_{sc} = 520\ °R$, and $Z_{sc} = 1$, one arrives at:

$$B_g = 0.0283 \frac{ZT_a}{p} \tag{2.21}$$

where:

B_g = volume factor of gas, cf/scf,
Z = deviation factor,
T_a = absolute temperature, °R,
p = pressure, psia.

EXAMPLE 2.5: WHAT IS THE ACTUAL VOLUME OF THE GAS, IF ITS VOLUME MEASURED AT STANDARD CONDITIONS IS 1.2 MSCF. OTHER DATA ARE IDENTICAL TO THOSE OF EXAMPLE 2.4.

Solution

The volume factor of the gas, from Eq. (2.21):

$$B_g = 0.0283 \times 0.85(200 + 460)/1{,}200 = 0.013.$$

Actual volume is found from Eq. (2.6):

$$V(p, T) = B_g V_{sc} = 0.013 \times 1{,}200 = 15.6 \text{ cu ft.}$$

Gas Density

The fact that gas volume factor is an explicit function of state parameters allows a direct calculation of gas density at any conditions. Based on the definition of volume factor, gas density can be expressed from Eq. (2.7) and, after substituting the formula for gas volume factor, we get:

$$\rho(p,T) = \frac{\rho_{sc}}{B_g} = \frac{0.0764\gamma_g p}{0.0283ZT_a} = 2.7\gamma_g \frac{p}{ZT_a} \quad (2.22)$$

where:

$\rho(p,T)$ = gas density, lb/cf
p = pressure, psia,
Z = deviation factor,
T_a = absolute temperature, °R.

The above formula is used to find the actual density of natural gases at any pressure and temperature based on the knowledge of their specific gravities and deviation factors.

EXAMPLE 2.6: WHAT IS THE ACTUAL DENSITY OF THE GAS GIVEN IN THE PREVIOUS EXAMPLES, IF ITS SPECIFIC GRAVITY IS 0.75.

Solution

The deviation factor is 0.85, as found in Example 2.4.

By using Eq. (2.22) we get:

$$\rho(p,T) = 2.7 \times 0.75 \times 1,200/(0.85 \times 660) = 4.3 \text{ lb/cu ft.}$$

An alternate way of calculation is the use of the previously calculated B_g value. Gas density at standard conditions:

$$\rho_{sc} = 0.0764\ \gamma_g = 0.0764 \times 0.75 = 0.057 \text{ lb/cu ft.}$$

Gas volume factor was found in Example 2.5 to be equal to 0.013. This allows a direct calculation of actual gas density:

$$\rho(p,T) = \rho_{sc}/B_g = 0.057/0.013 = 4.4 \text{ lb/cu ft.}$$

2.2.4.2 Gas property correlations

Pseudocritical Parameters

In most cases, gas composition is unknown and pseudocritical parameters cannot be calculated by the additive rule. The correlation of **Hankinson, Thomas**, and **Phillips** [8] gives pseudocritical pressure and temperature in the function of gas specific gravity:

$$p_{pc} = 709.6 - 58.7\gamma_g \quad (2.23)$$

$$T_{pc} = 170.5 + 307.3\gamma_g \quad (2.24)$$

where:

p_{pc}, T_{pc} = pseudocritical pressure and temperature, psia and °R,
γ_g = gas specific gravity, –.

These equations are valid only for sweet natural gases, i.e., gases with negligible amounts of sour components.

Sour gases, like CO_2 and H_2S, considerably affect critical parameters of natural gas mixtures. These effects were investigated by **Wichert** and **Aziz** [9,10], who proposed a modified calculation method to account for the presence of sour components in the gas. They developed a correction factor, to be calculated by the following formula:

$$e = 120\left(A^{0.9} - A^{1.6}\right) + 15\left(B^{0.5} - B^{4}\right) \tag{2.25}$$

Factors A and B are mole fractions of the components $CO_2 + H_2S$ and H_2S in the gas, respectively.

This correction factor is applied to the pseudocritical parameters, determined previously assuming the gas to be sweet, by using the equations given below:

$$T_m = T_{pc} - e \tag{2.26}$$

$$p_m = p_{pc} \frac{T_m}{T_{pc} + B(1 - B)e} \tag{2.27}$$

where:

T_m = modified pseudocritical temperature, °R
p_m = modified pseudocritical pressure, psia

Deviation Factor

The use of the **Standing–Katz** chart (Fig. 2.4) is a generally accepted way to calculate deviation factors for natural gases. The inaccuracies of visual read off and the need for computerized calculations necessitated the development of mathematical models that describe this chart. Some of these procedures involve the use of simple equations and are easy to use, e.g., the methods of **Sarem** [11] or **Gopal** [12]. Others use equations of state and require tedious iterative calculation schemes to find deviation factors, such as **Hall** and **Yarborough** [13] or **Dranchuk et al.** [14]. A complete review of the available methods is given by **Takacs** [15].

The following simple equation, originally proposed by **Papay** [16], was found to give reasonable accuracy and provides a simple calculation procedure:

$$Z = 1 - \frac{3.52 p_{pr}}{10^{0.9813 T_{pr}}} + \frac{0.274 p_{pr}^2}{10^{0.8157 T_{pr}}} \tag{2.28}$$

where:

Z = deviation factor, –,
p_{pr}, T_{pr} = pseudo-reduced pressure and temperature.

EXAMPLE 2.7: FIND THE DEVIATION FACTOR FOR A 0.65 SPECIFIC GRAVITY NATURAL GAS AT P = 600 PSIA AND T = 100 °F WITH PAPAY'S PROCEDURE. THE GAS CONTAINS 5% CO_2 AND 8% H_2S.

Solution

The pseudocritical parameters from Eqs (2.23) and (2.24):

$$p_{pc} = 709.6 - 58.7 \times 0.65 = 671.4 \text{ psia.}$$

$$T_{pc} = 170.5 + 307.3 \times 0.65 = 370.2 \text{ °R.}$$

The parameters A and B in the **Wichert–Aziz** correction for sour components are:

$$A = (5+8)/100 = 0.13.$$

$$B = 8/100 = 0.08.$$

The correction factor is found from Eq. (2.25):

$$e = 120\left(0.13^{0.9} - 0.13^{1.6}\right) + 15\left(0.08^{0.5} - 0.08^{4}\right) = 120(0.159 - 0.038) + 15\left(0.238 - 4.09\times10^{-5}\right) = 14.52 + 3.57 = 18.09.$$

The corrected pseudocritical parameters are evaluated with Eqs (2.26) and (2.27):

$$T_m = 370.2 - 18.76 = 351.4\ ^{\circ}\mathrm{R}.$$

$$p_m = 570.8 \times 351.4/(370.2 + 0.08(1 - 0.08)18.76) = 235{,}719.1/371.5 = 634 \text{ psia}.$$

Reduced parameters can be calculated using Eqs (2.17) and (2.18):

$$p_{\mathrm{pr}} = 600/634.4 = 0.946,$$

$$T_{\mathrm{pr}} = (100+460)/351.4 = 1.594.$$

Finally, Z-factor is found by **Papay**'s equation (Eq. (2.28)):

$$Z = 1 - 3.52 \times 0.946/\left(10^{0.9813\times1.594}\right) + 0.274 \times 0.946^{2}/\left(10^{0.8157\times1.594}\right) = 1 - 0.091 + 0.012 = 0.921.$$

2.3 BASIC HYDRAULICS

2.3.1 INTRODUCTION

In sucker-rod pumping operations, just as in other phases of oil production, several different kinds of fluid flow problems are encountered. These involve vertical or inclined flow of a single-phase fluid or of a multiphase mixture in well tubing, as well as horizontal or inclined flow in flowlines. A special hydraulic problem is the calculation of the pressure exerted by the static gas column present in a rod pumped well's annulus. All the problems mentioned require that the engineer be able to calculate the main parameters of the particular flow, especially the pressure drop involved.

In this section, basic theories and practical methods for solving pipe flow problems are covered that relate to the design and analysis of sucker-rod-pumped installations. As all topics discussed have a common background in hydraulic theory, a general treatment of the basic hydraulic concepts is given first. This includes detailed definitions of (and relevant equations for) commonly used parameters of pipe flow.

2.3.2 BASIC PRINCIPLES

2.3.2.1 The general energy equation

Most pipe flow problems in petroleum production operations involve the calculation of the pressure along the flow path. Pressure being one form of energy, its variation over pipe length can be found from an energy balance written between two points in the pipe. The general energy equation describes the conservation of energy and states that the change in energy content between two points of a flowing fluid is equal to the work done on the fluid minus any energy losses. From this basic principle, the change in flowing pressure, dp, over an infinitesimal length dl can be evaluated as:

$$\frac{dp}{dl} = \frac{g}{g_c} \rho \sin \alpha + \frac{\rho v dv}{g_c dl} + \left(\frac{dp}{dl}\right)_f \tag{2.29}$$

This equation gives the pressure gradient dp/dl for the flow configuration of an inclined pipe shown in Fig. 2.5. Analysis of the right-hand-side terms shows that, in this general case, pressure gradient consists of three components, according to the three terms in the equation: hydrostatic, acceleration, and friction gradients.

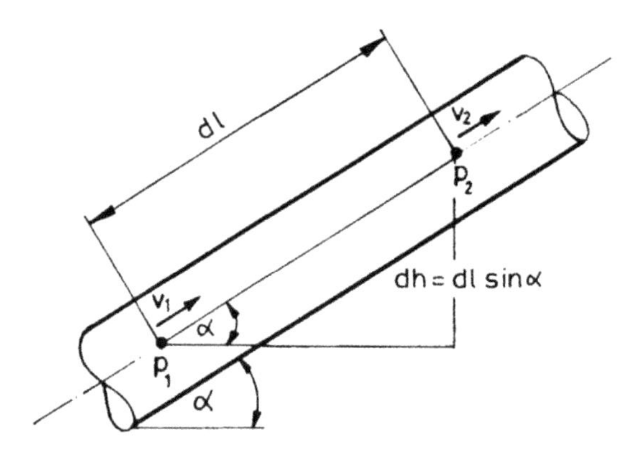

FIGURE 2.5

Flow configuration for the general energy equation.

The hydrostatic gradient, also known as elevation gradient, stands for the changes in potential energy. The acceleration gradient represents kinetic energy changes. The third component accounts for irreversible energy losses due to fluid friction. The relative magnitude of these components in the total pressure gradient depends on the number of flowing phases, pipe inclination, and other parameters [17].

2.3.2.2 Flow velocity

The velocity of flow is a basic hydraulic parameter and is usually calculated from the flow rate given at standard conditions. Using the definition of volume factor B to arrive at actual volumetric flow rate, fluid flow velocity is calculated from the next formula, which is valid for both gas and liquid flow:

$$v = \frac{q(p, T)}{A} = \frac{q_{sc} B}{A} \tag{2.30}$$

where:

v = flow velocity,
q_{sc} = volumetric fluid rate at standard conditions,
B = volume factor at prevailing conditions,
A = cross-sectional area of pipe.

In multiphase flow, only a fraction of pipe cross-section is occupied by one phase, because at least two phases flow simultaneously. Thus, the above formula gives imaginary values for the individual phases, frequently denoted as superficial velocities. Although actually nonexistent, these superficial velocities are widely used in multiphase flow correlations.

2.3.2.3 Friction factor

Frictional losses constitute a significant fraction of flowing pressure drops; thus their accurate determination is of utmost importance. For calculating friction gradient or frictional pressure drop, the **Darcy–Weisbach** equation is universally used:

$$\left(\frac{\mathrm{d}p}{\mathrm{d}l}\right)_f = f\frac{\rho v^2}{2g_c d} \tag{2.31}$$

where:

$(\mathrm{d}p/\mathrm{d}l)_f$ = frictional pressure drop, psi/ft,
f = friction factor, –,
$g_c = 32.2$, a conversion factor,
ρ = fluid density,
v = flow velocity,
d = pipe diameter.

The friction factor f is a function of the Reynolds number N_{Re} and the pipe's relative roughness and is usually determined from the **Moody** diagram.

2.3.2.4 Reynolds number

Due to the complexity of single-phase and, especially, multiphase flow, the number of independent variables in any hydraulic problem is quite large. One commonly used way to decrease the number of variables and to facilitate easier treatment is the forming of dimensionless groups from the original variables. Such dimensionless groups or numbers are often used in correlations, the most well-known one being the Reynolds number, defined below:

$$N_{\mathrm{Re}} = \frac{\rho v d}{\mu} \tag{2.32}$$

This equation gives a dimensionless number in any consistent set of units. Using the customary unit of cP for viscosity, it can be transformed to:

$$N_{\mathrm{Re}} = 124\frac{\rho v d}{\mu} \tag{2.33}$$

where:

N_{Re} = Reynolds number, –,
ρ = fluid density, lb/cu ft,
v = flow velocity, ft/s,
d = pipe diameter, in,
μ = fluid viscosity, cP.

2.3.2.5 Moody diagram

Friction factor, in general, was found to be a function of the Reynolds number and pipe relative roughness. Relative roughness is defined here as the ratio of the absolute roughness ε of the pipe inside wall to the pipe inside diameter:

$$k = \frac{\varepsilon}{d} \tag{2.34}$$

where:

k = pipe relative roughness, –,
ε = pipe absolute roughness, in,
d = pipe diameter, in.

Typical absolute roughness values are $\varepsilon = 0.0006$ in for new and $\varepsilon = 0.009$ in for used well tubing.

There are several formulae describing friction factors for different flow conditions. Most of them are included in the **Moody** diagram [18], which is a graphical presentation of **Darcy–Weisbach**-type f values. Use of this chart, given in Fig. 2.6, is generally accepted in the petroleum industry.

As seen in the Moody diagram, the friction factor is a different function of the variables N_{Re} and k in different ranges. In laminar flow (i.e., Reynolds numbers lower than 2,000–2,300) f varies with the Reynolds number only:

$$f = \frac{64}{N_{Re}} \tag{2.35}$$

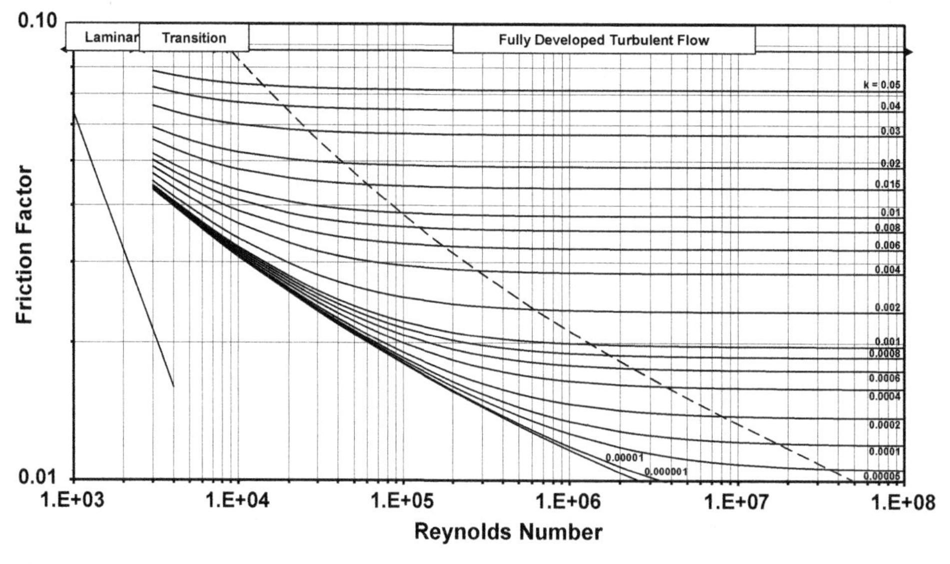

FIGURE 2.6

The **Moody** [18] diagram: friction factors for pipe flow.

On the other hand, in fully developed turbulent flow, for Reynolds numbers over 2,000–2,300 and for a rough pipe, friction factor is a sole function of the pipe's relative roughness. There exists a transition region between smooth wall flow and fully developed turbulent flow where the value of friction factor is determined by both N_{Re} and k values.

For computer applications the use of the **Colebrook** equation is recommended; it can be solved for friction factors in both the transition and the fully developed turbulent regions [19]. It is an implicit function and can be solved for f values using some iterative procedure, e.g., the **Newton–Raphson** scheme.

$$\frac{1}{\sqrt{f}} = 2\log\frac{3.7}{k} - 2\log\left(1 + \frac{9.335}{kN_{Re}\sqrt{f}}\right) \tag{2.36}$$

where:

f = friction factor, –,
k = pipe relative roughness, –,
N_{Re} = Reynolds number, –.

The inconvenience of an iteration scheme can be eliminated if an explicit formula is used to calculate friction factors. **Gregory** and **Fogarasi** [20] investigated several such models and found that the formula developed by **Chen** [21] gives very good results. This formula, as given below, very accurately reproduces the **Moody** diagram over the entire range of conditions. It does not necessitate an iterative scheme and thus can speed up lengthy calculations (e.g., multiphase pressure drop calculations) involving the determination of a great number of friction factors.

$$\frac{1}{\sqrt{f}} = -2\log\left(\frac{k}{3.7065} - \frac{5.0452}{N_{Re}}\log A\right) \tag{2.37}$$

where:

$$A = \frac{k^{1.1098}}{2.8257} + \left(\frac{7.149}{N_{Re}}\right)^{0.8981} \tag{2.38}$$

2.3.3 SINGLE-PHASE LIQUID FLOW

Liquids are practically incompressible; that is why flow velocity is constant in a pipe with a constant cross-section. With no change in velocity along the flow path no kinetic energy changes occur and the general energy equation has to be modified accordingly. Setting the right-hand second term to zero in Eq. (2.29) and substituting the expression of friction gradient from Eq. (2.31) in the term $(dp/dl)_f$:

$$\frac{dp}{dl} = \frac{g}{g_c}\rho\sin\alpha + f\frac{\rho v^2}{2g_c d} \tag{2.39}$$

This equation gives the pressure gradient for liquid flow in an inclined pipe. It can be solved for pressure drop in a finite pipe length l, and using customary field units will take the form:

$$\Delta p = \frac{1}{144}\rho l\sin\alpha + 1.294\times10^{-3} f\frac{l}{d}\rho v^2 \tag{2.40}$$

where:

Δp = pressure drop, psi,
ρ = flowing density, lb/cu ft,
l = pipe length, ft,
α = pipe inclination above horizontal, degrees,
v = flow velocity, ft/s,
d = pipe diameter, in.

In case of vertical flow $\alpha = 90°$, sin $\alpha = 1$, and pressure drop is calculated by:

$$\Delta p = \frac{1}{144}\rho l + 1.294 \times 10^{-3} f \frac{l}{d} \rho v^2 \tag{2.41}$$

For horizontal flow, there is no elevation change and inclination angle is zero; thus sin $\alpha = 0$. Therefore, total pressure drop for horizontal flow consists of frictional losses only, and can be expressed by using the formula:

$$\Delta p = 1.294 \times 10^{-3} f \frac{l}{d} \rho v^2 \tag{2.42}$$

All the pressure drop formulae discussed above contain the liquid flow velocity v, which is evaluated on the basis of Eq. (2.30) and, using field measurement units, can be written as:

$$v = 0.0119 \frac{q_l B}{d^2} \tag{2.43}$$

where:

v = flow velocity, ft/s,
q_l = liquid flow rate, STB/d,
d = pipe diameter, in,
B = volume factor, bbl/STB.

The procedures for calculating pressure drop in single-phase liquid flow are illustrated by presenting two example problems.

EXAMPLE 2.8: CALCULATE THE PRESSURE DROP OCCURRING IN A HORIZONTAL LINE OF LENGTH 4,000 FT WITH AN INSIDE DIAMETER OF 2″ FOR A WATER FLOW RATE OF 1,000 BPD. PIPE ABSOLUTE ROUGHNESS IS 0.00015″, WATER SPECIFIC GRAVITY IS 1.05. USE $B_W = 1$ AND $\mu = 1$ CP.

Solution

Find the flow velocity in the pipe by using Eq. (2.43):

$$v = 0.0119 \times 1{,}000/(1 \times 2^2) = 2.98 \text{ ft/s}.$$

Next, the Reynolds number is found with Eq. (2.33), substituting $\rho = \gamma_w \rho_{wsc}/B_w$:

$$N_{Re} = (124 \times 1.05 \times 62.4 \times 2.98 \times 2)/1 = 48{,}422.$$

Flow is turbulent, as $N_{Re} > 2{,}300$. Friction factor is read off from the Moody diagram at $N_{Re} = 48{,}422$ and at a relative roughness of $k = 0.00015/2 = 7.5 \times 10^{-5}$.

From Fig. 2.6:

$$f = 0.021.$$

The frictional pressure drop is calculated using Eq. (2.42):

$$\Delta p = 1.294 \times 10^{-3} \times 0.021\left(4,000 \times 1.05 \times 62.4 \times 2.98^2\right)/2 = 32 \text{ psi}.$$

EXAMPLE 2.9: CALCULATE THE WELLHEAD PRESSURE REQUIRED TO MOVE 900 BPD OF OIL INTO A 200 PSI PRESSURE SEPARATOR THROUGH A FLOWLINE WITH A LENGTH OF 2,000 FT AND A DIAMETER OF 2 IN. THE OIL PRODUCED IS GASLESS WITH A SPECIFIC GRAVITY OF 0.65 AND HAS A VISCOSITY OF 15 CP AND A B_O OF 1. THE FLOWLINE HAS A CONSTANT INCLINATION, THE SEPARATOR BEING AT A VERTICAL DISTANCE OF $H = 100$ FT ABOVE THE WELLHEAD.

Solution

The hydrostatic pressure drop is the first term in Eq. (2.40).

Using $\rho_o = \gamma_o\, \rho_{wsc}/B_o = 0.65 \times 62.4/1 = 40.6$ lb/cu ft, and $\sin \alpha = h/l = 100/2{,}000 = 0.05$,we get $\Delta p_{hydr} = (40.6 \times 2{,}000 \times 0.05)/144 = 28.2$ psi.

For calculating frictional pressure drop, flow velocity and Reynolds number are calculated by Eqs (2.43) and (2.33), respectively:

$$v = 0.0119\left(900/1 \times 2^2\right) = 2.68 \text{ ft/s}.$$

$$N_{Re} = 124(40.6 \times 2.67 \times 2)/15 = 1,799.$$

Flow is laminar, as N_{Re} is less than the critical value of 2,300. Friction factor for laminar flow is given by Eq. (2.35) as:

$$f = 64/1,799 = 0.036.$$

Frictional pressure drop is calculated using the second term in Eq. (2.40):

$$\Delta p_f = 1.294 \times 10^{-3} \times 0.036\left(2,000 \times 40.6 \times 2.67^2\right)/2 = 13.5 \text{ psi}.$$

Total pressure drop equals $28.2 + 13.5 = 41.7$ psi. The pressure required at the wellhead is the sum of separator pressure and flowing pressure drop:

$$p = 200 + 41.7 = 241.7 \text{ psi}.$$

2.3.4 MULTIPHASE FLOW

A full treatment of multiphase flow is beyond the scope of this book, so this section gives only a very basic overview of the topic.

Multiphase flow of oil, water, and gas is a very complex phenomenon and has been the subject of intensive research efforts. The main difficulties encountered in describing multiphase flow are associated with the following major problems: (1) liquid and gas concurrently flowing in a pipe may assume different geometrical arrangements, called flow patterns; (2) the flowing mixture is compressible; thus its density continuously changes with changes in pressure and temperature; and (3) besides fictional losses, a new kind of energy loss is also occurring. This is called slippage loss and is the result of the gas phase's slipping through the liquid due to the great specific gravity difference between phases.

Most multiphase correlations assume the flowing mixture to be homogeneous and use the general energy equation written for this hypothetical fluid. This approach reduces the number of equations to be solved and allows the application of methods developed for single-phase flow. Due to the complexity of multiphase flow, numerous empirical correlations have been developed so far.

For the purposes of sucker-rod pumping design and analysis, it is sufficient to limit the discussion of multiphase flow problems to the presentation of gradient curves. Gradient curves are families of graphs containing precalculated pressure traverses in horizontal or vertical pipes. They are developed by using a particular pressure drop correlation and by assuming several basic data as chart basis. The coordinate system generally used for plotting gradient curves has coordinates of pressure vs depth or pressure vs pipe length. This presentation permits a direct determination of pressure distribution in a given pipe.

Figure 2.7, taken from **Brown** [22], gives a typical vertical pressure gradient curve sheet. The basic data that are used to develop such families of curves are tubing size, liquid production rate, water cut, fluid properties, and temperature. Actual values of these parameters are displayed in the figure. Gradient curves of different sources use different assumed data, so care must be taken to find those curves that most closely fit the conditions at hand. The individual curves on a sheet of gradient curves represent different gas/oil or gas/liquid ratios. As seen in Fig. 2.7, changes in the gas/liquid ratio markedly change the shape of the pressure traverse curve.

The use of gradient curves is quite straightforward. First the sheet with the required pipe size and liquid rate is selected. If no such sheet is found, interpolation using the closest sheets available may be necessary. The next step is the selection of the pressure traverse curve that corresponds to the actual gas/liquid ratio. This particular curve represents the pressure distribution along tubing length and then can be used in subsequent calculations.

2.3.5 STATIC GAS COLUMN PRESSURE

2.3.5.1 Theoretical background

Generally, the casing–tubing annulus in a sucker-rod-pumped well is not sealed at the bottom, as is usual in flowing or gas-lifted wells. The absence of a packer enables well fluids to enter the annular space that acts as a natural gas/liquid separator. The separation of well fluids results in having a liquid and a gas column in the annulus. As the casing head is directly connected to the flowline at the surface, the pressure at the top of the annular gas column equals the tubing head pressure. This circumstance provides a very effective way to calculate bottomhole pressures from the wellhead pressure by adding the pressure exerted by the gas and liquid columns to the wellhead pressure.

The gas column present in a pumping well's annulus is considered to be in a semistatic state, as flow velocities in the annular space are most often negligible. This is due to the large cross-sectional areas involved and the relatively low gas flow rates. In such cases, frictional losses can be disregarded and the pressure distribution in the column is affected only by gravitational forces. The general energy equation for this case is written with the friction and acceleration gradient neglected, and it only contains the hydrostatic term. From Eq. (2.29):

$$\frac{dp}{dl} = \frac{g}{g_c} \rho \sin \alpha \tag{2.44}$$

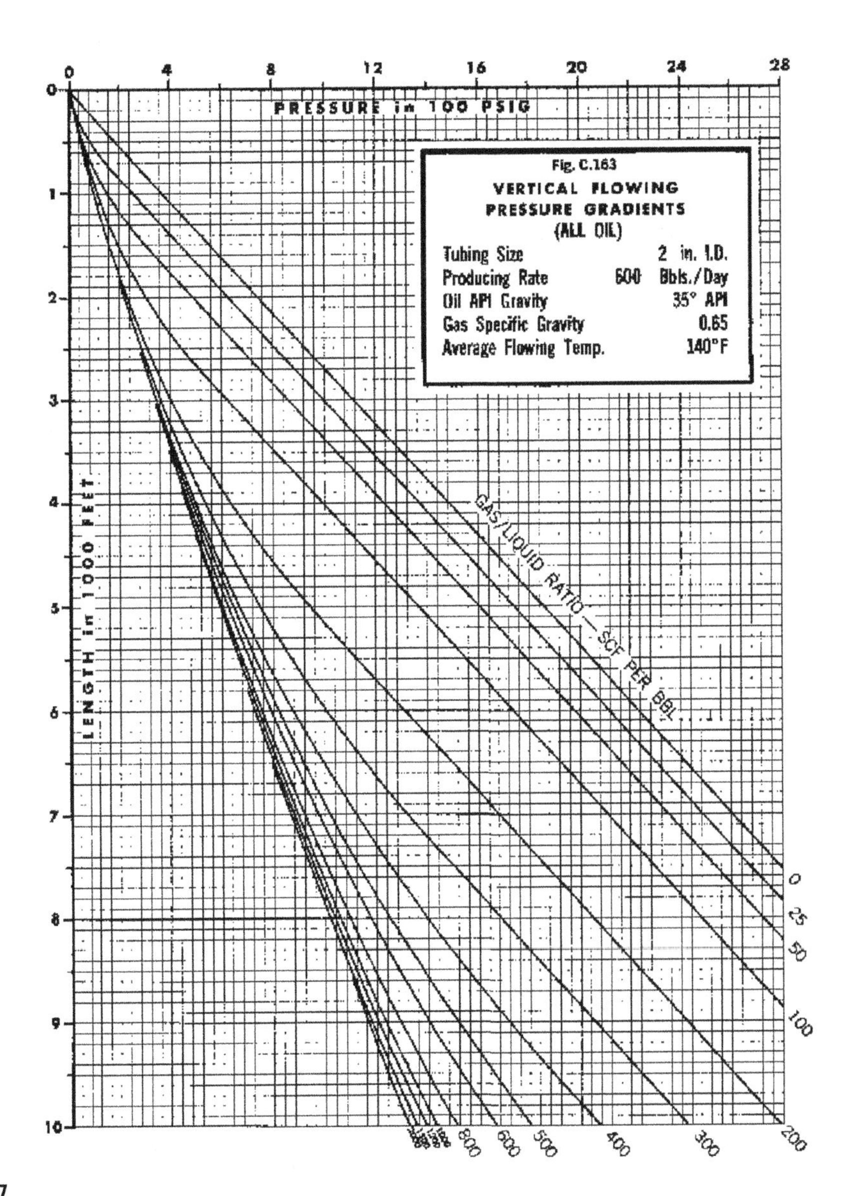

FIGURE 2.7

Example gradient curve sheet from **Brown** [22].

For an inclined pipe dl sin $\alpha = $ dh, where dh is the increment of vertical depth coordinate. Introducing this expression and using psi for the unit of pressure, we get:

$$\frac{\mathrm{d}p}{\mathrm{d}h} = \frac{1}{144}\rho_g \tag{2.45}$$

The actual density of gas, ρ_g, can be evaluated in the knowledge of its specific gravity and volume factor. Using Eqs (2.3) and (2.21), gas density can be written as:

$$\rho_g = \frac{\gamma_{gsc}}{B_g} = \frac{0.0764\gamma_g}{0.0283\frac{ZT_a}{p}} \tag{2.46}$$

Upon substitution of this expression into the original equation the following differential equation is reached:

$$\frac{dp}{dh} = 0.01877\gamma_g \frac{p}{ZT_a} \tag{2.47}$$

where:

dp/dh = gas pressure gradient, psi/ft,
γ_g = gas specific gravity, –,
p = pressure, psi,
Z = deviation factor, –,
T_a = absolute temperature, °R.

The above formula cannot be analytically solved, since it contains the empirical functions of deviation factor Z and actual temperature T. In order to develop a numerical solution, it is assumed that Z and T are constants for a small vertical depth increment Δh. Using these assumptions, Eq. (2.47) can already be solved between two points in the annulus as:

$$p_2 = p_1 \exp\left(0.01877\gamma_g \frac{\Delta h}{Z_{avg}T_{avg}}\right) \tag{2.48}$$

where:

p_1 = gas column pressure at depth h, psi,
p_2 = gas column pressure at depth $h + \Delta h$, psi,
Z_{avg} = average Z factor between points 1 and 2, –,
T_{avg} = average absolute temperature between points 1 and 2, °R,
Δh = vertical distance between points 1 and 2, ft.

Equation (2.48) provides a simple way for approximate calculations of gas column pressures. This involves a trial-and-error procedure, the main steps of which are detailed in the following. Note that the bottom pressure will be calculated using only one depth increment, i.e., Δh is set to the total height of the gas column.

1. Assume a value for the pressure at the bottom of the column, which is to be calculated.
2. Calculate an average pressure in the column.
3. From temperature distribution data, find an average temperature in the column.
4. Obtain a deviation factor for the above conditions.
5. Use Eq. (2.48) to find a calculated value of gas column pressure.

6. If the calculated and the assumed pressures are not close enough, use the calculated one for a new assumed pressure and repeat the calculations. This procedure is continued until the two values are sufficiently close.

EXAMPLE 2.10: LIQUID LEVEL IN THE ANNULUS OF A PUMPING WELLS IS AT A DEPTH OF 4,500 FT FROM THE SURFACE. FLOWING WELLHEAD PRESSURE IS 300 PSIA AND THE PRODUCED GAS IS OF 0.7 SPECIFIC GRAVITY. WELLHEAD TEMPERATURE IS 100 °F AND THE TEMPERATURE AT THE LIQUID LEVEL IS 180 °F.

Solution

Assume a p_2 of 330 psia at the bottom of the gas column. Average pressure in the column:

$$p_{avg} = (300 + 330)/2 = 315 \text{ psia.}$$

Average temperature is calculated as:

$$T_{avg} = (100 + 180)/2 = 140\ °\text{F} = 140 + 460 = 600\ °\text{R}.$$

The pseudocritical parameters of the gas are calculated by Eqs (2.23) and (2.24):

$$p_{pc} = 709.6 - 58.7 \times 0.7 = 669 \text{ psia,}$$

$$T_{pc} = 170.5 + 307.3 \times 0.7 = 386\ °\text{R}.$$

The pseudo-reduced parameters for the average conditions are found with Eqs (2.17) and (2.18):

$$p_{pr} = 315/669 = 0.47,$$

$$T_{pr} = 600/386 = 1.54.$$

Deviation factor Z is read off Fig. 2.4:

$$Z = 0.955.$$

Calculated p_2 is evaluated using Eq. (2.48):

$$p_2 = 300 \exp(0.01877 \times 0.7 \times 4,500/0.955/600) = 300 \exp(0.1032) = 332.6 \text{ psia.}$$

This is not equal to the assumed value of 330 psia; thus a new trial is made. The average pressure with the new assumed bottom pressure of 332.6 psia is calculated by:

$$p_{avg} = (300 + 332.6)/2 = 316.3 \text{ psia.}$$

Average and pseudo-reduced temperatures did not change; pseudo-reduced pressure is:

$$p_{pr} = 316.3/669 = 0.473$$

Using Fig. 2.4:

$$Z = 0.952$$

The new value for p_2:

$$p_2 = 300 \exp(0.01877 \times 0.7 \times 4,500/0.952/600) = 300 \exp(0.1035) = 332.7 \text{ psia.}$$

Since this is sufficiently close to the assumed value of 332.6 psia, gas pressure at the given level is 332.7 psia.

The calculation scheme previously detailed has its inherent errors that are due to the use of only one depth step. The proper use of Eq. (2.48), however, requires that the assumption of a constant Z factor and a constant temperature be met in the given depth increment Δh. This can only be assured by dividing the total gas column length into sufficiently small sections. The smaller the depth increments taken, the smaller the variation of Z and T in the incremental column height and the more accurately the actual conditions are approximated. Therefore, to ensure maximum accuracy it is essential to use a stepwise iteration procedure with sufficiently small depth increments.

2.3.5.2 Computer solution

This calculation model easily lends itself to computer solutions. Figure 2.8 offers the flowchart of a program that follows the above procedure. The basic steps of the calculations are essentially the same as

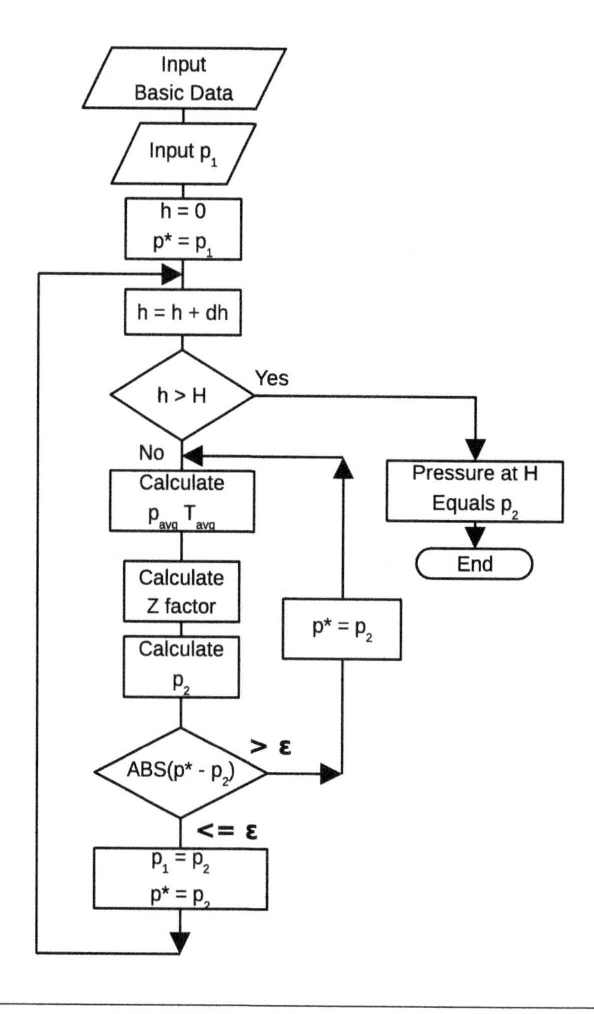

FIGURE 2.8

Flowchart for calculating pressure distribution in a static gas column.

detailed before, the only difference being that the total column length is divided into several increments. For the first increment, the pressure at the upper point equals the known surface pressure p_1. Starting from this, pressure p_2 valid at the lower point of the increment is calculated. Finding p_2 involves a trial-and-error procedure and using a first guess of $p^* = p_1$ is convenient. After the assumed value p^* and the calculated p_2 have converged, the next depth increment is taken. Pressure p_1 at the top of the next lower increment must equal the pressure at the bottom of the previous increment. The procedure is repeated for subsequent lower depth increments until reaching the bottom of the gas column. At the last increment, the final p_2 value gives the pressure at the bottom of the gas column.

EXAMPLE 2.11: USING THE DATA OF THE PREVIOUS EXAMPLE, CALCULATE THE PRESSURE EXERTED BY THE GAS COLUMN WITH THE DETAILED PROCEDURE. USE A DEPTH INCREMENT OF 450 FT.

Solution

Calculation results are contained in Table 2.1 The accuracy for subsequent p_2 values was set to 0.1 psia. As seen from the table, only two trials per depth increment were necessary. The final result of 332.5 psia compares favorably with the 332.7 psia calculated in the previous example.

Table 2.1 Calculation results for Example 2.11

H, ft	p_1, psia	p^*, psia	p_{avg}, psia	T_{avg}, °R	Z	p_2, psia
450	300.00	300.00	300.00	564	0.955	303.31
		303.31	301.65	564	0.955	303.31
900	303.31	303.31	303.31	572	0.956	306.60
		306.60	304.95	572	0.956	306.60
1,350	306.60	306.60	306.60	580	0.957	309.88
		309.88	308.24	580	0.957	309.88
1,800	309.88	309.88	309.88	588	0.958	313.15
		313.15	311.52	588	0.958	313.15
2,250	313.15	313.15	313.15	596	0.959	316.40
		316.40	314.78	596	0.958	316.40
2,700	316.40	316.40	316.40	604	0.960	319.64
		319.64	318.02	604	0.959	319.64
3,150	319.64	319.64	319.64	612	0.961	322.87
		322.87	321.26	612	0.960	322.87
3,600	322.87	322.87	322.87	620	0.962	326.09
		326.09	324.48	620	0.962	326.09
4,050	326.09	326.09	326.09	628	0.963	329.29
		329.29	327.69	628	0.963	329.29
4,500	329.29	329.29	329.29	636	0.964	332.47
		332.47	330.88	636	0.964	332.48

2.3.5.3 Universal gas gradient chart

Investigations of calculated pressure traverses in static gas columns have shown that pressure vs depth curves valid for a given gas specific gravity and a given surface pressure can be approximated very closely by straight lines. An explanation of this observation is that the effects on the gas density of the increase in temperature down the well and of the increasing pressure tend to offset each other. The use of average pressure gradients is, therefore, a viable approach to the calculation of gas column pressures.

Figure 2.9 presents a chart constructed using the above logic, where pressure gradients are plotted for different gas specific gravities in the function of surface pressure. A similar chart is given by **Brown** and **Lee** [23] that was developed by using a constant deviation factor, as found by the present author. Figure 2.9, on the contrary, was constructed with the proper consideration of gas deviation factors and can thus provide higher accuracy.

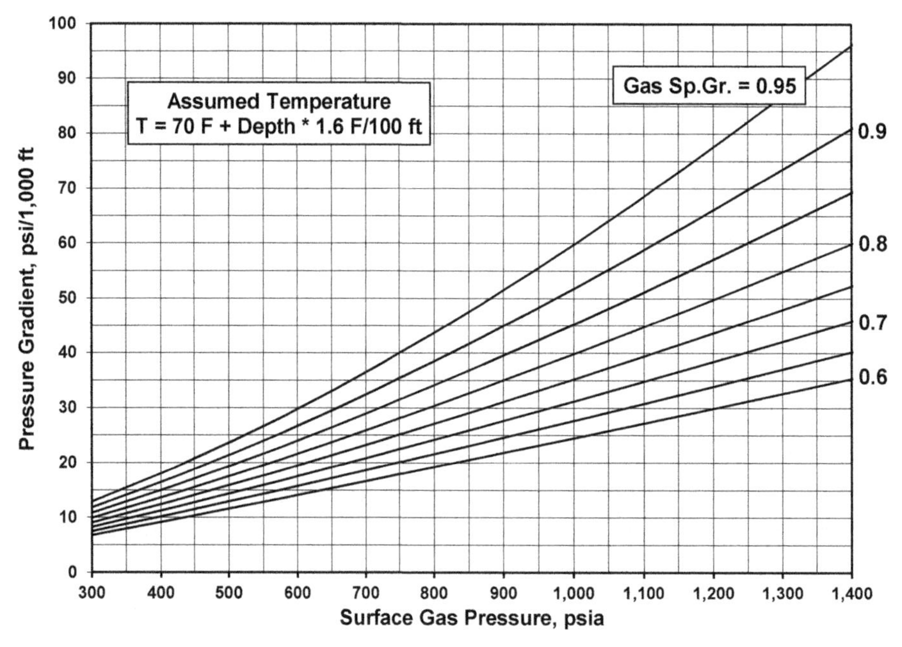

FIGURE 2.9

Static gas pressure gradient chart.

Appendix A contains a full-page copy of the above chart to be used for estimating annular gas column pressures.

EXAMPLE 2.12: USE FIG. 2.9 TO ESTIMATE GAS COLUMN PRESSURE FOR THE PREVIOUS EXAMPLE.

Solution

At $p_1 = 300$ psia surface pressure and a specific gravity of 0.7, pressure gradient is found as 7 psia/1,000 ft. Gas column pressure with this value:

$$p_2 = p_1 + \text{grad}\ H/1,000 = 300 + 74.5 = 331.5\ \text{psia}.$$

The discrepancy of this result from the value calculated in Example 2.11 is due to the difference of actual and chart basis temperatures.

2.4 INFLOW PERFORMANCE OF OIL WELLS

2.4.1 INTRODUCTION

The proper design of any artificial lift system requires an accurate knowledge of the fluid rates that can be produced from the reservoir through the given well. Present and also future production rates are needed to accomplish the following basic tasks of production engineering:

- selection of the right **type** of lift,
- detailed **design** of production equipment, and
- estimation of future well performance.

The production engineer, therefore, must have a clear understanding of the effects governing **fluid inflow** into a well. Lack of information may lead to overdesign of production equipment or, on the contrary, equipment limitations may restrict attainable liquid rates. Both of these conditions have an undesirable impact on the economy of artificial lifting and can cause improper decision making as well.

A well and a productive formation are interconnected at the **sandface**, the cylindrical surface where the reservoir is opened. As long as the well is kept shut in, sandface pressure equals reservoir pressure, and thus no inflow occurs to the well. It is easy to see that, in analogy to flow in surface pipes, fluids in the reservoir flow only between points having different pressures. Thus a well starts to produce when the pressure at its sandface is decreased below reservoir pressure. Fluid particles in the vicinity of the well then move in the direction of pressure decrease and, after an initial period, a **stabilized rate** develops. This rate is controlled mainly by the pressure prevailing at the sandface, but is also affected by a multitude of parameters, such as reservoir properties (rock permeability, pay thickness, etc.), fluid properties (viscosity, density, etc.), and well completion effects (perforations, well damage). These latter parameters being constant for a given well, at least for a considerable length of time, the only means of controlling production rates is the control of **bottomhole pressures**. The proper description of well behavior, therefore, requires that the relationship between bottomhole pressures and the corresponding production rates be established. The resulting function is called the well's **inflow performance relationship** (IPR) and is usually obtained by running **well tests**.

2.4.2 BASIC CONCEPTS

Darcy's Law

The equation describing filtration in porous media was originally proposed by **Darcy** and can be written in any consistent set of units as:

$$\frac{q}{A} = -\frac{k}{\mu}\frac{\mathrm{d}p}{\mathrm{d}l} \tag{2.49}$$

This formula states that the rate of liquid flow, q, per cross-sectional area, A, of a given permeable media is directly proportional to permeability, k, and the pressure gradient, $\mathrm{d}p/\mathrm{d}l$, and is inversely proportional to liquid viscosity, μ. The negative sign is included because flow takes place in the direction of decreasing pressure gradients. Darcy's equation assumes a steady-state, linear flow of a single-phase fluid in a homogeneous porous media saturated with the same fluid. Although these conditions are seldom met, all practical methods are based on Darcy's work.

Drainage Radius

Consider a well producing a stable fluid rate from a homogeneous formation. Fluid particles from all directions around the well flow toward the sandface. In idealized conditions, the drainage area (the area where fluid is moving to the well) can be considered a circle. At the outer boundary of this circle, no flow occurs and undisturbed reservoir conditions prevail. Drainage radius, r_e, is the radius of this circle and represents the greatest distance of the given well's influence on the reservoir under steady-state conditions.

Average Reservoir Pressure

The formation pressure outside the drainage area of a well equals the undisturbed reservoir pressure, p_R, which can usually be considered a steady value over longer periods of time. This is the same pressure as the bottomhole pressure measured in a shut-in well, as seen in Fig. 2.10.

Flowing Bottomhole Pressure

Figure 2.10 shows the pressure distribution in the reservoir around a producing well. In shut-in conditions, the average reservoir pressure, p_R, prevails around the wellbore, and its value can be measured in the well as Static bottomhole pressure (***SBHP***). After flow has started, bottomhole pressure is decreased and pressure distribution at intermediate times is represented by the dashed lines. In steady state, the well produces a stabilized liquid rate and its bottomhole pressure attains a stable value, p_{wf}. The solid line on Fig. 2.10 shows the pressure distribution under these conditions.

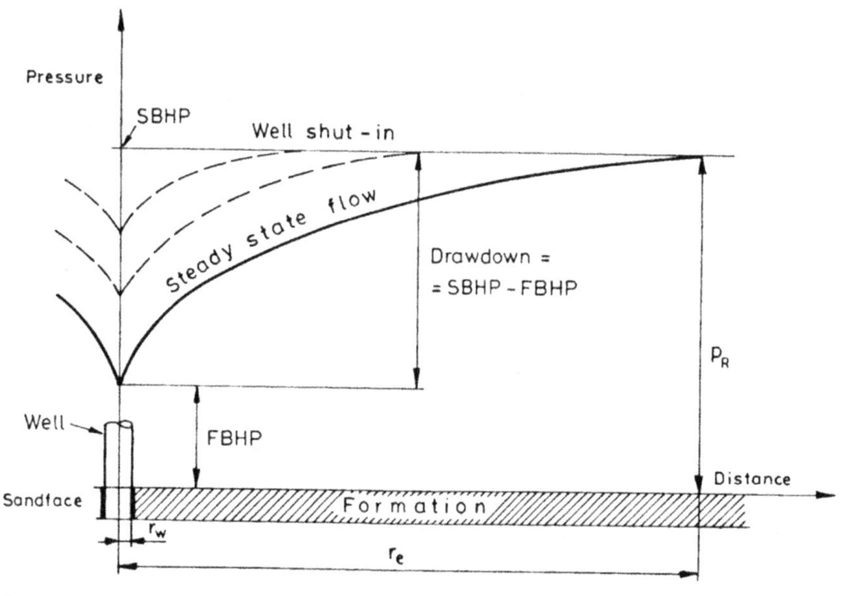

FIGURE 2.10

Pressure distribution in the formation around a well.

Pressure Drawdown

The difference between static and flowing bottomhole pressures is called pressure drawdown. This drawdown causes the flow of formation fluids into the well and has the greatest impact on the production rate of a given well.

2.4.3 THE PRODUCTIVITY INDEX MODEL

The simplest approach to describe the inflow performance of oil wells is the use of the **productivity index** concept. It was developed using the following simplifying assumptions:

- flow is **radial** around the well,
- a single-phase, **incompressible** liquid is flowing,
- permeability distribution in the formation is **homogeneous**, and
- the formation is fully saturated with the given liquid.

For the previous conditions, **Darcy's equation** can be solved to find the production rate of the well:

$$q = \frac{0.00708kh}{\mu B \ln\left(\frac{r_e}{r_w}\right)}\left(p_R - p_{wf}\right) \tag{2.50}$$

where:

q = liquid rate, STB/d,
k = effective permeability, mD,
h = pay thickness, ft,
μ = liquid viscosity, cP,
B = liquid volume factor, bbl/STB,
r_e = drainage radius of well, ft,
r_w = radius of wellbore, ft,
p_R = reservoir pressure, psi, and
p_{wf} = flowing bottomhole pressure, psi.

Most parameters on the right-hand side of the above equation are constant, which permits collecting them into a single coefficient called **productivity index**, *PI*:

$$q = PI\left(p_R - p_{wf}\right) \tag{2.51}$$

where:

q = liquid rate, STB/d,
PI = productivity index, STB/d/psi,
p_R = reservoir pressure, psi, and
p_{wf} = flowing bottomhole pressure, psi.

This equation states that liquid inflow into a well is directly proportional to **pressure drawdown**. It plots as a straight line on a pressure-vs-liquid flow rate diagram, as shown in Fig. 2.11. The endpoints of the line are the **average reservoir pressure**, p_R, at a flow rate of zero and the maximum **potential rate** at a bottomhole flowing pressure of zero. This maximum rate is the well's **absolute open flow**

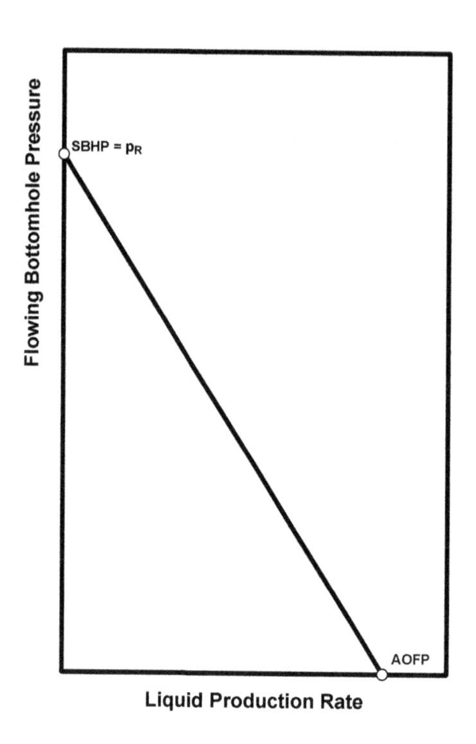

FIGURE 2.11

Well inflow performance according to the constant *PI* concept.

potential (AOFP) and represents the flow rate that would occur if flowing bottomhole pressure could be reduced to zero. In practice, it is not possible to achieve this rate, and it is only used to compare the deliverability of different wells.

The use of the PI concept is quite straightforward. If the average reservoir pressure and the productivity index are known, use of Eq. (2.51) gives the flow rate for any flowing bottomhole pressure. The well's *PI* can either be calculated from reservoir parameters or be measured by taking flow rates at various Flowing bottomhole pressures (*FBHPs*).

EXAMPLE 2.13: A WELL WAS TESTED AT P_{WF} = 1,500 PSI PRESSURE AND PRODUCED Q = 1,000 BPD OF OIL. SHUT-IN BOTTOMHOLE PRESSURE WAS P_{WS} = 2,500 PSI. WHAT IS THE WELL'S *PI* AND WHAT IS THE OIL PRODUCTION RATE AT P_{WF} = 1,000 PSI?

Solution

Solving Eq. (2.51) for *PI* and substituting the test data:

$$PI = q/\left(p_R - p_{\mathrm{wf}}\right) = 1,000/(2,500 - 1,500) = 1 \text{ bopd/psi}.$$

The rate at 1,000 psi is found from Eq. (2.51):

$$q = PI\left(p_{\mathrm{ws}} - p_{\mathrm{wf}}\right) = 1(2,500 - 1,000) = 1,500 \text{ bopd}.$$

2.4.4 INFLOW PERFORMANCE RELATIONSHIPS

2.4.4.1 Introduction

In many wells on artificial lift, bottomhole pressures below **bubble point pressure** are experienced. Thus, there is a **free gas** phase present in the reservoir near the wellbore, and the assumptions that were used to develop the PI equation are no longer valid. This effect was observed by noting that the productivity index was not a constant as suggested by Eq. (2.51). Test data from such wells indicate a downward-curving line, instead of the straight line shown in Fig. 2.11.

The main cause of a curved shape of inflow performance is the **liberation** of solution gas due to the decreased pressure in the vicinity of the wellbore. This effect creates an increasing gas saturation profile toward the well and simultaneously decreases the effective permeability to liquid. Liquid rate is accordingly decreased in comparison to single-phase conditions, and the well produces less liquid than indicated by a straight-line PI curve. Therefore, the constant PI concept cannot be used for wells producing below the bubble point pressure. Such wells are characterized by their **inflow performance relationship** (or IPR) curve, to be discussed in the following.

2.4.4.2 Vogel's inflow performance relationship correlation

Vogel used a numerical reservoir simulator to study the inflow performance of wells depleting solution gas drive reservoirs. He considered cases below bubble point pressure and varied pressure drawdowns, fluid properties, and rock properties. After running several combinations on the computer, Vogel found that all the calculated IPR curves exhibited the same general shape [24]. This shape is best approximated by a dimensionless equation given as follows, and is graphically depicted in Fig. 2.12.

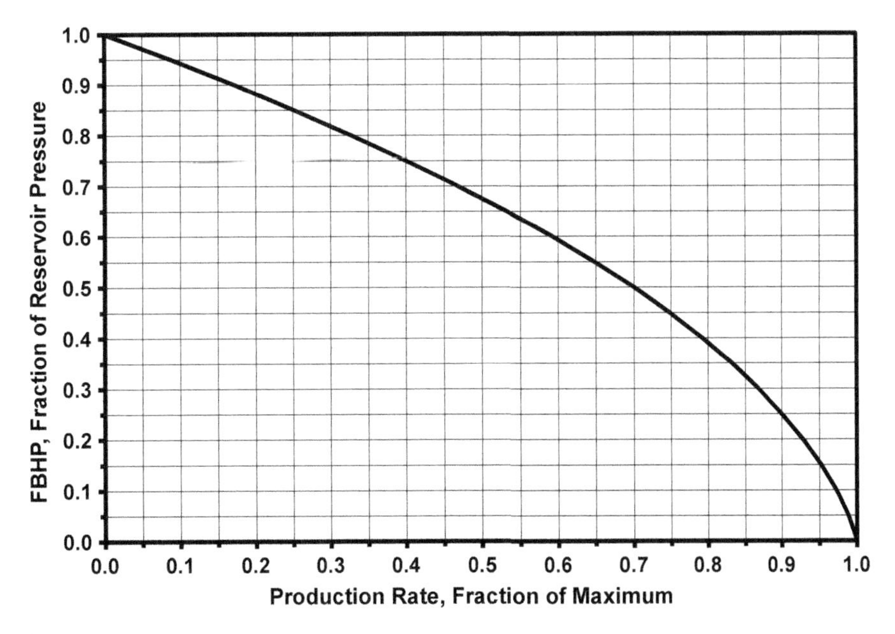

FIGURE 2.12

Vogel's [24] dimensionless inflow performance curve.

$$\frac{q}{q_{max}} = 1 - 0.2\frac{p_{wf}}{p_R} - 0.8\left(\frac{p_{wf}}{p_R}\right)^2 \quad (2.52)$$

where:

q = production rate at bottomhole pressure p_{wf}, STB/d,
q_{max} = maximum production rate, STB/d, and
p_R = average reservoir pressure, psi.

Although Vogel's method was originally developed for solution gas drive reservoirs, the use of his equation is generally accepted for other drive mechanisms as well. It was found to give reliable results for almost any well with a bottomhole pressure below the bubble point of the crude.

In order to use Vogel's method, reservoir pressure needs to be known, along with a single stabilized rate and the corresponding flowing bottomhole pressure. With these data it is possible to construct the well's IPR curve by the procedure discussed in the following example problem.

EXAMPLE 2.14: USING DATA OF THE PREVIOUS EXAMPLE FIND THE WELL'S MAXIMUM FLOW RATE (AOFP) AND CONSTRUCT ITS IPR CURVE, BY ASSUMING MULTIPHASE FLOW IN THE RESERVOIR.

Solution

Substituting the test data into Eq. (2.52):

$$1,000/q_{max} = 1 - 0.2(1,500/2,500) - 0.8(1,500/2,500)^2 = 0.592.$$

From the above equation the AOFP of the well:

$$q_{max} = 1,000/0.592 = 1,689 \text{ bopd}.$$

Now find one point on the IPR curve where p_{wf} = 2,000 psi using Fig. 2.12.
p_{wf}/p_R = 2,000/2,500 = 0.8, and from Fig. 2.12:

$$q/q_{max} = 0.32, \text{ and } q = 1,689 \times 0.32 = 540 \text{ bopd}.$$

The remaining points of the IPR curve are evaluated similarly.

Figure 2.13 shows the calculated IPR curve along with a straight line valid for PI = 1, as found in Example 2.13. Calculated parameters for the two cases are listed as follows:

	Max. rate	Rate at 1,000 psi
Vogel	1,689	1,330
Constant PI	2,500	1,500

Comparison of the preceding results indicates that considerable errors can occur if the constant PI method is used for conditions below the bubble point pressure.

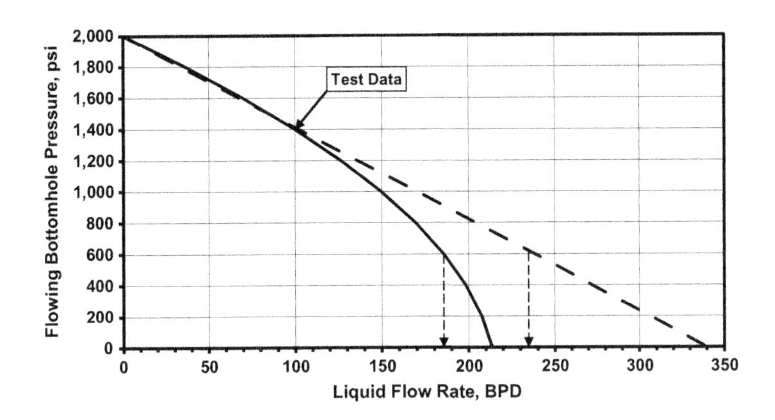

FIGURE 2.13

Comparison of IPR curves for Example 2.14.

2.4.4.3 The composite inflow performance relationship curve

The **Vogel** correlation, as discussed in the previous section, can be applied:

- if the well's flowing bottomhole pressures are below the bubble point pressure, and
- if only oil is produced.

The **composite IPR curve** introduced by **Brown** [25] eliminates these restrictions and provides a way to describe the well's inflow performance in a broad range of conditions.

It should be clear that inflow conditions at pressures greater than the bubble point pressure are described by the constant PI principle discussed previously. According to **Darcy's law**, however, the same constant productivity index should control the inflow conditions of a well producing water only. Thus, in wells with a water cut of 100%, inflow for any conditions is described by the productivity index alone.

Wells producing liquids with a water cut of less than 100% and with pressures lower than the bubble point pressure should have IPR curves somewhere between the curve valid for pure oil (the Vogel correlation) and the one valid for pure water production. The schematic description of such **composite IPR curves** is presented in Fig. 2.14.

The composite curves shown in the figure exhibit three distinct intervals:

1. Well inflow at pressures greater than the bubble point pressure is along a straight line having a slope equal to the *PI* (productivity index).
2. For liquid rates less than the maximum oil rate $q_{o\ \max}$, the slope of the curve is composed of the slope for the Vogel curve and the constant PI curve.
3. When the well produces liquid rates greater than its maximum oil rate $q_{o\ \max}$, a straight line is used because the production is mostly water.

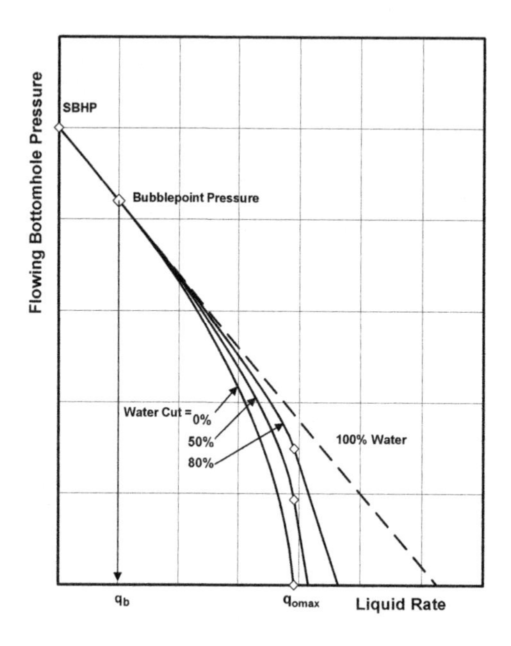

FIGURE 2.14

Schematic depiction of composite IPR curves.

In the following, the calculation of flowing bottomhole pressures for given liquid rates is described while using the **composite IPR** curve principle. This enables one to construct the well's IPR curve and to find *FBHPs* for any rate or to find the liquid rates belonging to any *FBHP* value. The construction of the IPR **curve** is done differently in each of the intervals introduced previously.

For water cuts of 100% or for liquid rates less than the rate valid at the bubble point pressure the following formula can be used:

$$FBHP = SBHP - \frac{q_l}{PI} \tag{2.53}$$

Flowing bottomhole pressures for water cuts less than 100% and liquid rates less than the well's maximum oil rate are found from:

$$FBHP = f_w\left(SBHP - \frac{q_l}{PI}\right) + 0.125 f_o\ p_b\left(-1 + \sqrt{81 - 80\frac{q_l - q_b}{q_{o\,max} - q_b}}\right) \tag{2.54}$$

Finally, for water cuts less than 100% and liquid rates greater than the well's maximum oil rate, *FBHP* is calculated from the expression:

$$FBHP = f_w\left(SBHP - \frac{q_{o\,max}}{PI}\right) - (q_l - q_{o\,max})\,\text{slope} \tag{2.55}$$

where:

$SBHP$ = static bottomhole pressure, psi,
PI = productivity index, bpd/psi,
q_l = liquid flow rate, bpd,
f_w = water cut, –,
f_o = produced oil fraction, –,
p_b = bubble point pressure, psi.

The previous formulas include the following unknown parameters:

PI, the slope of the straight portion of the IPR curve;
q_b, the liquid rate at the bubble point pressure;
$q_{o\,\max}$, the well's maximum oil rate;
slope, the slope of the IPR curve at liquid rates greater than $q_{o\,\max}$.

These parameters are evaluated from production well tests providing liquid rates at different *FBHPs*. Their determination, based on a single well test, depends on the relation of the actual *FBHP* and the well's *SBHP* to the bubble point pressure and is accomplished as detailed in the following.

Case One: *SBHP* and test pressure above the bubble point pressure

The well's *PI* is easily found from Eq. (2.51) as:

$$PI = \frac{q_{\text{test}}}{SBHP - FBHP_{\text{test}}} \tag{2.56}$$

The liquid rate valid at a bottomhole pressure equal to the bubble point pressure is calculated as given here:

$$q_b = PI\left(SBHP - p_b\right) \tag{2.57}$$

The maximum oil rate when producing 100% oil is found from:

$$q_{o\,\max} = q_b + \frac{PIp_b}{1.8} \tag{2.58}$$

The slope of the composite curve's linear section is evaluated from:

$$\text{slope} = \frac{f_w \frac{0.001 q_{o\,\max}}{PI} + 0.125 f_o p_b \left(-1 + \sqrt{81 - \frac{80\left(0.999 q_{o\,\max} - q_b\right)}{q_{o\,\max} - q_b}}\right)}{0.001 q_{o\,\max}} \tag{2.59}$$

Finally, the well's maximum liquid production rate is found from:

$$q_{l\max} = q_{o\,\max} + f_w \frac{SBHP - \frac{q_{o\,\max}}{PI}}{\text{slope}} \tag{2.60}$$

With the previous parameters known, flowing bottomhole pressures for any rate are evaluated from Eq. (2.53) through Eq. (2.55), resulting in the points of the composite IPR curve.

Case Two: *SBHP* above and test pressure below the bubble point pressure.

Since the point belonging to the production test does not fall on the linear portion of the IPR curve, the productivity index can only be determined by a trial-and-error procedure. The iterative determination of the proper *PI* value is facilitated by assuming the following initial value:

$$PI_{initial} = \frac{q_{test}}{f_o\left\{SBHP - p_b + \frac{p_b\left[1-0.2\frac{FBHP_{test}}{p_b}-0.8\left(\frac{FBHP_{test}}{p_b}\right)^2\right]}{1.8}\right\} + f_w(SBHP - FBHP_{test})} \quad (2.61)$$

Based on this *PI*, the calculation sequence detailed for **Case One** is followed and first the parameters q_b, $q_{o\ max}$, and *slope* are determined. Based on those, the flowing bottomhole pressure belonging to the test rate q_{test} is found. If the calculated value differs from the measured *FBHP*, a new iteration step is required. Usually, only a few iterations are needed to arrive at the proper *PI* value. Based on the final value, the well's **composite IPR curve** is evaluated from Eqs (2.53)–(2.55).

Case Three: *SBHP* and test pressure below the bubble point pressure

Again, iteration is required to arrive at the proper *PI* value and the use of the following initial value is advised:

$$PI_{initial} = \frac{q_{test}}{f_o\frac{SBHP\left[1-0.2\frac{FBHP_{test}}{SBHP}-0.8\left(\frac{FBHP_{test}}{SBHP}\right)^2\right]}{1.8} + f_w(SBHP - FBHP_{test})} \quad (2.62)$$

The calculation steps for the construction of the **composite IPR curve** are identical to those detailed for **Case Two,** with the following modification: in the relevant formulas the substitutions $p_b = SBHP$ and $q_b = 0$ must be used.

EXAMPLE 2.15: CALCULATE THE POINTS OF THE COMPOSITE IPR CURVE FOR A WELL THAT PRODUCES A LIQUID RATE OF 150 BPD AT A FLOWING BOTTOMHOLE PRESSURE OF 2,200 PSI AND HAS A STATIC BOTTOMHOLE PRESSURE OF 2,500 PSI. THE PRODUCED OIL'S BUBBLE POINT PRESSURE AT BOTTOMHOLE CONDITIONS IS 2,100 PSI AND THE PRODUCING WATER CUT EQUALS 60% ($F_W = 0.6$).

Solution

Since both the *SBHP* and the $FBHP_{test}$ values are less than the bubble point pressure, calculations follow those described for **Case One**.

The *PI* of the well is found from Eq. (2.56), and using the measured test data we get:

$$PI = 150/(2,500 - 2,200) = 0.5 \text{ bpd/psi}.$$

The liquid rate at the bubble point pressure is found from Eq. (2.57):

$$q_b = 0.5(2,500 - 2,100) = 200 \text{ bpd}.$$

The maximum oil rate, as found from Eq. (2.58), equals:

$$q_{o\ max} = 200 + 0.5 \times 2,100/1.8 = 783 \text{ bpd}.$$

The slope of the linear portion of the curve is evaluated from Eq. (2.59):

$$\text{slope} = \left\{0.6 \times 0.001 \times 783/0.5 + 0.125 \times 0.4 \times 2,100\left[-1 + (81 - 80(0.999 \times 783 - 200)/(783 - 200))^{0.5}\right]\right\} / (0.001 \times 783) = 8.22.$$

After these preliminary calculations, points on the three sections of the composite IPR curve can be evaluated. In the following, detailed calculations for only one point in each interval are presented.

Above the bubble point, take the liquid rate of 100 bpd and find the corresponding *FBHP* from Eq. (2.53) as follows:

$$FBHP = 2,500 - 100/0.5 = 2,300 \text{ psi}.$$

In the next interval, take a rate of 600 bpd to find the *FBHP* required from Eq. (2.54):

$$FBHP = 0.6(2,500 - 600/0.5) + 0.125 \times 0.4 \times 2,100\left\{-1 + [81 - 80(600 - 200)/(783 - 200)]^{0.5}\right\} = 1,212 \text{ psi}.$$

Finally, take a liquid rate $q_l = 820$ bpd, greater than the maximum oil rate, and find the corresponding *FBHP* from Eq. (2.55):

$$FBHP = 0.6(2,500 - 783/0.5) - (820 - 783)8.22 = 256.2 \text{ psi}.$$

2.5 THE BASICS OF NODAL ANALYSIS

System analysis of producing oil and gas wells (often called **nodal analysis**) is the latest addition to the petroleum engineer's arsenal of design and analysis tools. The methodology and calculation procedures developed in the last three decades are based on the recognition that the underground reservoir, the producing well, and the surface liquid and gas handling equipment constitute a complex, **interrelated** system. Accordingly, any process in any element of the system entails changes that occur not only in the given part but also in the system as a whole. This section introduces the basic principles of system analysis as adapted to the description of producing well behavior.

2.5.1 INTRODUCTION

Petroleum fluids found in an underground reservoir move through a complex system to reach their destinations on the surface. This system is called the **production system** and comprises the following main components: the **reservoir**, the producing **well**, the surface **flowline**, and the **separator**. Some of these can further be divided into smaller elements; for example, the well, besides the tubing string, may contain safety and/or gas lift valves, as well as other components. The production system is thus a system of **interconnected** and **interacting** elements that all have their own specific performance relationships, but each, in turn, also depends upon and influences the other elements. In order to produce fluids from the well, all components of the system must work together. Thus the solution of any fluid production problem requires that the production system be treated as a complete entity.

The outlines of this principle were first given by **Gilbert** [26], the father of production engineering, in the 1950s. He described the interaction of the reservoir, the well, and the wellhead choke and proposed a system-oriented solution for determining the production rate of a flowing well. The practical use of Gilbert's ideas was limited, mainly due to the limitations of the methods available in his time for modeling the performance of the system's elements. During the last decades, however, research into the behavior of the individual hydraulic elements of oil and gas wells has been very intensive. As a result of this progress, there exist today several different theories, calculation procedures, and design procedures that reliably model the performance of the elements of a production system. Good examples for this are the numerous correlations available for calculating pressure traverses in vertical and horizontal pipes.

The wide selection of available calculation models and the advent of computers, which eased the burden of the necessary calculations, led to the reappearance of Gilbert's ideas in the early 1980s [27,28]. The new contributions aim at the **numerical simulation** of the production system's hydraulic behavior, but also enable the **optimization** of the system to produce the desired flow rate most economically.

The system analysis methods and procedures mentioned above were named "**nodal analysis**" by **K. E. Brown**, and the term has generally been accepted. A full treatment of nodal analysis principles has been given by **Beggs** [29]. The application of this theory to flowing and artificially lifted oil wells can have immediate practical and economic advantages.

2.5.2 THE PRODUCTION SYSTEM

The **production system** of any oil or gas well comprises part of the reservoir, the system transporting well fluids to the surface, and the surface separation equipment. These components have their own performance relationships describing their behavior under different flow conditions. The formation is characterized by the laws of flow in porous media, whereas in most of the other components single or multiphase flow in pipes takes place. Accordingly, a proper description of the total system's behavior cannot be achieved without a thorough investigation of each component's performance. The different calculation models developed for the characterization of the system's components, therefore, provide a firm foundation for system analysis.

Consider a simple flowing oil well with only some of the components described above. The schematic drawing of this simplified case is shown in Fig. 2.15. Fluids in the formation flow to the well from as far as the boundary of the drainage area, 1. After entering the well through the sandface, vertical flow in the tubing starts at the well bottom, 2. In the tubing string vertical or inclined tubing flow takes place up to the wellhead, 3. A surface choke bean is installed at 4, which is also the intake point to the flowline. Horizontal or inclined flow in the flowline leads into the separator, 5.

The points in the flowing well's production system designated by the numbers in Fig. 2.15 constitute the "**nodes**" of the system analysis theory. They separate the different components of the system: the formation, the tubing string, the flowline, etc. Between any two node points, flowing pressure decreases in the direction of the flow and the pressure drop can be calculated based on the characteristics of the given component. In this way, the performance of any component is established as the variation of the flow rate with the pressure drop across the given component. This task is accomplished through the application of the different pressure drop calculation models valid for the different kinds of hydraulic components.

Two node points deserve special consideration: the boundary of the drainage area (point 1) and the separator (point 5). These points constitute the two **endpoints** of the production system and their pressures are considered constant for longer periods. Formation pressure at the outer boundary of the drainage area of the well, of course, changes with the depletion of the reservoir, but for production engineering purposes involving short and medium time periods, this change can be **disregarded**.

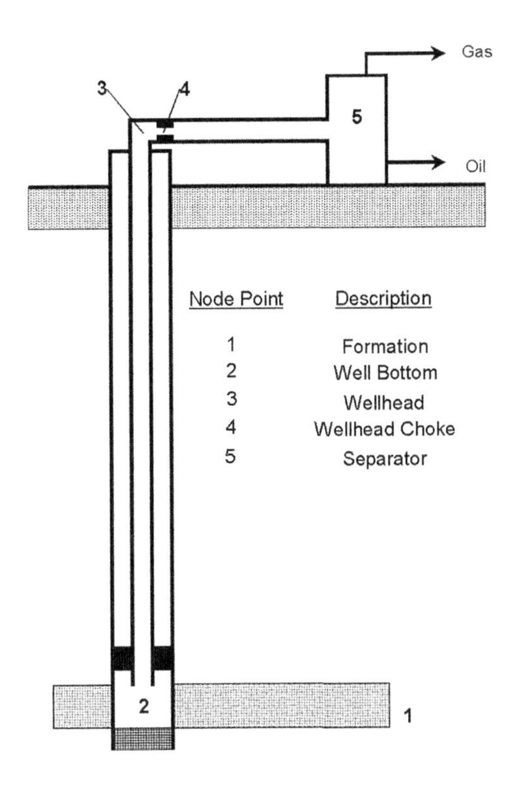

FIGURE 2.15

The production system of a flowing well with the node points.

Separator pressure, at the same time, is usually set for the **whole life** of the field and is held constant. Thus, the two endpoint pressures, i.e., the average reservoir pressure and the separator pressure, can duly be considered **constant** values for the purposes of system analysis.

Any oil or gas well's production system can be divided into its components using appropriately placed nodes. A typical sucker-rod-pumped case is illustrated in Fig. 2.16. The unique features of a common sucker-rod pumping well are the absence of a packer in the well and the connection of the annulus at the casing head to the flowline, usually through a check valve. Due to the annular space being open both downward and at the wellhead, two paths are available for the fluid to move up the hole. One of these is the tubing string, where fluids are lifted to the surface by the downhole pump. The other path open to fluids is in the annulus, where liquid rises to a stationary level, above which a gas column exists. Since the two flow paths are connected at the bottom (node 2), it follows from the hydrostatic law of communicating vessels that the pressure exerted at node 2 by each subsystem must be equal. The result of the requirement is that, contrary to flowing wells, there are two ways to calculate bottomhole pressures in pumping wells: one through the tubing string and the other in the annulus. The dynamic liquid level, therefore, is an indication of the well's actual bottomhole pressure. This observation is of great importance to the system analysis of rod-pumped wells, as will be seen in a later chapter.

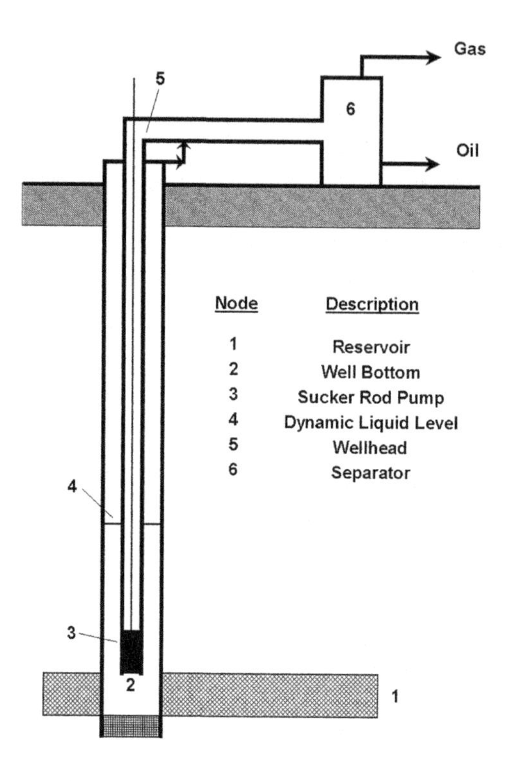

FIGURE 2.16

Nodal system of a typical sucker-rod-pumped well.

2.5.3 BASIC PRINCIPLES

One of the many objectives of system analysis is the determination of the **flow rate** of a given production system. The solution of this problem is illustrated here through the example of a flowing well.

As discussed before, in connection with Fig. 2.15, an oil well can be considered a **series-connected** hydraulic system made up of its components, which are bracketed by the appropriately placed **nodes**. Evaluation of the total system's performance permits the following conclusions to be made:

- Mass flow rate throughout the system is **constant**, although phase conditions change with changes in pressure and temperature.
- Pressure **decreases** in the direction of flow because of the energy losses occurring in the various system components.
- At node points, input pressure to the next component **must** equal the output pressure of the previous component.
- System parameters being **constant** for considerable periods are: the **endpoint** pressures at the separator and in the reservoir; the wellbore and surface **geometry** data (pipe diameters, lengths, etc.); and the **composition** of the fluid entering the well bottom.

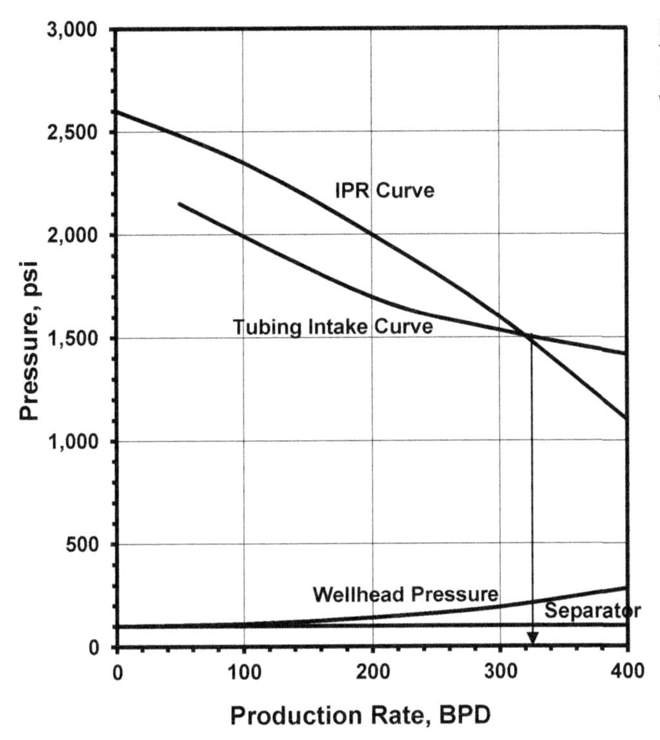

FIGURE 2.17

Example system analysis for a flowing oil well after [30].

Taking into account these specific features of the production system, a procedure can be devised to find the flow rate at which the system will produce. This starts with dividing the system into two **subsystems** at an appropriately selected node called the **"solution node."** The next step is to find pressure vs rate curves for each subsystem. These functions are constructed started from the node points with known pressures at the separator and at the well bottom. The intersection of the two curves gives the **cooperation** of the subsystems and thus the desired rate.

A simple example is shown in Fig. 2.17 [30]. The well is a low producer and has no surface or downhole chokes installed. The well's production system is divided at node 2 with one subsystem consisting of the flowline and the tubing string, and the other being the formation. The pressure-vs-rate diagram of the formation is the familiar **IPR curve**. The other curve is constructed by summing the separator pressure and the pressure drops in the flowline (wellhead pressure curve) and by further adding to these values the pressure drops occurring in the tubing string. The resulting curve is the **tubing intake pressure** vs production rate. The total system's rate is found at the intersection of this curve with the IPR curve and is 320 bpd in the present example.

The same procedure can be followed starting from a different node position, but it will give the same result. This shows that system analysis allows a flexibility to study different situations.

REFERENCES

[1] Gould TL. Vertical two-phase steam-water flow in geothermal wells. JPT August 1974:833–42.
[2] Standing MB. A pressure-volume-temperature correlation for mixtures of California oils and gases. API Drill Prod Pract 1947:275–86.
[3] Lasater JA. Bubble point pressure correlation. Trans AIME 1958;213:379–81.
[4] Vazquez M, Beggs HD. Correlations for fluid physical property prediction. Paper SPE 6719 presented at the 52nd annual fall technical conference and exhibition of SPE, Denver, Colorado. October 9–12, 1977.
[5] Beal C. The viscosity of air, natural Gas, crude oil and its associated gases at oil field temperatures and pressures. Trans AIME 1946;165:94–112.
[6] Chew J, Connally Jr CH. A viscosity correlation for gas-saturated crude oils. Trans AIME 1959;216:23–5.
[7] Standing MB, Katz DL. Density of natural gases. Trans AIME 1942;146:140–9.
[8] Hankinson RW, Thomas LK, Phillips KA. Predict natural gas properties. Hydrocarbon Processing April 1969:106–8.
[9] Wichert E, Aziz K. Compressibility factor for sour natural gases. Can J Chem Eng April 1971:267–73.
[10] Wichert E, Aziz K. Calculate Z's for sour gases. Hydrocarbon Processing May 1972:119–22.
[11] Sarem AM. Z-Factor equation developed for use in digital computers. OGJ July 20, 1959:64–6.
[12] Gopal VN. Gas Z-factor equations developed for computer. OGJ August 8, 1977:58–60.
[13] Yarborough L, Hall KR. How to solve equation of state for Z-factors. OGJ February 18, 1974:86–8.
[14] Dranchuk PM, Purvis RA, Robinson DB. Computer calculations of natural gas compressibility factors using the standing and katz correlation. In: Inst. Of petroleum technical series No. IP 74-008; 1974.
[15] Takacs G. Comparing methods for calculating Z-factor. OGJ May 15, 1989:43–6.
[16] Papay J. Change of technological parameters in producing gas fields. in Hungarian Proc OGIL Bp 1968: 267–73.
[17] Bradley HB, editor. Petroleum engineering handbook. Society of Petroleum Engineers; 1987 [Chapter 34].
[18] Moody LF. Friction factors for pipe flow. Trans ASME 1944;66:671.
[19] Fogarasi M. Further on the calculation of friction factors for use in flowing gas wells. J Can Petr Techn April–June 1975:53–4.
[20] Gregory GA, Fogarasi M. Alternate to standard friction factor equation. OGJ April 1, 1985:120–7.
[21] Chen NH. An explicit equation for friction factor in pipe. Ind Eng Chem Fund 1979;18:296.
[22] Brown KE. The technology of artificial lift methods, vol. 3a. Tulsa OK: PennWell Books; 1980.
[23] Brown KE, Lee RL. Easy-to-use charts simplify intermittent gas lift design. WO February 1, 1968:44–50.
[24] Vogel JV. Inflow performance relationships for solution-gas drive wells. JPT January 1968:83–92.
[25] Brown KE. The technology of artificial lift methods, vol. 4. Tulsa OK: PennWell Books; 1984.
[26] Gilbert WE. Flowing and gas-lift well performance. API Drill Prod Pract 1954:126–57.
[27] Proano EA, Mach JM, Brown KE. Systems analysis as applied to producing wells. Mexico City: Congreso Panamericano de Ingeniera del Petroleo; March 1979.
[28] Mach JM, Proano EA, Brown KE. A nodal approach for applying systems analysis to the flowing and artificial oil and gas wells. Society of petroleum engineers paper no. 8025; 1979.
[29] Beggs HD. Production optimization using nodal analysis. 2nd ed. OGCI Publications; 2003.
[30] Takacs G, Szilas AP, Sakharov VA. Hydraulic analysis of producing Wells. in Hungarian Koolaj es Foldgaz May 1984:129–36.

CHAPTER

SUCKER-ROD PUMPING SYSTEM COMPONENTS AND THEIR OPERATION

3

CHAPTER OUTLINE

Sucker-Rod Pumping Handbook. http://dx.doi.org/10.1016/B978-0-12-417204-3.00003-0

3.1 INTRODUCTION

The individual components of a sucker-rod pumping system can be divided into two major groups: surface and downhole equipment. The main elements of a common installation are shown in Fig. 3.1.

The **surface equipment** includes the following:

The **prime mover** provides the driving power to the system and can be an electric motor or a gas engine.

The **gear reducer** or gearbox reduces the high rotational speed of the prime mover to the required pumping speed and, at the same time, increases the torque available at its slow speed shaft.

The **pumping unit**, a mechanical linkage that transforms the rotary motion of the gear reducer into the reciprocating motion required to operate the downhole pump. Its main element is the **walking beam**, which works on the principle of a mechanical lever.

The **polished rod** connects the walking beam to the sucker rod string and ensures a sealing surface at the wellhead to keep well fluids within the well.

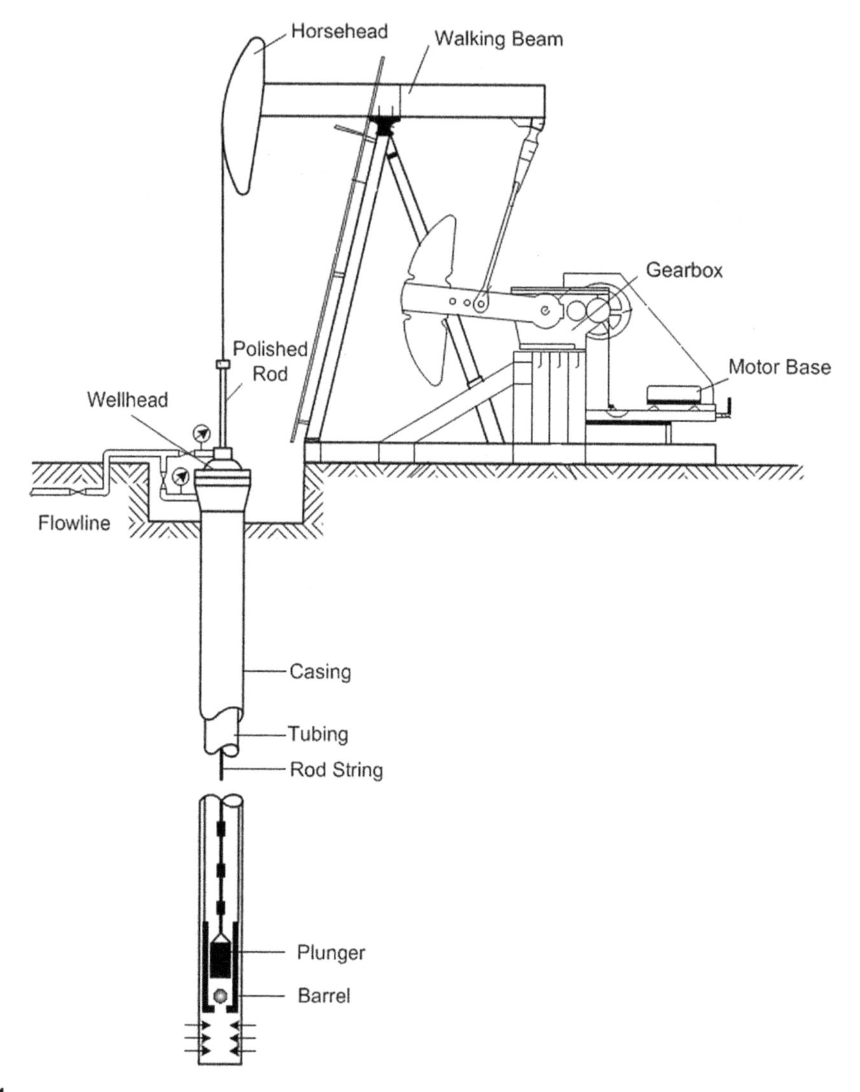

FIGURE 3.1

The components of a sucker-rod pumping system.

The wellhead assembly contains a **stuffing box** that seals on the polished rod and a **pumping tee** to lead well fluids into the flowline. The casing–tubing annulus is usually connected, through a check valve, to the flowline.

The **downhole equipment** includes the following:

The **rod string**, composed of sucker rods, run inside the tubing string of the well. The rod string provides the mechanical link between the surface drive and the subsurface pump.

The **pump plunger**, the moving part of a usual sucker rod pump, is directly connected to the rod string. It houses a ball valve, called **traveling valve**, which, during the upward movement of the plunger, lifts the liquid contained in the tubing.

The pump barrel or **working barrel** is the stationary part (cylinder) of the subsurface pump. Another ball valve, the **standing valve**, is fixed to the working barrel. This acts as a suction valve for the pump, through which well fluids enter the pump barrel during upstroke.

Chapter 3 presents a review of the different elements in the sucker-rod pumping system and includes detailed descriptions of the individual parts. The discussion is started from the well bottom and the different types of subsurface pumps are treated first. The features and operational principles of standard pumps and their structural parts are explained. Ancillary downhole equipment, such as tubing anchors and downhole gas separators, are covered.

The most vital part of the pumping system is the rod string, because its trouble-free operation is critical to the performance of the whole system. According to its importance, great emphasis is laid on the treatment of the sucker-rod string in Section 3.5. The theories of proper mechanical design of rod strings are fully described; all available design procedures are detailed and illustrated by presenting example problems. Information on the evaluation and prevention of rod string failures concludes this section.

The next most important section (Section 3.7) deals with the constructional and operational details of pumping units. The available pumping unit types are described based on their geometrical similarities and detailed procedures are presented for the calculation of their kinematic parameters. At the end of Chapter 3, gear reducers and the different types of prime movers are covered.

3.2 SUBSURFACE PUMPS

3.2.1 INTRODUCTION

The subsurface pumps used in sucker-rod pumping work on the **positive displacement principle** and are of the cylinder and piston type. Their basic parts are the **working barrel** (cylinder), the **plunger** (piston), and two **ball valves**. The valve affixed to the working barrel acts as a suction valve and is called the **standing valve**. The other valve contained in the plunger acts as a discharge valve and is called the **traveling valve**. These valves operate like check valves and their opening and closing during the alternating movement of the plunger provides a means to displace well fluids to the surface.

Before a detailed review of the different pump types, it is important to have a basic understanding of a sucker-rod pump's operation. The discussion of the **pumping cycle**, i.e., the basic period of the pump's operation, is presented in connection with Fig. 3.2. This figure depicts a common sucker-rod pump with the plunger moving inside a stationary barrel. The barrel is connected to the lower end of the tubing string, while the plunger is directly moved by the rod string. The positions of the barrel and the plunger, as well as the operation of the standing and traveling valves, are shown at the two extreme positions of the up- and downstroke. For simplicity of description, pumping of an incompressible fluid, i.e., liquid, is assumed.

At the start of the **upstroke**, after the plunger has reached its lowermost position, the traveling valve closes due to the high hydrostatic pressure in the tubing above it. Liquid contained in the tubing above the traveling valve is lifted to the surface during the upward movement of the plunger. At the same time, the pressure drops in the space between the standing and traveling valves, causing the standing valve to open. Wellbore pressure drives the liquid from the formation through the standing

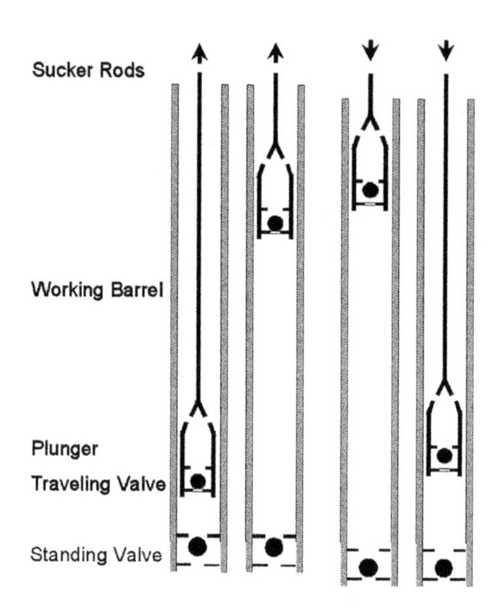

FIGURE 3.2

Schematic description of the pumping cycle.

valve into the barrel below the plunger. Lifting of the liquid column and filling of the barrel with formation liquid continues until the end of the upstroke. It is important to note that during the whole upstroke, the full weight of the liquid column in the tubing string is carried by the plunger and the rod string connected to it. The high pulling force causes the rod string to stretch due to its elasticity.

After the plunger has reached the top of its stroke, the rod string starts to move downwards. The **downstroke** begins, the traveling valve immediately opens, and the standing valve closes. This operation of the valves is due to the incompressibility of the liquid contained in the barrel. When the traveling valve opens, liquid weight is transferred from the plunger to the standing valve, causing the tubing string to stretch. During downstroke, the plunger makes its descent with the open traveling valve inside the barrel filled with formation liquid. At the end of the downstroke, the direction of the rod string's movement is reversed and another pumping cycle begins. Liquid weight is again transferred to the plunger, causing the rods to stretch and the tubing to return to its unstretched state.

This elementary description of the pumping cycle has shown that the sucker-rod pump operates like any single-acting **piston** pump. The most significant difference between a pump used at the surface and a sucker-rod pump lies in the way the piston, or plunger, is driven. In surface pumps, e.g., mud pumps, the rod connecting the driving mechanism to the piston is quite short and its length does not change considerably during operation. Therefore, piston stroke length equals the stroke imposed by the driving mechanism. The subsurface pump's plunger, on the other hand, is operated by a **string** of rods, the length of which can be several thousands of feet. Due to its **elastic** behavior, this long string periodically stretches and recoils, which makes the plunger's movement complex and complicated to predict. Plunger stroke length, therefore, cannot be readily found from the stroke length measured at the surface.

The pumping cycle, as described in Fig. 3.2, assumes **idealized** conditions to prevail:

- Single-phase **liquid** is produced, and
- The barrel is **completely** filled with well fluids during the upstroke.

If any of these conditions are not met, the operation of the pump can seriously be affected. All problems occurring in such situations relate to the changes in **valve action** during the cycle. As already mentioned, both valves are simple **check valves**, which open or close according to the relation of the pressures above and below the valve seat. The valves, therefore, do not necessarily open and close at the two extremes of the plunger's travel. As a consequence, the **effective plunger stroke length**, i.e., the part of the stroke used for lifting well fluids, can be less than total plunger stroke length.

In case well fluids in the barrel contain some **free gas** at the start of the downstroke, the traveling valve remains **closed** as long as this gas is compressed to a pressure sufficient to overcome the liquid column pressure above it. Part of the stroke is taken up by the **gas compression** effect and effective plunger stroke length is accordingly reduced. A similar problem occurs at the start of the upstroke when pumping gassy fluids. Just before the upstroke starts, the gas–liquid mixture occupying the space between the standing and traveling valves is at the hydrostatic pressure of the liquid column in the tubing. When the plunger begins its upward movement with the traveling valve closed, this high-pressure mixture starts to expand, allowing only a gradual pressure decrease below the plunger. This effect delays the **opening** of the standing valve until the pressure above the valve drops to wellbore pressure. The fraction of plunger travel during this process can considerably reduce the stroke length available for the barrel to fill up with liquids. Thus the plunger's **effective** stroke length is decreased again.

These effects can be combined with the **incomplete filling** of the pump barrel during the upstroke. This usually occurs when pump capacity is higher than the inflow rate to the well. All of these conditions can lead to a considerable reduction of the plunger stroke available for lifting well fluids and, consequently, to decreased pumping rates. The design of a rod-pumping installation, therefore, must take into consideration the actual downhole conditions at the pump.

3.2.2 BASIC PUMP TYPES

The two principal categories of sucker-rod pumps are the **tubing** and the insert or **rod pump**. Their basic difference is in the way the working barrel is installed in the well.

In a **tubing pump**, Fig. 3.3, the working barrel is an integral part of the tubing string. It is connected to the bottom of the tubing and run to the desired depth with the tubing string. This construction allows using a barrel diameter slightly less than the tubing inside diameter. Its main disadvantage is that the barrel can only be serviced by **pulling** the entire tubing string.

Below the barrel of a tubing pump, a **seating nipple** is mounted into which the **standing valve** can be locked. After the barrel and the tubing string are already in the well, the plunger with the traveling valve is run on the rod string. The standing valve is attached to the bottom of the plunger by a standing **valve puller** during installation. The standing valve is lowered into the seating nipple, where it locks either mechanically or by the use of friction cups. The standing valve puller is then disengaged, and the plunger is raised to its working position. Removal of the standing valve is also possible with the use of the valve puller. This eliminates the need for pulling the tubing string to repair the standing valve.

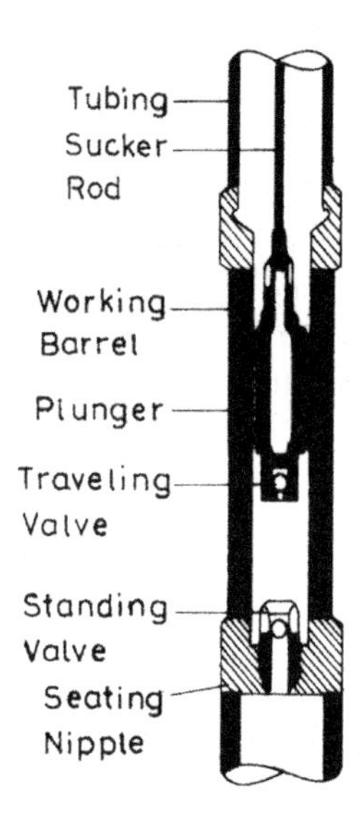

FIGURE 3.3

The basic parts of a tubing pump.

The rod or **insert pump**, contrary to a tubing pump, is a complete **pumping assembly**, which is run into the well on the sucker-rod string. This assembly contains the working **barrel** (also called barrel tube), the **plunger** inside the barrel, and both standing and traveling **valves** (see Fig. 3.4). Only the seating nipple has to be run with the tubing string to the desired pumping depth. Then the pump assembly is run on the rod string, and a mechanical or cup-type hold-down is used to lock it in place. The standing valve of a rod pump is a part of the barrel.

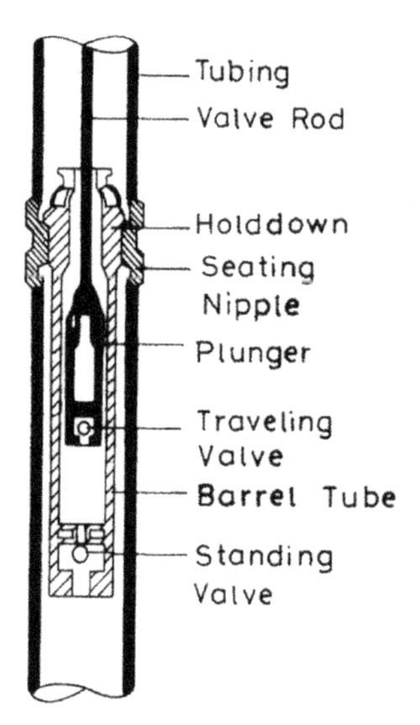

FIGURE 3.4

The basic parts of a rod pump.

The **casing pump** is a variation of the rod pump, which is used in wells without a tubing string. The pump assembly is seated in a **packer**. This type of installation is generally used in wells with high production capacities, because the size of the pump that can be run is limited by the size of the casing only.

The selection of the proper pump type to be used in a given installation is based on several factors. As a rule, tubing pumps can be used to move larger liquid volumes than rod pumps. The biggest disadvantage of tubing pumps is that the whole tubing string must be pulled should the working barrel fail. Several operational problems, such as gas production, sand, and corrosion, should also be considered to make the final decision.

3.2.3 API PUMPS

3.2.3.1 Classification of API pumps

Most sucker-rod pumps used in the world petroleum industry conform to the specifications of the **American Petroleum Institute** (API). The pumps standardized in **API Spec 11AX** [1] have been classified and given a letter designation by API. The designation is a two-letter code for **tubing** pumps, and a three-letter code for **rod** pumps. Explanation of these letter codes follows.

- The first letter refers to the basic type of pump:
  - **T** for tubing pumps and
  - **R** for rod pumps.
- The second letter stands for the type of **barrel**, whether it is a heavy- or a thin-wall barrel; tubing pumps come in heavy-wall barrel only. Different code letters are used for pumps with **metal** plungers and for pumps with **soft-packed** plungers:

Metal Plungers	Soft-Packed Plungers
• **H** for heavy wall,	**P** for heavy wall,
• **W** for thin wall,	**S** for thin wall.
• **X** for heavy wall, internally threaded.	

- The third letter is used for rod pumps only and signifies the location of the seating assembly. The seating assembly or **hold-down** is always at the bottom of a traveling-barrel pump; other rod pumps can be seated at the top or at the bottom, as given below:
  - **A** for top hold-down (top anchor),
  - **B** for bottom hold-down (bottom anchor), and
  - **T** for traveling barrel, bottom hold-down.

For example, the pump designated as **TP** stands for a tubing pump with heavy wall and a soft-packed plunger; an **RWB** pump is a rod pump with a thin-wall stationary barrel, metal plunger, and a bottom anchor.

3.2.3.2 Characteristics of API pumps

In the following, the main features of API pumps are detailed, along with a list of relative advantages and disadvantages [1–3].

3.2.3.2.1 Tubing pumps

Tubing pumps are the oldest types of sucker-rod pumps, with a simple and rugged construction, and always come with a **heavy** wall barrel. Their inherent advantage over other pump types is the relative large pumping **capacity** due to the large barrel sizes. A schematic drawing of a tubing pump in the upstroke position is presented in Fig. 3.5. The figure shows a pump with a metal plunger, designated by the API code of **TH**; the same pump with a soft-packed plunger is coded **TP**.

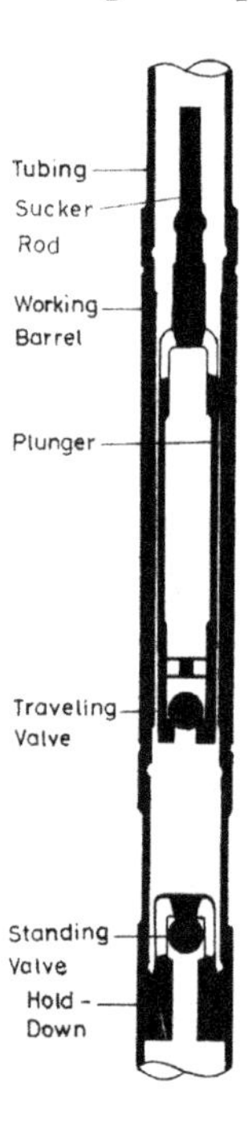

FIGURE 3.5

Cross-section of an API TH tubing pump.

The relative **advantages** of tubing pumps are as follows:

- They provide the **largest** pump sizes in a given tubing size, with barrel inside diameter (ID) usually only 1/4 in smaller than tubing ID. These large barrels allow more fluid volume to be produced than with any other type of pump.
- Tubing pumps have the **strongest** pump construction available. The barrel is an integral part of the tubing and can thus withstand high loads. The rod string is directly connected to the plunger without the necessity of a **valve rod**, making the connection more reliable than in rod pumps.
- The tubing pump is usually less expensive than rod pumps due to its **fewer** parts.
- The large valve sizes result in low pressure losses in the pump; production of **viscous** fluids is also possible.

The main **disadvantages** of tubing pumps are listed below.

- Workover operations usually require the tubing to be **pulled**. High pump repair costs are the greatest drawbacks in the use of tubing pumps.

- Tubing pumps perform poorly in **gassy** wells. The relatively large dead space, the space between the standing valve and the traveling valve at the bottom of stroke, causes poor valve action and low pump efficiency.
- Lifting depth can be limited by the large fluid **loads** associated with the large plunger areas and the use of high-strength sucker rods may be needed. At greater depths, excessive plunger stroke loss is expected due to large amounts of rod and tubing stretch.

3.2.3.2.2 Stationary barrel top anchor rod pumps

Figure 3.6 shows the cross-section of an **RHA** pump during upstroke. Its heavy-wall working **barrel** is held in place at the top of the pump assembly, a **preferred** seating arrangement for the majority of pumping installations. The plunger of the **RHA** pump is made of metal. Other pumps in this category are **RWA**, with a thin-wall barrel and metal plunger, and **RSA**, with a thin-wall barrel and a soft-packed plunger.

Advantages of these pumps include:

- The top **hold-down** is recommended in **sandy** wells because sand particles cannot settle over the seating nipple due to the continuous washing action of the fluids pumped. Therefore, the pump assembly usually does not get **stuck** and can easily be pulled should it need servicing.
- When pumping **gassy** fluids from wells with low fluid levels, this pump performs well because the standing valve is submerged deeper in well fluids than in the case of bottom hold-down pumps.
- A **gas anchor** can be directly connected to the pump barrel when free gas is present.
- If a long barrel length is required, the top hold-down gives better **support** to the pump assembly than a bottom hold-down. Barrel movement can also be less with a limited rubbing action of the barrel against the tubing.

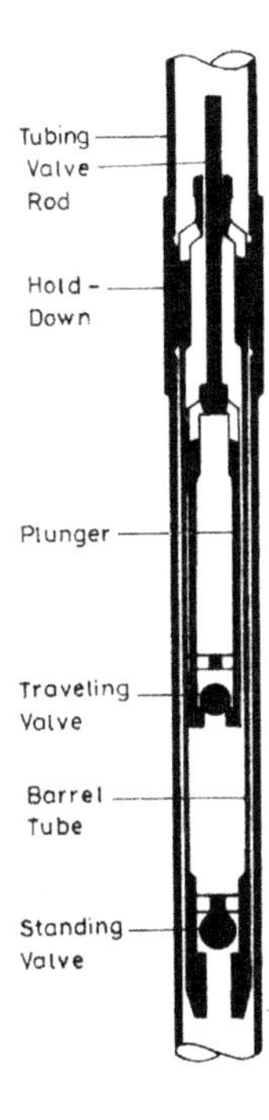

FIGURE 3.6

Cross-section of an API RHA rod pump.

Disadvantages of these pump types are:

- Due to the top anchor position, the outside of the barrel is at **suction** pressure, while the inside is exposed to the high **hydrostatic** pressure of the liquid column in the tubing. The big pressure **differential** across its wall can deform or even **burst** the barrel, especially if it is of the thin-wall type, limiting the use of **RWA** pumps to about 5,000 ft.
- On the downstroke, the barrel is under high **tensile loads** due to the weight of the liquid column being supported by the standing valve. The mechanical strength of the barrel, therefore, limits the **depth** at which such pumps can be used.
- The valve rod can **wear** by rubbing against its guide and can be the weak link in the rod string.
- Compared to a traveling-barrel pump, this pump has more **parts** and a higher initial cost.

3.2.3.2.3 Stationary barrel bottom anchor rod pumps

The cross-section of an **RHB** pump during upstroke is shown in Fig. 3.7. This should usually be the **first** pump to consider for **deep** well service. The working barrel is fixed to the tubing at the bottom of the pump assembly, which has definite advantages in deep wells. **RHB** pumps have metal plungers and heavy-wall barrels, **RSB** pumps have a thin-wall barrel and a soft-packed plunger. The **RWB** pump with a thin-wall barrel is the most popular type of all sucker-rod pumps. The recently introduced **RXB** pumps are similar to **RWB** pumps but have internally threaded heavy-wall barrels with larger outside diameters that do not pass through the seating nipple; this is the reason they come in bottom anchor versions only.

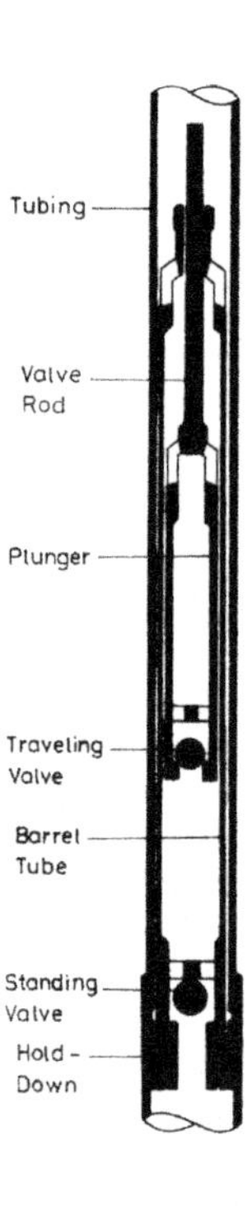

FIGURE 3.7

Cross-section of an API RHB rod pump.

The main **advantages** of these pumps are listed below.

- The outside of the barrel is always at the **hydrostatic** pressure exerted by the liquid column in the tubing. Thus, the pressure **differential** across the barrel wall is much less than that in a top anchor-type pump, making the barrel less prone to mechanical damage. Consequently, this pump can be used to greater depths than other rod pumps.
- Use of this pump is advisable in wells with low fluid **levels** because it can be run very close to well bottom, the deepest point of the pumping assembly being the seating nipple.
- The standing valve is usually **larger** than the traveling valve and this feature ensures a smooth intake to the pump. Foaming tendency of well fluids is also reduced.
- In deviated wells the barrel can **pivot** above the seating nipple, which reduces wear.

Disadvantages are the following:

- During downtime, or in intermittent operation, **sand** or other solid particles can settle on top of the plunger, which can be stuck in the barrel when the pump is started again.
- The annulus between the tubing and the barrel can fill up with **sand** or other solids, preventing the pulling of the pump.
- The valve rod can be a **weak** point compared to the rod string.
- Pump **cost** is higher than for traveling-barrel pumps due to more parts.

3.2.3.2.4 Traveling-barrel rod pumps

The operation of any piston pump is based on the **relative** movement between the piston and cylinder. From this follows that the same pumping action is achieved in a rod pump if the plunger is stationary and the **barrel** moves. The traveling-barrel rod pumps operate on this principle and have the plunger held in place while the barrel is moved by the rod string. The position of the anchor or hold-down is invariably at the **bottom** of the pump assembly.

Traveling-barrel rod pumps are versatile and can be used in normal, sandy, and corrosive wells. Figure 3.8 gives a cross-section of an **RHT** pump. The plunger is attached to the bottom hold-down by a short hollow **pull tube**, through which well fluids enter the pump. The **standing valve**, situated on top of the plunger, is of a smaller size than the traveling valve. Thin-wall pumps are designated **RWT** and those with a soft-packed plunger **RST**.

Traveling-barrel pumps are widely used when **sand** production is a problem; their main **advantages** are listed below.

- The traveling barrel keeps the fluid in **motion** around the hold-down, preventing sand or other solids from **settling** between the seating nipple and the hold-down. Therefore, pulling of the pump assembly is usually trouble-free.
- This pump is recommended for **intermittent** pumping of sandy wells, because sand cannot get between the plunger and the barrel during shutdowns.
- The connection between the rod string and the traveling barrel is **stronger** than that between the valve rod and the rod string in stationary barrel pumps.
- It's a **rugged** construction with fewer parts than stationary barrel pumps, with a lower **price**.

Disadvantages of traveling-barrel pumps are:

- The size of the standing valve is **limited** because it has to fit into the barrel. This relatively smaller valve offers high **resistance** to fluid flow, allowing gas to break out of solution, causing poor pump operation in gassy wells.

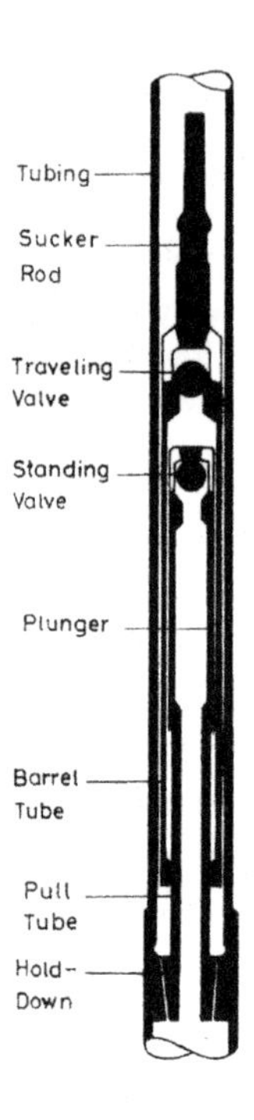

FIGURE 3.8

Cross-section of an API RHT rod pump.

- In deep wells, the high hydrostatic pressure acting on the standing valve on the downstroke may cause the pull tube to **buckle** and excessive **wear** can develop between the plunger and barrel. This limits the length of the barrel that can be used in deep wells.
- Pumping of highly **viscous** fluids is not recommended because the small standing valve can cause excessive pressure drops at the pump intake.

3.2.3.3 Specification of pump assemblies

In order to completely specify a sucker-rod pumping assembly, the American Petroleum Institute proposed the use of a multicharacter designation in **API Spec 11AX** [1]. This designation is used worldwide and orders on sucker-rod pumps conforming to it are generally accepted. The complete **designation** contains several groups specifying the different parts of the pump assembly.

The first numeric group defines the nominal tubing size the given pump is supposed to operate in.

Code	Tubing Size, in
15	1.9
20	2 3/8
25	2 7/8
30	3 1/2
40	4 1/2

The second group is a three-figure code that specifies the pump bore size; this refers to the **inside** diameter of the barrel bore, which is practically equal to the plunger's outside diameter. Non-API pumps with larger diameters are also manufactured for use in larger tubing sizes.

Code	Pump Size, in
106	1 1/16
125	1 1/4
150	1 1/2
175	1 3/4
178	1 25/32
200	2
225	2 1/4
250	2 1/2
275	2 3/4
375	3 3/4

The third group is the API letter designation of the pump, explained before. The next group is a single-letter code that refers to the type of **seating** assembly, which can be a mechanical (designated by the letter **M**) or a cup-type (designated by the letter **C**) hold-down. The last numeric group of the designation refers to the length of the pump: measurements are listed in feet; the first number gives the **barrel** length, the second number gives the nominal **plunger** length, and the last two give the lengths of barrel **extensions**. Extensions can be added to both ends of the barrel to prevent some operational problems.

To order a specific pump, the additional information required by the manufacturer is the specification of the materials for the barrel and the plunger, plunger fit, and valve material.

EXAMPLE 3.1: DESCRIBE THE SUCKER-ROD PUMP HAVING THE FOLLOWING DESIGNATION:

25-175 RWAC 10-3-1-1

Solution

Using the previous descriptions the above designation is resolved as:

Tubing size is 2 7/8 in nominal; pump bore size is 1 3/4 in. The pump is a stationary, thin-wall barrel rod pump with a cup-type top hold-down assembly and a metal plunger. The barrel of the pump is 10 ft long, plunger length is 3 ft, and upper and lower barrel extensions are 1 ft long.

3.2.4 STRUCTURAL PARTS

The sucker-rod pump's **nominal size** is the exact **inside** diameter of the barrel. Plunger outside diameter, in the case of metal plungers, differs only slightly from the ID of the barrel used. **Plunger fit**, i.e., the clearance between the plunger and barrel, is in the order of a few **thousands** of an inch. Specifications for sucker-rod pump components are contained in **API Spec 11AX** [1].

3.2.4.1 Barrels

Working barrels are the largest and most expensive components of downhole pumps; they are lengths of cold-drawn metal tubing, machined and honed, with the inside wall **polished** to allow for a smooth movement of the plunger. In the latest API publications, only **one-piece barrels** are specified, but up to the end of the 1970s, **liner barrels** were also standard. A liner barrel consisted of an outer jacket containing one or more **sectional** liners of 1ft length each.

Depending on the threads applied to both ends, **pin-end** or **box-end** barrels are available. The wall **thickness** of barrels and barrel tubes is different for the heavy- and the thin-wall versions. **Heavy-wall** usually means a wall thickness of 3/16 in or greater; **thin-wall** barrels are about 1/8 in thick. Heavy-wall barrels have **pin** ends (Fig. 3.9), thin barrels are manufactured with **box** ends (Fig. 3.10). The only exception is the heavy-wall **RXB** pump, which has a barrel with box end internal threads and has a larger outside diameter than the **RW** pump. Because of this, it can be used only with a bottom hold-down, since the barrel cannot run through the seating nipple.

Barrels are manufactured in standard lengths of up to 24 ft; longer pumps are usually made up of several barrels joined together, like in the case of a record 100 ft pump [4].

Selecting the **length** of the pump barrel is very important and must be the first step in designing a sucker-rod pump. The barrel must be sufficiently long to contain the **plunger** with its **valve(s)** and the maximum anticipated plunger **stroke** length. Since plunger stroke length cannot directly be determined, the **maximum** polished rod stroke length available on the pumping unit is assumed. In order to estimate the real plunger stroke, it is customary to add a spacing **allowance** to the polished rod stroke length. The spacing allowance for **steel** rods is usually 24 in for depths less than 4,000 ft; at higher

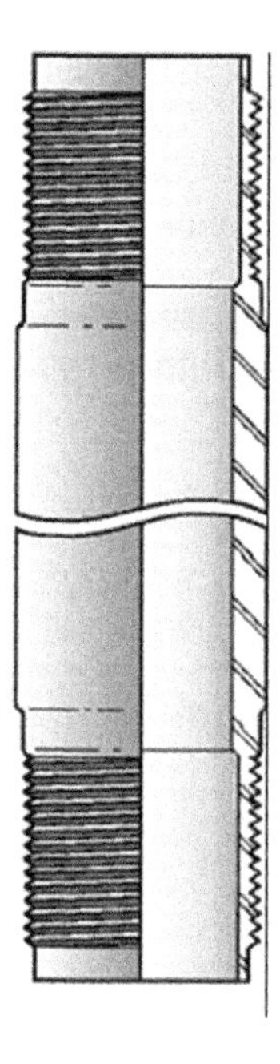

FIGURE 3.9

Heavy-wall barrel with pin end.

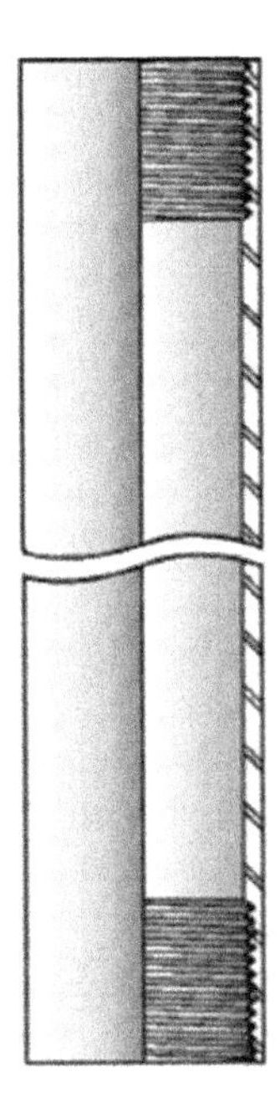

FIGURE 3.10

Thin-wall barrel with box end.

depths 6 in per 1,000 ft is used. For **fiberglass** rod strings, because of their much greater elasticity than for steel, at least three times greater allowance is required and the use of computer programs to size the barrel is recommended.

For adverse conditions like sand and corrosion problems, **plating** of the inside surface of barrels is used. **Chrome** plating provides a very hard surface against **abrasion** damage. **Nickel carbide** barrels have carbon steel as base metal on which nickel containing very small particles of carbide is plated. This solution combines the advantages of a **strong** structure due to the carbon steel base metal and an excellent resistance to most kinds of **corrosion** (including CO_2, H_2S, and chlorides) guaranteed by the nickel carbide plating.

3.2.4.2 Plungers

The earliest types of plungers used in sucker-rod pumps were the **soft-packed** ones using cups made of a resilient material to seal on the barrel wall. They can safely be used in both tubing and rod pumps and offer high corrosion resistance to well fluids. The strength of the sealing cups, however, limits the application of such plungers to **moderate-depth** wells. In deeper wells, usually below 5,000 ft, the high hydrostatic pressure above the plunger causes excessive **slippage** of produced fluids past the plunger so that pump displacement is drastically reduced. Deeper wells, therefore, require a more effective seal and a closer fit between barrel and plunger, which can only be achieved if a **metal-to-metal** sealing system is used. All-metal plungers, or simply called **metal plungers**, are the only solution in deep wells.

Metal plungers are manufactured in different versions, with **plain** or **grooved** outside surfaces. Grooved plungers are at an advantage when the well produces some **sand**. Sand or other solid particles can be trapped in the grooves, thus preventing them from scoring the barrel and plunger. When a plain plunger is used, on the other hand, a solid particle can travel the whole length of the plunger, damaging both the barrel and the plunger surfaces. A small score on these polished surfaces results in increased liquid **slippage** past the plunger because of the great pressure **differential** across the plunger. Pump volumetric efficiency and liquid production can thus be greatly reduced.

Just like barrels, plungers also are manufactured in **pin-end** and **box-end** types. Figure 3.11 illustrates the most popular pin-end **plain** plunger. Box-end plungers (as shown in Fig. 3.12) can have a smaller wall thickness than pin-end plungers and offer less resistance to flow, need less material, and are usually less expensive. A grooved plunger is shown in Fig. 3.13.

The **surface** of metal plungers may be **plain** steel, **chrome** or **nickel** plated, or **sprayed** metal. Plain metal is used for normal conditions; **chrome**-plated plungers are recommended for **abrasive** conditions without H_2S corrosion. **Sprayed** metal plungers are the most popular and have a flame-sprayed coat of a nickel-based alloy; they are economical and are used in moderately abrasive wells.

On the upstroke, the radial **clearance** between the barrel's inside wall and the plunger is subjected to a high pressure **differential**. High hydrostatic pressure, due to the weight of the liquid column in the tubing, acts from above the plunger, and a much smaller pump intake pressure acts from below. Liquid from above the plunger, therefore, is constantly forced to **slip back** past the plunger. In order to maintain a high pumping efficiency, the rate of this **slippage** must be kept at a minimum. This is the reason why only a very small clearance is allowed in a sucker-rod pump. Generally available **plunger fits** range from 0.001 to 0.005 in, measured on the diameter, and are designated as $-1 \ldots -5$ fits. The selection of the plunger fit to be used is based on actual well conditions, but liquid **viscosity** is the

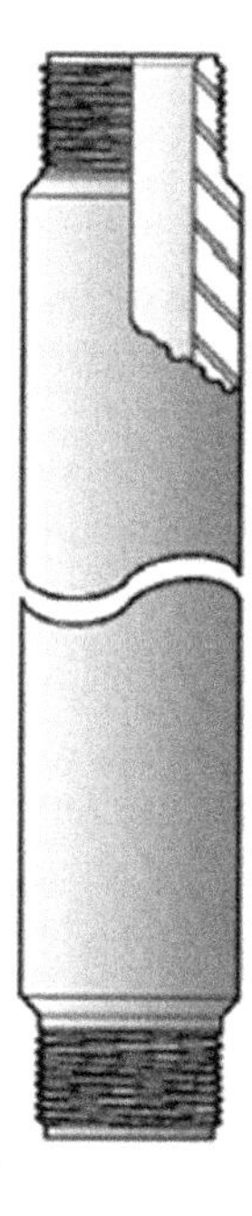

FIGURE 3.11

Pin-end plain metal plunger.

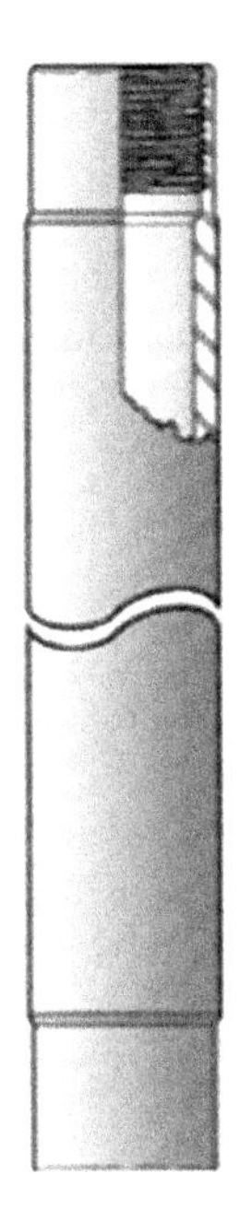

FIGURE 3.12

Box-end plain metal plunger.

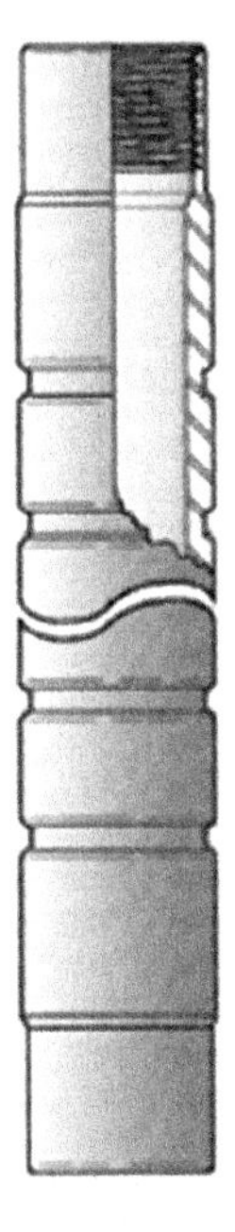

FIGURE 3.13

Box-end grooved metal plunger.

governing factor. The higher the viscosity, the less the slippage past the plunger with the same fit, because viscous oils offer more resistance to flow than lighter ones. High-viscosity crudes can efficiently be pumped with a −5 fit plunger, whereas low viscosity crudes may need a −1 fit to be used.

The general rule for selecting the **length** of metal plungers is as follows:

- for pumping depths less than 3,000 ft, a plunger length of 3 ft is sufficient,
- for wells between 3,000 ft and 6,000 ft depth, plunger length should be 3 ft plus 1 ft/1,000 ft of lift, and
- in wells deeper that 6,000 ft, 6-ft-long plungers are recommended.

Soft-packed plungers consist of an inside mandrel onto which a series of **packing** elements are installed, with metal **wear rings** placed on top and on bottom. The packing elements can be either **cup**- or **ring**-shaped and are made of a synthetic material with an increased hardness for increased well depths. The design and construction of soft-packed plungers is not standardized; Fig. 3.14 illustrates typical plungers [3]. The cups in a **cup** plunger (Fig. 3.14a) expand on the upstroke to form an effective seal on the barrel and contract on the downstroke to move freely in well fluids. They are well suited for pumping crudes with little **sand** content. Packing rings in **ring** plungers (Fig. 3.14b) are less effective than cups but wear slower and more uniformly. **Combination** plungers (Fig. 3.14c) employ both cups and rings to provide a sealing effect. Such plungers are very effective when fine-grain sand is present.

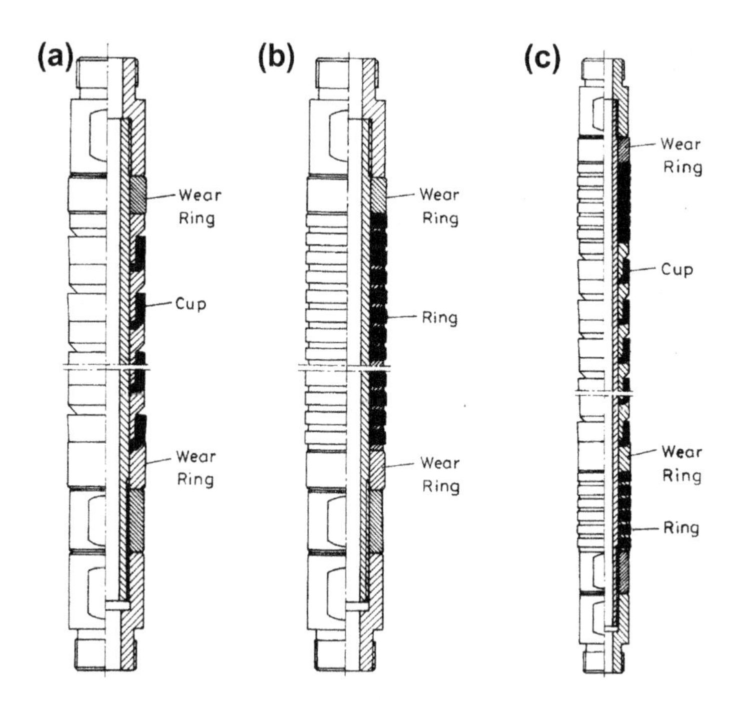

FIGURE 3.14

Soft-packed plungers by Oilwell Div [3]. (a) cup type, (b) ring type, (c) combination type.

Soft-packed plungers are recommended where the **lubrication** quality of the well stream is poor or where sand or abrasive-laden fluids tend to **stick** metal plungers. They are economical to use in relatively **shallow** wells. Other advantages are their lower **price** and the easy way they can be **repaired** at the well site by changing the packing elements.

When selecting the **length** of soft-packed plungers, the general rule of thumb is to have 4–6 cups for every 1,000 ft of well depth.

3.2.4.3 Pump valves

3.2.4.3.1 API valves

Valves are considered the **heart** of the sucker-rod pump, because an efficient pumping operation mostly depends on the proper action of the **standing** and **traveling** valves. API valve assemblies are simple check valves and operate on the **ball-and-seat** principle (Fig. 3.15). **Seats** are machined, precision ground, and finished from corrosion- and erosion-resistant metals. They are usually **reversible** and can be used on both sides. The metal **balls** are precision finished and each ball-and-seat combination is **lapped** together to provide a perfect seal; the matched set is finally **vacuum tested.** Because they come in sets, none of them can be used with a part coming from another set; if the ball or seat is permanently damaged a **new** valve **assembly** must be used.

Highly reliable sealing action is required between the ball and the seat because of the very high **differential** pressures across the valve during pumping. Small initial imperfections on the sealing surfaces or later damages due to abrasion or corrosion cause an increased liquid **slippage** and a rapid deterioration of valve action due to fluid cutting.

Balls and seats are available in different metals; stainless **steel** or **alloy** materials are used where abrasion and corrosion are low. **Tungsten carbide** is another reliable material, but it cannot be used in wells with H_2S and CO_2 content; under such conditions nickel carbide balls and seats are better. Exotic ceramics are also gaining popularity. **Zirconia** ceramic balls and seats, for example, are chemically inert, are totally resistant to corrosion and abrasion, and provide a longer life than metal valve parts.

The **traveling** valve can be placed either above or below the plunger. The top position is recommended in wells producing little or no gas. Its advantage is that the plunger itself is not loaded by the weight of the fluid column, eliminating plunger stretch. Valve placement below the plunger reduces the **dead space** between the standing and traveling valves, a desired feature when gassy fluids are pumped. Pumps with this valve arrangement provide favorable volumetric efficiencies and are less prone to **gas locking**. Both traveling and standing valves can be **doubled** to extend the service life of pumps. Especially in sandy wells, two valves in series ensure a much higher level of safety against the cutting action of fluid jets.

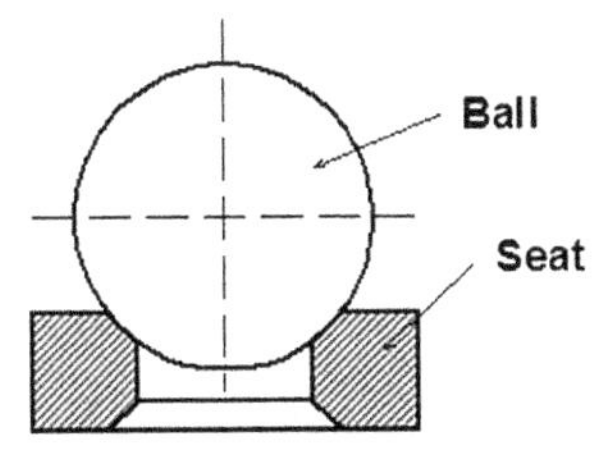

FIGURE 3.15

Valve seat and ball.

Table 3.1 Main Data of Standard Valves Used in Subsurface Pumps

		Ball Size	
Pump Size, in	**Seat Bore, in**	**Std. in**	**Altern. in**
1 1/16	0.460	0.625	
1 1/4	0.550	0.750	0.688
1 1/2	0.670	0.938	0.875
1 3/4	0.825	1.125	1.000
2	0.960	1.250	1.125
2 1/4	1.060	1.375	1.250
2 1/2	1.310	1.688	1.500
2 3/4	1.310	1.688	1.500
3 3/4	1.700	2.250	2.000

The API specification on pump valves and cages (**API Spec 11AX** [1]) leaves several design options open to manufacturers. The diameter of the seat bore, shape of seat seal surface, cage construction, and metallurgical processes are among those factors that vary from manufacturer to manufacturer and can considerably affect valve performance. Care has to be taken to select those products that provide optimum pump operation under the specific conditions of the actual well [5].

The main data of pump **valves** offered by several manufacturers are presented in Table 3.1. As shown, most manufacturers offer **alternate**, smaller ball sizes, too, for pumping viscous fluids; these provide a greater **clearance** between the ball and the valve cage and offer less resistance to flow. The dimensional data refer to **traveling valves** used in stationary barrel pumps and are identical for standing valves in traveling-barrel pumps. Standing valves in stationary-barrel pumps or traveling valves in traveling-barrel pumps are one size larger or smaller, respectively.

3.2.4.3.2 The Petrovalve

Traditional ball-and-seat valves in subsurface pumps have several fundamental operational troubles:

- Due to the extremely great number of pumping cycles per day, the repeated **impact** of the two parts against each other can severely damage the sealing surfaces.
- The usual cage constructions, intended to reduce the impact damage, usually restrict the flow **area** available through the valve and can be the source of pumping inefficiencies due to low fillage of the pump barrel.
- In deviated wells valve wear is **uneven** and can lead to premature failure.

The **Petrovalve** design [6,7], shown in Fig. 3.16, tries to overcome these hindrances and provides an increased flow area as well as a more efficient and longer-lasting positive seal. Instead of a ball, the closing element is a **hemisphere** whose movement is restricted in one dimension only. This is achieved by the action of the **stem** protruding on both sides from the hemisphere. An upper and a lower **guide** ensure that the valve stem and the closing element move perpendicularly to the valve seat's sealing surface; the closing element is always kept centered in the housing, both in the open and in the closed position. The results of this are: (1) when open, flow is **uniformly** distributed around the closing element, (2) during closing the closing element falls **directly** on the seat, and (3) the valve operates

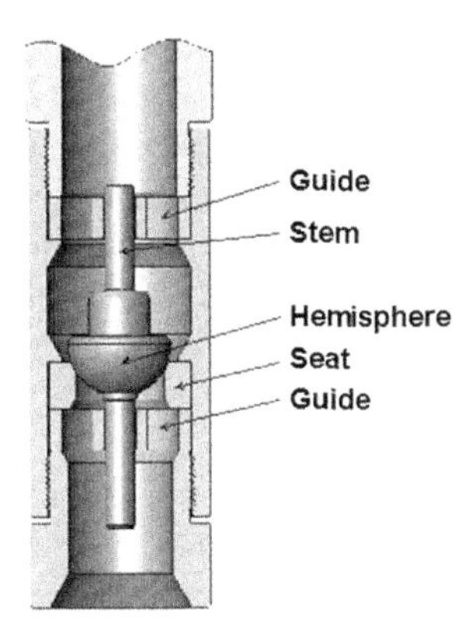

FIGURE 3.16

The "Petrovalve" design of pump valves.

identically at **any** inclination angle. Compared to a ball-and-seat-type valve, the closing element always falls precisely on the seat and its damage due to **rattling** is eliminated. Due to its construction the Petrovalve provides, depending on its size, a flow area 27–112% greater than an API valve of the same size; this has definite advantages when pumping highly viscous crudes.

A modification of the Petrovalve called the **Gas Breaker** is a **mechanical** traveling valve that automatically prevents **gas lock** situations. The operation of the **Gas Breaker** is shown in Fig. 3.17 when the plunger is close to the bottom of its stroke. The valve itself is a **Petrovalve** with a

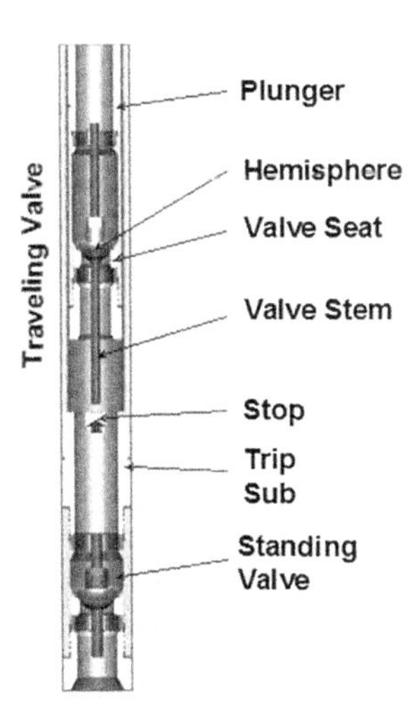

FIGURE 3.17

The "Gas Breaker" mechanical traveling valve.

longer stem, which hits a stop built in the **trip sub** installed just above the standing valve when the plunger reaches the bottom of the stroke. In a gas-locked pump, the traveling valve would not open at the bottom of the stroke, but the **Gas Breaker** automatically opens at the end of each pumping cycle, allowing the gas compressed in the barrel to be released to the tubing string. For proper operation a **slight** tap must be felt during the plunger **spacing** operation; after that, rods must be picked up by about 1/2 in; this ensures that the valve opens at the end of each downstroke. Like the normal Petrovalve, the **Gas Breaker** valve also operates perfectly at any inclination angle.

3.2.4.4 Valve cages

During valve operation, the ball is periodically **seated** and **unseated** on the valve seat. The high hydrostatic pressure at pump depth causes the ball to hit the seat with high-impact forces. In cases where the ball's movement is not restricted, it may **rattle**, i.e., move off the center line of the seat bore when lifted. Then, during closing, the ball may hit one side of the seat only, which results in excessive **wear** to both the seat and the ball. To decrease valve damage and improve performance, **valve cages** are used that **guide** and **restrict** the ball's movement. The role of valve cages is to restrict the **lateral** and **vertical** travel of valve balls, while providing the least possible flow **restriction** across the valve assembly.

The API specifies **open** and **closed** valve cages to be used in standing and traveling valves. Open and closed cages in **standing** valves are shown in Figs 3.18 and 3.19, respectively; the same for **traveling** valves are depicted in Figs 3.20 and 3.21.

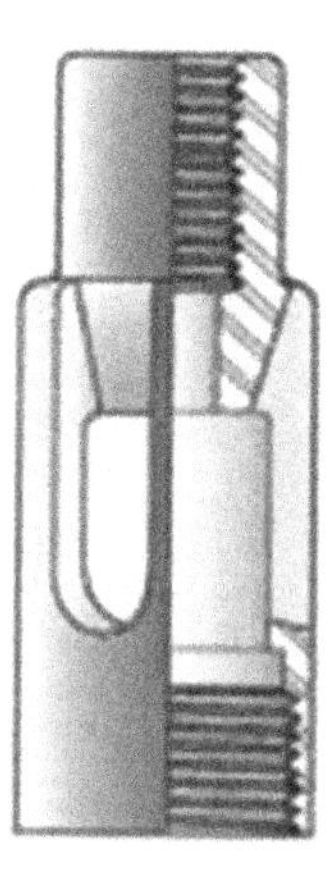

FIGURE 3.18

Open API cage for standing valves.

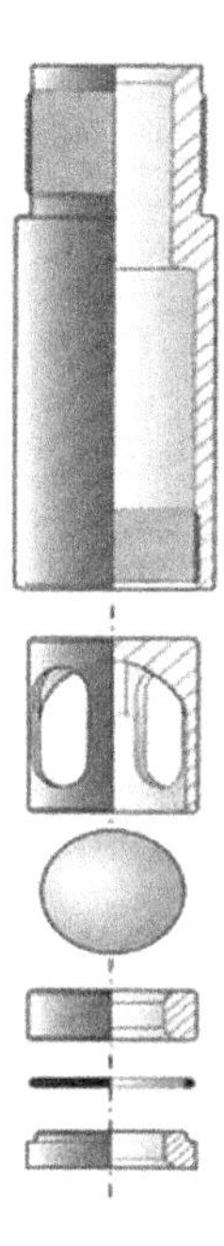

FIGURE 3.19

Closed API cage for standing valves.

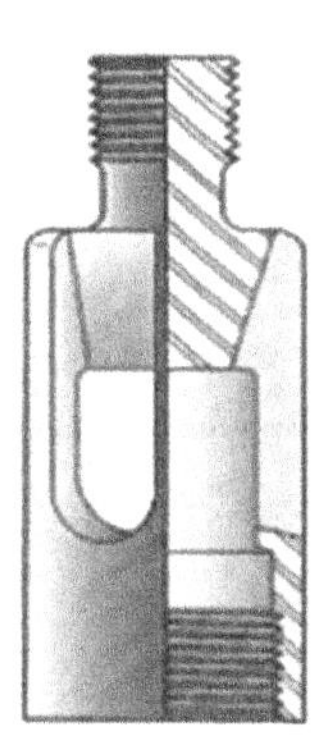

FIGURE 3.20

Open API cage for traveling valves.

The main problems with **conventional** API valve cages are that:

- their constructional length is relatively long, making the **dead space**, i.e., the space between the two valves at the bottom-most position of the plunger, undesirably great, and
- they usually present a **restriction** on the liquid flow passing through the valve.

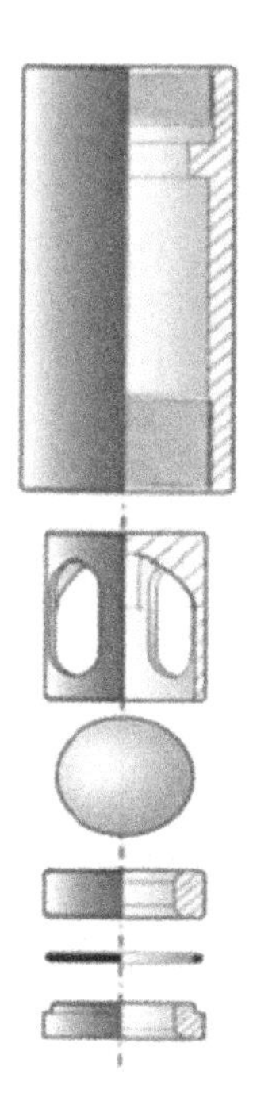

FIGURE 3.21

Closed API cage for traveling valves.

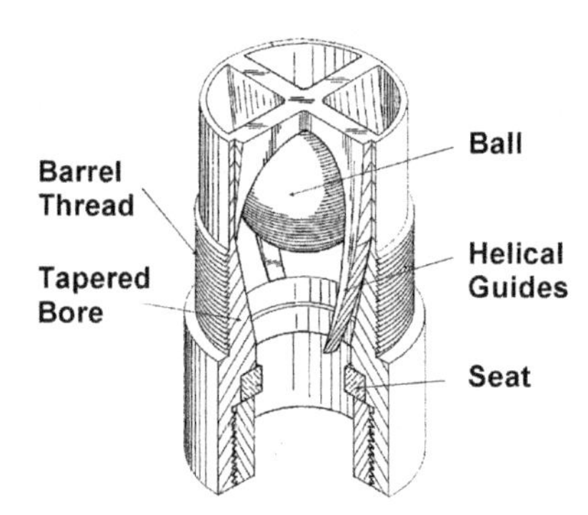

FIGURE 3.22

Construction of the HIVAC valve cage.

These disadvantages are mostly eliminated by the **HIVAC** (high volume and compression) cage design shown in Fig. 3.22 [8] that offers extremely high flow capacities (up to five times higher) as compared to conventional models. The most important features of the HIVAC cage are as follows.

- Since the total **length** of the cage is much reduced, the pump's **dead space** is decreased and its tolerance to free gas production is greatly enhanced.

- The bore above the valve seat is **tapered** and its cross-section increases upward. This ensures that the flow area around the rising ball increases with an increased pumping rate and that the pressure drop across the valve is kept at a low level.
- The ball's **guides** are helically twisted and thus impart a circular motion to the fluid as it flows through the valve; this feature keeps the solids **suspended** in the fluid for a longer period and reduces sanding problems.

The HIVAC cage can be used in either the **traveling** or the **standing** valve or both, but a standing valve with this cage alone can bring about significant increases in pumping rates. Another beneficial feature is the much less sensitivity to gas locking when producing gassy well streams [9].

3.2.4.5 Hold-downs (pump anchors)

Hold-downs or **anchors** are used to **affix** the stationary part of a rod pump, be it either the barrel or the plunger, to the tubing string. The other function of the hold-down is to prevent well fluids from flowing back from the tubing by **sealing** off fluid column pressure from bottomhole pressure. The **hold-down** is run into a **seating nipple** previously installed in the tubing, where it is mechanically locked or held down by friction forces. The operation of the pump imparts vertical forces on the stationary member of the pump assembly, and these forces are directly transferred to the hold-down. The anchoring mechanism has to withstand these forces; otherwise, the pump **unseats** and stops producing.

Hold-downs can be either mechanical or cup-type; above a temperature of about 250 F only mechanical devices can be used. The **mechanical hold-down** positively locks in the seating nipple's recess by a **spring** action. Sealing is ensured by a seating **ring** that forms a metal-to-metal seal on the inside of the seating nipple. Figures 3.23 and 3.24 show bottom-anchor and top-anchor mechanical hold-downs, respectively. These are especially recommended in deep wells because the mechanical lock makes them less likely to unseat. **Cup-type anchors** provide the required hold-down force by mechanical friction only. Molded plastic cups (Fig. 3.25) are utilized to form a friction seal in the seating nipple. They offer less resistance to pump **unseating** than mechanical-lock anchors.

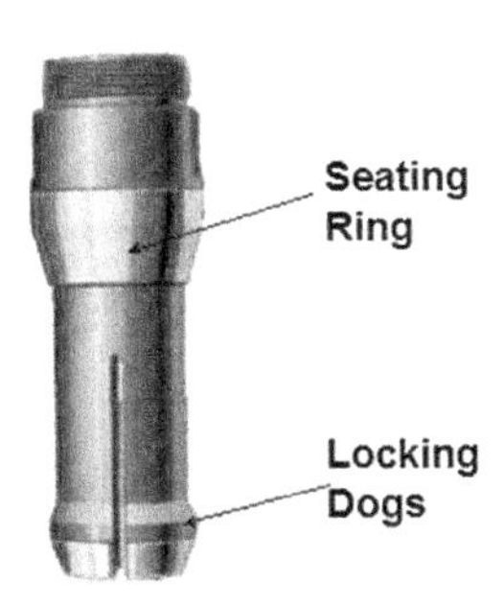

FIGURE 3.23

Bottom-anchor mechanical hold-down.

FIGURE 3.24

Top-anchor mechanical hold-down.

FIGURE 3.25

Cup-type hold-down.

3.2.5 SPECIAL PUMPS

In addition to standard API pumps, there is a multitude of nonstandard sucker-rod pump types available from different manufacturers. Some structural parts of such special pumps still **conform** to API specifications, making pump **repairs** easier and less expensive. Most special pumps provide features not available in standard pumps and are recommended in wells with operational problems like **gas interference**, **sand-laden** or highly **viscous** fluids, etc. High production rates also require special pumps to be used; one example for a high-volume pump is the **casing** pump.

In the following sections, the most important types of special pumps will be described. Although other pumps are also available, the ones presented here were chosen to be representative for the general types available from several manufacturers.

3.2.5.1 Top-and-bottom-anchor rod pump

As discussed before, both the top- and the bottom-anchor positions had definite **advantages** in stationary barrel rod pumps. The advantages are combined in this pump, which is basically the well-known stationary barrel rod pump but has **two** hold-downs installed. It may be designated as **RHAB** or **RWAB**, depending on the barrel wall used. Figure 3.26 shows a cross-section of this pump type in the downstroke position. The bottom hold-down is usually of the **mechanical** type and provides most of the necessary hold-down force. The upper or top hold-down is a **cup-type** one, its main function being to ensure a seal on top of the pump.

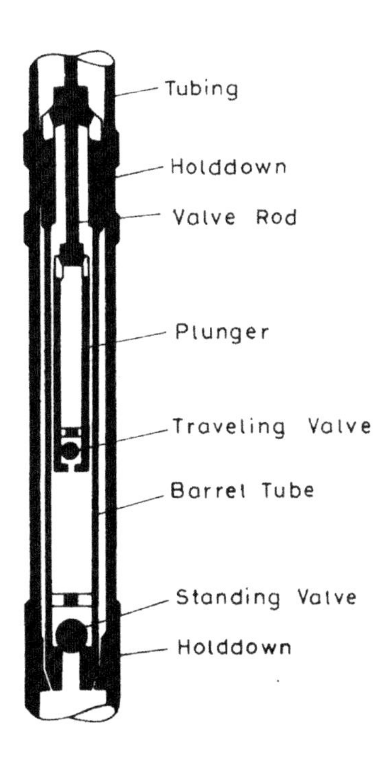

FIGURE 3.26

Cross-section of a top-and-bottom-anchor rod pump.

Dual-anchor rod pumps are recommended if a **long** pump is required because the two hold-downs ensure an increased support to the barrel. The barrel's outside is not subjected to high pressure due to the sealing effect of the top hold-down, thus preventing barrel damage. When pumping **sandy** crudes, the top hold-down eliminates the possibility of sand **settling** around the barrel tube and the sanding-in of the pump. Additionally, the outer surface of the barrel is protected against the corrosive effects of well fluids. Its main disadvantage is the relative high **cost** due to the two hold-downs required and the need to prepare a special tubing section where the pump is set.

3.2.5.2 Ring-valve pump

The use of a sliding **sand valve** on the valve rod of a rod pump is an old practice to prevent sand from **settling** between the barrel and the plunger during downtime. Later it turned out that pumps equipped with such valves also ensure very efficient pumping operations in **gassy** wells [10,11]. The sand valve is usually called a **sliding** or **ring valve**, hence the name of the pump. It is sometimes also called a **two-stage pump**, because the ring valve facilitates a two-stage pumping action, as described below.

Figure 3.27 illustrates the operation of a ring-valve rod pump. The ring valve is essentially a **check** valve situated in an open cage on top of the barrel. It can slide up and down on the valve rod, to which it is closely fitted. When the plunger starts its downward travel, the ring valve immediately **closes** due to the fluid column pressure above it. At the same time, the fluid load is **transferred** to the ring valve and the traveling valve is **unloaded**. As the plunger travels downwards, pressure in the space above the traveling valve quickly drops and this helps the valve to open shortly after the downstroke begins.

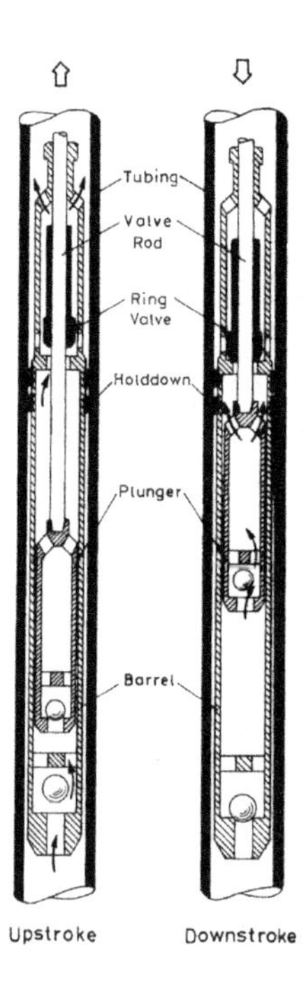

FIGURE 3.27

The operation of a ring-valve rod pump.

When the upstroke starts, the ring valve is still closed and the plunger compresses the gas–liquid mixture in the barrel, soon lifting the ring valve off its seat. For most of the upstroke the pump operates just like any other **stationary** barrel rod pump, because the ring valve is open and offers no resistance to upward fluid flow.

The ring valve action just described is extremely beneficial for pumping fluids with free **gas**. In such cases, conventional pumps can have very low efficiencies and can easily be **gas locked** or pounding fluid. A gas lock is a situation when the traveling valve stays closed throughout the pumping cycle, and no fluid is produced to the surface. A **fluid pound** occurs on the downstroke when the plunger approaches the liquid level in the partially filled barrel with the traveling valve still closed. Both of these conditions can be attributed to a single cause: the traveling valve cannot open at the **beginning** of the downstroke due to the **compressibility** of the fluid contained in the barrel. The ring valve, by **unloading** the traveling valve at the start of the downstroke, enables the latter to open at the proper time and thus eliminates the occurrence of fluid pound or gas-locked conditions. At the start of the **upstroke**, the ring valve remains seated due to fluid load and opens only when the gas–liquid mixture above the plunger is compressed. This two-stage pumping process helps to prevent **gas locking** of the pump.

The ring valve is applicable to all versions of stationary barrel rod pumps. **Advantages** of the pumps equipped with ring valves include:

- more complete **fillage** of the pump barrel and increased **net** plunger stroke length when producing gassy fluids,
- reduction in rod string **loading** due to the elimination of fluid pound; lower sections of the rod string always remain in tension, and
- the range of rod stresses is reduced and rod **life** increases.

The ring-valve pump's **disadvantage** is that it can suffer from fluid pound on the upstroke. This can occur when the barrel is only partially filled with liquid and the ring valve does not open at the start of the upstroke. Then, during upstroke, the liquid column above the plunger can pound on the ring valve. Application of a **charger valve** [12] eliminates this pound, too, by dumping some fluid on the plunger from above the charger valve, near the bottom of the stroke.

3.2.5.3 Three-tube pump

All sucker-rod pumps suffer from **abrasion** of the closely fitted barrel and plunger when liquids containing sand or other abrasives are pumped. **Abrasion wear** is less if plunger **fit** is increased, but this entails a drastic drop in the volume pumped. One very efficient way to reduce wear and still maintain a good pump operation is the use of a **three-tube pump**. The efficiency of a three-tube pump does not depend on plunger fit because sealing of the moving parts is ensured by **hydraulic** means.

Figure 3.28 shows the operation of a **three-tube pump**. It is similar in design to a **traveling-barrel** bottom-anchor API rod pump. The barrel tube with a standing valve is surrounded by two **concentric** traveling tubes that are fastened together and have a traveling valve (TV) on top of them. The inner tube acts as a plunger and has another traveling valve at its bottom; the outer tube can be considered a traveling barrel. These tubes are **loosely** fitted to each other, with about 0.015 in clearance between them. This is about **three times** as much as the largest fit between metal plungers and barrels. An important feature is a small **bore** near the top of the plunger tube.

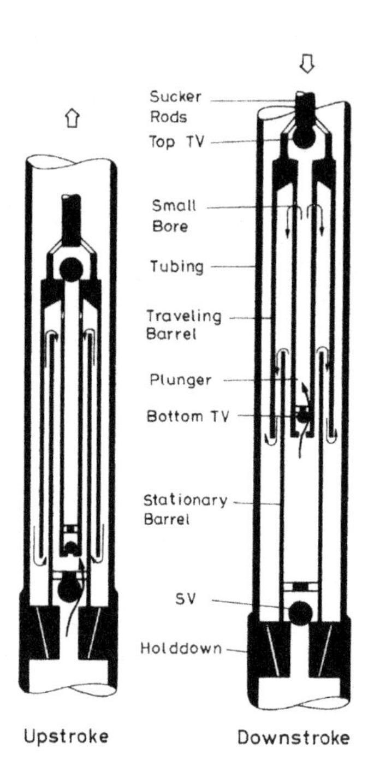

FIGURE 3.28

The operation of a three-tube rod pump.

The operation of a three-tube pump involves several stages. On the upstroke, both traveling valves are **closed**, and the standing valve is open. Because of the loose fit between the three tubes, well fluids can **leak** through the annular spaces from the high-pressure area in the tubing towards the low-pressure area below the bottom traveling valve. As the tubes are comparatively long, large **pressure drops** develop in the clearances, thus reducing the **leakage** of well fluids to a minimum.

When the downstroke begins, both traveling valves open and the standing valve closes. Now, well fluids are displaced from above the standing valve over the bottom traveling valve. A small fraction of this fluid **escapes** through the small bore at the top of the plunger tube, as well as through the loose **fit** between the plunger and the stationary barrel, and passes the clearance between the two barrels to enter the tubing. Just like on the upstroke, the leak rates are **small** again, but the direction of flow is **reversed**. Fluids present in the annular areas of the tubes are, therefore, **recirculated** during theup- and downstrokes instead of being actually displaced into the tubing. This effect provides the necessary **sealing** between the loosely fitted tubes and facilitates maintenance of an effective pumping action.

The three-tube pump is ideally suited to produce **sand-laden** fluids because the use of loosely fitted tubes results in negligible **wear** of the working parts. Wells after hydraulic fracturing can be very efficiently **cleaned** out with three-tube pumps.

The three-tube pump **combines** the advantages of both a traveling- and a stationary-barrel rod pump. The traveling barrel keeps the fluid in motion around the bottom hold-down and eliminates pump sticking due to sand settling. The top traveling valve prevents sand from **settling** inside the pump during downtime. In wells with exceptionally high bottomhole **temperatures**, the problems associated with the unequal thermal **expansion** of barrels and plungers in conventional pumps are eliminated because of the loose fits employed. The **disadvantage** of three-tube pumps is that deeper wells require longer tubes to be used, in order to reduce pump slippage due to the higher fluid column pressures.

3.2.5.4 Pumps with hollow valve rods

In deep wells, the **valve rod** in stationary-barrel rod pumps tends to **buckle** during the downstroke. Buckling is extremely severe in fluid pound conditions when high compressive loads act on the valve rod's two ends, causing excessive **wear** of the rod on the rod guide and even on the inside of the tubing. The use of a **tube**, usually a standard pull tube, instead of the valve rod is recommended. The **hollow valve rod** is much **stronger** than the solid one, and it has the advantage of greatly reducing or totally eliminating **buckling** problems. This is why several manufacturers offer stationary-barrel rod pumps equipped with pull tubes to extend the service life of pumps in deeper wells.

A pump was developed that combines the benefits of the hollow valve rod with those of the **ring valve** or two-stage pump [13]. The pump is called the **two-stage hollow valve-rod pump** and was developed for pumping high-gas–oil ratio wells. The pump is also capable of producing a well from below a **packer**, when all fluids (liquid and gas) are led through the pump.

The operation of a two-stage hollow valve-rod pump is explained in Fig. 3.29. The pump's basic features are the use of a **second** (top) traveling valve (which is the equivalent of the sand or ring valve

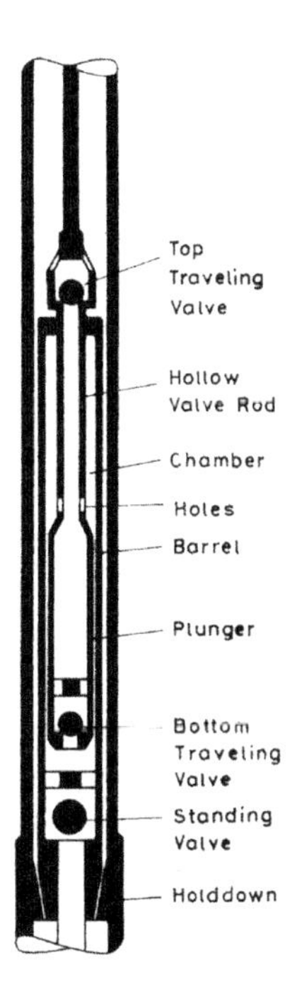

FIGURE 3.29

Cross-section of the two-stage hollow valve-rod pump.

used in ring-valve pumps), a ported **hollow** valve rod, and a **chamber** above the plunger. Other parts are similar to those of a stationary-barrel rod pump. The sequence of the pump's operation is discussed below, by assuming a gas–liquid mixture to be pumped.

At the start of the **upstroke** the bottom traveling valve in the plunger is closed and carries the fluid load. The standing valve opens and well fluids enter the barrel below the ascending plunger. As the upstroke proceeds, the plunger displaces the fluids contained in the **chamber** and forces them through the holes in the hollow valve rod and up to the top traveling valve. During this process, the gas–liquid mixture is **compressed**, because the volume displaced from the chamber is considerably more than the inside volume of the hollow valve rod. The elevated pressure prevents further **gas liberation** from the liquid phase and opens the top traveling valve. At the top of the upstroke, upward fluid flow through the top traveling valve stops and the valve ball returns to its seat.

As soon as the plunger starts its downward travel, the chamber space starts to **increase**, creating a pressure drop. This decrease in pressure helps the **top** traveling valve to **close** and the bottom traveling valve to open. The standing valve is closed and the plunger displaces the fluids from the barrel into both the chamber and the hollow valve rod. In case a single-phase liquid is produced, this plunger displacement is more than the increase in the chamber's volume, which results in some liquid displacement into the tubing. The situation is changed if a gas–liquid mixture is present in the barrel. Then, due to the **compressibility** of the mixture, fluids are **compressed** in the chamber and a much lesser volume, if any, is discharged in the tubing.

This pump is manufactured in two **versions**, the only difference between the two being the way the hollow valve rod is fitted to the top of the barrel. If a **loose** fit is used, the pump provides good operation in wells with both **gas** and **sand** problems, although the pump's compression ratio is reduced. Pumps can tolerate high sand concentrations because fluid movement around the valve rod ensures a continuous **removal** of sand particles from the chamber situated above the plunger.

The other version of the two-stage hollow valve-rod pump (**Gas Chaser** from Harbison-Fischer) [11] has a perfect seal between the valve rod and the barrel, thus providing a high compression ratio in the pump's chamber. Well fluids, therefore, are compressed in **two** stages: first on the **downstroke** from the barrel into the chamber, then on the **upstroke** from the chamber to the tubing. The pump's net compression ratio is the product of the individual ratios and can thus reach extremely high values. This feature explains why these pumps can operate with very high gas–liquid ratios; even when run on a **packer** with all well fluids entering the pump, they can handle large volumes of **gas** and do not suffer from **gas locking**.

3.2.5.5 Oversized tubing and casing pumps

Oversized tubing pumps are necessary when liquid rates greater than those achievable using normal tubing pumps are required. They use plungers of much greater sizes than the inside diameter of the tubing; because of this the complete pump **assembly** must be attached to the bottom of the tubing string. Connection of the plunger to the bottom of the rod string is facilitated by using an **on-and-off tool** that allows connecting and disconnecting of the plunger's valve rod. This tool also allows pulling of the rod string without retrieving the pump assembly. A basic problem with oversized pumps is the difficulty of pump **repairs** because of the need for pulling the tubing string to get access to the pump assembly. Operation of these big pumps significantly increases the loads on the rod string because of the high pump loads and requires proper rod string sizes and taper designs.

Casing pumps are high-capacity **rod** pumps run in wells **without** a tubing string. The pump barrel is normally set on top of a **packer**, causing the complete well stream pass through the pump; there is no annulus to provide gas **separation**. In case gas production is low, the big-size pumps generally used can produce very high rates. The application of casing pumps has its **limitations**: the casing string may be **damaged** by the rods moving inside the casing without a tubing string, and **fishing** operations of a broken rod string are difficult because of the large diameter of the casing.

3.2.5.6 Stroke-through pumps

In case of substantial **sand** or **solid** content in the produced liquid, the operating conditions of rod pumps are improved if so-called **stroke-through pumps** are used. These pumps are regular pumps with a slight **modification** of the barrel extensions that have **larger** inside diameters than the barrel itself. The plunger is designed to **stroke out** of the regular barrel both at the top and at the bottom of its stroke length. This results in a **cleaning** action that prevents **buildup** of sand or solids on the unswept part of the barrel and on the end of the plunger.

3.2.5.7 The Farr plunger

The **Farr plunger** was developed to radically improve the performance of sucker-rod pumps in wells producing considerable amounts of **sand** [14–16]. It was found that the main source of sand problems in standard pumps is the connection between the sucker rod and the plunger. The **connector** is a conical piece of equipment with a maximum outside diameter (OD) at its bottom of about 0.06 in less than the plunger OD. The **gap** between the connector and the barrel is thus much **greater** (at least 30 times) than the normal **fit** (0.002–0.007 in) between the plunger and the barrel. This situation creates the following problems if pumping fluids with sand content:

- During the upstroke, sand suspended in the produced fluid is **directed** (funneled) into and collects in the **gap** between the connector and the pump barrel, eventually reaching the much tighter space between the plunger and the barrel; see Fig. 3.30.
- When the well is shut down, which occurs frequently if the well is on time cycle control, sand will **settle** on the connector and the wedge-shaped space around the connector, filled up with sand, can prevent the movement of the plunger when pumping is resumed.

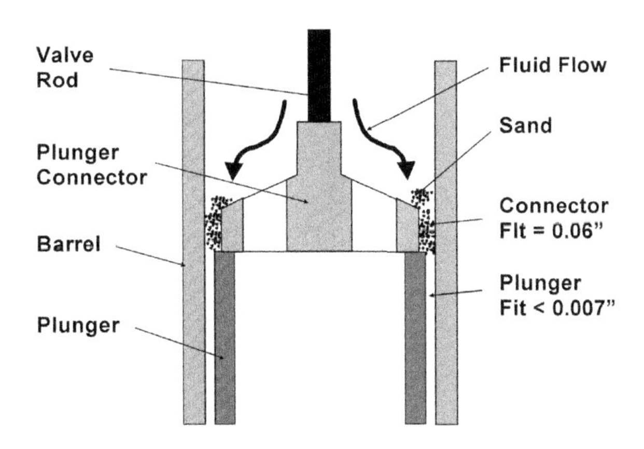

FIGURE 3.30

Sand movement during the upstroke in a standard downhole pump.

The **Farr plunger**, as shown in Fig. 3.31, has its rod–plunger connection at the **bottom** of the plunger (just above the traveling valve), eliminating the greater **gap** between the connector and the barrel present in standard pumps. The other modification involves the shape of the plunger's top, which is now tapering **inward**. The results of these changes are the following operational features of the Farr plunger, as illustrated in Fig. 3.31:

- With the plunger on the upstroke in the pumping cycle, the tapered top of the plunger directs sand-laden fluid **away** from the plunger–barrel **fit**, preventing sand particles from entering there.
- During system shutdowns, sand particles settling out from suspension in the fluid fall in the **plunger** and collect there instead of in the gap between the plunger and the barrel. When the unit is turned on, the first pumping cycle will **flush** out the plunger and return the sand in suspension with the produced fluid.

Long-term field applications of the **Farr plunger** have proved that sucker-rod pumps with these plungers last **two to six times** longer than standard pumps in wells producing sand. Economic advantages are the result of several factors: less workover costs, less production losses, and lower repair costs.

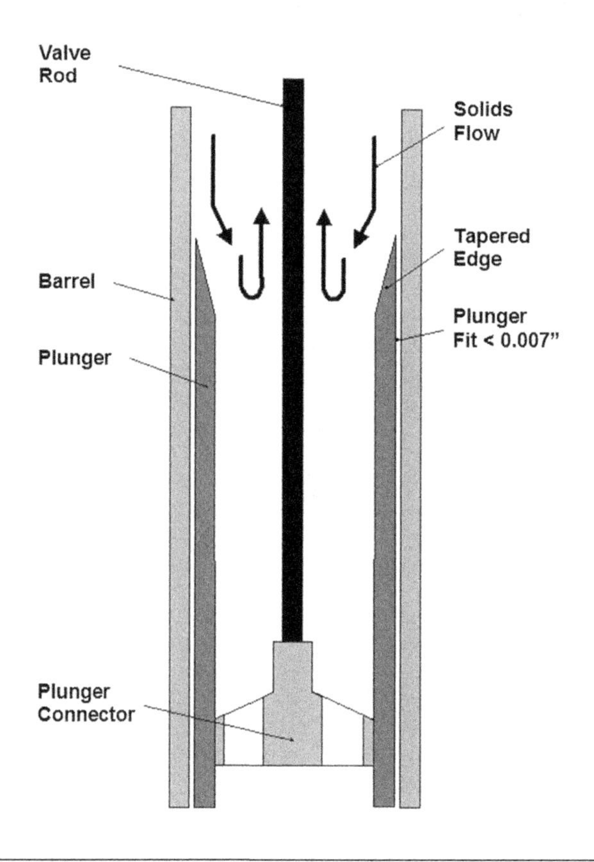

FIGURE 3.31

Construction of the Farr plunger.

3.2.5.8 Other pump types

The **tapered-barrel** or **variable-slippage pump** provides a simple solution to prevent gas locks and helps the traveling valve to open on time [17]. As shown in the schematic drawing in Fig. 3.32, the top part (or a separate section) of the barrel is **tapered** outward. When the plunger approaches the top of the upstroke the clearance between the plunger and the barrel progressively increases. This increases the liquid slippage past the plunger and any gas contained in the barrel can escape, eliminating a gas-locked condition. As the plunger starts its downstroke, the traveling valve opens immediately because pressures are equalized above and below it. The disadvantage of this solution is the reduced pumping efficiency due to the increased slippage past the plunger, but that is justified by the pump's ability to prevent gas locks and fluid pounds.

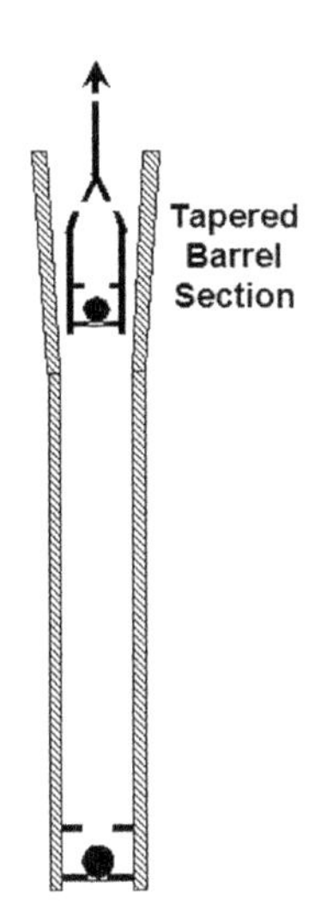

FIGURE 3.32

Sucker-rod pump with tapered barrel.

The **Panacea pump**, according to its name (Greek: **cure-all**) solves **two** problems at the same time: it breaks **gas**-locked conditions and eliminates the damage of **sand** in the pump [18]. The pump contains **two** barrels connected to each other by a lengthened **coupling** with a larger inside diameter; no other modification to the pump is necessary. Based on the required plunger stroke length, the lengths of the two barrels and the plunger are selected in such a way that the plunger passes completely through the coupling on both the up- and the downstroke. Meeting this requirement calls for proper pump spacing.

The operation of the Panacea pump in a gassy well is illustrated in Fig. 3.33. When the plunger descends past the large-diameter coupling situated between the two barrels, some **gas** from below can **bypass** the plunger as it moves through the cavity formed by the coupling; this gas is then released in the tubing. On the upstroke, in turn, some **liquid** will flow down below the plunger as the plunger rises through the coupling. These relatively small amounts of bypassed gas and liquid can return pump operation to normal during consecutive pumping cycles, even in a completely gas-locked pump. The pump was successfully used in wells with a wide range of pump setting depths and plunger stroke lengths.

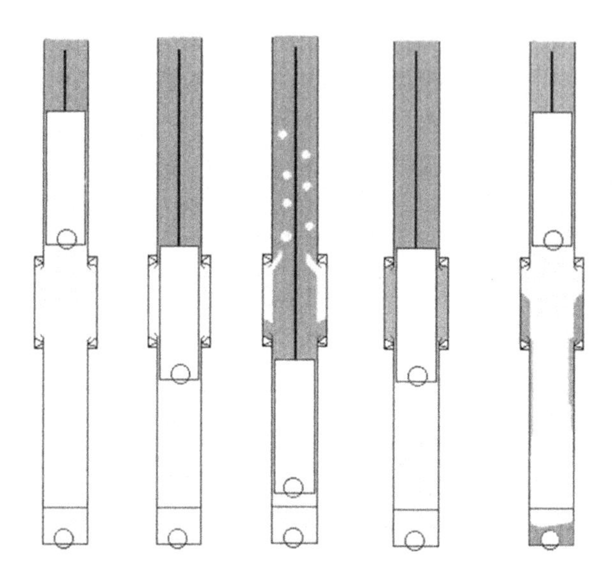

FIGURE 3.33

Operation of the "Panacea pump."

In wells producing high concentrations of **sand**, the outside of the Panacea pump's plunger is continuously **wiped** all along its entire length. This is because the plunger completely passes through the large-diameter coupling and any sand grains left in the cavity are **washed** away by the well fluid on each stroke. Since sand particles cannot get between the barrel and the plunger, their damage to the pump is much reduced. As shown, the simple addition of a large-diameter coupling in the middle of the barrel greatly increases the applicability of the sucker-rod pumps in adverse conditions.

The **Sand-Pro pump** developed for producing solids-laden well fluids contains **two** plungers connected by a pull tube. The upper plunger's main function is to **handle** the produced sand; for this reason it is soft-packed and contains soft valve cups for sealing on the barrel's inside wall. This plunger does not contribute to fluid lifting and has a low pressure differential acting on it. Fluid lifting is performed by the **lower** plunger with a sprayed metal surface and a **close** fit to the barrel. The two plungers are connected with a normal **pull tube**, which is ported at its lower end; see Fig. 3.34. The pull tube is surrounded by a pipe, called **sand shield**, fixed to the top of the lower plunger and open from above; this arrangement creates an annulus around the pull tube.

During the operation of the pump sand and solid particles get past the upper plunger without doing much damage there and get in the **annulus** between the pull tube and the sand shield. This way most of the solid particles produced by the well do not reach the lower plunger and cannot damage the pump. On the downstroke, sand collected in the annulus migrates through the pull tube's ports inside the plunger and is **mixed** with the fluid flow to be lifted on the upstroke along with the fluid produced. The

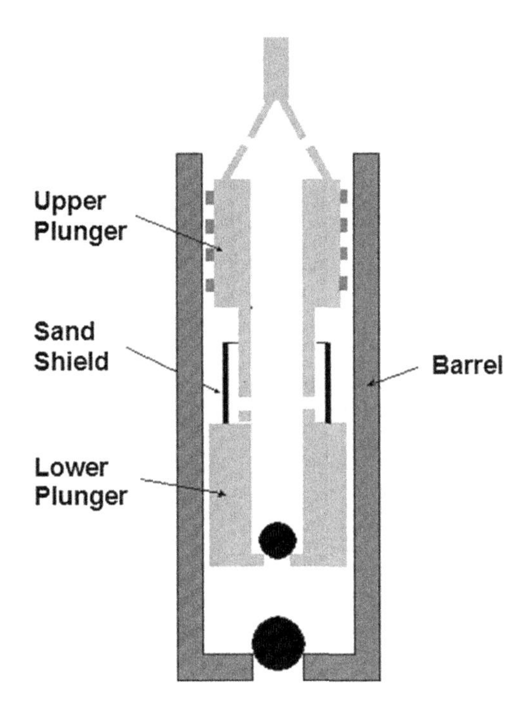

FIGURE 3.34

Construction of the "Sand-Pro pump."

Sand-Pro pump's main advantages are the drastically increased pump life and the reduced workover cost in fields producing higher amounts of solids [19,20].

The **Samson pump**, originally patented by **Skillman** [21], has a plunger much **longer** than the barrel; the plunger always sticks out of the barrel. The sealing surface between the plunger and the barrel is not constant during the pumping cycle as in conventional pumps, where it equals the surface of the plunger. In the Samson pump the **length** of the seal **varies** and is at a maximum at the bottom of the stroke when it is several times **greater** than in a usual sucker-rod pump. As the plunger moves upward, the seal length decreases and reaches its minimum at the top of the stroke; good design ensures this minimum to be sufficient for proper operation of the pump. Figure 3.35 shows the position of the plunger relative to the barrel at the bottom and at the top of the plunger stroke, respectively.

The advantageous features of the Samson pump are the direct consequences of the great and variable pressure-sealing surface provided by the construction of the pump. One of these features is the decreased **slippage** through the plunger–barrel fit that can be as small as one third of that in conventional pumps of the same dimensions. As a consequence, plungers with greater fits can be used without an increase in fluid slippage and the pump can handle solids much better; friction in the pump decreases and equipment life increases.

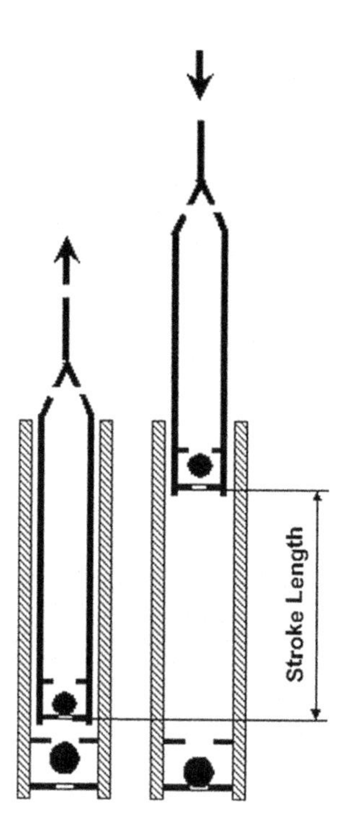

FIGURE 3.35

Operation of the "Samson pump."

Extensive field trials proved that the pump's main application area can be the pumping of sand-laden fluids. On the upstroke solid particles suspended in the fluid are **washed** away from the top of the plunger–barrel interface, unlike in conventional pumps where they are directed into the fit between the barrel and the plunger, as shown in Fig. 3.30. On the downstroke the plunger is **wiped** free of solids as it falls into the barrel. During downtimes, especially in wells on intermittent pumping, solids cannot settle within the barrel as they do in other pumps and resuming the operation of the pump happens without problems; sticking of the pump is eliminated [22].

The **Loc-No Plunger** [11] works like a mechanically operated traveling valve and can be used in stationary-barrel rod pumps to prevent gas locking of the pump in gassy wells. As shown in Fig. 3.36, this device replaces the normal plunger plus the traveling valve and has a freely sliding **sleeve** that acts similarly to a traveling valve. At the start of the upstroke the sleeve, due to the friction between its outside and the barrel, **slides** downward to mechanically seal on the drop and to provide the positive seal required for the pumping action. During the upstroke well fluids can thus enter the barrel space because of the pressure differential created by the plunger.

As soon as the downstroke starts, the sleeve immediately moves off the drop because the frictional force acting on it changes direction, and the traveling valve opens at once. Any gas collected at the top

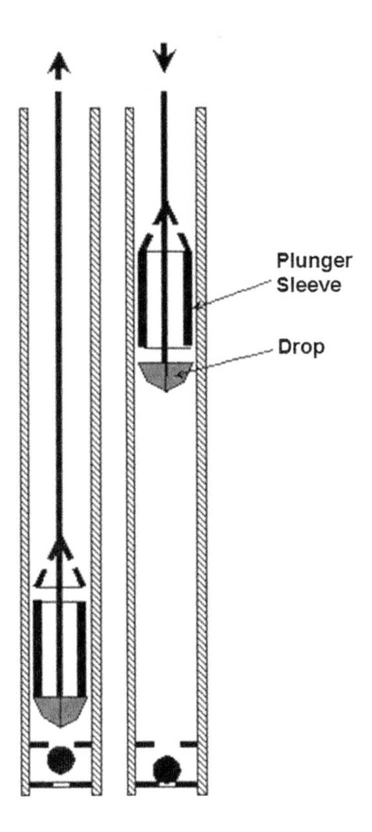

FIGURE 3.36

Operation of the "Loc-No plunger."

of the barrel can pass through the plunger and fluid load is instantly transferred to the traveling valve; gas lock is automatically eliminated. The Loc-No plunger can be used in any stationary-barrel pump in gassy wells, but its application is limited to wells shallower than 5,000 ft, mainly because of the imperfect sealing action between the sliding sleeve and the drop.

3.2.6 PUMP SELECTION

3.2.6.1 Pump size and type

The selection of the proper sucker-rod pump for a given installation is a complicated task and many aspects must be considered, such as well geometrical data, special pumping conditions, etc. Table 3.2 helps in the selection of pump **sizes** for different pump types and contains the **maximum** tubing, pump, and rod sizes that can fit common casing strings; S.H. indicates when sucker rods with slim-hole couplings only can be used.

Table 3.2 Maximum Tubing, Pump, and Rod Sizes That Can Fit Common Casing Strings

Casing Size, in	Max. Tubing Size, in	Max. Rod Size, in	Max. RW Pump, in	Max. RXB Pump, in	Max. RH Pump, in	Max. TH Pump, in	Max. Oversize Pump, in
2 7/8	1 1/2	5/8 SH	1 1/4			1 1/2	2
3 1/2	2 1/16	3/4 SH	1 1/4	1 1/4		1 1/2	2
4	2 3/8	7/8 SH	1 1/2	1 1/2	1 1/4	1 3/4	2 1/4
4 1/2	2 7/8	1 SH	2	2	1 3/4	2 1/4	2 3/4
5	2 7/8	1 SH	2	2	1 3/4	2 1/4	2 3/4
5 1/2	3 1/2	1 1/8	2 1/2		2 1/4	2 3/4	3 3/4
6 5/8	3 1/2	1 1/8	2 1/2		2 1/4	2 3/4	3 3/4
7	4 1/2	1 1/8	3 1/4		2 3/4	3 3/4	4 3/4
7 5/8	4 1/2	1 1/8	3 1/4		2 3/4	3 3/4	5 3/4

The different types of sucker-rod pumps behave differently under **special** operating conditions such as high liquid rates, gas interference, and sand production. Table 3.3 provides a general comparison of the available pump **types** according to a classification of special conditions [23] and allows one to evaluate the merits of each. As seen, the group with the most advantageous features is the stationary-barrel, top-hold-down rod pump type that includes API designations **RHA**, **RWA**, and **RSA**.

3.2.6.2 Allowable pump setting depths

The mechanical strength of sucker-rod pumps, especially in deep wells, must be checked and the stresses generated under actual operating conditions must be kept under control. In this respect the most important part of the pump is the working **barrel**; this being a piece of pipe, the procedure to follow is very similar to selecting pipes for different pressure conditions. Loads and stresses in the barrel come from two sources: (1) **radial** loads caused by the difference between inside and outside pressures, and (2) **axial** loads acting on the barrel's metal cross-sectional area. Accordingly, the three possible modes of loading on the working barrel are as follows:

- **Burst loading** occurs when pressure inside the barrel is greater than the outside pressure.
- **Collapse loading** is caused by a pressure differential due to high outside and low inside pressures.
- **Axial loading** is the result of tensile loads acting on the barrel's metal cross-section.

Since these loadings occur **concurrently**, the working barrel fails when the mechanical stress caused by **any** of the above loads exceeds the allowable stress of the barrel material. Due to constructional differences in pump designs, the **significance** of the three loading conditions varies with the type of the pump; consequently, barrels of different pumps typically fail under different conditions. Typical failure modes of API pumps are listed in Table 3.4.

As shown, working barrels of **bottom-anchor** rod pumps typically **collapse** during the **upstroke** because of the high hydrostatic pressure acting on their outside while the pressure inside the barrel is low. They are not affected by burst loading because inside and outside pressures are quite similar during the **downstroke**; the possibility of tensile failures is also low because axial loads in

Table 3.3 The Merits of Different Pump Types under Special Conditions According to Petterson et al. [23]

	Special Operating Conditions									
Pump Type	**High Rate**	**Deep Well**	**Low Fluid Level**	**Low Speed**	**Intermittent Pumping**	**Gas**	**Sand**	**Corrosion**	**Scale**	**Paraffin**
Stationary barrel top hold-down rod pump	☑	X	☑☑	☑	☑	☑☑	☑☑	☑☑	☑	☑
Stationary barrel bottom hold-down rod pump	☑	☑☑	☑	☑	☑	☑	X	X	☑	☑
Traveling barrel bottom hold-down rod pump	☑	☑	X	☑	☑☑	X	☑	X	☑	☑
Three-tube pump	X	X	X	X	☑☑	X	☑☑	X	☑	☑
Stroke-through pump						X	☑☑		☑☑	☑☑
Tubing pump	☑☑	X	N/A	☑	N/A	X	☑	☑☑	☑	☑
Casing pump	☑☑	X	N/A	☑	N/A	X	X	X	X	X

☑ = good; ☑☑ = very good; X = not recommended; N/A = not applicable.
Deep well = deeper than 7,000 ft.

Table 3.4 Typical Failure Modes of API Pumps

Pump Type	Failure Mode
Rod pumps	
Stationary barrel, top anchor	Burst or axial
Stationary barrel, bottom anchor	Collapse
Traveling barrel, bottom anchor	Collapse
Tubing pumps	Burst or axial

the barrel wall are taken up by the bottom anchor. The barrels of **top-anchor** rod pumps or **tubing** pumps, on the other hand, are not loaded during the upstroke because both the inside and outside pressures are equal. On the **downstroke**, however, there are high **burst** and **axial** loads on the barrel because the high hydrostatic pressure in the tubing acts inside the barrel and on the standing valve. Since the pressure outside the barrel is equal to pump intake pressure, which is usually very low, the danger of the barrel's failure due to burst is high. At the same time, the barrel is stretched due to the high axial load on the standing valve and can suffer a tension failure. In these cases, therefore, **both** burst and axial failures are possible. Axial failure in tubing pumps, of course, can be eliminated completely by anchoring the tubing string when all axial loads are transmitted to the casing string.

Loads from the three main sources previously described vary with the setting **depth** of the pump because the pressures acting on the working barrel change with pumping depth and the density of the fluid pumped. This makes it possible to find allowable pump setting depths (ASDs) for different pump types and operating conditions [24]. The **API RP 11AR** publication [25] presents the relevant formulae and tabulates calculated ASD values for different available barrel materials. Calculations assume a safety factor of 1.25 (25% overdesign), a service factor of $SF = 1$ (corrosion or other problems require the use of $SF < 1$), actual dimensional data of API pump components, the endurance limits of different metals, and pumping of water. For illustration, Table 3.5 contains allowable pump setting depths (ft) for low-carbon steel materials and usual API pump types and sizes according to **Hein and Loudermilk** [24].

Table 3.5 Allowable Pump Setting Depths (ft) for Low Carbon Steel Materials According to Hein and Loudermilk [24]

	Pump Size, in						
Pump Types	1.25	1.50	1.75	2.00	2.25	2.50	2.75
RWA, RSA	6,394	5,520		3,732		3,183	
RWB, RSB, RWT	16,936	14,705		9,727		6,362	
RHA	8,321	8,818	6,749		4,876		
RHB, RHT	27,148	24,249	21,897		18,323		
TH, TP			10,019		7,763		6,262

3.2.6.3 Pumps for gassy wells

As already discussed, the valves of the sucker-rod pump operate properly only if an **incompressible** liquid occupies the barrel space. In this case, during the upstroke the standing valve opens as soon as the plunger starts its upstroke while on the downstroke the traveling valve opens as soon as the plunger starts to move down. Under these conditions the **full** stroke length of the plunger is utilized for liquid production. The situation changes dramatically when **gassy** fluids are pumped because of the **compressibility** of the fluid that gets into the barrel. The operation of the valves will not follow the required behavior and their opening and closing will be **delayed**, resulting in lost plunger stroke length and in reduced pumping rates. A detailed description of the operation of sucker-rod pumps handling gassy fluids follows.

3.2.6.3.1 The effects of compression ratio

When the upstroke starts, the gas–liquid mixture contained in the space between the standing and traveling valves is at the **hydrostatic** pressure of the liquid column in the tubing. This pressure equals the **discharge** pressure of the pump that keeps the standing valve in the **closed** position. The plunger begins its upward movement with a closed traveling valve so the physical **volume** available for the gas–liquid mixture increases. This, in turn, creates the **expansion** of the mixture and a gradual reduction of pressure below the rising plunger. The standing valve, being a simple check valve, can open only when the pressure above it drops below the pump intake pressure available from the wellbore. Since the expansion of the mixture takes up part of the plunger's upstroke, the delayed opening of the standing valve reduces considerably the stroke length available for the barrel to fill up with liquids. The plunger's **effective** stroke length, as well as its liquid output, is **decreased**.

During the **downstroke**, the operation of a pump producing gassy fluids is very similar to the behavior just described. As the downstroke starts, the standing valve closes but the traveling valve remains **closed** as long as the gas–liquid mixture in the barrel is compressed to a pressure sufficient to overcome the liquid column pressure above it. Compression of the gas–liquid mixture occurs while the plunger moves downward and the fraction of the stroke required for this to happen reduces the **effective** plunger stroke length and the pumping rate.

In the worst scenario, single-phase **gas** occupies the barrel space and the valves of the pump do not open at all; the pump's only action is the compression and expansion of this gas; the pump is **gas locked**. No liquid reaches the surface and the energy efficiency of the pumping system goes down to zero. Since this is a highly undesirable kind of operation, in the following we investigate the behavior of sucker-rod pumps when pumping gassy fluids and the ways of avoiding gas-locked situations.

Figure 3.37 shows the conditions at the start and at the end of the **upstroke** in a gas-locked pump where both the traveling and standing valves stay closed for most of the length of the plunger's stroke. At the start of the upstroke gas pressure below the plunger equals the pump discharge pressure, p_d, and the volume available is the so-called unswept volume or **dead space**, V_{usw}, in the pump. At the top of the stroke the pressure is assumed to drop to the pump intake pressure, *PIP*, so that the standing valve can open. The expanded volume of the gas is the sum of the unswept volume and the swept volume; this latter is identical to pump **displacement**. Assuming an **isothermal** change, the relationship between gas volume and pressure is according to the formula $p\,V = \text{const.}$ and, using the parameters just described, we receive the following expression:

$$p_d V_{usw} = PIP(V_{sw} + V_{usw}) \tag{3.1}$$

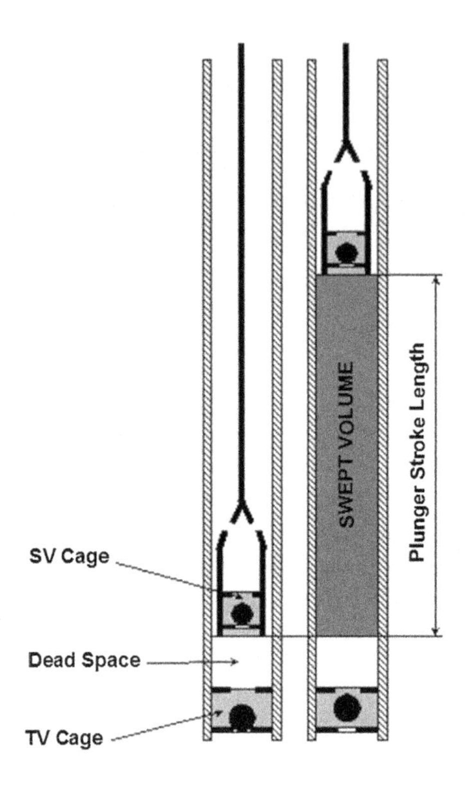

FIGURE 3.37

Illustration of the dead space in a downhole pump.

The formula can be modified to:

$$\frac{p_d}{PIP} = \frac{(V_{usw} + V_{sw})}{V_{usw}} = CR \tag{3.2}$$

where:

p_d, PIP = pump discharge and intake pressures, psi,
V_{sw}, V_{usw} = swept and unswept volumes in barrel, cu in, and
CR = the pump's compression ratio, –.

The ratio of pressures on the left-hand side of the formula must be corrected because they act on different surfaces due to the **bevel** on the traveling valve's seat. The correction, based on the appropriate seat diameters (see Fig. 3.38), is:

$$\frac{p_d}{PIP}\frac{\frac{D_2^2\pi}{4}}{\frac{D_1^2\pi}{4}} = \frac{p_d}{PIP}\frac{D_2^2}{D_1^2} = \frac{p_d}{PIP}VSR \tag{3.3}$$

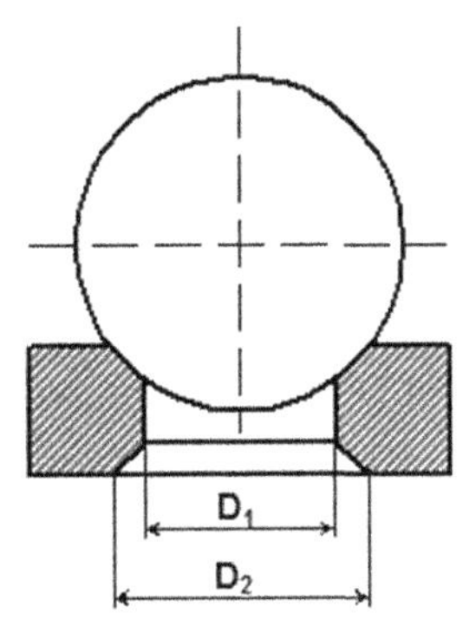

FIGURE 3.38

Dimensions of a beveled traveling valve.

The final formula becomes, after substituting this in Eq. (3.2):

$$\frac{p_d}{PIP} = \frac{CR}{VSR} \tag{3.4}$$

where:

VSR = valve seat area ratio, –.

The formula expresses the conditions under which gas lock can be **prevented**; if the two sides of the expression **match**, then at the end of the upstroke the pressure in the barrel decreases to the pump intake pressure and the standing valve can open. Exactly the same formula can be derived for the plunger's downstroke when the pump compresses the gas in the barrel to the discharge pressure that opens the traveling valve. In general, the left-hand side of Eq. (3.4) represents the **required** compression ratio for the actual pumping conditions and the right-hand side represents the **available** compression ratio from the pump used. Gas locking is prevented when the compression ratio of the pump is equal to or greater than the required ratio; then the pump can even operate with gas being pumped only.

Let's investigate the different terms figuring in Eq. (3.4) and their ranges in usual sucker-rod pumps [26]. The pump's compression ratio, CR, as shown in Eq. (3.2), is a function of the pump's swept and un-swept volumes. The swept volume, V_{sw}, equals the barrel volume corresponding to the plunger stroke length (see Fig. 3.37). The unswept volume, also called the **dead space**, on the other hand, may have the following components:

- Spaces in traveling and standing valve **cages**. Manufacturers usually do not publish these data, but cages of different designs can have widely varying dead spaces.
- Dead space due to improper selection of the valve rod's **length**. The ideal length for the valve rod is found when the distance between the traveling and standing valves during assembly is about 1/2 in at the bottom position of the plunger.
- Inherent dead spaces in the barrel of different pump types. **Thin**-wall rod pumps have the **least** amount of dead space; heavy-wall rod pumps have much more because they need the use of extension **couplings**. Tubing pumps are the **worst** because of their larger diameters and the standing valve puller attached to the bottom of the plunger.

- The plunger's **internal** volume is part of the dead space if the traveling valve is located at the **top** of the plunger.
- Additional dead space caused by improper **spacing** of the plunger. The plunger must be spaced so that rods are run in the well until the plunger tags on the standing valve; then it must be raised just to remove the tag.

According to its definition (Eq. 3.2) the compression ratio, *CR*, depends on the pump **displacement** (the swept volume) as well: the greater the displacement for the same dead space, the greater the value of *CR*. This is clearly seen in Fig. 3.39, where *CR*s of different types of pumps are given for various plunger stroke lengths; longer strokes represent greater pump displacements and, consequently, higher compression ratios. The relatively poor performance of tubing pumps can also be observed.

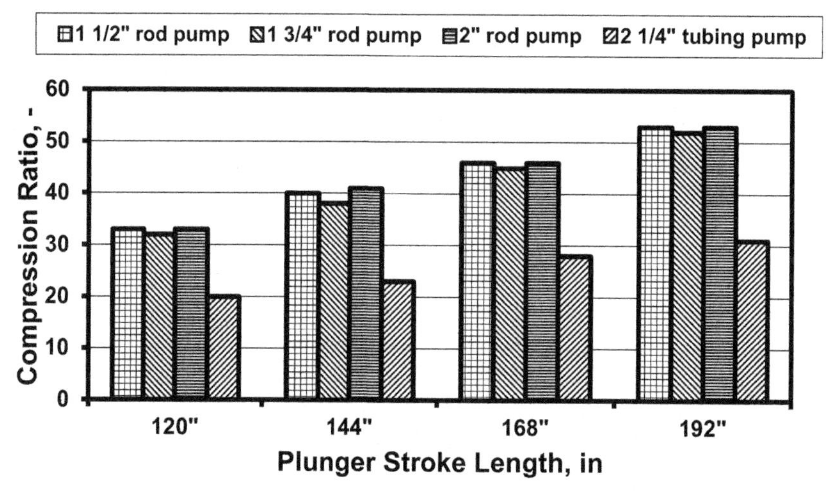

FIGURE 3.39

Typical compression ratios of API sucker-rod pumps.

Finally, the valve seat area ratio, *VSR*, represents the effect of **beveling** of the traveling valve seat. Although manufacturers seldom publish this piece of data, the use of an average value of $VSR = 1.16$ is recommended [26].

Checking gas-lock conditions is accomplished by using Eq. (3.4) and following the calculation steps given here:

1. Based on the type and size of the pump, find the dead space in the pump at ideal pump spacing, and the valve seat area ratio, *VSR*.
2. Correct the dead space for actual (usually greater than ideal) spacing.
3. Find the pump displacement and calculate the pump's compression ratio, *CR*.
4. Calculate pump intake and discharge pressures. *PIP* is found from the annular liquid level in the casing; p_d is equal to tubing head pressure plus the hydrostatic column pressure of the produced liquid at pump setting depth.

5. Using PIP or p_d as a known variable, solve Eq. (3.4) for the other variable. Compare the calculated value to the actual value found in step 4.
6. Gas lock is possible if $PIP_{\text{calculated}} > PIP_{\text{actual}}$ or $p_{d\ \text{calculated}} < p_{d\ \text{actual}}$.

EXAMPLE 3.2: DETERMINE WHETHER A 1.75 IN RWA PUMP SET AT 4,500 FT WITH A STROKE LENGTH OF 120 IN CAN WORK WITHOUT GAS LOCKING IF THE FLUID LEVEL ABOVE THE PUMP IS 300 FT. LIQUID GRADIENT IN THE TUBING AND IN THE ANNULUS IS 0.35 PSI/FT, TUBING HEAD PRESSURE IS 150 PSI. ASSUME TWO DIFFERENT SPACINGS OF THE PLUNGER AT 1 IN AND 5 IN

Solution

The pump's dead space was found as 9 cu in, and the average valve seat area ratio of $VSR = 1.16$ is used.

Additional dead space for the different spacings is found from the pump size.

For 1-in spacing, $1\ (1.75^2\ \pi)/4 = 2.4$ cu in must be added and the unswept volume becomes:

$$V_{usw} = 9 + 2.4 = 11.4\,\text{cu in.}$$

For 5-in spacing, $5\ (1.75^2\ \pi)/4 = 12$ cu in must be added and the unswept volume becomes:

$$V_{usw} = 9 + 12 = 21\,\text{cu in.}$$

Pump displacement is found from the plunger stroke length and the pump diameter:

$$V_{sw} = 120\left(1.75^2\pi\right)/4 = 288.6\ \text{cu in.}$$

Compression ratios for the two cases from Eq. (3.2) are:

$$CR_1 = (11.4 + 288.6)/11.4 = 26.3.$$

$$CR_2 = (21 + 288.6)/21 = 14.7.$$

PIP is found from the annular liquid level above the pump:

$$PIP = 300 \times 0.35 = 105\ \text{psi.}$$

Pump discharge pressure, p_d, is calculated as the sum of the tubing head and the hydrostatic pressures:

$$p_d = 150 + 4{,}500 \times 0.35 = 1{,}725\ \text{psi.}$$

Use the actual PIP and solve Eq. (3.4) for pump discharge pressure:

$$p_d = PIP\ \ CR/VSR.$$

For the first case the pump discharge pressure is $p_d = 105 \times 26.3/1.16 = 2{,}380$ psi, which is greater than the actual discharge pressure (1,725 psi), so the pump does not get gas locked because it can compress even gas.

For the second case $p_d = 105 \times 14.7/1.16 = 1{,}330$ psi, which is less than the actual discharge pressure (1,725 psi), so the pump will become gas locked if a great amount of gas gets into the barrel.

3.2.6.3.2 Conclusions

Based on the previous discussions on compression ratios in sucker-rod pumps, several recommendations may be given to **prevent** gas lock situations when producing wells with higher gas rates. As will be seen, all proposed measures aim at **maximizing** the compression ratio of the pump by reducing the amount of total dead space in the pump.

- Try to use thin-walled rod pumps with longer strokes.
- Use plungers with traveling valves situated at the bottom of the plunger.
- Reduce the spacing of the plunger to 1 in.
- Anchoring the tubing string is recommended because it allows proper spacing of the pump.
- Use valve cages with a minimum of dead space.
- Avoid the use of double valve arrangements because they increase the size of the dead space.

3.2.6.4 Selection of materials

Pump parts are usually available in various materials with different chemical and physical properties. The proper material selection for the different components of a pumping assembly must have a high priority, because it directly influences the **service life** of subsurface pumps. The choice of materials should be based on the proper consideration of the following five factors:

- **corrosiveness** of the well fluids,
- **abrasive** particles present in well fluids,
- mechanical **strength** required,
- **compatibility** of different materials, and
- overall **costs**.

Corrosion is attributed to the presence of **salt water** and/or corrosive **gases** in the well stream. Such gases are **hydrogen sulfide** (H_2S), **carbon dioxide** (CO_2), and **oxygen** (O_2). The forms of corrosion damage are different in different environments and include, among others, **pitting**, i.e., local material loss on the surface and **galvanic** corrosion of two dissimilar metal parts. Other corrosion types are **hydrogen embrittlement** and **sulfide stress cracking** when sour gases are present. All these forms of corrosion result in a gradual or sudden **failure** of the affected parts and can considerably reduce the service life of subsurface pumps. Different metals offer different resistances to the various kinds of corrosion.

Abrasion is always present when fluids containing **solid** particles are pumped. The basic abrasive material in pumping operations is the **sand** coming from productive formations. Severe damage is expected where closely fitted moving parts are involved. As abrasion works on the surface of structural parts, **hardening** or **plating** of the critical surfaces reduces the harm done.

All materials preliminarily selected for their corrosion and abrasion resistance qualities should be checked for mechanical **strength**. The working stresses in the given pump component must be considered before a final selection is made. Very often, especially in sour environments, the metals with suitable corrosion resistance can turn out to be not **strong** enough for the purpose. In these cases, another material must be selected, or another pump type is considered which imparts less loading on the structural part studied. The best material, therefore, is found only by trial and error.

When selecting materials for the individual parts making up a complete pumping assembly, the **compatibility of metals** must also be considered. Incompatibility problems can result in **electrolysis** between two dissimilar metal parts joined together, because salt water or other fluids usually act as electrolytes. Another cause of incompatibility is the difference in the **thermal expansion** of two metal parts. A closely fitted plunger, for example, can get stuck in a barrel if the barrel's material expands less than the plunger material.

Last but not least, material selection is a question of **costs** as well. In order to find the optimum solution, not only initial equipment costs but rather the total costs over a longer period of time must be considered. For estimating total costs, the anticipated **frequency** of failures and the costs of **repair** and

maintenance jobs have to be assumed. In conclusion, field experience is a deciding factor in selecting the proper materials, but manufacturers' recommendations must also be taken into account.

After this basic discussion of the factors affecting material selection, the most important metals used in subsurface pump parts are listed below, with a short description of their basic properties, after [2,27].

Carbon Steels—provide no protection against **corrosion** or **abrasion**. If plated with chrome, abrasion resistance is excellent, and metal spraying offers moderate abrasion resistance.

Alloy Steels—regular alloys are fairly **resistant** to corrosion but not to **abrasion**. Hardening, plating, or metal spraying increases abrasion resistance.

Stainless Steels—are usually not recommended in **abrasive** fluids; corrosion resistance is moderate to excellent.

Brass—is easily eroded by abrasive well streams, unless plated. Different brass alloys have moderate to high corrosion resistance.

Monel—excellent for **corrosive** environments, but no resistance to abrasion unless plated with a hard surface.

Cobalt Alloys—high corrosion and abrasion resistance. Used for valve **balls** and **seats** and as a coating for high-wear surfaces.

Carbides—excellent against corrosion and abrasion, **tungsten carbide** is mainly used for balls and seats.

Ceramics—Zirconia ceramic materials are extremely resistant to both corrosion and abrasion and are used for balls and seats.

3.3 TUBING ANCHORS

As discussed in Section 3.2.1, the tubing string is subjected to a varying load during the pumping cycle. On the upstroke it is not loaded, since the plunger carries the weight of the liquid in the tubing. However, this load is transferred to the tubing at the start of the **downstroke**. Owing to this variable load, a freely suspended tubing string periodically **stretches** and **recoils** during the pumping cycle. It is quite easy to see that tubing stretch reduces the plunger stroke **length** available for lifting of fluids, resulting in a reduction of pump displacement. This phenomenon has been recognized for many years, and the **anchoring** of the tubing string to the casing became standard practice.

However, these effects do not completely describe the behavior of the tubing string in a pumping well. It was found by **Lubinski**, based on his investigations of rotary drill pipe, that the lower portion of a freely suspended tubing string is subject to **helical buckling** [28]. Buckling is present during the upstroke only, and causes the tubing to wrap around the rod string below a certain depth, called the **neutral point**. The shape of the buckled tubing is a helical surface that is constrained on the inside by the rod string and on the outside by the casing. Figure 3.40 shows a cross-section of the tubing in a pumping well during the up- and downstroke. On the upstroke, the rod string is fully under tension and **straight**, due to the fluid load carried by the traveling valve. The tubing string, being unloaded, contracts and moves upward to its unstretched position. At the same time, a compression force arises in the lower portion of the tubing, which causes it to **buckle**. This buckling is most pronounced immediately above the pump and diminishes up the hole until the neutral point, above which tension due to tubing weight overcomes the compressive force. When the downstroke starts, fluid load is transferred to the standing valve and the tubing returns to its fully stretched position.

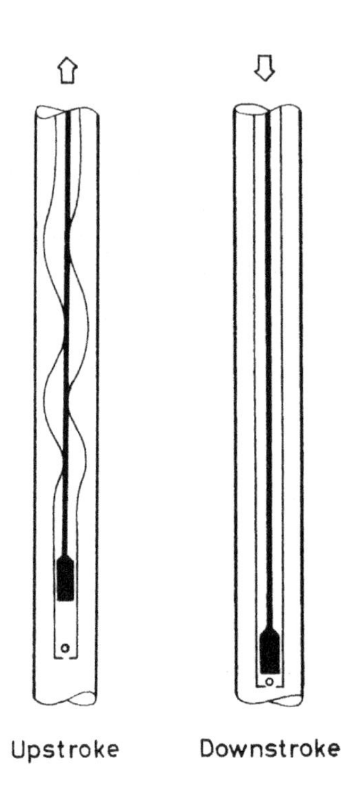

FIGURE 3.40

The movement of a freely suspended tubing string during the pumping cycle.

The variable movement of the tubing string causes several operational **problems**, all of which are the result of the excessive **friction** between the rods and the tubing. The friction is more pronounced if the well stream contains **abrasives**, which increase the wear on the metal surfaces. Greater wear can also accelerate corrosion damage, so an effective corrosion inhibition may be needed. The main detrimental effects of tubing **buckling** are:

- The rods **wear** on the inside of the tubing, which can lead to rod and tubing failures. The casing string also can suffer from mechanical wear.
- The loads on the pumping unit increase because of the increased **frictional** forces between the rods and the tubing.
- More **torque** and **power** are required at the surface to drive the pumping unit.

In early production practice, a simple **hook-wall packer** was used to anchor the tubing. The hook-wall packer can be considered a compression anchor that relies on a **compressive** force for holding properly, preventing downward tubing movement but allowing upward movement. It follows from the operational principle that the compression anchor can **walk** up the hole and will fix the tubing to the casing string at the **uppermost** position of the tubing's movement. As seen in Fig. 3.41, the tubing is still **buckled** on the upstroke, its deflection being limited by the straight, heavily loaded rod string. The situation is even worse during downstroke, when the tubing, although loaded by the

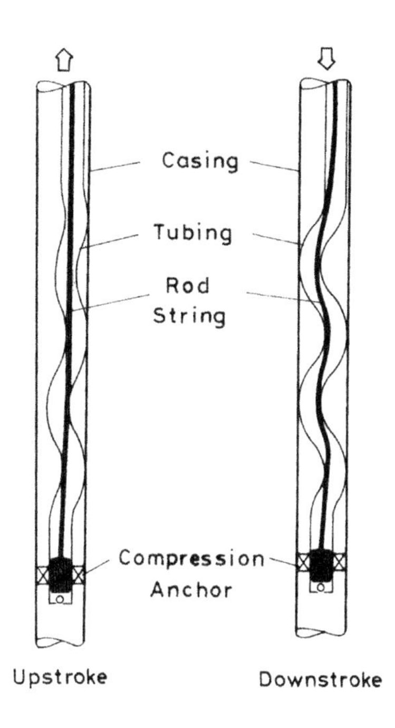

FIGURE 3.41

The operation of a compression anchor in a pumping well.

weight of the fluid column, cannot straighten out because of the anchor. Buckling is more severe, since the tubing's lateral movement is only limited by the casing and not by the rod string that is no longer under great tension and thus offers little resistance. These unfavorable features made the compression anchor, in the majority of cases, **impractical** to use. The only advantages of a compression anchor are the relatively easy retrieval and the ability to act as a tubing catcher, should the tubing part.

A better way to anchor the tubing string is to use a **tension-type anchor**. This can be set at any depth in the casing and does not allow any upward movement. Its holding power is derived from a **tension** force on the tubing. The required force or pickup to be applied to the tubing depends on well conditions and must be calculated before setting the anchor. This anchor completely eliminates tubing **buckling** and additional troubles associated with it. Its only **drawbacks** are that it is more difficult to retrieve, and it does not prevent a parted tubing from falling into the well.

The tubing **anchor–catcher** (TAC), shown in Fig. 3.42, offers the advantages of both the tension and the compression anchors; its slips, when set, do not allow any vertical movement. It is set by left-hand rotation and retrieved by an opposite rotation. When set, the anchor–catcher permits the proper tensioning of the tubing and completely prevents buckling. The required surface pull on the tubing string is calculated from the tension that would exist in the tubing during the downstroke, were the tubing freely suspended in the well [29,30]. In case the tubing parts, it is not allowed to fall to the bottom. By preventing any vertical movement of the tubing, the anchor–catcher eliminates

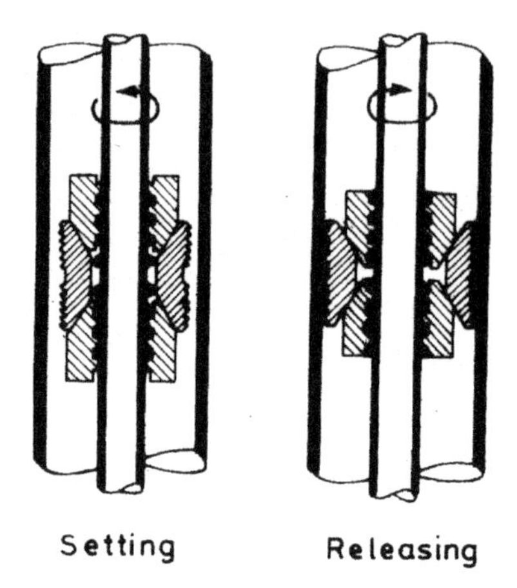

FIGURE 3.42

The principle of operation of a tubing anchor–catcher (TAC).

wear on rod, tubing, and casing strings. Its use increases the effective stroke length of the plunger and thus can increase pump **displacement** as well. The loading of the rod string, as well as the pumping unit, is also reduced. The TAC must be set at a depth close to the pump to eliminate tubing movement but never below the perforations, to prevent the accumulation of solid material on top of it.

3.4 DOWNHOLE GAS SEPARATORS (GAS ANCHORS)

3.4.1 INTRODUCTION

Sucker-rod pumps, like all piston pumps working on the positive displacement principle, are designed to pump an **incompressible** liquid phase. These pumps, therefore, provide very efficient fluid lifting operations in wells where negligible amounts of **free gas** are present at pump depth. However, the majority of oil wells produce, along with the liquid phases, some gas that can exist at the pump intake either as **free** or as **dissolved** gas. Gas that is free at suction pressure drastically reduces the pump's volumetric efficiency because pump valves do not operate properly during the pumping cycle. On the **upstroke**, the standing valve's opening is **delayed**. In addition to this, the barrel only **partially** fills with liquid, since gas occupies some of the barrel space. On the **downstroke**, the traveling valve is kept **closed** for a portion of its downward travel by the fluid load from above. All these effects decrease the plunger's **effective** stroke length and cause a considerable reduction of pump displacement. In extreme cases a **gas lock** can also develop, which completely stops the pumping action.

The detrimental effects of pumping a gaseous mixture with a sucker-rod pump give rise to a multitude of operational problems. These include low pump **efficiencies**, increased pump and rod **failures** due to fluid and gas pounding, and **losses** in liquid production. Accordingly, lifting costs get higher and production economy is considerably decreased. This is why **gas interference** is considered one of the biggest enemies facing the production engineer working in sucker-rod pumping.

There are two main areas of **remedial** measures to improve pumping operations in gassy wells:

- the greatest possible amount of free gas is **separated** downhole and is not allowed to enter the pump, and
- such pumps are selected that minimize the detrimental effects of free gas.

The second case often involves the use of **special** pumps that have been discussed in Section 3.2.5; the first case is discussed in the following.

3.4.1.1 Free gas volume calculations

The presence of free gas at the pump suction depends on the **pump intake pressure** (*PIP*), the thermodynamic properties of the well fluids, and the well temperature at suction conditions. The case of direct gas production, usually a result of gas coning around the well, should also be mentioned but will not be included in the following. Depending on the produced crude's bubble point pressure, two cases are possible:

1. For *PIP* values higher than or equal to bubble point pressure, **no free gas** enters the pump, and
2. At *PIP* values lower than the bubble point pressure, a progressively greater portion of **solution gas** evolves from the crude oil as free gas.

The "in situ" free gas volumetric rate present at pump suction conditions is found from the well's production gas–liquid ratio (*GLR*) and the amount of gas, R_s, still in solution at the pump intake pressure, *PIP*, as given here:

$$q'_g = \frac{q_l}{1 + WOR}[GLR(1 + WOR) - R_s]B_g \tag{3.5}$$

where:

q'_g = in situ volumetric gas rate at pump suction, ft^3/d,
q_l = the well's liquid production rate, STB/d,
WOR = production water–oil ratio, –,
GLR = production gas–liquid ratio, scf/STB,
R_s = solution gas-oil ratio (GOR) at pump intake pressure and temperature, scf/STB, and
B_g = gas volume factor at pump intake pressure and temperature, ft^3/scf.

Knowledge of the free gas rate allows the determination of the gas **content** of the well stream entering the subsurface pump **without** a gas separator, an important parameter when evaluating the operating conditions of the pump. It can be calculated as the ratio of the gas rate to the total fluid rate using the following formula:

$$Gas\% = \frac{\frac{q'_g}{5.614}}{q_l\left(\frac{B_o}{1+WOR} + B_w\frac{WOR}{1+WOR}\right) + \frac{q'_g}{5.614}} 100 \tag{3.6}$$

where:

$Gas\%$ = in situ gas content, %,
q'_g = in situ volumetric gas rate, ft^3/d,
q_l = the well's liquid production rate, STB/d,
WOR = production water–oil ratio, –, and
B_o, B_w = oil and water volume factors, bbl/STB.

The sucker-rod pump can operate without any problems if the calculated gas content is around a few percentages, but gas contents above 10–20% indicate **gassy** conditions; at considerably higher values, severe problems including **gas locking** can occur.

EXAMPLE 3.3: FIND THE IN SITU FREE GAS VOLUME AT THE PUMP INTAKE CONDITIONS FOR THE DATA GIVEN

Oil Rate, STB/d	50	*PIP*, psi	100
Water Rate, STB/d	60	Oil API	34
Produced GLR, scf/STB	50	Water *SpGr* –	1.03
Temperature, F	150	Gas *SpGr* –	0.60

Solution

Oil specific gravity is found from Eq. (2.4):

$$\gamma_o = 141.5/(131.5 + 34) = 0.855.$$

The *WOR* from the liquid rates is $WOR = 60/50 = 1.2$.
The solution gas–oil ratio at suction conditions is found from Eq. (2.10):

$$y = 0.00091 \times 150 - 0.0125 \times 34 = -0.288, \text{ and}$$

$$R_s = 0.6\left(100/18/10^{-0.288}\right)^{1.205} = 10.5 \text{ scf/STB}.$$

Gas volume factor calculations necessitate the calculation of the critical parameters of the gas from Eqs (2.23) and (2.24):

$$p_{pc} = 709.6 - 58.7 \times 0.6 = 674 \text{ psia}.$$

$$T_{pc} = 170.5 + 307.3 \times 0.6 = 355 \text{ R}.$$

Pseudoreduced parameters and the deviation factor are calculated from Eqs (2.17) and (2.18) and Eq. (2.28), respectively:

$$p_{pr} = (100 + 14.7)/674 = 0.170.$$

$$T_{pr} = (150 + 460)/355 = 1.719.$$

$$Z = 1 - 3.52 \times 0.170/10^{0.9813 \times 1.719} + 0.274 \times 0.170^2/10^{0.8157 \times 1.719} = 0.895.$$

Volume factor of gas is calculated from Eq. (2.21):

$$B_g = 0.0283 \times 0.895(300 + 460)/(100 + 14.7) = 0.135 \text{ cu ft/scf}.$$

Volume factors of water and oil are found from Eqs (2.8) and (2.11):

$$B_w = 1.0 + 1.21\text{E-4}(150\text{-}60) + 1\text{E-6}(150 - 60)^2 - 3.33\text{E-6} \times 100 = 1.018 \text{ bbl/STB}.$$

$$F = 10.5(0.600/0.855)^{0.5} + 1.25 \times 150 = 220.7.$$

$$B_o = 0.972 + 1.47\text{E-4}\ \ 220.7^{1.175} = 1.045 \text{ bbl/STB}.$$

The free gas volumetric rate at suction conditions is calculated from Eq. (3.5) as follows:

$$q'_g = (50 + 60)/(1 + 1.2)\ [50(1 + 1.2) - 10.5]\ 0.135 = 670 \text{ cu ft/d}.$$

The gas content of the well stream reaching the sucker-rod pump can be found from Eq. (3.6):

$$Gas\% = (670/5.614)/\{(50 + 60)\ [1.045/(1 + 1.2) + 1.018 \times 1.2/(1 + 1.2)] + (670/5.614)\} = 0.513 = 51.3\%.$$

As seen, the gas–liquid mixture at the pump intake contains more than 51% of free gas, so the conditions are very detrimental to sucker-rod pumping.

3.4.1.2 Basics of downhole gas separation

Almost all downhole gas separators operate on the principle of **gravitational** separation, because the use of other separation methods (like centrifugal separation) is difficult in a rod-pumped well. Thus the force of **gravity** is utilized to separate the gas, usually present in the form of small gas **bubbles**, from the liquid phase. In order to understand the separation process, the typical behavior of free gas bubbles contained in the well stream should be discussed first. Gas bubbles submerged in a moving liquid are subjected to two basic forces: **buoyancy** and viscous **drag**. During steady-state flow conditions these forces are in balance and the gas bubbles tend to **rise** relative to the liquid with a definite terminal upward velocity. This velocity is a function of bubble **size** and the **viscosity** of the surrounding liquid. Smaller bubbles have lower rising velocities and velocity decreases in more viscous liquids; a value of 0.5 ft/s is generally used for average conditions, i.e., average-sized bubbles (about 1/4 in diameter) moving in low-viscosity liquids.

If the well stream is led into a space of sufficient capacity, then liquids, being denser than gas, flow downward due to gravity, but gas bubbles tend to rise. As long as downward liquid velocity is lower than the terminal velocity of the gas bubbles, the resultant gas velocity is directed upward and the gas phase continuously **rises** compared to the liquid phase. High liquid velocities, on the other hand, result in the gas bubbles being taken along with the liquid and no separation of the phases takes place. For an effective gas separation, therefore, flow velocity of the liquid must be kept below the typical bubble rise velocity of 0.5 ft/s. This requirement can only be met if the cross-sectional **area** available for liquid flow is properly selected by considering the liquid production rate of the well. Downhole gas separators work according to this principle: they force the liquid phase to have a velocity lower than 0.5 ft/s by properly selecting the space available for liquid flow.

The separation capacity of any kind of gravitational separator used in sucker-rod pumped wells is usually evaluated by using the **Clegg** correlation [31,32], which gives an **estimate** for the *GLR* of the gas–liquid mixture leaving the separator and reaching the pump, which is connected downstream of the separator:

$$GLR'_{ing} = C\ PIP^{2/3}\ v_{sl}^{0.5} \tag{3.7}$$

where:

GLR'_{ing} = dimensionless *GLR* ingested by the pump, –,
C = separator-type coefficient, –,
PIP = pump intake pressure, psia, and
v_{sl} = liquid superficial velocity in separator, ft/s.

The formula contains the coefficient C, which varies with the type of gas anchor, as given here:

Gas Anchor Type	Coefficient
Natural or cup type	$C = 0.028$
Packer type	$C = 0.036$
Poor-boy type	$C = 0.1$

The superficial liquid velocity is found from the following expression, which calculates the downward velocity of liquid flow in the separator body.

$$v_{sl} = 9.36 \times 10^{-3} \frac{q_l}{A} \left[\frac{B_o}{1 + WOR} + B_w \frac{WOR}{1 + WOR} \right] \tag{3.8}$$

where:

v_{sl} = superficial liquid flow velocity, ft/s,
q_l = the well's liquid production rate, STB/d,
WOR = production water–oil ratio, –,
B_o, B_w = oil and water volume factors, bbl/STB.
A = downflow cross-sectional area in separator, in^2.

The volumetric rate of gas **ingested** by the pump after separation is easily found from the **Clegg** correlation (Eq. 3.7) by multiplying GLR'_{ing} by the in situ liquid rate of the well:

$$q'_{ing} = 5.614 GLR'_{ing} q_l \left[\frac{B_o}{1 + WOR} + B_w \frac{WOR}{1 + WOR} \right] \tag{3.9}$$

where:

q'_{ing} = in situ volumetric gas rate ingested by the pump, cu ft/d,
GLR'_{ing} = dimensionless GLR ingested by the pump, –,
q_l = the well's liquid production rate, STB/d,
WOR = production water–oil ratio, –,
B_o, B_w = oil and water volume factors, bbl/STB.

The knowledge of the **ingested** amount of gas allows the determination of the gas **content** of the fluid leaving the gas separator and reaching the subsurface pump by using Eq. (3.6). If this value is less than the one calculated without gas separation, then the use of the gas anchor is **justified** because the separator reduces the amount of free gas entering the subsurface pump. Otherwise the gas separator is inefficient in removing free gas from the well stream.

3.4.2 GAS SEPARATOR TYPES

3.4.2.1 The natural gas anchor

Downhole gas **separators** used in sucker-rod pumping are often called **gas anchors**. The simplest and most **efficient** gas anchor is the **natural gas anchor,** shown in Fig. 3.43. It utilizes the lower section of the casing–tubing annulus as a **sump** for natural gas separation. Its most important feature is

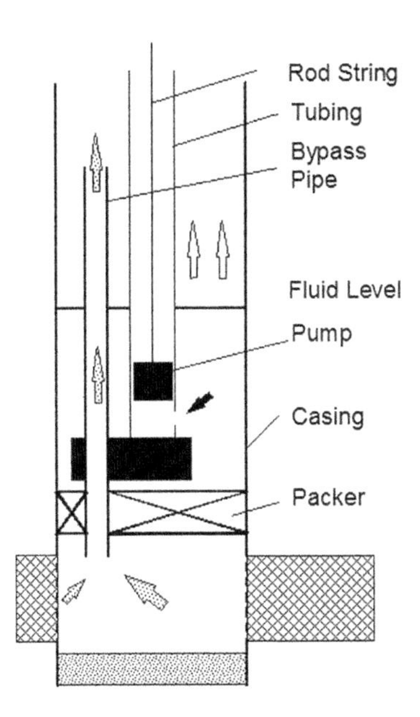

FIGURE 3.43

Construction of the natural gas separator.

that the pump is set at least 10 ft **below** the level of the lowermost casing perforations. The sucker-rod pump's suction is thus at a greater depth than the sandface where formation fluids enter the well. Well fluids, liquids and gas, must therefore move **downward** to enter the pump and gas separation takes place in the annulus between the casing and the tubing strings. As the annulus is kept open at the surface, the separated gas can leave the well, ensuring an undisturbed separation process.

Since liquid flow takes place between the casing and the tubing strings, the cross-sectional **area** available for flow is the **largest** possible and liquid velocity, as a consequence, is low. As long as the superficial liquid velocity is below 0.5 ft/s (the terminal rise velocity of average-sized gas bubbles) the natural gas anchor provides a **perfect** gas separation and no free gas enters the pump. Because the maximum possible cross-sectional area is utilized for liquid flow in this separator, the natural gas anchor provides the **maximum** separation capacity over any other gravitational gas separator.

Table 3.6 gives the maximum liquid rates [33] that can be handled by natural gas anchors in **annuli**. The rates are based on the assumption that as long as liquid downward velocity in the annulus is below the terminal **rise** velocity of gas bubbles, all gas coming from the formation is **separated** and will rise to the surface. As seen from the table, greater annular cross-sections associated with smaller separator sizes inside the same casing (as in modified natural gas anchors) provide perfect separation of produced gas at higher liquid production rates.

The requirements for using a **natural** gas anchor are:

- The well should be drilled to a depth **below** the productive formation to provide a downhole sump. This lower section of the well should be **clean** enough to provide a trouble-free pump operation.
- Casing size should be sufficiently **large** to ensure that downward velocity of the liquid flow does not exceed the rising velocity of gas bubbles.
- The pressure kept at the casinghead should be sufficiently **low** to allow the separated gas to flow freely to the surface.

Table 3.6 Maximum liquid rates handled by natural gas anchors, according to McCoy et al. [33]

Casing Size, in	Separator Size, in	Liquid Capacity, blpd
	Natural Gas Anchors	
7	3 1/2	1,150
7	2 7/8	1,335
7	2 3/8	1,440
5 1/2	2 7/8	635
5 1/2	2 3/8	740
4 1/2	2 7/8	305
4 1/2	2 3/8	410
	Modified Natural Gas Anchors	
5 1/2	1 1/2	820
4 1/2	1 1/4	520

3.4.2.2 The modified natural gas anchor

As already discussed, the lower the liquid downward velocity in the gas separator, the more efficient is the separation of the phases. For a given daily liquid rate, flow velocity is inversely proportional to the cross-sectional area available to flow; increasing this area improves the operation of the separator. This is the principle behind the **modified** natural gas anchor (Fig. 3.44), where a small-diameter

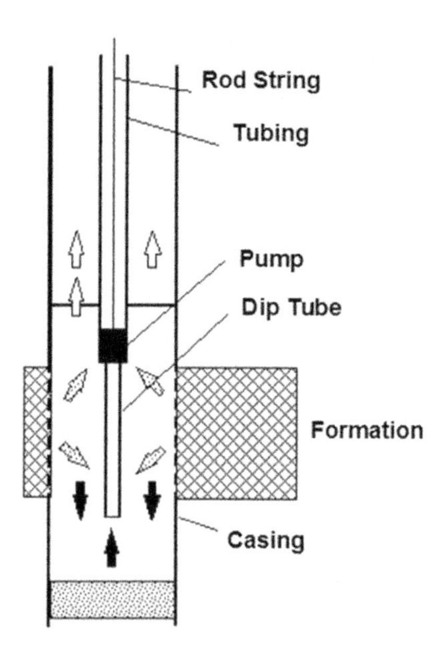

FIGURE 3.44

Construction of the modified natural gas separator.

(maximum 1.5 in) **dip tube** is installed below the pump. The pump is run just above the casing perforations but the dip tube must reach down below the perforations. In order to minimize frictional pressure losses in the dip tube, its length should be 10–20 ft at most. Compared to the original natural anchor, the modified version provides a greater flow area for the liquid and can achieve greater separation efficiencies if the diameter of the casing is small.

3.4.2.3 Packer-type gas separators

If for some reason the pump cannot be set below the casing perforations, it is set above the perforations and the use of a gas separator is a must in wells producing substantial gas volumes. According to their gas separation efficiency, **packer-type gas separators** come next to natural gas anchors. These are all set above a packer or contain diverter cups so that the complete well stream containing free gas must flow through the separator, which utilizes the casing annulus for gas separation. An excellent overview of the various versions of available types is given by **Clegg** [32].

3.4.2.3.1 Parallel-type gas anchor

Figure 3.45 illustrates the construction and operation of a **parallel**-type downhole gas separator [34]. Multiphase mixture is directed from below the packer through a small bypass pipe extending over the fluid level and then falls back in the annulus. The relatively large cross-sectional area of the casing–tubing annulus ensures low downward liquid velocities and thus permits gas to efficiently **separate** from the liquid. Pump suction is located deep below the fluid level where practically no gas phase is present. Packer-type gas anchors should be run directly above the perforations to ensure that pump suction is **below** the fluid level. In case of rapidly changing well conditions, this type of gas anchor may need frequent alterations and can turn out to be impractical. Another limitation is the relatively high investment cost and the sensitivity to failures caused by mechanical problems.

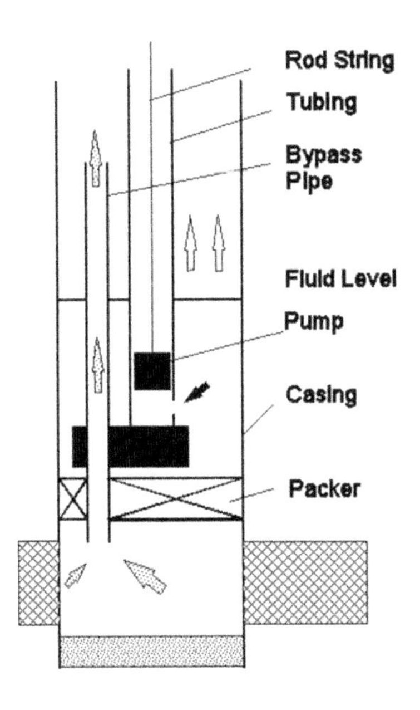

FIGURE 3.45

Parallel-type packer gas separator.

EXAMPLE 3.4: CALCULATE THE EFFECT OF USING A PARALLEL PACKER-TYPE GAS ANCHOR IN THE WELL GIVEN IN EXAMPLE 3.3. THE WELL'S CASING STRING HAS AN INTERNAL DIAMETER OF 4.9 IN, THE TUBING OUTSIDE DIAMETER IS 2.375 IN, AND A BYPASS PIPE OF 2 IN OD IS USED

Solution

The separator's coefficient is 0.036 (packer gas anchor); the cross-sectional area in the annulus available for liquid flow is:

$$A = \pi/4\left(4.9^2 - 2.375^2\right) - \pi/4 \times 2^2 = 11.3\ \text{in}^2.$$

The superficial liquid flow velocity is evaluated from Eq. (3.8):

$$v_{sl} = 9.36\text{E-}3(50 + 60)/11.3[1.045/(1 + 1.2) + 1.018 \times 1.2/(1 + 1.2)] = 0.094\ \text{ft/sec}.$$

The **Clegg** correlation (Eq. 3.7) gives the GLR of the gas–liquid mixture leaving the separator:

$$GLR'_{ing} = 0.036(100 + 14.7)^{2/3}0.094^{0.5} = 0.26.$$

The rate of ingested gas is calculated from Eq. (3.9):

$$q'_{ing} = 5.614 \times 0.26(50 + 60)\ [1.045/(1 + 1.2) + 1.018\ 1.2/(1 + 1.2)] = 166\ \text{cu ft/d}.$$

Compared to the free gas volume coming from the formation, the gas volume entering the pump is reduced from 670 to 166 cu ft/d. The gas content of the mixture leaving the separator can be found from Eq. (3.6) as follows:

$$Gas\% = (160/5.614)/\{(50 + 60)\ [1.045/(1 + 1.2) + 1.018 \times 1.2/(1 + 1.2)] + (160/5.614)\} = 0.207 = 20.7\%,$$

and the separator removed 75% of the gas from the mixture entering the well.

3.4.2.3.2 Concentric-type gas anchor

Another version of the packer-type gas separator is called **concentric** [35] because it contains two concentric tubes, as shown in Fig. 3.46. Well fluid from below the packer enters through the seven

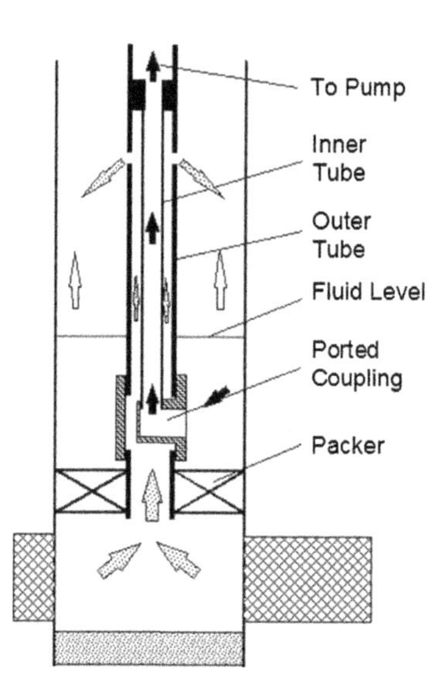

FIGURE 3.46

Concentric-type packer gas separator.

holes in the ported coupling and rises in the annulus between the inner and outer tubes, and finally falls into the casing annulus through ports in the outer tube. Gas separation takes place in the space between the casing string and the separator body; as a result, the fluid above the packer contains a negligible amount of free gas. Gasless liquid is sucked in by the sucker-rod pump situated above the separator through the inner tube and the ported coupling.

Total length of this separator is 40 ft and it is available in different outside diameters to match different casing sizes [36].

EXAMPLE 3.5: COMPARE THE USE OF THE CONCENTRIC ANCHOR TO THE PARALLEL GAS ANCHOR DISCUSSED IN EXAMPLE 3.4, IF THE OUTSIDE DIAMETER OF THE SEPARATOR IS 2.62 IN

Solution

The cross-sectional area in the annulus available for liquid flow is:

$$A = \pi/4\ \left(4.9^2 - 2.62^2\right) = 13.5\ \text{in}^2.$$

The superficial liquid flow velocity is evaluated from Eq. (3.8):

$$v_{sl} = 9.36\text{E} - 3\ (50 + 60)/13.5\ [1.045/(1 + 1.2) + 1.018 \times 1.2/(1 + 1.2)] = 0.079\ \text{ft/sec}.$$

The **Clegg** correlation (Eq. 3.7) gives the *GLR* of the gas—liquid mixture leaving the separator:

$$GLR'_{ing} = 0.036(100 + 14.7)^{2/3} 0.079^{0.5} = 0.239.$$

The rate of ingested gas is calculated from Eq. (3.9):

$$q'_{ing} = 5.614 \times 0.239(50 + 60)[1.045/(1 + 1.2) + 1.018 \times 1.2/(1 + 1.2)] = 152\ \text{cu ft/d}.$$

The gas content of the mixture leaving the separator can be found from Eq. (3.6) as follows:

$$Gas\% = (152/5.614)/\{(50 + 60)[1.045/(1 + 1.2) + 1.018 \times 1.2/(1 + 1.2)] + (152/5.614)\} = 0.193 = 19.3\%,$$

which is less than the value received from the parallel-type gas anchor; this indicates that the concentric gas anchor is more efficient than the parallel one.

3.4.2.3.3 The diverter gas separator

This recently developed downhole gas separator [37] provides the **same** separation efficiency as the natural gas anchor and can be used in wells where the pump cannot be run below the perforations. Its main feature is the use of flexible elastomer **cups**, instead of a packer, that seal on the inside of the casing and **divert** the well stream vertically through an inner tube; see Fig. 3.47. Liquid and gas are discharged from the inner tube into the annulus, where separation of gas from the liquid takes place. Gasless liquid is sucked in by the sucker-rod pump from above the rubber cups and reaches the pump through the annulus between the inner and outer tubes. The separator is relatively **short**, being only 5 ft tall.

The advantages of the diverter separator over the parallel type of packer separator are: (1) with elimination of the bypass pipe, the total cross-sectional area of the annulus is utilized for gas separation, and (2) the rubber cups provide easy setting and retrieval as compared to the mechanical problems that may arise when a packer is used. However, practical limitations related to its light construction prevent the separator's use with a packer or a tail pipe; this is the reason why it is not widely used.

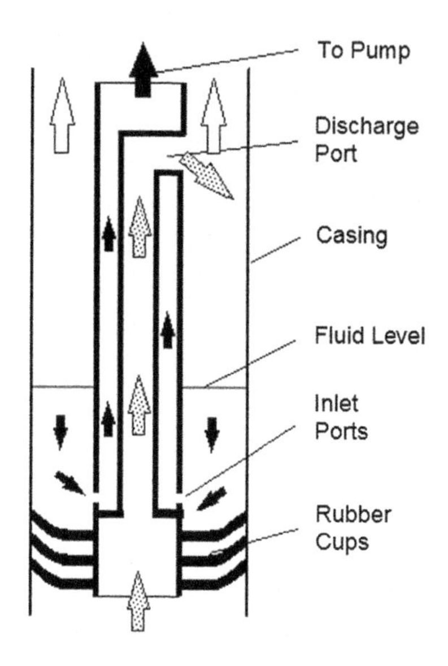

FIGURE 3.47

The diverter gas separator.

3.4.2.3.4 The seating nipple separator

Based on the experiences with the diverter separator, the **seating nipple separator** is a much heavier construction very similar to the concentric-type separator [38]. The separator is set on a **packer** or is equipped with diverter **cups** so that well fluids coming from the formation are directed into the separator. It includes the seating nipple for the downhole pump so that the pump can be located within inches of the liquid in the casing annulus surrounding the separator. As shown in Fig. 3.48, well fluid

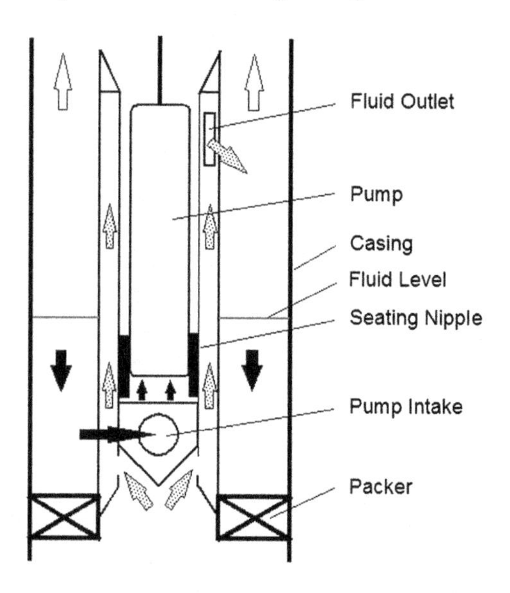

FIGURE 3.48

The seating nipple gas separator.

rises through a concentric annulus **inside** the separator to a single outlet from where it is discharged almost horizontally to impinge on the casing wall; as a result, liquid falls down on top of the packer. Separation of gas from the liquid takes place in the annulus between the **casing** and the separator **body**. Since the outside diameter of the separator is about the same as the tubing OD, **maximum** flow area is available and liquid flow velocity is low, maximizing the efficiency of gas separation. Liquid with a low gas contamination is sucked in by the downhole pump because the integral seating nipple is at the bottom of the separator. This arrangement ensures that the pressure drop is **negligible** as the liquid flows from the separator to the pump suction. All these features make it possible to achieve separation efficiencies and flow capacities **equal** to those of the natural gas anchor even when the pump cannot be set below the perforations.

3.4.2.3.5 The use of tail pipes

It is a quite common practice to run below the downhole pump a longer section of tubing (called a **tail pipe**) in conjunction with a packer down to the depth of the perforations, especially when the pump is set a considerable distance above the formation. This solution can also be used in **horizontal** wells where the packer is set in the well's vertical section and the tail pipe allows reaching the horizontal section. The use of a tail pipe is very advantageous in decreasing the flowing bottomhole pressure and thereby increasing the well's liquid production **rate**. The explanation of this feature is related to the behavior of vertical multiphase flow at low liquid rates. Under such conditions the dominant component of multiphase pressure drop is due to **gas slippage** and this increases as pipe size increases. Therefore, the multiphase flowing gradient in the tail pipe, compared to that in the casing string, is lower because of the smaller diameter of the tail pipe. The lower pressure gradient results in a lower total pressure drop in the tail pipe and a lower flowing bottomhole pressure at the perforations. In wells with low liquid capacities, therefore, the well's liquid production rate can be increased by the utilization of a tail pipe connected to the downhole pump.

3.4.2.4 The collar-size gas separator

The **collar-size gas separator** [39,40] offered by the Echometer Company was introduced in 1998. The background of this innovative tool lies in the behavior of vertical multiphase flow in annuli. When the casing and the tubing strings are concentric, the distribution of gas bubbles in the liquid is **uniform** across the total flow cross-section. In **eccentric** annuli, however, gas bubbles will tend to flow in the **wider** side of the annulus and phase concentrations will not be uniform in the cross-section available for flow. Liquid concentration will be much higher where the radial space between the pipes is **smaller**, whereas in the wider side of the annulus gas concentration greatly increases, as shown in Fig. 3.49. This phenomenon is utilized in the construction of the gas separator by forcing the separator against

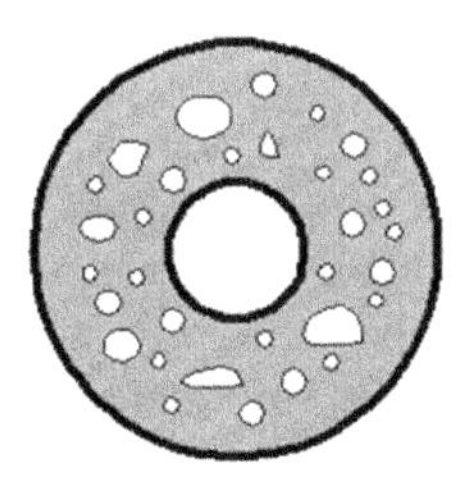
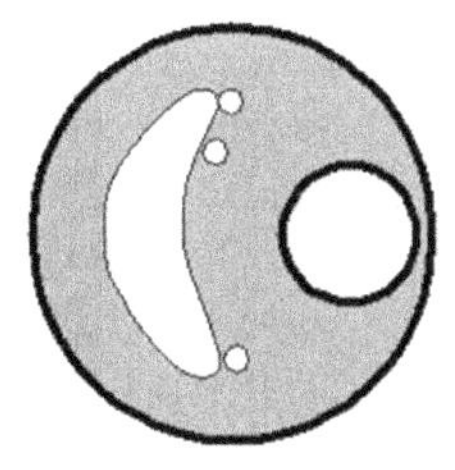
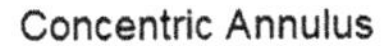

FIGURE 3.49

Flow patterns in vertical annuli.

the casing wall and by placing its intake ports at the narrower side of the eccentric annulus formed by the casing and the separator's body. This way, the mixture sucked in by the separator is much **richer** in liquid than in cases where the separator is concentrically located inside the casing.

The construction of the gas separator is illustrated in Fig. 3.50. The separator's body (the outer tube) is made from thin-wall pipe and has an outside diameter equal to the OD of the tubing collar. This feature ensures that the separator can lie on the low side of the casing string with or without a decentralizer spring to take advantage of the eccentric annulus thus created. The properly sized, thin-wall (1/8 in thickness) dip tube maximizes the internal area of the separator and facilitates low downward liquid velocities inside the separator, a prerequisite for an efficient gas separation. Gassy fluid enters the separator through two 1 1/4 in inlet ports and separated gas leaves the separator through two 1 1/4 in vent holes situated 5 in above the inlet ports. Field trials have shown that a separator length of about 5.5 ft is sufficient; in sand-producing wells one or two tubing joints may be used between the separator body and the bull plug to allow for solids accumulation.

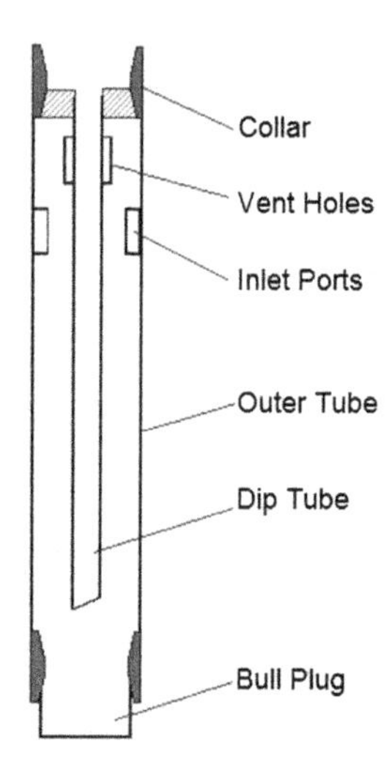

FIGURE 3.50

Construction of the collar-size gas separator.

Although the collar-size separator is very similar to a poor-boy separator, detailed later, it provides much **higher** separation efficiencies (1) because it is set eccentrically in the casing string by gravity, and (2) because of the much greater area between the outer tube and the dip tube. As shown in Fig. 3.49, the eccentric position ensures that the gas–liquid mixture entering the separator has a much reduced gas content. The bulk of free gas flowing in the wider side of the annulus rises to the annular liquid level and is released to the surface. Any free gas entering through the inlet ports is separated in the annulus between the outer tube and the dip tube, provided the liquid downward velocity is less that 0.5 ft/s; separated gas leaves the separator through the vent holes on the outer tube.

The collar-size gas separator has become very popular with thousands of installations worldwide today [41]. In most cases an average liquid rate increase of 25% is observed and the pumping time in intermittently produced wells decreases by about 50%.

Table 3.7 Liquid and Gas Handling Capacities of Collar-Size Gas Separators [42]

Tubing Size in	Separator OD in	Liquid Capacity bpd	4 1/2 in Casing	5 1/2 in Casing	7 in Casing
			Gas Capacity @ 1 atm, Mscf/d		
2 3/8	3.00	229	35	76	154
2 7/8	3.75	413	11	52	130
3 1/2	4.50	624		23	101
4	5	778			79
4 1/2	5.6	1,016			49

The **collar-size gas separator** is manufactured in different sizes; its liquid and gas capacities are given in Table 3.7 for different casing–tubing combinations [42]. The liquid capacities given represent the **maximum** pumping rates allowed for a **perfect** separation; the use of higher-capacity sucker-rod pumps rapidly deteriorates separation efficiency. Gas capacities are valid for **atmospheric** pump suction pressures assuming no liquid above the pump and continuous venting of casinghead gas to the atmosphere. Increased pressures boost the separator's ability to remove greater gas volumes proportional to suction pressure. At pump intake pressures above atmospheric pressure the gas capacities given in the table must be multiplied by the number of atmospheres, i.e., *PIP*/14.7 (*PIP* in psi units).

3.4.2.5 The poor-boy gas separator

The **least** efficient and lowest-capacity (maximum 50 bpd liquid) gas anchor is the **poor-boy** anchor, shown in Fig. 3.51. It consists of a **mud anchor**, which is a bull plugged perforated pipe section, and a

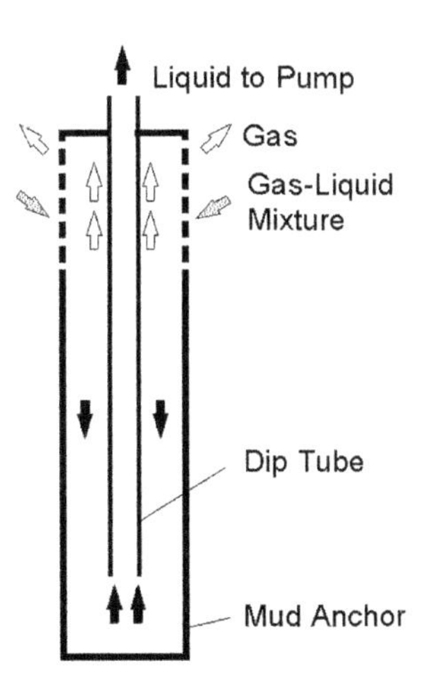

FIGURE 3.51

Poor-boy gas anchor (downhole gas separator).

dip tube inside the mud anchor. The gas anchor is run **immediately** below the pump, pump suction being connected to the dip tube, and is immersed in well fluids below the annulus fluid level. The larger gas bubbles in the well's annular volume rise directly to the surface of the fluid level. Thus, the mixture entering the gas anchor through the intake ports contains bubbles of **medium** to **small** size only. This mixture is drawn by pump suction pressure downward in the annular space between the mud anchor and the dip tube. During this downward travel, gas bubbles, due to their lower gravity, **rise** to the top of the anchor where they can **escape** into the casing annulus. Liquids descending in the anchor body become increasingly gas free, and the liquid entering the pump through the dip tube will contain a reduced amount of free gas.

Poor-boy gas anchors should be properly sized for the actual operating conditions. The basic requirement of proper gas separation is, like in the case of any gravitational separator, that downward velocity of the liquids in the anchor body be less than the rising velocity of gas bubbles. Assuming a given liquid rate, liquid velocity is decreased by increasing the inside diameter of the anchor, which is limited by the casing size of the well. The dip tube should extend about 4 ft below the lowest perforation on the mud anchor. Other design factors include dip tube size (usually 1 in), total area of inlet ports, and anchor volume. Anchor volume should be sufficiently large to assure a proper **retention time** for the separation to occur; a rule of thumb is two times the pump displacement per stroke. A detailed procedure for designing poor-boy gas anchors can be found in [43].

Poor-boy gas anchors are typically made from conventional oilfield tubing, perforated subs, etc., having thick walls, and are therefore, not very well suited for constructing gas separators because the area available for gas separation is limited. Liquid and gas capacities of poor-boy separators, as compared to other separators, are low; typical values are given in Table 3.8 [44].

Table 3.8 Liquid Handling Capacities of "Poor-Boy" Gas Anchors, According to McCoy et al. [44]

Size and description	Dip tube size, in	Liquid capacity, bpd
3 1/2 perforated tubing sub	1.25	260
2 7/8 Perforated tubing sub	1.00	177
2 7/8 Perforated tubing sub	1.25	134
2 3/8 Perforated tubing sub	0.75	121
2 3/8 Perforated tubing sub	1.00	94
2 3/8 Perforated tubing sub	1.25	51
2 3/8 Perforated tubing sub & 1.5 in pump	1.76	37

Modified poor-boy gas anchors are used to improve the separation capacity and efficiency of poor-boy anchors; they have outside diameters **greater** than the tubing size normally used. This modification increases the cross-section available for fluid flow and reduces the downward liquid velocity inside the separator; gas separation efficiency is thus enhanced.

EXAMPLE 3.6: INVESTIGATE THE USE OF A POOR-BOY GAS ANCHOR IN THE WELL DISCUSSED IN EXAMPLE 3.3, IF THE INSIDE DIAMETER OF THE SEPARATOR IS 3 IN WITH A DIP TUBE OUTSIDE DIAMETER OF 1 IN

Solution

The cross-sectional area in the annulus available for liquid flow is:

$$A = \pi/4(3^2 - 1^2) = 6.3\ \text{in}^2.$$

The superficial liquid flow velocity in the separator body is evaluated from Eq. (3.8):

$$v_{sl} = 9.36\text{E-}3(50 + 60)/6.3[1.045/(1 + 1.2) + 1.018 \times 1.2/(1 + 1.2)] = 0.169\ \text{ft/sec}.$$

The **Clegg** correlation (Eq. 3.7) gives the *GLR* of the gas–liquid mixture leaving the separator as $C = 0.1$ for the poor-boy separator:

$$GLR'_{ing} = 0.1(100 + 14.7)^{2/3}\ 0.169^{0.5} = 0.97.$$

The rate of ingested gas is calculated from Eq. (3.9):

$$q'_{ing} = 5.614 \times 0.97(50 + 60)\ [1.045/(1 + 1.2) + 1.018 \times 1.2/(1 + 1.2)] = 617\ \text{cu ft/d}.$$

The gas content of the mixture leaving the separator can be found from Eq. (3.6) as follows:

$$Gas\% = (617/5.614)/\{(50 + 60)\ [1.045/(1 + 1.2) + 1.018 \times 1.2/(1 + 1.2)] + (617/5.614)\} = 0.492 = 49.2\%,$$

and the poor-boy separator removed only 7.9% of the gas from the well stream. As seen in the previous examples, this value is about 10% of the gas that packer-type separators could remove; this shows the very low efficiency of the poor-boy separator.

3.4.3 CONCLUSIONS

An extensive research project conducted at the University of Texas at Austin [45–47] involved the evaluation of most types of gas anchors available today. Using a full-scale experimental setup to model the conditions in a well's annulus, different types of gas anchors were tested for their liquid and gas handling capacities. Many of the study's conclusions were in **disagreement** with widely acknowledged rules of thumb and assumptions used in the industry for years. The most important findings are the following:

- If the pump intake can be set below the casing perforation, no gas separator is needed. Natural gas separation provides 100% of separation efficiency as long as the superficial liquid velocity in the annulus is below 0.5 ft/s.
- The recommended **length** of the gas anchor or gas separator is often overestimated; these studies found that a length of 5.5–6 ft is sufficient.
- The old recommendation [13] of running the gas separator deep below the bottom perforations (at least 30 ft) is wrong, since both laboratory and field observations indicated that a minimum distance of 1–2 ft is enough for a **complete** separation of the produced gas.
- Previous studies indicated that proper separation requires great entry ports with a total area equal to about four times the separator annulus area. This fact was also disproved and it was shown that increasing the total entry area did not increase the separator efficiency.

3.5 THE SUCKER-ROD STRING

3.5.1 INTRODUCTION

The sucker-rod string is the most vital part of the pumping system, since it provides the **link** between the surface pumping unit and the subsurface pump. It is a peculiar piece of mechanical equipment and has almost **no analogies** in man-made structures, being several thousand feet long and having a maximum diameter of slightly more than one inch. The behavior of this perfect **slender bar** can have a fundamental impact on the efficiency of fluid lifting and its eventual failure leads to a total loss of production. Therefore, a properly designed rod string not only assures good operating conditions but can considerably reduce total production costs as well.

The rod string is composed of individual **sucker rods**, connected to each other, until the required pumping depth is reached. Early pumping installations used wooden poles, usually made of hickory, which had steel end-fittings to facilitate joining of the rods. As average well depths increased and greater rod strength was required, the all-steel sucker rods appeared, around the beginning of the twentieth century. These are solid steel **bars** with forged **upset ends** to accommodate male or female **threads**, a design that has remained unchanged. The most important improvements in sucker-rod manufacturing methods over the years were the application of **heat treating** to improve corrosion resistance, better **pin constructions**, and the use of **rolling** instead of cutting for making the necessary threads. Steel rods different from the solid type were also made available, like the hollow sucker rod or rod tube, the continuous rod, and the flexible rod.

Steel rods have some common drawbacks: first, their relatively high **weight**, which increases the power needed to drive the pump; and second, their high **susceptibility** to corrosion damage in most well fluids. Both of these problems are eliminated by the use of the latest addition to the arsenal of rod pumping, the **plastic** sucker rods. The utilization of **fiberglass**-reinforced plastic materials in rod manufacture decreases total rod string weight, improves corrosion resistance, and has other benefits as well. Due to their numerous advantages, fiberglass sucker rods are increasingly favored by operators; their market share in the US reached over 10% in 1985 [48].

3.5.2 SOLID STEEL RODS

3.5.2.1 API specifications

The most widely used sucker rod type is the solid steel rod that has been standardized by the **American Petroleum Institute** since 1926. According to the latest **API Spec. 11B** [49], sucker rods come in 25 or 30 ft lengths; rod body diameters range from 5/8 to 1 1/4 in with 1/8-in increments. The usual construction of sucker rods is the one-piece rod that has upset ends with the necessary threaded connections. The weaker three-piece rods, seldom used, had their separate end connectors screwed onto the two ends of the rod body and could be used to limited well depths only.

The ends of sucker rods are **hot forged** and later **machined** and **threaded**. For a long time, either pin (male thread) or box (female thread) connections could be used, as illustrated in Fig. 3.52, but the latest editions of **API Spec. 11B** specify pin ends only.

The construction and nomenclature of rod ends is given in Fig. 3.53, where the short square section, called **wrench flat**, facilitates the use of power tongs for a proper makeup. Usually, rods with two pin ends are manufactured and a coupling is installed at one end in the factory.

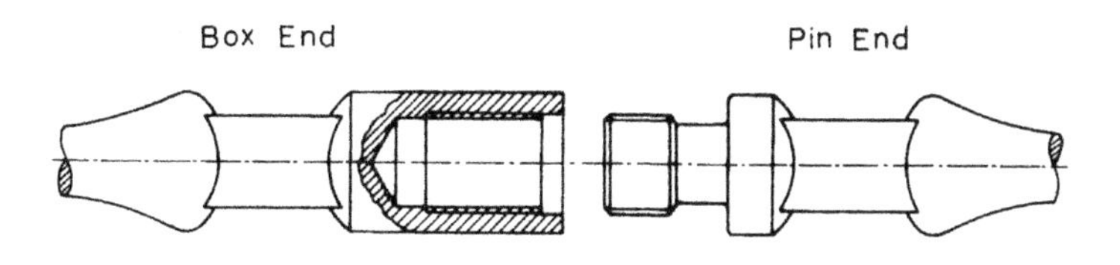

FIGURE 3.52

Sucker-rod box and pin ends.

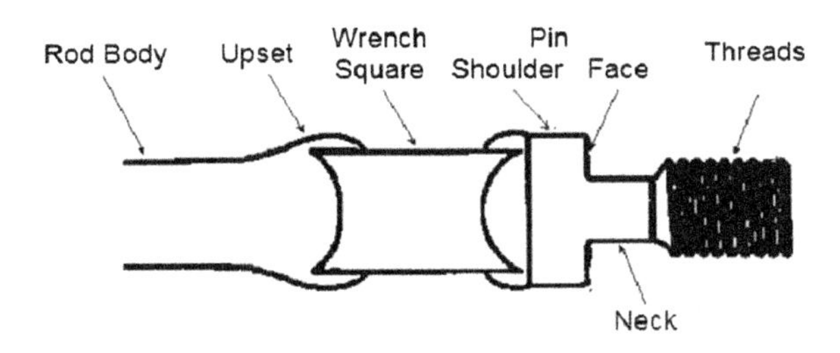

FIGURE 3.53

Construction details of a sucker-rod pin end.

The **physical parameters** of API sucker rods can be derived from their body diameter, d_r (in), as detailed in the following.

- **Metal area**, A_r (sq in), is found from the rod body diameter as $A_r = d_r^2\ \pi/4$.
- **Average rod weight in air**, w_r (lb/ft), must include the effects of sucker-rod joints; usually 9% is added to the weight calculated from the body diameter. Using the average density of steel (487.5 lb/cu ft), average rod weight in air is found as $w_r = 1.09 \times 487.5\ A_r/144$.
- **Corrected cross-sectional area**, A_c (sq in), represents the **average** cross-sectional area (taking into account the additional volume represented by rod body **upsets** and the **couplings**) of a longer rod section and can be calculated from a simple mass balance written for a 1-ft-long rod section:

$$A_c \frac{\rho}{144} = w_r \tag{3.10}$$

Solving the equation for A_c and substituting $\rho = 487.5$ lb/cu ft for the density of the steel material, we get the following expression:

$$A_c = 0.295\ w_r \tag{3.11}$$

where:

A_c = corrected cross-sectional area of sucker rod, sq in, and
w_r = average rod weight in air, lb/ft.

- **The elastic constant** of the rod, E_r (in/lb/ft), represents the stretch in inches of a 1-ft-long rod under a load of 1 lb. It is defined as $E_r = 12/A_r/E$; where $E = 3.1$ E7 psi is the steel material's average Young's modulus.

Table 3.9 Calculated Physical Data of API Sucker Rods

Rod Size, in	Metal Area, sq in	Weight in Air, lb/ft	Corrected Area, sq in	Elastic Constant, in/(lb ft)
5/8	0.307	1.135	0.335	1.262 E−6
3/4	0.442	1.634	0.482	8.762 E−7
7/8	0.601	2.224	0.656	6.437 E−7
1	0.785	2.904	0.857	4.929 E−7
1 1/8	0.994	3.676	1.084	3.894 E−7
1 1/4	1.227	4.538	1.339	3.154 E−7

Table 3.9 contains the calculated values of the rod data just described for available sucker-rod sizes.

A cross-sectional view of a sucker rod **coupling** (box and box) is given in Fig. 3.54. Couplings normally have the same size box threads at both ends, but couplings with different threads to connect different rod sizes are also available and are called **subcouplings**.

Dimensional data of **couplings** are also presented in **API Spec 11B** [49]; their length is 4 in, except for sucker rod size of 1 1/8 in that are 4.5 in long. Couplings for regular service are designated **full-size** couplings and those with a reduced outside diameter are called **slimhole** couplings. When using slimhole (SH) couplings, the ratio of the metal area in the coupling to the metal area of the rod body is much lower than the same ratio in the case of full-size couplings. Therefore, slimhole couplings are more **severely** loaded than the regular ones and are often the **weak** link in the rod string. Smaller tubing sizes in the well, however, may exclude the use of regular couplings. The minimum tubing sizes required for the different sizes of full-size and slimhole couplings are given in Table 3.10.

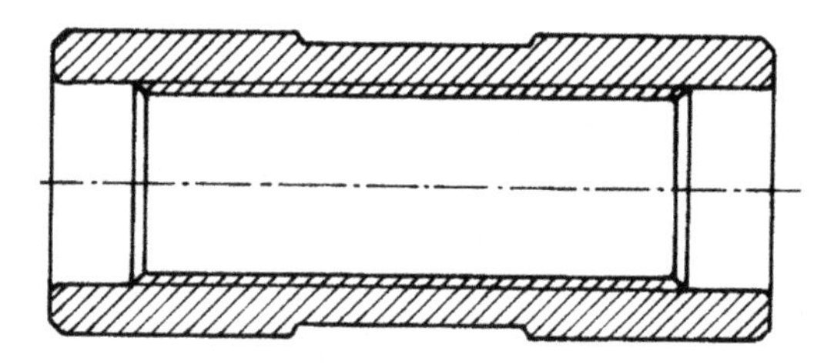

FIGURE 3.54

Sucker-rod coupling, box-to-box type.

Table 3.10 Dimensions of Full-Size and Slimhole Sucker-Rod Couplings

	Full Size		Slim Hole	
Rod Size, in	OD, in	Tubing Size, in	OD, in	Tubing Size, in
5/8	1 1/2	2 1/16	1 1/4	1.99
3/4	1 5/8	2 3/8	1 1/2	2 1/16
7/8	1 13/16	2 7/8	1 5/8	2 3/8
1	2 3/16	3 1/2	2	2 7/8
1 1/8	2 3/8	3 1/2	–	–

3.5.2.2 Sucker-rod joints

The operation of the sucker-rod pumping system presupposes the **integrity** of the rod string, which, in turn, depends on the proper functioning of each sucker-rod **joint** in the string. Therefore, it is of utmost importance for an engineer to understand the mechanism of these joints and to investigate the impact of downhole conditions on their operation. It must be clear that a **hand-tight** pin-to-coupling connection will cause, due to the periodic movement of the string, a gradual **unscrewing** of the joint and the rods will eventually part. If the tightness of the threaded connection is increased, the danger of unscrewing is limited, but a much greater problem, material **fatigue**, will start to become more pronounced. **Fatigue** is caused by repeated loading and unloading of the pin–coupling contact due to the varying well loads during the pumping cycle.

Both the rod pin and the coupling are highly **susceptible** to material fatigue, because the threads on both parts act as **stress raisers**. These are known to have a very detrimental effect on the endurance of metal parts. The deterioration of the joint can be further enhanced by downhole **corrosion**, because corrosive fluids can easily enter the threads, and by **bending** forces, which arise when the pin shoulder face separates from the coupling face. All the above effects influence the safe life of sucker-rod joints and lead to the fact that 99% of fatigue breaks in rod pins are caused by improper rod makeup.

3.5.2.2.1 API connections

The cross-section of a properly made-up (tightened) sucker-rod **joint**, showing the distribution of mechanical stresses in the metal parts, is given in Fig. 3.55, modified after **Hardy** [50]. The pin face is in contact with the coupling face and the high torque applied during makeup imposes a very high contact force between the two parts. This force acts on the metal areas of both the pin and the coupling and gives rise to high local stresses. The stress in the coupling is **compressive,** while that in the rod pin is **tensile**. After the rods are run in the well, the pumping **loads** directly act on the joint, putting an additional stress on the pin and the coupling.

Rod loads during the pumping cycle are always **tensile** loads and are caused by rod weight on the downstroke, and by rod weight plus fluid load on the upstroke. This additional tensile loading influences the total stresses in the pin and in the coupling in different ways. The stresses in the **pin** are **additive**; thus an increase in net tensile stress is observed. The amount of stress increase was found to be equal to the rod load divided by the combined cross-sectional area of the pin and the coupling [51]. Taking into account the dimensional data of standard pins and couplings, this value is less than **half** of the stress that would occur were the same load imposed on the pin area only. Consequently, the proper amount of prestressing considerably reduces the range of stress present in the sucker-rod pin.

The upper part of the coupling, on the other hand, is under **compression** because of the prestressing caused by proper makeup. Pumping loads, therefore, **reduce** the compression stresses in this area. A properly made-up sucker-rod joint, however, should originally have **compressive** stresses that are considerably **greater** than the maximum **tensile** stress expected from well loads. Therefore, the upper section of the coupling is always under compression, which keeps the coupling face firmly in **contact** with the pin shoulder face. In the lower sections of the coupling, compression is gradually decreasing until a neutral point is reached, below which tension takes over. The maximum tensile stress occurs in the middle sections of the coupling and is imposed by the pumping load alone.

The above discussion has shown that as long as a firm **contact** between the pin and the coupling faces is maintained, pumping loads are **carried** by the whole joint. The proper makeup of rods, therefore, has a crucial role in **preventing** rod failures. The most significant positive effect of

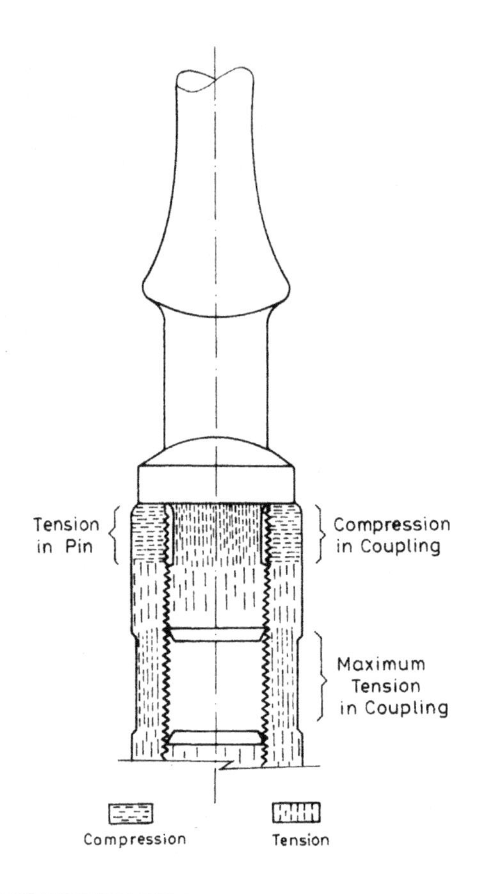

FIGURE 3.55

The distribution of stresses in a sucker-rod joint.

prestressing due to proper makeup on rod string performance is that it minimizes the **fatigue** loading in rod pins. This is the result of several factors:

- the **range** of stress is significantly reduced, as previously detailed,
- **bending** of the pin is eliminated, and
- **corrosion** damage is greatly reduced because of the tight thread connections.

Rods are made up using pneumatic or hydraulic **power tongs** that exert a high **torque** on the joint. Although power tongs usually permit the measurement of the applied torque, the torque itself is not a good **indicator** of stress levels in the joint. This is because the actual torque is influenced by a host of variables, such as the **finish** on the contact surfaces, thread **lubrication**, etc. Most of the torque is absorbed by friction on both the mating thread elements and the pin–coupling contact area. Thus, the **useful** torque that produces the necessary prestress in the pin amounts to only about 10% of the total applied. These are the reasons why the API recommendation **RP 11BR** [52] prescribes the **circumferential displacement** method for the accurate determination of stress levels in sucker-rod joints.

The **circumferential displacement** method, recommended to ensure a **proper** rod makeup, is accomplished as follows. After the pin and coupling are mated and made up to a **hand-tight position**, a vertical line is scribed across the joint. The joint is then finally made up with the power tong until the **distance** of the two vertical lines, measured on the circumference, equals the amount specified in tables published in **API RP 11BR** [52]. The recommended **displacement** values vary with the various rod sizes and grades. Because the contacting surfaces of the joints are smoothed out when the rods are first run into a well, subsequent makeups (when using used rods) require less torque to be applied. The difference in recommended circumferential displacements, however, is only significant for Grade D rods.

Several **improvements** in sucker-rod joint design have been made throughout the years. Of these, two deserve to be mentioned. The **undercut pin** first appeared in API specifications in the early 1960s, and is shown, along with the older-style pin, in Fig. 3.56. The undercut thread has long been used in other industries to decrease or even totally eliminate **fatigue** failures [53]. The **undercut**, or stress-relieving, section is a relatively long, slender **neck** on the pin with a diameter slightly less than the minor diameter of the threads. Its main functions are:

- the proper makeup torque causes more **elongation** in the undercut pin, thus ensuring the tightness of the joint under operating conditions,
- the neck is more **flexible** than the threaded section and allows for more **bending** without overstressing the pin, should the connecting faces separate, and
- stress **concentration** is reduced in the critical neck area because there are no threads there.

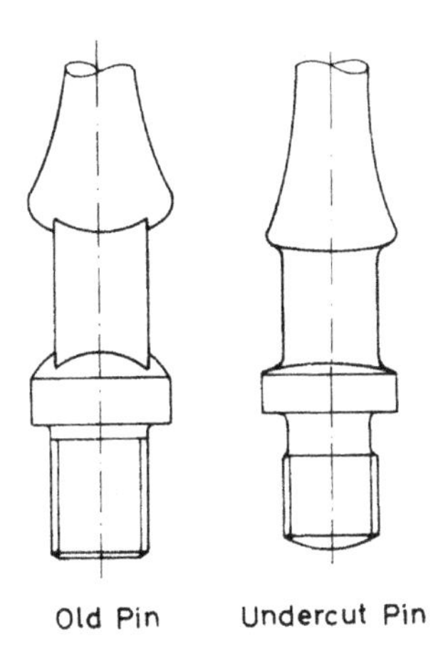

FIGURE 3.56

The older and the undercut API sucker-rod pins.

Threads on pins and couplings can be formed by several methods, but after the undercut pin became widely used, **rolling** is the preferred manufacturing process. **Thread rolling** is a **cold-forging** method using hardened dies that force the metal to flow into the required thread form. The older-style cut thread and the rolled thread are compared in Fig. 3.57, where cross-sections of both are displayed.

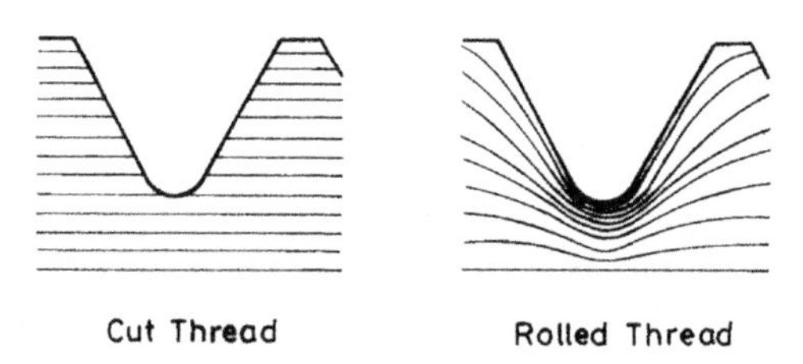

FIGURE 3.57

Cross-sections of a cut and a rolled thread.

The cutting process works by **removing** material from the metal and leaves the metal fibers exposed at the thread surfaces. The sharp V-cut at the bottom is an excellent **stress raiser** that accentuates fatigue at this highly stressed point. Thread rolling, on the other hand, does not cut the metal fibers, but **compresses** and expands them to the desired shape. The fibers follow the thread shape and are compressed at the root, giving maximum strength to that critical area. Rolled thread surfaces are superior in **finish** over cut threads, and greatly reduce friction during rod makeup. The elimination of surface imperfections prevents the occurrence of **stress raisers** that induce fatigue failure. The service **life** of sucker-rod joints is, therefore, greatly extended by thread rolling.

3.5.2.2.2 Premium connections

The API sucker-rod connection discussed previously was first standardized in 1926 and it served the industry well without major modifications until the 1970s; see **API Spec. 11B** [49]. In the last decades, however, as rod stresses in deeper and deeper wells became higher and higher, an inherent **weakness** of the three-part (two pins and a coupling) connection caused an increased number of rod **failures**. The majority of these failures are due to material **fatigue** in the sucker-rod joint, making the joint the **weakest** point in the rod string.

Most failures in the API sucker-rod joint can be explained by the **limitations** of the sucker-rod thread design, which are:

- The **gap** existing between the rod and the coupling threads allows relative **movement** (even if at a minimum scale) of the two connected parts, especially at high dynamic axial loads.
- During makeup and normal operations the connecting threads are permanently **deformed**, which easily leads to the formation of small cracks and a final fatigue break.
- The stress distribution is not **uniform**, with high stress concentrations at some places, especially at the **first** engaged thread.

After the industry realized the inherent weakness of the standard API sucker-rod joint, different solutions were offered to address the problem. In the following discussions, the most important developments in this area are presented, but it is still to early to tell which will be generally accepted. The emergence of the various methods proves, however, that sucker-rod pumping still has a high potential for producing relatively high liquid rates from deeper wells.

Improved 7/8 in Rod Connections

Experience in an Argentinean oilfield proved that most connection failures occurred in the 7/8 in taper sections. An investigation showed that, compared to other rod sizes, the pin of the 7/8 in rod is **weaker** because its cross-sectional area, in relation to the rod body area, is less than for **any** other rod. This observation led to the development of a non-API rod design, and 7/8 in rods with the same pins as used on 1 in rods are manufactured by the **Tenaris** Company [54]. The use of the **bigger** pin has increased the metal cross-sectional area at the pin undercut, where most of the failures regularly occur. Manufacturing these rods required a completely new **forging** process to create the pin ends.

The improved 7/8 in rod connections proved to be very successful in wells with high liquid rates and depths of more than 6,500 ft; connection failures in 60-rod strings were practically eliminated. Experience showed this solution to be **superior** to the use of high-strength rod materials.

Four-Part Modified API Connections

After an extensive experimental investigation of the sucker-rod joint involving finite element analysis, **Carstensen** invented a four-part connection system [55]. The system can be applied to **any** new or used API sucker rod and coupling and dramatically **improves** their load-carrying capacities. Figure 3.58 shows a cutaway of the new joint, which is made up of four parts: the two pins, one coupling, and a **torque button** squeezed between the pins.

The four-part connection is created from standard API rods and couplings by following the next process [56], most of which is executed in a workshop:

1. Original pins are **machined** so that the distance from the shoulder to the pin end is **precisely** uniform and the pin ends are perfectly **flat** and at right angles to the axis of the rod.
2. The coupling is also machined to a **precise** overall length so that its ends are flat and parallel to each other.
3. The coupling is made up on one of the pins while the advancement of the pin related to the coupling end is closely **monitored** by a precision indicator. Torque is applied to the joint until a predetermined **distance** between the pin end and the coupling end face is reached. This operation ensures that the rod's neck is **prestretched** sufficiently to ensure a highly reliable connection.
4. The **torque button**, usually made of a different material than the rod to prevent **galling**, is installed and locked in place against the pin end. This concludes the **preparations** in the workshop and ensures that the connections made at the rig will always break out first.
5. At the rig, after applying some thread lube, the other rod is made up to a hand-tight position. Now the **gap** between the torque button and the end of the new pin is exactly the same as the stretch previously applied to the other side of the coupling.

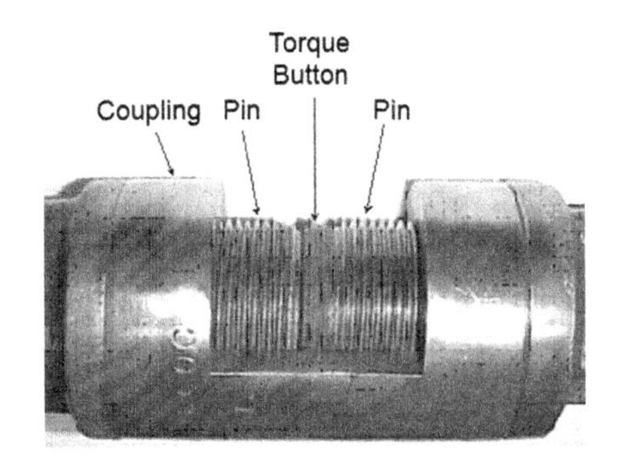

FIGURE 3.58

The four-part modified API connection.

6. The new rod is driven further into the coupling until the pin end **contacts** the torque button. By this time both pins are **prestretched** to exactly the **same** level.
7. Further torque applied to the connection ensures that the **middle** portion of the coupling is also stretched to the required level.

The most important **benefits** of the four-part connection are:

- The threads of the pin and the coupling are firmly **locked** and no relative **movement** between them is possible. In an API connection, the two pin ends are unsupported and the cyclic loading during pumping operations generates **microscopic** movements, leading to permanent deformations and eventual **fatigue** failures.
- The load-bearing **metal** area is increased at the middle sections of the coupling.
- The connection is much more **rigid**, as compared to the API joint, and dynamic loads are transferred to the rod body without much bending.

In conclusion, the new construction greatly increases the **life** of the sucker-rod joint due to the much reduced **fatigue** loading involved. Since both the pin and the coupling are highly prestressed, the same pumping loads resulting in tension stresses cause lower stress **fluctuations**. The **range** of stress decreases and the material will fail under fatigue after a much higher number of **cycles**. The inventor claims a **six-fold** increase in fatigue life over that of API joints in sucker rod-pumped wells involving tension loads only. If used in progressive cavity pumping (PCP) installations where a combination of tension and torsion loads occur in the rod string, an increase of 2.5 times was observed.

The four-part sucker-rod connection is manufactured under the trade name **PRO/KC** by Weatherford [57].

Premium Connection by Tenaris

The Argentinian sucker-rod manufacturer, **Tenaris**, conducted an extensive development program with the objective of improving the **reliability** of sucker-rod joints. Their final solution was a completely **new thread form** [58,59] that substantially improves the fatigue life of the joint. The main features of the new premium connection (see Fig. 3.59) are the following:

- Threads on the pin and the coupling are **tapered** and have a **trapezium** profile.
- Threads have **flank-to-flank** contact when engaged with diametrical interference.
- Stresses in the rod pin after proper makeup are **low**.

The new thread form completely eliminates the **gap** between the pin and coupling threads as well as their relative **motion**, the basic **weakness** of the API threads. **Looseness** of the joint is thus much

FIGURE 3.59

Pin end of Tenaris premium sucker rods.

reduced, with only minor permanent deformations occurring in the pin and coupling threads. Because of this, the occurrence of small initial **cracks** is mostly eliminated and the fatigue life of the joint greatly exceeds that of the conventional API connections. Due to these improvements, the stress distribution in the joint is more uniform, without severe stress concentrations.

Figures 3.60 and 3.61 compare the **geometry** and the typical **stresses** in the conventional and the new connections, respectively. As seen, the API thread form allows **movement** between the pin and the coupling, which is completely **eliminated** in the new design. The shading of the figures illustrates the distribution of **stresses**, with darker shades representing greater stresses; as seen, the new connection design has more **uniform** stresses in the joint.

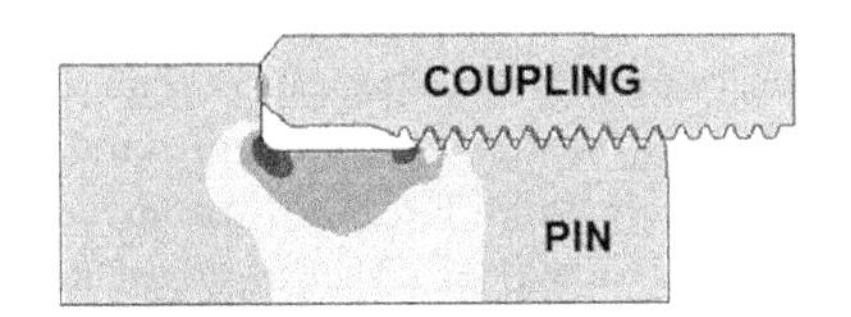

FIGURE 3.60

Geometry of and stress distribution in an API connection.

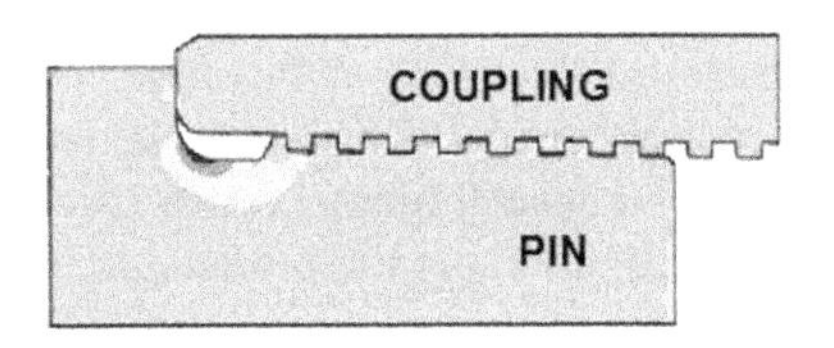

FIGURE 3.61

Geometry of and stress distribution in a premium connection.

Based on fatigue tests in the laboratory and field applications in several countries, the main **benefits** of the new premium connections can be summarized as follows.

- Due to the high fatigue resistance of the joint, stress ranges **greater** than those for Grade D rods are allowed, and the Goodman diagram must be modified accordingly.
- Since the number of joint failures dramatically decreases, **workover** costs drop considerably.
- The higher load capacity of the rods with the new connection allows production of liquid rates from **deep** wells where only ESP units could be used before.
- The higher strength of the premium connections allows the use of **lighter** rod strings and a consequent reduction in energy requirements.
- **Corrosion** problems are much reduced if, instead of using high-strength rod materials that are not efficient to combat corrosion, rods made of weaker grades but having the new premium joints are used.
- Due to the greater **integrity** of the new connection, it can be used in PCP applications where it can withstand greater torque loadings than normal rods.

3.5.2.3 Rod materials

API Materials

The material of steel sucker rods normally has an **iron** content of more than 90%. Alloying elements are added to increase **strength** and **hardness**, to improve the effects of metallurgical treatments, and to combat corrosion. Steels used for rod manufacture fall into two broad categories, **carbon steels** and **alloy steels**. Carbon steels contain carbon, manganese, silicon, phosphorus, and sulfur only; alloy steels contain other elements as well. Different manufacturers offer different compositions under various trade names. The compositions and mechanical properties of standard API sucker-rod materials (called **rod grades**) are given in **API Spec. 11B** [49]. **Grade C** rods made of **AISI 1536** material and having a minimum tensile strength of 90,000 psi are the cheapest, but their applicability is limited to noncorrosive environments and average pumping loads. The chrome–molybdenum alloy **Grade D** material (115,000 psi minimum tensile strength) allows for higher operating stresses but is limited to wells with average corrosion without H_2S. **Grade K** is a special nickel–molybdenum alloy used in mildly corrosive fluids and has a minimum tensile strength of 90,000 psi.

Non-API Materials

The continual efforts for increasing the strength of sucker rods have long indicated that prestressing the rod surface can bring about favorable features [60]. The high-strength **Electra** (Weatherford **EL**) sucker rods are made with a manufacturing process that includes full-length induction **hardening** that results in a special **binary** construction of the rod material. As shown in Fig. 3.62, the **core** of the rod is covered, all along the rod length, by an outer **case** with a thickness of about 5–8% of the rod diameter. This outer case is, as a consequence of the induction heat treatment, in a permanent extreme

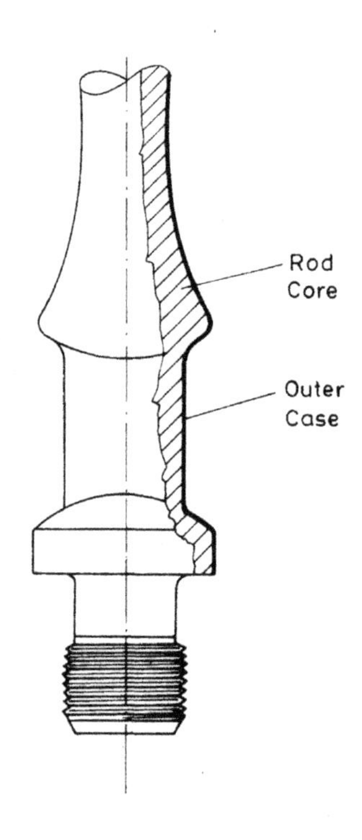

FIGURE 3.62

The binary construction of an Electra sucker rod.

compression with an average stress of as high as 120,000 psi. Since this stress is well above the highest possible tensile stresses that can ever be imposed on the rod under the most severe conditions, the outer case of the rod will never be subjected to **tensile** stresses. Material **fatigue** is mostly **eliminated**, because the rod case is under compression during the whole pumping cycle and the range of stress, mainly responsible for fatigue failures, can be ignored. The maximum allowable working stress of these rods is specified as 50,000 psi, regardless of the actual stress range in the rods due to well loads. Electra rods require extra **care** during field handling and running because of their exceptionally hard surfaces.

There are now more non-API **high-strength rods** available than ever before. These rods permit pumping of more fluid from greater depths or to get more production without overloading the rods. **Norris** was the first company to come up with this type of rod (**Norris 97**), which is manufactured differently from other rod types. Their minimum tensile strength is $T_a = 140{,}000$ psi, much higher than any of the API rod materials. Other rods in this category are **LTV HS** and **Trico 66**. All these rod materials are much **harder** than API **Grade D**, and consequently are more **sensitive** to handling and corrosion damage. In corrosive wells, therefore, the application of high-strength rod materials requires an effective corrosion inhibition program.

High-strength rod materials, except the Weatherford **EL**, are susceptible to stress **fluctuations** and material **fatigue** but can be subjected to larger stress ranges without becoming overloaded. For designing rod strings made up from these high-strength rods, a stress range diagram different from the original modified **Goodman** diagram must be used, as discussed later.

3.5.3 OTHER TYPES OF STEEL RODS

3.5.3.1 Continuous rods

Many downhole problems of rod pumping are inherently associated with the existence of the **joints** in the sucker-rod string, because these are subjected to severe operational conditions and are the **weakest link** in the system. The main **problems** related to the existence of sucker-rod connections in the rod string are:

- Joints can be **loosened** due to improper initial makeup and by operational problems like fluid pound conditions.
- **Fatigue** failure of rod connections is the commonest cause of workover operations.
- In deviated wells couplings can **erode** the tubing because side loads are concentrated at them, intensifying their rubbing action on the tubing wall.
- The larger outside diameters of the connections along the rod string (typically every 25 ft) may **restrict** the pumping rate because of the higher flowing pressure losses. This effect is important when producing higher rates through smaller tubing strings.
- The falling velocity of rods during the downstroke is decreased by the **piston effect** of the connections and the string may **float**, especially when pumping heavier fluids. This may cause heavy dynamic loads in the pumping unit, as well as compression loads in lower rod sections.

A **continuous rod string** without any joints, therefore, totally eliminates rod pin and coupling failures and increases rod string life. Such rods have long been available and are called either **continuous** rods, **coiled** rods, or **Corods** [61]. This technology was developed in the early 1970s in

Table 3.11 Mechanical Data of Corods Available from a Major Manufacturer

Corod No. -	Nominal Rod Size, in	Metal Area, sq in	Weight in Air, lb/ft	Elastic Constant, in/(lb ft)
#2	12/16	0.441	1.50	0.934 E−6
#3	13/16	0.518	1.76	0.796 E−6
#4	14/16	0.601	2.04	0.687 E−6
#5	15/16	0.690	2.35	0.599 E−6
#6	16/16	0.785	2.67	0.526 E−6
#7	17/16	0.886	3.01	0.466 E−6
#8	18/16	0.994	3.38	0.415 E−6

Canada to reduce well service costs but did not gain general acceptance mainly because of the special **handling** requirements and the associated high workover costs. Serviceability issues, however, are mostly eliminated today and the popularity of continuous rods is rapidly increasing.

The original continuous rods, **Corods**, are made from the same materials as solid rods, and have a **semi-elliptical** cross-section. Their nominal size is the outside diameter of a solid rod with the same cross-sectional area. Available sizes from one major manufacturer range from 12/16 to 18/16 in in 1/16 in increments; their main mechanical data are given in Table 3.11 [62]. The **elliptical** shape allows long rod strings to be **spooled** onto special transport reels without any permanent damage to the rods. The previously assembled total rod string, consisting of sections with different sizes **welded** together, is run into the well using a special service rig.

The special features of **Corods** result in improved performance in deviated wells where mechanical friction between the rod and tubing strings is usually excessive and can lead to premature failure. The use of **Corods** decreases mechanical wear of rods as well as the tubing inside wall because of two effects:

- As compared to conventional sucker rods that contact the tubing inside wall at one point only in a given cross-section, Corods have at least two contact points, as shown in Fig. 3.63. The figure compares 1 in slimhole sucker rods with 1 in (#6) Corods in different tubing sizes.
- If examined lengthwise, due to the relatively great outside diameter difference between the sucker-rod body and the couplings, conventional rods contact the tubing inside wall along the length of the couplings only, that is, 4–4 1/2 in, while Corods have a continuous contact.

The combined result of the two effects just described is that contact **areas** about 150 times greater can be expected when using Corods. If similar side loads are assumed, it is clear that mechanical wear will be much reduced and rod and tubing life will be substantially lengthened when using Corods instead of conventional sucker rods [63].

The main **advantage** of the application of **Corods** is the elimination of rod pins and couplings from the string; other advantages are as follows:

- Rod **fall** problems are eliminated; pumping speed can be increased.
- **Larger** rod sizes may be used in the same size of tubing or a smaller tubing size may be used for the same rod size.

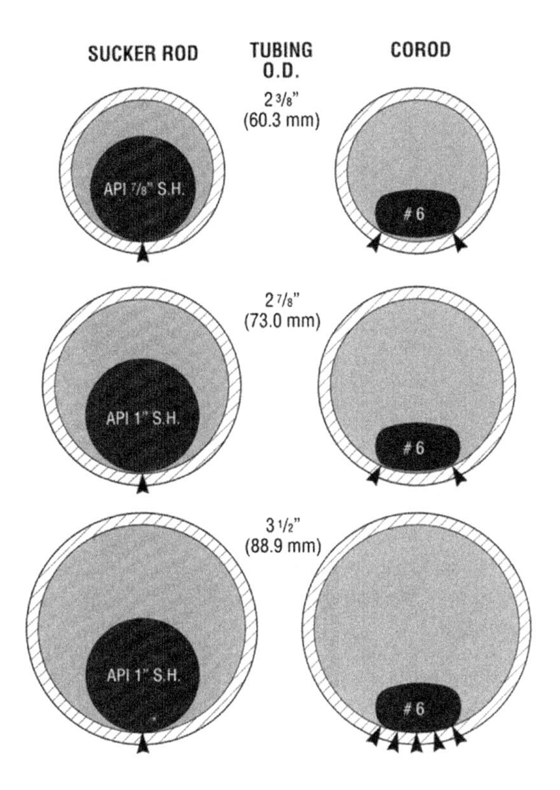

FIGURE 3.63

Comparison of rod–tubing contacts for coupled rods and Corods.

- In deviated wells downhole **friction** between the tubing and the rod string is decreased because the contact surfaces are **bigger**; tubing and rod lives increase.
- Rod stresses and surface torque requirements are **reduced** and it may be possible to pump larger volumes from the same well.

Their **limitations** include the need for a special rig for well **servicing** and a need for special **welding** procedures for field repairs.

Many drawbacks of using early generations of **Corods** are eliminated by the latest technology utilizing **coiled rods** of **round** cross-section. These are usually offered in four sizes, 13/16 in, 7/8 in, 1 in, and 1 1/8 in (usually designated as #3R, #4R, #6R, and #8.5R sizes), and are delivered to the well site on 18 ft reels fitted on trailers. Their **advantages** are very similar to the original **Corods** and they can be run in smaller tubing sizes, as shown in Fig. 3.64, where Corods and conventional (conv.) sucker rods of 7/8 in and 1 in sizes are shown in different sizes of tubing. As seen, the Corods provide much greater cross-sectional area for fluid flow than normal sucker rods of the same size.

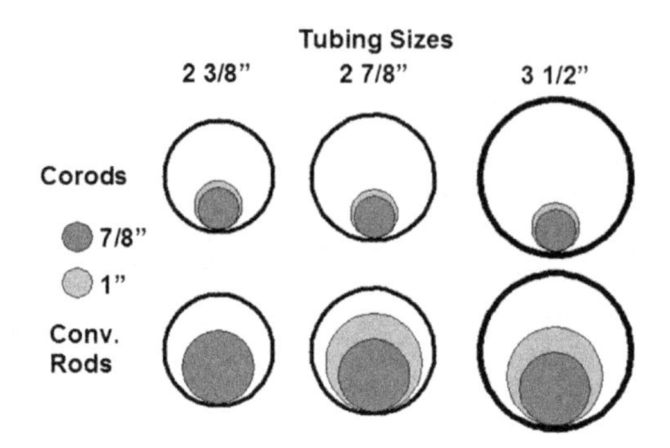

FIGURE 3.64

Fluid flow areas for coupled rods and round Corods.

The **service** requirement of using coiled rods was greatly reduced [64] by two inventions:

- A special **injector head** is used to run or pull the coiled rod string. The injector head has universal applications: can be used on workover rigs, may be truck-mounted, etc.
- Welding of rods is facilitated by a **portable**, gas-fired butt-welding unit that can produce reliable welded connections in the oilfield.

The many advantages of **coiled rods** account for their increasing popularity; the number of units has doubled in the last five years with a total of over 10,000 installations worldwide.

3.5.3.2 Other rod types

Another type of **continuous** sucker rod is the **Flexirod**, which is a special kind of wire **rope** [65] developed in the early 1960s. The rope is made up of 37 individual wires of high-strength steel and has an outside nylon **jacket** that provides corrosion and abrasion resistance. Flexirods have sizes ranging from 5/8 to 1 in, with 1/8 in increments [66], and have less than **half** the weight of the same solid rods. Their use requires several **modifications** in the pumping system: special **pumps** are to be run in the well to keep the string in tension during the entire pumping cycle, and a **hollow** polished rod is to be used.

Hollow sucker rods or **sucker tubes** can be used to advantage in **slimhole** completions where the well is cased with a tubing string [67]. Since fluid lifting takes place **inside** the hollow rod string, no tubing is needed in the well. The hollow rods require special **wellhead** configurations, consisting of a hollow polished rod and a flexible hose connected to the flow line. In wells with high sand production, these rods prevent sand **settling** problems. Injection of corrosion inhibitors can also be efficiently solved with the use of hollow rods. Their application is restricted to lower fluid rates, because large fluid volumes involve great pressure losses inside the tubes. The design and analysis of hollow rod pumping installations is similar to those of a solid rod installation [68].

3.5.4 FIBERGLASS SUCKER RODS

Steel sucker rods have two inherent **disadvantages** when used in deep corrosive wells: one is their great **weight,** and the other is that they are susceptible to severe **corrosion** damage. A long steel rod string is heavily loaded by its **own** weight, and even a small load on the plunger can result in higher than allowed stresses. High stresses combined with the corrosive action of well fluids rapidly increase rod failure frequencies. Because of these problems, low-weight, corrosion-resistant, nonmetal materials have definite application potential in sucker rod manufacture. A widely used such material is glass-reinforced plastic, **fiberglass** for short. The first field test with fiberglass sucker rods was performed in 1975 [69]. They became commercially available in 1977 and have evolved into the highly reliable products known today [70,71].

Fiberglass sucker rods consist of a plastic rod body and two steel end **fittings** bonded to both of the rod's ends. The end connectors have standard API pin threads (Fig. 3.65) and their special construction allows for high gripping forces that resist the forces pulling out the rod body from the fitting. The body is a **composite** of a thermosetting **resin** and of 1.5 million parallel glass **fibers**, each 15 micron thin. The individual fibers have an extremely high tensile strength and, depending on the resin/glass ratio achieved during manufacture, the final rods can have a tensile strength of 110–180,000 psi. Compared to steel, fiberglass rods are about **25% stronger** and weigh only **one third** of the respective steel rods. When subjected to an axial force, they **elongate about four times** as much as steel. Such an excessive stretch **prohibits** the use of an all-fiberglass string. Therefore, steel rods are run below the fiberglass ones to increase the total weight, but the usual combinations of fiberglass/steel rods still weigh only about one half as much as an all-steel string.

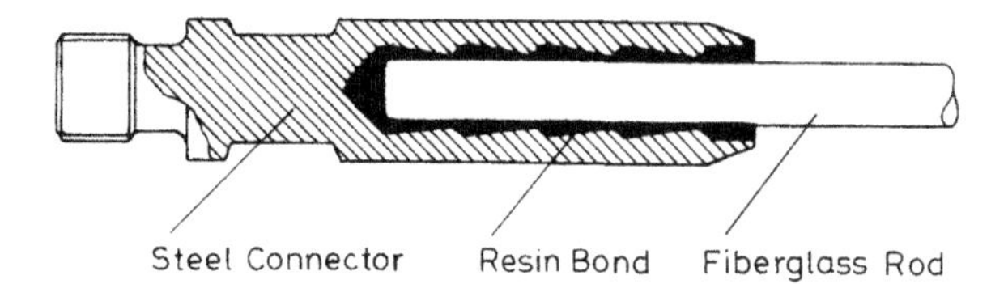

FIGURE 3.65

The construction of an end connector for fiberglass sucker rods.

Specifications for fiberglass rods are published in **API Spec. 11B** [49] (earlier in **API Spec. 11C** [72]). These rods are classified based on the rod body's **diameter**, the operating **temperature** rating, and the grade of end fittings. Nominal sizes are 3/4 in, 7/8 in, 1 in, and 1 1/4 in; standard lengths are 25, 30, and 37.5 ft. Rods have one size smaller standard **API connections** than their nominal size, except 1 1/4 in rods, which have 1 in connections. Since rod strength decreases with increasing temperatures, the safe operational temperature must be specified by the manufacturer. Average weights and elastic properties of rods vary with the different manufacturers due to differences in the manufacturing, the resins used, etc. The actual value of the modulus of elasticity has a direct impact on the stretch properties of the available fiberglass rods [73]; average values range from 6.3 to 7.2×10^6 psi as compared to 30×10^6 psi of steel rods. Table 3.12 contains basic data of fiberglass rods available from one manufacturer and is valid for rods 37.5 ft long with an elastic modulus of $E = 6.3 \times 10^6$ psi.

Table 3.12 Basic Data of Fiberglass Sucker Rods Available from a Major Manufacturer

Rod Size, in	Rod Area, sq in	Weight in Air, lb/ft	Elastic Constant, in/(lb ft)
3/4	0.442	0.48	4.308 E−6
7/8	0.601	0.64	3.168 E−6
1	0.785	0.80	2.425 E−6
1 1/4	1.227	1.29	1.552 E−6

The **benefits** of using fiberglass rods are numerous and can be summed up as follows:

- The most important advantage is that production rates can be **increased** by running a heavy steel rod section below the fiberglass rods and utilizing the greater **stretch** of fiberglass to increase downhole pump stroke. This effect is maximized if the pump is run with a pumping speed near the string's resonant frequency.
- In addition to increased well production, **operating** costs are usually reduced because of the lower total **weight** of the rod string and the reduced power requirement.
- Polished rod **loads** and energy requirements decrease and smaller pumping units can be used.
- The fiberglass rod string offers excellent resistance to **corrosion** and reduces workover costs.

Limitations of the use of fiberglass sucker rods are given in the following:

- The service life of these rods is heavily reduced at high well **temperatures**.
- They do not tolerate **compression** loading that must be avoided.
- Because of the relative softness of the rod material, they are prone to mechanical **damage** during handling and running/pulling operations.
- They are more **expensive** than steel rods, but the benefits usually justify the increased investment costs.

3.5.5 THE DESIGN OF SUCKER-ROD STRINGS

3.5.5.1 Introduction

The weight of the sucker-rod string is **distributed** along its length, and any section has to carry at least the weight of all the rods below it. This fact suggests that the ideal sucker-rod string would be a **continuous taper** from top to bottom. Since such a shape is practically impossible to achieve, one tries to approach the ideal construction by designing **tapered** strings with sections of increasing diameters toward the surface. For shallow wells, **straight** rod strings made up from one rod size only are also used, but deeper wells inevitably require the application of **tapered** strings.

In order to simplify the identification of strings made up of different rods, tapered strings are designated using standard **API code** numbers. The taper code system is different for the **coupled** (conventional) and the **continuous** rod types. For coupled rods with a 1/8 in increment in available rod sizes, the first numeral of the code refers to the **largest** rod size, and the second numeral refers to the **smallest** rod size in the string (both expressed in **eighths** of an inch). For example, a three-taper string composed of 1 in, 7/8 in, and 3/4 in rods is designated with the code **86**. The continuous rod sizes are available in 1/16 in increments, and their code is expressed in sixteenths of an inch, without the first numeral. Thus a rod string made up of 1 in, 15/16 in, and 7/8 in continuous rods is designated as **64**: since the largest rod in the string is 16/16 in, the first numeral of the code is a 6, and the second numeral is 4 because the smallest rod size is 14/16 in.

Composite fiberglass–steel rod strings are usually designated by a **four-digit** code. The first digit of the code gives the **size** of the fiberglass rods in eighths of an inch; the next two digits represent the **percentage** of fiberglass rods in the string, and the last digit refers to the **steel** rod size in eighths of an inch. For example, a rod string made up of 80% of 1 in fiberglass and 20% of 7/8 in steel rods is designated by the code **8807**.

A properly designed rod string should provide failure-free pumping operations for an extended period of time. Rod string design aims at the determination of:

- the rod **sizes** to be used in the string,
- the **lengths** of the individual taper sections, and
- the rod **material** to be used.

In order to find an ideal solution, detailed design calculations should be performed, and the actual well conditions should be properly taken into account. The two **basic** problems in sucker-rod string design are:

- how rod **loads** are to be calculated, and
- what principle to use for the determination of taper **lengths**.

At the time of designing, of course, the anticipated rod loads are not known, and they also depend on the taper lengths that are about to be determined. Therefore, one has to rely on approximate calculations to find probable rod loads that will occur during pumping. To decrease the effects of uncertainties in the design, it is customary to assume that **water** is pumped and that the fluid level is at the pump setting depth, i.e., the well is **pumped off**. Both of these assumptions increase the calculated rod loads and improve the safety of the rod string design.

3.5.5.2 Rod loads

The most basic property of any sucker-rod string is its **elastic behavior**, which is responsible for the complexity of its operation. The forces that excite the string at its two ends (at the surface through the polished rod, at the lower end by the subsurface pump) produce elastic **force waves** that travel in the rod material with the speed of the **sound**. These waves are of different magnitude and phase, and their interference and reflection can greatly affect the actual forces that occur in any rod section. Due to the complexity of describing these force waves, most rod string design procedures disregard the rod loads arising from such effects.

The possible rod loads during a complete pumping cycle at any depth in the string can be classified into the following groups:

- **Weight** of rods. This force is **distributed** along the string. At any section, it is equal to the weight of the rods below the given section. It is positive for both the up- and the downstroke. Here, and in the following, the load is said to be positive if it is directed downwards.
- **Buoyancy** force. This force always **opposes** the rod weight and is equal to the hydraulic lift caused by immersing the rods into the produced liquid. It is customary to handle the sum of the rod weight and buoyant force by using specific rod weights in fluid.
- **Fluid load** is a **concentrated** force acting at the bottom of the string only during upstroke, and equals the force resulting from the net hydrostatic pressure of the fluid lifted, acting on the area of the pump plunger. It is always positive.
- **Dynamic** loads are the results of changes in acceleration, during the pumping cycle, of the moving masses (rods and fluid column). The magnitude and direction of these forces is constantly changing during the pump stroke, but generally, dynamic loads result in a positive net load for the upstroke and a negative load for the downstroke.
- **Friction** forces are of two kinds: (1) fluid friction and (2) mechanical friction. Fluid is moving with the rods during upstroke and against the rods on the downstroke. Mechanical friction forces oppose the movement of rods; they are positive during upstroke and negative during downstroke.

An examination of these forces during a complete pumping cycle shows that the rod string is exposed to a **cyclic loading**. Although the upper rods are always in **tension**, the tension level considerably increases during the upstroke due to the load of the fluid lifted, the dynamic loads, and the friction forces. The downstroke load consists of only the buoyant weight of the rods minus dynamic loads and friction forces. Thus, the loading of the sucker-rod string is **pulsating tension**, which must be accounted for in its mechanical design. This is why the string has to be designed for **fatigue** endurance, as done in most of the present-day procedures.

3.5.5.2.1 True and effective loads

As shown previously, the most important components of the rod string load are the **gravitational** and **buoyant** forces, so a detailed treatment of those is given in the following. Since both of these forces are axial loads, and slender rods are extremely vulnerable to compression, one has to thoroughly investigate the variation of such forces during the pumping cycle. For a first approach, let us examine only the **static loads** during the **downstroke** of the pumping cycle.

When calculating the buoyancy forces acting on the rod string, two formulations of the **Archimedes principle** can be used:

- buoyancy forces acting on a submerged body can be calculated from the volume of the fluid **displaced**, or
- buoyancy forces can be defined as forces acting on each taper's cross-sectional **area** exposed to hydrostatic pressure.

Using the first approach and assuming negligible wellhead pressure, the **wet** weight of the sucker-rod string immersed in the produced fluid is found as the difference between the weight of the string in air and the weight of the fluid volume displaced by the string. This is the load measured during the downstroke and is usually designated as the rod string's **buoyant weight**, W_{rf}; it can be expressed using the volume of the string, V, as follows:

$$W_{rf} = V(\rho_{st} - \rho_l) = V\ \rho_{st}\left(1 - \frac{\rho_l}{\rho_{st}}\right) \tag{3.12}$$

Let us define the density of the produced liquid, ρ_l, by the density of water and the liquid specific gravity, and introduce the density of steel, ρ_{st}, in the formula:

$\rho_l = 62.4\ SpGr$ (lb/cu ft), and

$\rho_{st} = 487.5$ (lb/cu ft).

After substitution of these into Eq. (3.12) we get the following formula for the buoyant weight of a tapered sucker-rod string, valid at the top of the string:

$$W_{rf} = W_r(1 - 0.128\ SpGr) \tag{3.13}$$

where:

W_{rf} = buoyant weight of the rod string, lb,

W_r = total weight of the rod string in air, lb, and
$SpGr$ = specific gravity of the produced liquid.

In case there is a **significant** pressure on the wellhead due to flow line pressure, this pressure is transmitted in the fluid filling up the tubing string and its effect **decreases** the surface weight of the string. It is easy to see that wellhead pressure acts on the cross-sectional area of the bottom taper, A_1, then on the difference between the cross-sectional areas of the next and the bottom taper, $A_2 - A_1$, etc. as follows:

$$WHP(A_1 + (A_2 - A_1) + (A_3 - A_2) + (A_4 - A_3) + \ldots) = WHP\ A_{top} \tag{3.14}$$

Equation (3.13) is then modified to include the effect of wellhead pressure on the surface load of the rod string:

$$W_{rf} = W_r(1 - 0.128\ SpGr) - WHP\ A_{top} \tag{3.15}$$

where:

W_{rf} = buoyant weight of the rod string, lb,
WHP = wellhead pressure, psi, and
A_{top} = the corrected cross-sectional area of the top taper section, sq in
$SpGr$ = specific gravity of the produced liquid.

The treatment of the buoyant forces just described does not permit the calculation of the **distribution** of rod loads; this is only possible if buoyant forces are found from the liquid hydrostatic **pressure** and the exposed cross-sectional **areas** of the tapers. A detailed derivation is facilitated with the help of Fig. 3.66, where the schematic arrangement of a three-taper rod string is depicted with the

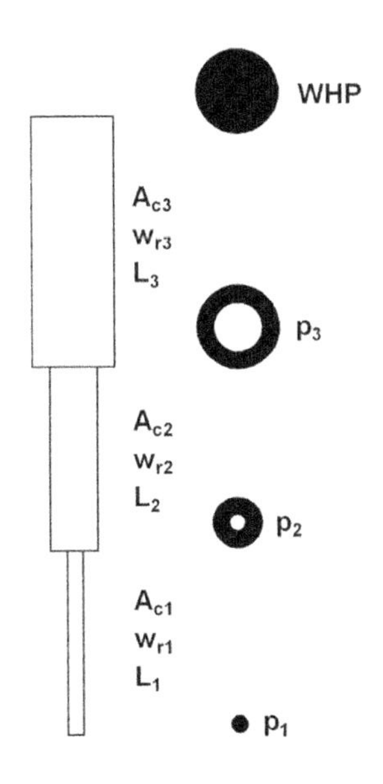

FIGURE 3.66

Schematic arrangement of a three-taper rod string.

required parameters of each taper, shown along with the hydrostatic pressures and projected rod areas. Static loads at the bottom and the top of each taper (denoted by subscripts B and T, respectively) will be calculated based on the balance of forces acting on each taper section.

Two kinds of loads will be calculated:

True loads are found from actual forces acting on the rod section. These represent **actual** forces in the rod material and are used for finding mechanical **stresses** in rod sections.

Effective loads, on the other hand, are **imaginary** forces that indicate the occurrence of **buckling** (see more detail in the next section) in the rod string; buckling occurs only when the effective load becomes negative. The definition of effective load is presented by the next formula [74,75]:

$$F_{\text{eff}} = F_{\text{true}} + p\,A_c \tag{3.16}$$

where:

F_{eff} = effective load, lb,
F_{true} = true mechanical load, lb,
A_c = the corrected cross-sectional area of the rod at the given depth, sq in, and
p = hydrostatic pressure at the same depth, psi.

Loads in Taper #1

At the **bottom** of the taper the hydrostatic pressure equals:

$$p_1 = WHP + 0.433\,SpGr(L_1 + L_2 + L_3)$$

Rod loads are calculated as:

$$F_{\text{true }1B} = -p_1\,A_{c1}$$

$$F_{\text{eff }1B} = F_{\text{true }1B} + p_1\,A_{c1} = 0$$

At the **top** of the taper the hydrostatic pressure equals:

$$p_2 = WHP + 0.433\,SpGr(L_2 + L_3)$$

Rod loads are calculated as:

$$F_{\text{true }1T} = F_{\text{true }1B} + L_1\,w_{r1}$$

$$F_{\text{eff }1T} = F_{\text{true }1T} + p_2\,A_{c1}$$

Loads in Taper #2

At the **bottom** of the taper the hydrostatic pressure equals:

$$p_2 = WHP + 0.433\,SpGr(L_2 + L_3)$$

Rod loads are calculated as:

$$F_{\text{true }2B} = F_{\text{true }1T} - p_2(A_{c2} - A_{c1})$$

$$F_{\text{eff }2B} = F_{\text{true }2B} + p_2\,A_{c2}$$

At the **top** of the taper the hydrostatic pressure equals:

$$p_3 = WHP + 0.433\ SpGr\ L_3$$

Rod loads are calculated as:

$$F_{\text{true }2T} = F_{\text{true }2B} + L_2\ w_{r2}$$

$$F_{\text{eff }2T} = F_{\text{true }2T} + p_3\ A_{c2}$$

Loads in Taper #3

At the **bottom** of the taper the hydrostatic pressure equals:

$$p_3 = WHP + 0.433\ SpGr\ L_3$$

Rod loads are calculated as:

$$F_{\text{true }3B} = F_{\text{true }2T} - p_3(A_{c3} - A_{c2})$$

$$F_{\text{eff }3B} = F_{\text{true }3B} + p_3\ A_{c3}$$

At the **top** of the taper the hydrostatic pressure equals the wellhead pressure, WHP, and the loads are calculated as follows:

$$F_{\text{true }3T} = F_{\text{true }3B} + L_3\ w_{r3}$$

$$F_{\text{eff }3T} = F_{\text{true }3T} + WHP\ A_{c3}$$

Calculation of the static loads during the **upstroke** is straightforward: the **fluid load** on the downhole pump's plunger has to be added to each load component calculated previously. Fluid load on the plunger is calculated from the net lift, represented by the **dynamic liquid level**, and the area of the plunger, as follows:

$$F_o = \left(WHP + 0.433\ SpGr\ L_{\text{dyn}}\right) A_p \tag{3.17}$$

where:

F_o = fluid load on plunger, lb,
WHP = the wellhead pressure, psi,
$SpGr$ = the specific gravity of the produced liquid, –,
L_{dyn} = the dynamic liquid level in the annulus, ft, and
A_p = the cross-sectional area of the pump plunger, sq in.

Exhibit 3.1 and Fig. 3.67 contain calculated **static** loads for an example three-taper rod string. As shown, true loads (solid curve) in the lower taper show compression forces but the effective loads in the same taper (dashed curve) are all positive, indicating that no buckling occurs in that taper.

CALCULATION OF STATIC TRUE AND EFFECTIVE LOADS ON A TAPERED SUCKER-ROD STRING with No External Forces on the Bottom of the Rods

Input Data

WHP (psi) =	100
SpGr =	1.00
Pump Size (in) =	1.75
Pump Setting Depth (ft) =	7,000
Dyn. Liquid Level (ft) =	3,000

Rod String Data

Taper	Size	Length	w_r	A_c
	in	ft	lb/ft	in^2
top	1	1,971	2.904	0.857
middle	7/8	2,220	2.224	0.656
bottom	3/4	2,809	1.634	0.482

Calculated Loads

Depth	Place		Weight in Air	Downstroke TRUE	Downstroke EFF.	Upstroke TRUE	Upstroke EFF.
ft			lbs	lbs	lbs	lbs	lbs
0	1	Top	15,251	13,217	13,303	16,582	16,668
1,971	1	Bottom	9,527	7,493	8,310	10,858	11,675
1,971	7/8	Top	9,527	7,685	8,310	11,050	11,675
4,191	7/8	Bottom	4,590	2,748	4,004	6,113	7,369
4,191	3/4	Top	4,590	3,081	4,004	6,446	7,369
7,000	3/4	Bottom	0	-1,509	0	1,856	3,365

EXHIBIT 3.1

Calculated static true and effective loads in a sample tapered rod string.

Investigation of the distribution of static rod loads allows the following conclusions to be made:

- In spite of the different approaches to the **Archimedes principle**, true and effective loads at the top of the rod string are **identical** for negligible wellhead pressures, with only minor differences for nonzero wellhead pressure values.
- The **slope** of the load distribution lines, i.e., the change in load for a unit depth increment, is different for the two cases. For **true** loads it is equal to the weight of rods in **air**, w_r, whereas for **effective** loads it equals the **buoyant weight** of the rods, w_{rf}.

It must be noted that previous discussions are valid for **static conditions** only when no **external** forces act on the rod string. In actual cases, however, several external forces may occur at the bottom of the string and along its length; these can considerably change the variation of rod loads during both the up- and the downstroke. Possible external forces on the rod string are:

- Viscous **drag** force opposing the movement of the plunger in the barrel,
- The force caused by the pressure drop in the **fluids** flowing through the pump valves,

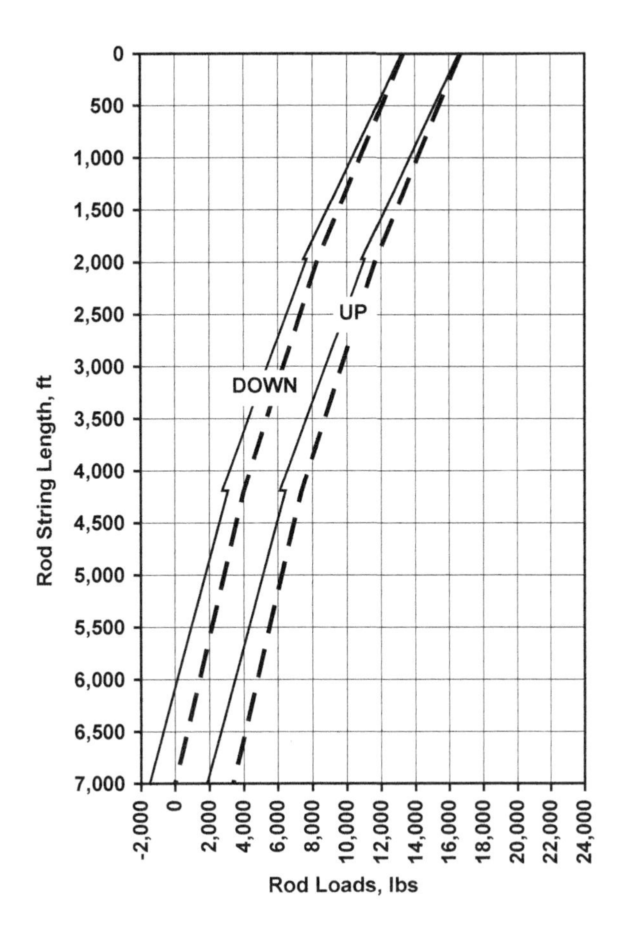

FIGURE 3.67

Distribution of static loads along the length of a sample rod string.

- Other **dynamic** forces due to pump malfunctions (e.g., gas or fluid pound),
- Viscous **drag** forces (fluid friction) along the rod string, and
- **Fluid** inertial forces.

Since rod loads, except on the surface, cannot normally be measured, their most accurate determination is possible through the solution of the damped **wave equation** (see a later chapter) that gives the distribution of loads along the rod string. These loads are different from the static loads calculated by the formulae detailed previously. Figure 3.68 compares for the previous example the static and the dynamic effective loads for a pumping speed of 10 strokes per minute (SPM) and assuming no external loads except fluid damping forces. As seen, actual effective loads during the downstroke may reach negative values; this is a prime indication of eventual rod buckling conditions.

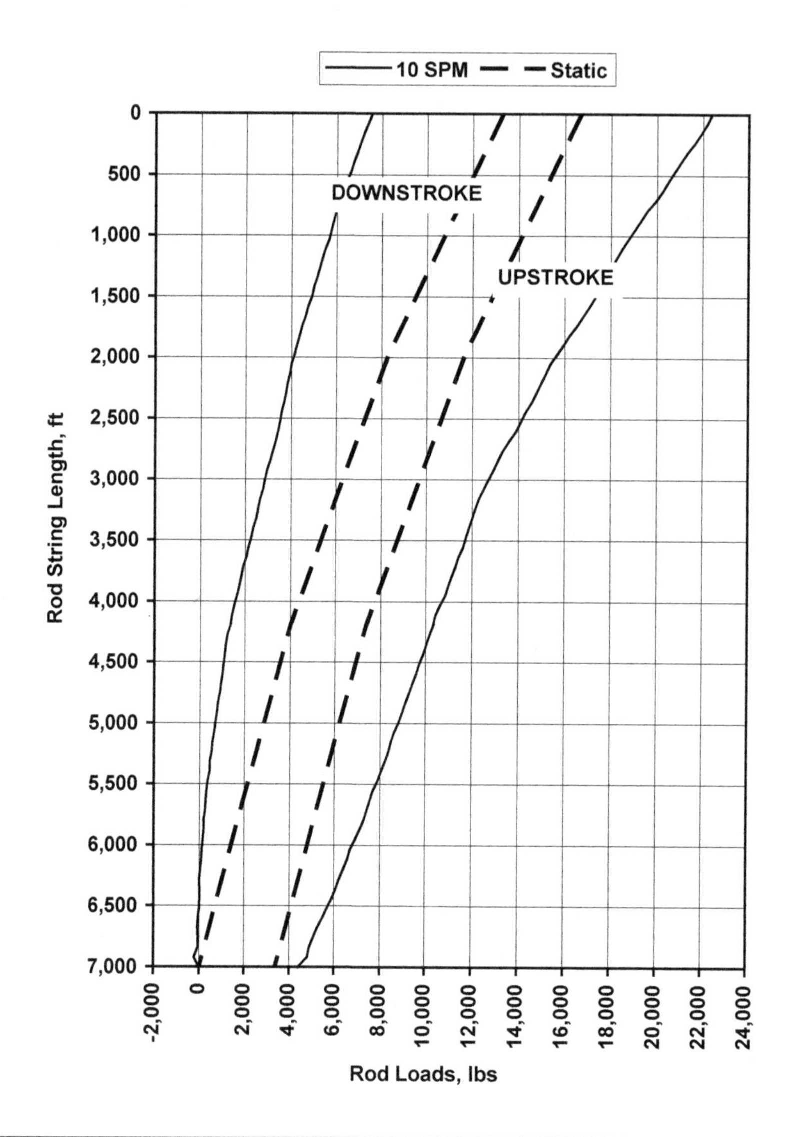

FIGURE 3.68

Distribution of predicted loads along the length of a sample rod string.

3.5.5.2.2 Buckling of the Rod String

The sucker-rod string, as a mechanical structure, is an absolute **slender** column because its length-to-diameter ratio (aka **slenderness** ratio) is extremely great. For a single 1 in rod the slenderness ratio is (25 ft × 12)/1 in = 300, much greater than 200, the approximate lower limit for slender steel columns; complete rod strings, of course, have even greater ratios. The sucker-rod string, as a slender column, is prone to **buckling**: the unstable deflection that occurs when a column is subjected to **compressive** stresses above a critical level. When buckled, the rod string assumes a **sinusoidal** or **helix** ("corkscrewed") shape inside the tubing below the **neutral point** where its loading changes from tension to compression. The high **bending** stresses present in the buckled rod string can lead to:

- immediate **failure** (in case the maximum stress surpasses the elastic limit of the steel), or
- a shortened fatigue **life** of the string (in case stresses are within the elastic operational range). Rods then undergo complete cycles of compression to tension; this kind of loading is more detrimental to rod life than the normal pulsating tension.

The deflected rod string contacts the tubing inside wall at several places and high side forces between the tubing and the rods cause severe mechanical **wear** of both components. Clear indications of buckling conditions, therefore, are frequent tubing **leaks** and rod breaks. This is the reason why buckling of the rod string is never allowed and specific measures to prevent its occurrence are usually taken.

In order to find where and when buckling is a problem, the load conditions of the rod string have to be evaluated, as done in the previous section. It must be clear that no part of the string is under compression during the **upstroke**, so it never buckles at that time. On the **downstroke**, however, **buoyancy** acting on the bottom taper plus any **external** forces (e.g., friction) acting at the pump are of **compressive** nature; there could be some compressive forces higher up the string as well. Therefore, the possibility of rod buckling must be evaluated only for the **downstroke** portion of the pumping cycle, with the main emphasis on the **bottom** taper.

It has been proved before by **Lubinski** and others [76,77] that buoyancy forces **do not** generate buckling; this is the reason why **effective** instead of true loads should be calculated when checking for buckling tendencies in the rod string. On the other hand, external forces can lead to buckling; an evaluation of the **external forces** occurring during the **downstroke** [78,79] resulted in the following general conclusions:

- Viscous **drag** in the plunger–barrel fit for low- to medium-viscosity fluids and normal pumping speeds is usually **negligible** but may be substantial for more viscous liquids.
- Pressure **drop** across the traveling valve can be **substantial** in cases when higher pumping speeds and smaller valves are used; the resulting compressive force arising on the plunger area can **buckle** the bottom rods.
- Mechanical (Coulomb) friction between the rods and the tubing can greatly contribute to compressive loads in the string, especially in deviated wells.

Buckling of the sucker-rod string is a special case among the many versions of buckling of **slender** elastic **columns** first investigated by the famous Swiss mathematician **Euler** in the eighteenth century. The bottom of the string is considered **fixed** at the bottomhole pump with no lateral movement possible; the upper end, however, is **free** to move laterally inside the tubing string. With its bottom fixed, the weight of the rod alone can create a compressive force that can initiate **buckling**. Under these conditions, the minimum length (also called **critical** length) of a rod whose weight creates sufficient compressive load to buckle the rod is found from the **Euler formula** [80], valid for the conditions just described:

$$L_{cr} = \sqrt[3]{\frac{0.795}{144} \pi^2 \frac{E\,I}{w_{rf}}} \tag{3.18}$$

where:

L_{cr} = critical length, ft
$E = 3.0 \times 10^7$, modulus of elasticity of the steel material, psi,

$I = (d^4 \pi)/64$, moment of inertia of the rod's cross-section, in^4, and
w_{rf} = specific rod weight in fluid, lb/ft.

Equation (3.18) gives the **critical length**, measured from the bottom of a rod taper section, at which **buckling** due to rod weight is initiated. In case the length of the rod under compression is **greater** than this value, the rod is expected to buckle.

The **critical load** belonging to buckling conditions is easily found from Eq. (3.18) after substituting the position of the **neutral** point as $L = F_{cr}/w_{rf}$:

$$F_{cr} = \sqrt[3]{\frac{0.795}{144}\pi^2 E\, I\, w_{rf}^2} \tag{3.19}$$

where:

F_{cr} = critical load that initiates buckling, lb.

Critical buckling lengths and loads for steel sucker-rod sizes, assuming they are submerged in **water**, are contained in Table 3.13.

Table 3.13 Critical Buckling Lengths and Loads for Steel Sucker-Rod Sizes

Rod Size, in	Critical Length, ft	Critical Force, lbs
5/8	23.1	22.9
3/4	26.1	37.2
7/8	28.9	56.2
1	31.6	80.2
1 1/8	34.2	109.8
1 1/4	36.6	145.5

From the previous description of sucker-rod buckling the following **conclusions** can be drawn:

- Standard API rods are subjected to buckling at very **low** compressive loads; see Table 3.13. For example, a 1 in rod buckles under a force of less than 100 lb.
- Bigger-diameter rods resist greater buckling forces. This is the reason why stronger than conventional rods called **sinker bars** are used when buckling is suspected.
- Both the critical length and critical load are independent of the **strength** of the rod material.

3.5.5.2.3 Using sinker bars to prevent buckling

One possible way to prevent buckling of the bottom sucker-rod taper section is to run heavy rods, called **sinker bars**, below the bottom rod taper. The length of the sinker bar section must be such that the weight of this section should overcome the compressive load acting at the bottom of the string. This ensures that the **neutral point** with zero axial load falls at the top of the sinker bar section; the bottom rod taper will not be subjected to compression. Although sinker bars (due to their greater diameter) have greater critical lengths and can resist higher compression loads than sucker rods, they can also be subject to buckling and must be **checked** to find if they can withstand the compression valid at the bottom of the sucker-rod string. If buckled, a bigger size of sinker bars must be selected and the whole process is repeated.

The necessary **length** of the sinker bar section to overcome the compression load depends on the weight and size of the selected bars and can be found according to the principle just described. The required formula is:

$$L = \frac{F_{ext}}{w_{rf}} \tag{3.20}$$

where:

L = required sinker bar section length, ft
F_{ext} = external compressive force, lb, and
w_{rf} = buoyant weight of the selected sinker bar, lb/ft.

After the length of the necessary sinker bar section is found, the section must be **checked** for buckling tendency using Eq. (3.18). If the calculated critical length of the sinker bar section is less than the length selected, a stronger sinker bar must be chosen. After the proper sinker bar size is found, the individual taper lengths of the rod string must be appropriately corrected.

EXAMPLE 3.7: USING THE ROD STRING GIVEN IN EXHIBIT 3.1, CHECK THE CONDITIONS FOR ROD BUCKLING AND FIND THE NECESSARY SIZE AND LENGTH OF SINKER BARS TO PREVENT ROD BUCKLING IF A COMPRESSIVE LOAD OF 100 LB ACTS AT THE BOTTOM OF THE STRING DURING THE DOWNSTROKE DUE TO AN EXCESSIVE PRESSURE DROP ACROSS THE TRAVELING VALVE

Solution

The bottom taper is 3/4 in, the critical force for this size from Table 3.13 is 37.2 lb; since this is much less than the actual compressive load of 100 lb, the taper will buckle.

Let's investigate the use of a 1 1/4 in sinker bar section at the bottom of the string. The sinker bar's average weight is $w_r = 4.172$ lb/ft, its corrected cross-sectional area is $A_c = 1.231$ sq in. The required length of the sinker bar section is calculated from Eq. (3.20):

$$L = 100/[w_r(1 - 0.128\ SpGr)] = 100/[4.172(1 - 0.128 \times 1)] = 27.5\ \text{ft}.$$

To find the critical length of the selected sinker bar section, first its buoyant weight and moment of inertia are calculated:

$$w_{rf} = w_r(1 - 0.128\ SpGr) = 4.172(1 - 0.128 \times 1) = 3.638\ \text{lb/ft, and}$$

$$I = \pi d^4/64 = \pi 1.25^4/64 = 0.120\ \text{in}^4.$$

Now the critical length is found from Eq. (3.18), using $E = 3.0 \times 10^7$ psi for the modulus of elasticity for steel:

$$L_{cr} = \left[(0.795 \times \pi^2 \times 3.0 \times E7 \times 0.12)/3.638/144\right]^{1/3} = 37.8\ \text{ft}.$$

By selecting a sinker bar section of 25 ft (one rod) and adding it to the bottom of the string, the sinker bar section will not buckle because its critical length (37.8 ft) is bigger, as just calculated. The length of the bottom taper must be corrected to accommodate the 25 ft of sinker bars.

The static loads in the rod string are displayed in Exhibit 3.2, which contains true and effective loads in the different sections of the rod string. As seen, the effective compressive load on the pump plunger is now carried almost completely by the sinker bar section, but, as calculated earlier, this taper is strong enough to withstand the compression without buckling.

CALCULATION OF STATIC TRUE AND EFFECTIVE LOADS ON A TAPERED SUCKER-ROD STRING with SINKER BARS

Input Data

WHP (psi) =		100
SpGr =		1.00
Pump Setting Depth (ft) =		7,000
External Load on Plunger (lbs) =		100
Use Sinker Bars:	Size (in) =	1 1/4
	w_r (lb/ft) =	4.172
	A_c (in^2) =	1.231

Sinker Bar Section Design

Required Length of SB (ft) =	27	
Selected Length of SB (ft) =	25	
Critical Length for SB (ft) =	38	design OK
Critical Load for SB (lbs) =	137	

Rod String Data

Taper	Size	Length w/o SB	Length w. SB	w_r	A_c
	in	ft	ft	lb/ft	in^2
top	1	1,971	1,971	2.904	0.857
middle	7/8	2,220	2,220	2.224	0.656
bottom	3/4	2,809	2,784	1.634	0.482
Sinker Bar	1 1/4		25	4.172	1.231

Calculated Loads

Depth	Place		Weight in Air	Downstroke TRUE	Downstroke EFF.
ft			lbs	lbs	lbs
0	1	Top	15,314	13,172	13,258
1,971	1	Bottom	9,591	7,449	8,266
1,971	7/8	Top	9,591	7,640	8,266
4,191	7/8	Bottom	4,653	2,703	3,959
4,191	3/4	Top	4,653	3,036	3,959
6,975	3/4	Bottom	104	-1,513	-9
6,975	SB	Top	104	-3,850	-9
7,000	SB	Bottom	0	-3,954	-100

EXHIBIT 3.2

Static true and effective rod loads for Example 3.7.

It is important to note that the calculations just described involved **static** loads only and dynamic loads during normal pumping operations can considerably differ; buckling conditions will thus change. The proper detection as well as prevention of sucker-rod buckling, therefore, depends on the accurate knowledge of dynamic rod loads; this can only be attained from the solution of the **wave equation**, detailed in a later chapter of this book.

3.5.5.3 Fatigue endurance limits

3.5.5.3.1 Steel rods

It is well known that most of the sucker rod breaks are **fatigue** failures, which occur at rod stresses well **below** the ultimate tensile strength or even the **yield strength** of the steel used. **Material fatigue** is a plastic tensile failure, due to repeated stresses, which starts at some **stress raiser** (a surface imperfection like a nick or a corrosion pit) on the surface of the rod, and slowly **progresses** at right angles to the direction of the stress, i.e., across the rod material. The load-carrying cross-section is thus progressively reduced until the remaining metal area is overloaded and breaks. Such failures occur in most mechanical parts subjected to repeated stresses.

The design of any equipment subjected to cyclic loading must consider the nature of the loading and the appropriate **fatigue endurance limit** of the material. The fatigue endurance limit of a given steel material, in a broad sense, is that maximum stress level at which the equipment will operate under cyclic loading conditions for a minimum of **10 million** complete cycles. If the steel will withstand this number of cycles, it can be expected to tolerate these stress levels for an extended period of time. The endurance limit can only be determined **empirically** by running a series of experiments under controlled conditions. Steels can have different endurance limits depending on the nature of loading (e.g., **tension–compression**, or in case of rod strings, **pulsating tension**); thus the actual value of endurance limit is primarily controlled by the type of loading. Other important effects include the surface imperfections (notches, nicks, etc.) and the nature of the operating environment.

API Rod Materials

Fatigue endurance limits for steels under cyclic tension–compression loads (as in steel structures) have been long established. The fatigue endurance diagram (called the **Goodman diagram**), however, could not be used for designing rod strings because of the differences in the **nature** of rod string loading and those used in the experiments. **A. A. Hardy**, working with an API study group, proposed the following **modifications** [81,82] to the original Goodman diagram:

- the maximum tension stress allowed should be **less** than the yield strength, as loading of the steel beyond plastic deformation changes the material properties,
- **compression** cannot be allowed in rod strings, as this can cause **buckling** (to which slender rods are especially vulnerable) and premature failure, and
- an additional **safety factor** to account for the **corrosiveness** of the environment, generally called service factor, was to be included.

With some minor alterations, this **modified** Goodman diagram was **adopted** by the **API** [52] and is still in use today. The fatigue endurance limit of the steel rod material, called the **allowable stress** in this API publication, can be calculated by the following formula:

$$S_a = SF\left(\frac{T_a}{4} + 0.5625\ S_{\min}\right) \tag{3.21}$$

where:

S_a = fatigue endurance limit (allowable stress), psi
SF = service factor, –
T_a = minimum tensile strength of the rod material, psi
$S_{\min}$ = minimum rod stress, psi.

The term *SF* is called **service factor**, and its use allows for an additional **safety** factor in the string design when **corrosive** fluids are pumped. Its value is best established from field records because it varies with the nature of well fluids and the effectiveness of corrosion treatments. Generally accepted service factors for different environments and rod grades are listed in Table 3.14, taken from **Brown** [83].

Table 3.14 Generally Accepted Service Factors, According to Brown [83]

	Service Factors	
Environment	**Grade C**	**Grade D**
Noncorrosive	1.00	1.00
Salt water	0.65	0.90
Hydrogen sulfide	0.50	0.70

Although the rod **body** is usually the limiting factor in rod string design, **slimhole** couplings considerably reduce the string's strength due to their reduced cross-sectional areas. Therefore, if slimhole couplings are used, the allowable stresses calculated from Eq. (3.21) must be reduced by applying a **derating factor**. Values of the slimhole coupling derating factors for different rod sizes and grades are given in Table 3.15, based on data from **Bradley** [48].

Table 3.15 Slimhole Coupling Derating Factors Based on Data from Bradley [48]

	Derating Factors for Rod Grades		
Rod Size, in	**K**	**C**	**D**
5/8	–	0.97	0.77
3/4	–	–	0.86
7/8	0.93	0.88	0.69
1	–	–	0.89

The modified Goodman diagram for the API rod grades and the high-strength E-rods is shown in Fig. 3.69. The **allowable stress** values, as plotted, were calculated using Eq. (3.21) for a service factor of 1.0. As discussed in Section 3.5.2.3, the high-strength **E rods** can sustain a constant allowable stress. In addition, these rods do not necessitate the use of service factors to increase the safety of design in corrosive wells.

Figure 3.70 illustrates a **nondimensional** presentation of the modified Goodman diagram, which is constructed by plotting the maximum stresses against the minimum stresses, both expressed as a **percentage** of the material's minimum tensile strength. This diagram can be **universally** used for all API rod grades and enables the analysis of different rod grades on the same diagram.

The safe loading **limits** on the modified Goodman diagram (see Fig. 3.70) are above the $S_{max} = 0$ line (as below this value the rods are in **compression**, which was not included), and below the S_a (allowable stress) line valid for the given service factor. If maximum rod stress values, plotted against

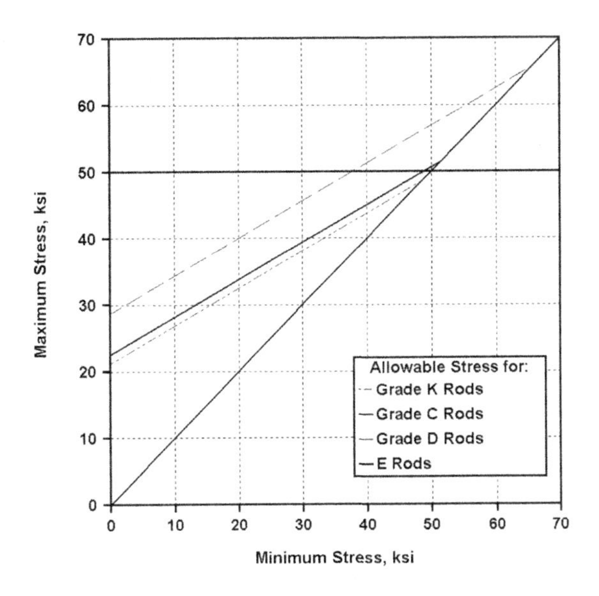

FIGURE 3.69

The modified Goodman diagram for various steel sucker-rod grades.

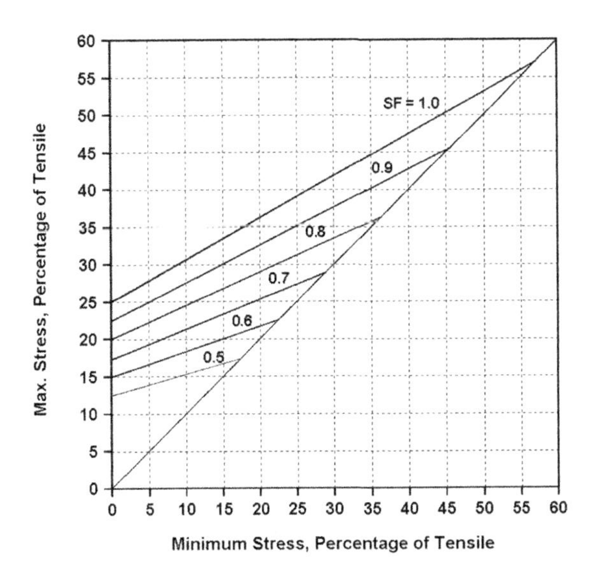

FIGURE 3.70

A universal modified Goodman diagram.

the appropriate minimum stresses, fall inside these limits, the rod string design is considered a **safe** one, allowing for failure-free continuous operation. Therefore, the aim of any string design method is to keep the **stresses** in the different taper sections within these safe operating **limits**.

It is also apparent from the diagram that, as minimum stress increases, the **range** of allowable stresses, i.e., the difference between maximum and minimum stresses, decreases. This means that when sucker rods operate under **high** stress levels, the minimum stresses must be kept fairly **high**. Fortunately, this is the situation in deep wells, where the weight of the long rod string constitutes the **major** part of the maximum rod load. For the same reasons, in **medium**-depth wells pumping with **large** plungers, resulting in fairly low minimum and high peak loads, the maximum rod stress may have to be **limited** in order not to **overload** the string.

High-Strength Non-API Rod Materials

The manufacturers of **Norris 97**, **LTV HS,** and **Trico 66** rods claim that their rods can handle wider ranges of stresses than regular API rods because of their greater strength. They recommend the following modification to Eq. (3.21) to calculate the maximum allowable stress in the rod material, where the minimum tensile strength of these rod materials equals $T_a = 140{,}000$ psi:

$$S_a = SF\left(\frac{T_a}{2.8} + 0.375\ S_{min}\right) \tag{3.22}$$

where:

S_a = fatigue endurance limit (allowable stress), psi
SF = service factor, –
T_a = minimum tensile strength of the rod material, psi
S_{min} = minimum rod stress, psi.

Norris was the first company to come up with this type of rod (**Norris 97**) and was also the first to use this method of stress analysis. The other companies, after developing rods similar to Norris 97 rods, recommend the use of the same stress range diagram.

Rods with Tenaris Premium Connections

As described in Section 3.5.2.2.2, the revolutionary new **thread** form developed by **Tenaris**, if used on sucker rods made of any conventional material, dramatically increases the rod string's fatigue life. This improvement is due to the fact that the **majority** of rod failures normally occur in the **joints**. Thanks to the exceptional strength of the new connection, rods can withstand much higher stress ranges than the strongest API material. The safe operating range of these rods, therefore, exceeds that of the rods made of **Grade D** material. Their maximum allowable stresses are found from the modified Goodman diagram utilizing the following formula and $T_a = 125{,}000$ psi as the tensile strength:

$$S_a = SF\left(\frac{T_a}{2.3} + 0.375\ S_{min}\right) \tag{3.23}$$

where:

S_a = fatigue endurance limit (allowable stress), psi
SF = service factor, –
T_a = minimum tensile strength of the rod material, psi
S_{min} = minimum rod stress, psi.

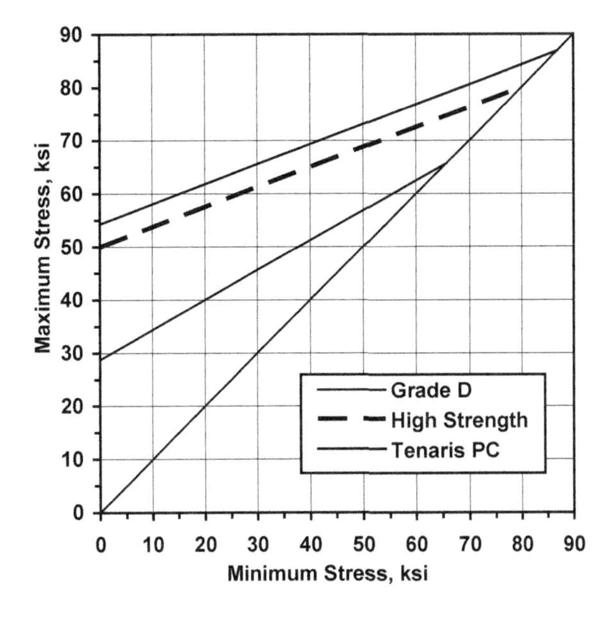

FIGURE 3.71

Comparison of the modified Goodman diagram valid for different rod materials.

Figure 3.71 presents a **comparison** of the modified Goodman diagram for **Grade D**, the high-strength materials, and for rods with **Tenaris** connections. As shown, the use of the new rod materials and premium connections greatly increases the stress range allowed in the sucker-rod string and can thus sufficiently raise the lifting **capacity** of rod-pumping installations.

EXAMPLE 3.8: THE SIZE OF THE TOP SECTION IN A TAPERED SUCKER-ROD STRING IS 1 IN, ROD MATERIAL IS GRADE D, AND THE WELL FLUID IS SALT WATER. THE MAXIMUM AND MINIMUM ROD LOADS WERE MEASURED AS 26,300 LB AND 15,700 LB, RESPECTIVELY. CHECK WHETHER THE ROD SECTION CAN SAFELY BE OPERATED UNDER THESE CONDITIONS, IF (1) FULL-SIZE AND (2) SLIMHOLE COUPLINGS ARE USED

Solution

The minimum and maximum rod stresses are calculated with the cross-sectional area of the 1 in rod found in Table 3.9:

$$S_{max} = 26,300/0.785 = 33,500 \text{ psi}$$

$$S_{min} = 15,700/0.785 = 20,000 \text{ psi}$$

The value of the service factor is found as 0.90 from Table 3.14, and the slimhole derating factor equals 0.89, taken from Table 3.15.

First, the full-size coupling case is checked, when no derating factor is to be applied to the allowable stress. The allowable stress is found from Eq. (3.21), where the minimum tensile strength of Grade D rods is 115,000 psi:

$$S_a = 0.90(115,000/4 + 0.5625 \times 20,000) = 36,000 \text{ psi}$$

Since this value is higher than the peak stress, the rod section is safe.
In the case of slimhole couplings, the allowable stress is reduced by using the derating factor:

$$S_a = 0.89 \times 36,000 = 32,040 \text{ psi}$$

In this case, the maximum rod stress is higher than the allowed stress and the same rod section is likely to fail if slimhole couplings are used.

3.5.5.3.2 Fiberglass rods

The fatigue life behavior of **fiberglass** sucker rods **completely** differs from that of steel rods. The basic difference is that fiberglass rods have a **finite** life under cyclic loading conditions. The failure-free life of these rods varies not only with the **range** of stress but also with the operating **temperature**. The higher the operating stress range in the rods, the lower the rod life that can be expected. Rod strength also decreases with an increase in operating **temperature**. These are the reasons why **API Spec. 11B** [49] (earlier **API Spec. 11C** [72]) requires that manufacturers supply a basic stress range **diagram** for the fiberglass rods they offer. This basic diagram should be constructed for a cycle number of 7.5 millions (equal to 1.8 years of operation at a pumping speed of 8 SPM) and an operating temperature of 160 F. In addition to these data, allowable range modifiers should be supplied for other cycle numbers (5, 10, 15, and 30 million) and temperatures. These modifiers are used like derating factors to calculate the allowable rod stresses for different required service lives and operating temperatures.

The stress range diagram defined above is the Goodman diagram for fiberglass sucker rods and is used for the same purposes as the modified Goodman diagram for steel rods. Figure 3.72 gives this

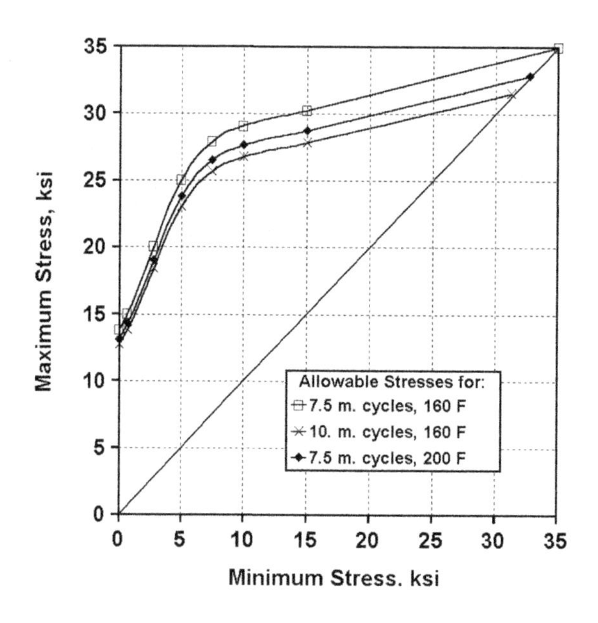

FIGURE 3.72

Goodman diagram (stress range diagram) for Fiberflex fiberglass sucker rods, using data from [84].

diagram for **Fiberflex** rods [84]. These rods have a maximum allowable static load of 35,000 psi at a temperature of 160 F. It can be observed that an increase in temperature or in cycle number reduces the allowable stresses and allows them to fall below the rod's rating. Some representative modifiers for these rods are the following: 85% for 15 million cycles, 80% for 30 million cycles, 95% for 200 F, and 80% for 240 F operating temperatures.

EXAMPLE 3.9: FIND THE SAFE MAXIMUM STRESS THAT CAN BE ALLOWED IN A FIBERGLASS ROD THAT IS SUBJECTED TO A MINIMUM STRESS OF 3,400 PSI AT 160 F. THE PUMPING SPEED IS 9 SPM AND A FAILURE-FREE OPERATION OF 3 YEARS IS DESIRED

Solution

The allowable maximum stress at a minimum stress level of 3,400 psi = 3.4 ksi at the rated conditions is read off from Fig. 3.72 as 22,000 psi.

The total number of cycles in 3 years:

3 years × 365 days/year × 1,440 min/day × 9 cycles/min = 14.2 million cycles.

Using the allowable stress modifier of 85%, valid for 15 million cycles, gives an allowed maximum stress of:

22,000 psi × 0.85 = 18,700 psi.

If the maximum stress in the given fiberglass rod is below this value, the rod can be expected to operate without fatigue failures for a period of 3 years.

3.5.5.4 Early design methods

The early rod string design methods all utilized the simplifying assumption that the string was exposed to a simple **tension** loading. Their goal was to keep the maximum rod stress at a value based on a percentage of the tensile strength of the rod material. Several different procedures were developed for the design of tapered rod strings. The one proposed in the **Bethlehem Handbook** [85], which was later adopted by the API in the earlier editions of **RP 11L**, gained wide acceptance. This involved a **balanced** design, based on static loading, by setting the **maximum** stress at the **top** of each section **equal**, and then selecting a material that could safely handle this stress.

Using this principle, different formulae can be derived for the calculation of rod taper percentages. Some are listed below, taken from **Zaba** [67]:

$$5/8'' - 3/4''\text{combinations}: \quad R_1 = 76.67 - 7.08\, d^2 \tag{3.24}$$

$$3/4'' - 7/8''\text{combinations}: \quad R_1 = 79.44 - 4.32\, d^2 \tag{3.25}$$

$$7/8'' - 1''\text{combinations}: \quad R_1 = 80.70 - 2.92\, d^2 \tag{3.26}$$

$$3/4'' - 7/8'' - 1''\text{combinations}: \quad R_1 = 67.24 - 6.96\, d^2 \tag{3.27}$$

$$R_2 = 15.60 + 3.82\, d^2 \tag{3.28}$$

where:

R_1 and R_2 are the length percentages of the rod taper sections, started from the pump, and
d = diameter of the pump plunger, in.

The simplified design procedures gave reasonable rod life for **shallower** depths, but, as well depths (and consequently rod stresses) increased, the old methods of design became inadequate. It was learned that with this design method practically all rod breaks were **fatigue** failures. These were occurring at operating stresses well below the tensile strength or the yield strength of the rod steel, causing designers to realize that a proper rod string design must include the **cyclic** nature of rod loading.

Although the above design theory is **outdated** and is seldom used today, it is still usable when applied to rod string designs made up of the high-strength **EL rods**. As discussed in Section 3.5.2.3 and shown in Fig. 3.69, the maximum allowed stress in these rods is a **constant** value, regardless of the actual range of stresses. Owing to this behavior of **EL** rods, string designs that set the **peak** stresses equal at the top of each rod section can be used successfully and will result in an optimum rod construction. Since the allowed rod stress for **EL** rods is 50,000 psi, a value that is considerably higher than the allowable stresses for API rod materials (see Fig. 3.69), **fewer** tapers with **smaller** rods are required and the total rod string weight is reduced. This can reduce torque and horsepower requirements on the pumping unit.

EXAMPLE 3.10: DESIGN AN API 65 TAPER ROD STRING USING EL RODS FOR THE FOLLOWING CASE: PUMP SETTING DEPTH 5,000 FT, PUMP SIZE 2.5 IN

Solution

From Eq. (3.24):

$$R_1 = 76.67 - 7.08 \times 2.5^2 = 32.42\%$$

and the length of the 5/8 in taper section:

$$32.42 \times 5,000/100 = 1,621 \text{ ft.}$$

The length of the 3/4 in section is 5,000−1,621=3,379 ft.
Rod weights in air, using data from Table 3.9:

$$W_{5/8} = 1,621 \times 1.135 = 1,840 \text{ lb},$$

$$W_{3/4} = 3,379 \times 1.634 = 5,521 \text{ lb}.$$

The fluid load on the plunger is found from:

$$F_o = 0.34 \ \mathrm{d^2L} = 0.34 \times 2.52 \times 5,000 = 10,625 \text{ lb}.$$

Minimum rod stresses are calculated based on the rod weight alone and using the cross-sectional areas found in Table 3.9:

$$S_{\min\ 5/8} = 1,840/0.307 = 5,990 \text{ psi},$$

$$S_{\min\ 3/4} = 1,840 + 5,521/0.442 = 16,650 \text{ psi}.$$

The maximum stresses are calculated based on the fluid load plus rod weight:

$$S_{\max\ 5/8} = (10,625 + 1,840)/0.307 = 40,600 \text{ psi},$$

$$S_{\max\ 3/4} = (10,625 + 1,840 + 5,521)/0.442 = 40,690 \text{ psi}.$$

Since both stress values are below the 50,000 psi rating of the **EL** rods, this string is properly designed. The calculated minimum and maximum stresses, if plotted on the **Goodman** diagram, would clearly fall above the maximum allowable stresses of even the strongest API rod material. This shows the superior strength of the **EL** rods.

3.5.5.5 Designs considering fatigue

3.5.5.5.1 West's method

It was shown by several investigators that previous designs, based on equal **maximum** stress at the top of each taper section, generally resulted in the **overloading** of lower rod sections. **Eickmeier** [86] observed this phenomenon, but it was **West** [87,88], who gave an improved string design to overcome the problem. To correct this inherent error of previous procedures, **West** developed a taper design that attains the **same** amount of **safety** for every taper section. His design goal is to have the same **ratio** of maximum stress to allowable stress in each taper. This means that the rod stresses, plotted on the modified **Goodman** diagram, will lie on the same $SF=$ **constant** line. Rod strings designed this way will have the **same** safety factor included in every taper, and will not have any **weak** points.

West calculated the maximum loads in the tapers as the sum of the rod weight in air, fluid load on plunger, and a dynamic force. He disregarded **buoyancy** effects to compensate for friction forces that are usually unknown but tend to act opposite the buoyant force. His approach to **dynamic** forces was to use the **Mills acceleration factor** method, which gives reasonable predicted loads for small pumps and medium pumping depths. His formulae to calculate rod loads for a single section are:

$$F_{\min} = L\, w_r (2 - f) \tag{3.29}$$

$$F_{\max} = F_o + L\, w_r f \tag{3.30}$$

where:

$F_{\min}$, $F_{\max}$ = minimum and maximum polished rod loads, lb
L = length of the taper section, ft
w_r = average weight of the taper section, lb/ft
F_o = fluid load on plunger, lb
f = Mills acceleration factor, –

$$f = 1 + \frac{S\,N^2}{70,500}$$

S = polished rod stroke length, in
N = pumping speed, SPM.

The design of taper lengths involves an **iterative** procedure, since the ratio of maximum to allowable stress, R, cannot be found in any section without the knowledge of the taper **lengths**, which are to be calculated. Figure 3.73 is a flowchart of the design procedure, which starts with the calculation of the required **ratio** of maximum and allowable stresses. This can be written for the entire string by assuming an **average** rod string weight:

$$R = \frac{F_o + L_{\text{total}}\, w_{\text{avg}}\, f}{A_n \frac{T_a}{4} + 0.5625\, L_{\text{total}}\, w_{\text{avg}}\, (2 - f)} \tag{3.31}$$

where the parameters not defined above are:

R = ratio of maximum and allowable stress, –
L_{total} = total rod string length, ft
w_{avg} = average rod string weight, lb/ft
A_n = cross-sectional area of the top rod, sq in.

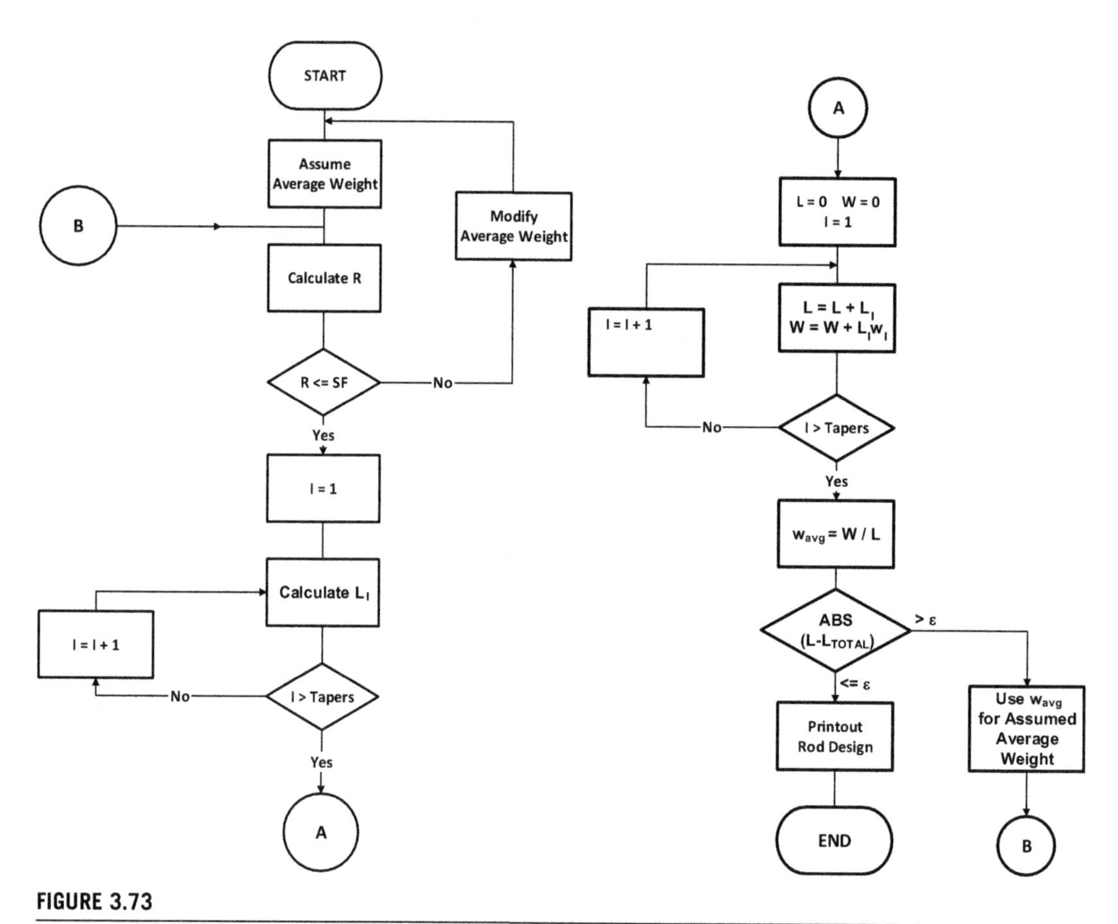

FIGURE 3.73

Flowchart of calculating rod string taper lengths with West's design procedure.

The calculated value of R is checked with the required **service factor**. If $R > SF$, a new average string weight is assumed and the calculation of R is repeated. In case no valid R is found, a stronger rod material must be used.

On the basis of the final value of R, taper **lengths** can readily be calculated to fulfill the requirements of the design method, i.e., to have the same ratio of maximum and allowable stresses in all tapers. The final formula to be used for the calculation of individual taper lengths:

$$L_i = \frac{F_{\max(i-1)} - R\left[A_i \frac{T_a}{4} + 0.5626\, F_{\min(i-1)}\right]}{w_i\ \left[0.5625\, R(2-f) - f\right]} \tag{3.32}$$

where:

L_i = length the ith rod taper, ft,
$F_{\max(i-1)}$ $F_{\min(i-1)}$ are the max. and min. rod loads in the previous rod section, lb,
A_i = cross-sectional area of the actual taper section, sq in,
T_a = minimum tensile strength of the rod material, psi,
w_i = average rod weight in air of the actual taper, lb/ft.

If the sum of calculated L_i taper lengths differs from the required pump setting depth, L_{total}, the process is **repeated** with a new **assumed** value of average rod weight based on the lengths just calculated. This procedure is repeated until convergence is found.

West also suggested a procedure to account for the **compression** that is present in the bottom taper(s). He developed a formula to determine the length and weight of rods in compression, and proposed starting the string design from the neutral point so determined.

EXAMPLE 3.11: CALCULATE THE REQUIRED TAPER LENGTHS USING THE FOLLOWING WELL DATA:

Pump setting depth	6,000 ft	Plunger size	2 in
PR stroke length	120 in	Pumping speed	6 SPM
Rod grade	API D	Service factor	0.9
API rod code	86	Liquid *SpGr*	1.0

Solution

Exhibit 3.3 contains the calculation results.

R O D S T R I N G D E S I G N R E S U L T S

Rod String Weight: 13906 [lbs] Avg. Rod Weight: 2.32 [lb/ft]

Size	Taper Lengths		Rod Stresses			SF
			Min.	Max.	Allow.	
[in]	[%]	[ft]	[psi]	[psi]	[psi]	[-]
1	35.5	2131	16639	29220	34298	0.7667
7/8	40.0	2400	12079	27253	31990	0.7667
3/4	24.5	1469	5093	24240	28453	0.7667

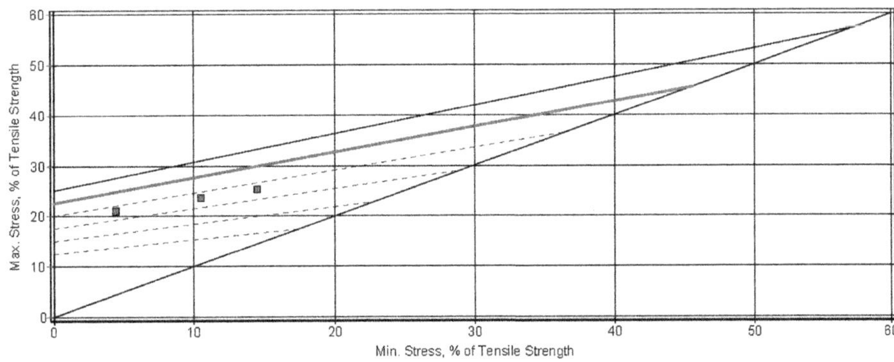

EXHIBIT 3.3

Sucker-rod string design results for Example 3.11.

EXAMPLE 3.12: DESIGN THE ROD STRING FOR THE SAME CONDITIONS AS IN EXAMPLE 3.11 BUT USE A COROD STRING

Solution

Exhibit 3.4 contains the calculation results. As seen, an **API 52** four-taper Corod string has the same loading as the coupled one calculated in the previous example but it is lighter by almost 3,000 lb. The Corod string, therefore, provides more energy-efficient operations and will probably have a longer service life because of the lack of couplings in the string.

ROD STRING DESIGN RESULTS

Rod String Weight: 11225 [lbs] Avg. Rod Weight: 1.87 [lb/ft]

Size	Taper Lengths		Rod Stresses			SF
			Min.	Max.	Allow.	
[in]	[%]	[ft]	[psi]	[psi]	[psi]	[-]
15/16	21.9	1316	15296	29133	33619	0.7799
7/8	23.3	1396	12755	28018	32332	0.7799
13/16	25.0	1499	9630	26647	30750	0.7799
3/4	29.8	1789	5694	24920	28758	0.7799

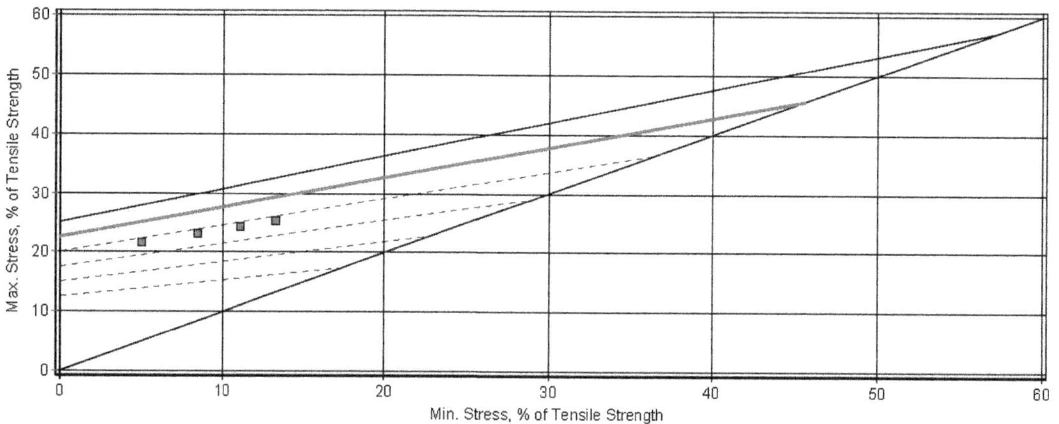

EXHIBIT 3.4

Sucker-rod string design results for Example 3.12.

3.5.5.5.2 Neely's method

Neely introduced the concept of **modified stress** and defined it as follows:

$$S_{mod} = S_{max} - 0.5625\ S_{min} \tag{3.33}$$

where:

S_{mod} = modified stress, psi
S_{max} = maximum rod stress, psi
S_{min} = minimum rod stress, psi.

The modified stress simply means the **intercept** at the vertical axis of the modified Goodman diagram (MGD) of a line that can be drawn starting from the actual $S_{min} - S_{max}$ point of the taper at hand, and having a **slope** parallel to the $SF = 1$ line. **Neely's** design procedure [89] attempts to **equalize** this modified stress at the **top** of each taper section.

Neely used the **buoyant** rod weight in calculating rod stresses and also used a new procedure to approximate **dynamic** forces. He proposed a simple empirical correlation to calculate the maximum dynamic force. It was a function of pumping speed, fluid load, stroke length, and a simplified spring constant for the rod string. The actual dynamic loads in the individual rod tapers were calculated based on the following assumptions:

- upstroke and downstroke dynamic forces are **equal**,
- dynamic forces decrease **linearly** with string length, starting from the surface maximum value calculated with his empirical correlation and diminishing to zero at plunger depth.

EXAMPLE 3.13: USING THE DATA OF EXAMPLE 3.11, DESIGN A ROD STRING WITH NEELY'S METHOD

Solution

Exhibit 3.5 presents detailed calculation results.

R O D   S T R I N G   D E S I G N   R E S U L T S

Rod String Weight: 13933 [lbs]   Avg. Rod Weight: 2.32 [lb/ft]

Size	Taper Lengths		Rod Stresses			SF
			Min.	Max.	Mod.	
[in]	[%]	[ft]	[psi]	[psi]	[psi]	[-]
1	36.7	2203	12590	29149	22067	0.8135
7/8	37.6	2254	7569	26250	21993	0.7953
3/4	25.7	1543	1696	22987	22033	0.7739

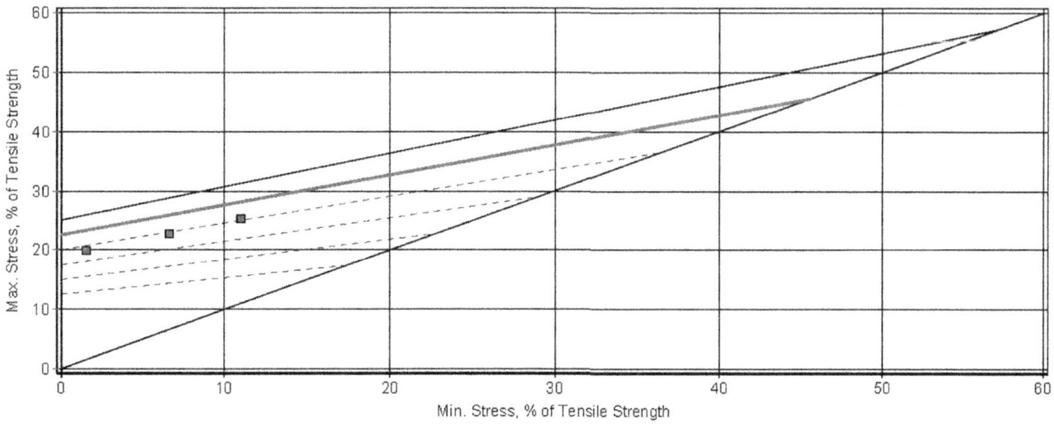

EXHIBIT 3.5

Sucker-rod string design results for Example 3.13.

3.5.5.5.3 API taper designs

In 1976, the **American Petroleum Institute** adopted the rod string design method proposed by **Neely**, and included rod **percentages** calculated by this procedure in the later editions of **RP 11L** [90]. The tables published in **RP 11L** gained wide acceptance, and thousands of rod strings have been installed since then that used these recommended taper lengths. Taper percentages in **RP 11L** are presented for different API taper combinations and are the sole function of the pump size used.

The use of the API taper lengths, of course, saves the **time** of a detailed string design, but the wide availability of personal computers today has eliminated the need to resort to such a **shortcut** method. The need for carrying out actual rod designs is further amplified by some unapparent **limitations** of the **RP 11L**. The basic assumptions used by **Neely**, and used but not shown [91] in the **API RP 11L**, to arrive at the published taper lengths are listed in Table 3.16.

Table 3.16 Basic Assumptions Used for the Calculation of Taper Percentages Published in API RP 11L after Neely [91]

Largest Rod Size, in	String Length, ft	Stroke Length, in	Pumping Speed, SPM
3/4	4,000	54	23.7
7/8	8,000	120	11.6
1	8,000	120	11.6
1 1/8	12,000	192	8.8
1 1/4	12,000	192	8.8

Rod designs for operating conditions, which are different from those presented in the above table, may require taper percentages that are significantly **different** from those listed in **API RP 11L**. Figure 3.74, presented by **Gault** [92], compares API tapers with those that can be calculated using

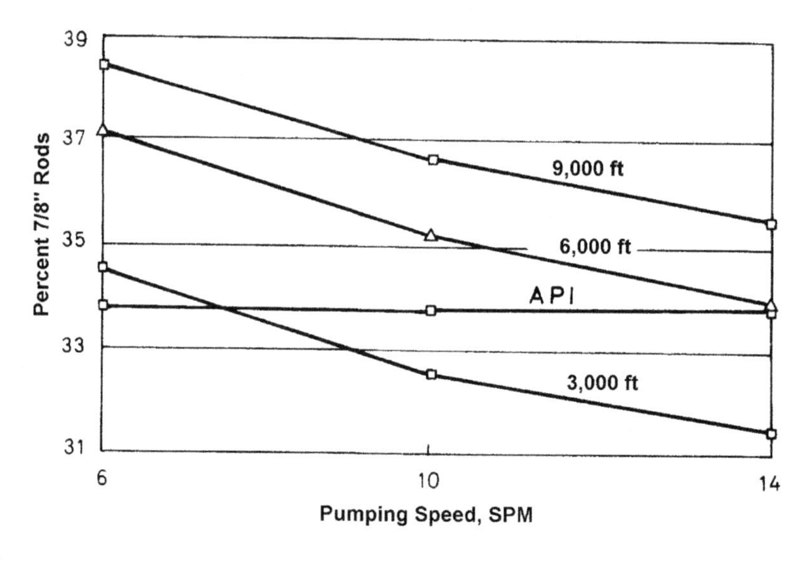

FIGURE 3.74

Comparison of API taper lengths with those calculated by Neely's method, after **Gault** [92].

Neely's procedure, for a 1.5 in plunger and 7/8–3/4 in rods. The API taper percentages, as shown, do not vary with well depth or pumping speed. The **differences** are apparent, and they lead to the conclusion that an **accurate** rod string design should be based on **actual** pumping conditions.

3.5.5.5.4 The Gault–Takacs method

The string design method developed by **Gault and Takacs** [93] overcomes the drawbacks of previous procedures, giving a more theoretically **sound** design method while requiring only **moderate** computational time. The **goal** of the proposed design is to have the same degree of **safety** in every taper section. The actual service factors will all be the **same**, and a rod string designed this way will be subjected to a **uniform** level of fatigue loading all along the string.

Of the rod loads considered, the calculation of **dynamic** loads is thoroughly revised, and a new procedure is proposed. For systems operating within the limitations of the **RP 11L**, calculation of the surface loads is made using the **RP 11L** procedure [90]. Thus, the calculated rod loads not only include the loads usually considered in design procedures but also reflect the effects of force wave reflections that take place in the string. This approach to rod load calculations, therefore, permits the inclusion of those forces that were **never** before considered in rod string design.

The other basic improvement is the way the **distribution** of dynamic forces along the string is calculated. Dynamic forces are the result of rod string **acceleration**. They have to be **proportional** to the **mass** being moved and must have **different** magnitudes during the up- and downstroke. It is a logical assumption to **distribute** the dynamic forces along the rod string in proportion to the distribution of rod **mass**. This means that the dynamic load calculated at the top of any taper section will be a function of the total mass of the rods below this level.

The **polished rod** loads, which reflect the sum of forces acting at the top of the rod string, are calculated by the **RP 11L** procedure, as given below:

$$PPRL = W_{\mathrm{rf}} + F_o + F_{\mathrm{du}} \tag{3.34}$$

$$MPRL = W_{\mathrm{rf}} - F_{\mathrm{dd}} \tag{3.35}$$

where:

$PPRL$ = maximum polished rod load, lb
$MPRL$ = minimum polished rod load, lb
W_{rf} = buoyant rod string weight, lb
F_o = fluid load on plunger, lb.

The surface (maximum) **dynamic** load components, F_{du} for the upstroke and F_{dd} for the downstroke, can be calculated from the knowledge of $PPRL$ and $MPRL$ values:

$$F_{\mathrm{du}} = PPRL - W_{\mathrm{rf}} - F_o \tag{3.36}$$

$$F_{\mathrm{dd}} = W_{\mathrm{rf}} - MPRL \tag{3.37}$$

The rod **loads** in the top section of the ith taper can now be written up by considering the distribution of dynamic forces, as detailed before:

$$F_{\max(i)} = F_o + \sum_{j=1}^{i} w_j\, L_j\, (1 - 0.128\, SpGr) + \frac{F_{\mathrm{du}}}{W_r} \sum_{j=1}^{i} w_j\, L_j \tag{3.38}$$

$$F_{\min(i)} = \sum_{j=1}^{i} w_j L_j (1 - 0.128\ SpGr) + \frac{F_{dd}}{W_r} \sum_{j=1}^{i} w_j L_j \tag{3.39}$$

where:

$F_{\max(i)}$; $F_{\min(i)}$ = maximum and minimum rod loads, lb
L_j = length of the jth taper, ft
w_j = average rod weight of the jth taper, lb/ft
$SpGr$ = specific gravity of the produced fluid, –
W_r = total rod string weight in air, lb.

The rod **stresses** are calculated based on these equations, and after some mathematical derivations, the following formulae are developed:

$$S_{\max(i)} = \frac{1}{A_i} \left\{ F_o + \left[\frac{F_{du}}{W_r} + 1 - 0.128\ SpGr \right] \sum_{j=1}^{i} w_j L_j \right\} \tag{3.40}$$

$$S_{\min(i)} = \frac{1}{A_i} \left\{ \left[1 - 0.128\ SpGr - \frac{F_{dd}}{W_r} \right] \sum_{j=1}^{i} w_j L_j \right\} \tag{3.41}$$

where:

A_i = cross-sectional area of rods in the ith taper, sq in.

Equations (3.40) and (3.41) can be combined into a single formula that is valid for a given taper section and does not contain the length of the taper:

$$S_{\max(i)} = \frac{F_o}{A_i} + S_{\min(i)} \frac{\frac{F_{du}}{W_r} + 1 - 0.128\ SpGr}{1 - 0.128\ SpGr - \frac{F_{dd}}{W_r}} \tag{3.42}$$

Equation (3.42) shows that, for a particular string design, the calculated $S_{\max} - S_{min}$ rod stress values for any taper in the string, if plotted on the modified Goodman diagram, will lie on a **straight** line. These lines for different tapers are **parallel** to each other, as their tangents are constant for a given design.

Starting from an **arbitrarily** selected string design, i.e., from the knowledge of taper lengths, the **RP 11L** calculations can be applied to find polished rod loads *PPRL* and *MPRL*. These values, in turn, can be used to determine an actual **service factor** for the top rod section. Given the value of $S_{\max}$ and $S_{\min}$, this can be calculated by using Eq. (3.21), which describes the allowable stress for the rod material:

$$SF_{act} = \frac{S_{\max}}{\frac{T_a}{4} + 0.5625\ S_{\min}} \tag{3.43}$$

where:

SF_{act} = actual service factor, –
$S_{\max}$ = maximum stress in taper, psi,

S_{min} = minimum stress in taper, psi,
T_a = minimum tensile strength of rod material, psi.

Using this value, the equation of the $SF = SF_{act}$ line can be written up, that goes through the point that corresponds to the top taper. According to the design goal set forth previously, the other tapers in the string should have the **same** service factor; thus their respective $S_{max} - S_{min}$ points must lie on this very same line. It was shown above, however, that these points must lie on the straight lines described by Eq. (3.42) as well. Thus, the **simultaneous** solutions of Eq. (3.42) and the allowable stress line valid at the calculated SF_{act} value will satisfy the requirements of the design procedure.

The above solution will give the following values for the required minimum rod stresses, S_i^*, of the individual tapers:

$$S_i^* = \frac{SF_{act}\frac{T_a}{4} - \frac{F_o}{A_i}}{\frac{\frac{F_{du}}{W_r}+1-0.128\ SpGr}{1-0.128\ SpGr-\frac{F_{dd}}{W_r}} - 0.5625\ SF_{act}} \tag{3.44}$$

To find the actual taper lengths satisfying the conditions of the design method, one has to **substitute** the S_i^* values just calculated into Eq. (3.41), which gives the minimum rod stress as a function of taper length. Solving that equation for the ith taper length, one arrives at the following **final** formula:

$$L_i = \frac{1}{w_i}\left(\frac{A_i\ S_i^*}{1 - 0.128\ SpGr - \frac{F_{dd}}{W_r}} - \sum_{j=1}^{i-1} w_j\ L_j\right) \tag{3.45}$$

The above equations enable one to calculate, starting from an **arbitrary** rod string design, the required taper lengths that **satisfy** the goal set forth in the design method. These taper lengths usually represent only an **intermediate** design, and the flowchart in Fig. 3.75 describes the iterative procedure to be followed.

First, a rod **grade** is selected for the string, which determines the **tensile** strength to be used. Then, an **initial** set of taper lengths is assumed for the actual pump setting depth and the API code of the string to be designed. With these data, it is now possible to find the polished rod **loads** by using the **RP 11L** calculation model. The SF_{act} valid for the top of the string is also determined.

The next problem is to find, for every taper, the S_i^* value that satisfies the design goals. Using these values, the required L_i lengths of the tapers can be obtained. These taper lengths represent an **intermediate** string design, and a check is made on their **total** length. If the total length differs from the pump setting **depth** desired, further **iterations** are required. In such cases, taper lengths are adjusted to give the required total length, and the calculations are repeated with this new assumed string design.

After having reached the necessary total rod string length, a final **check** is made on the value of SF_{act}. The reason for this is that the design procedure sets no **limit** to service factor values; it only tries to set them **equal** in the tapers. In cases where the required SF is lower than the actual SF_{act}, a stronger rod material has to be selected. Then the whole design procedure is repeated with this new rod grade.

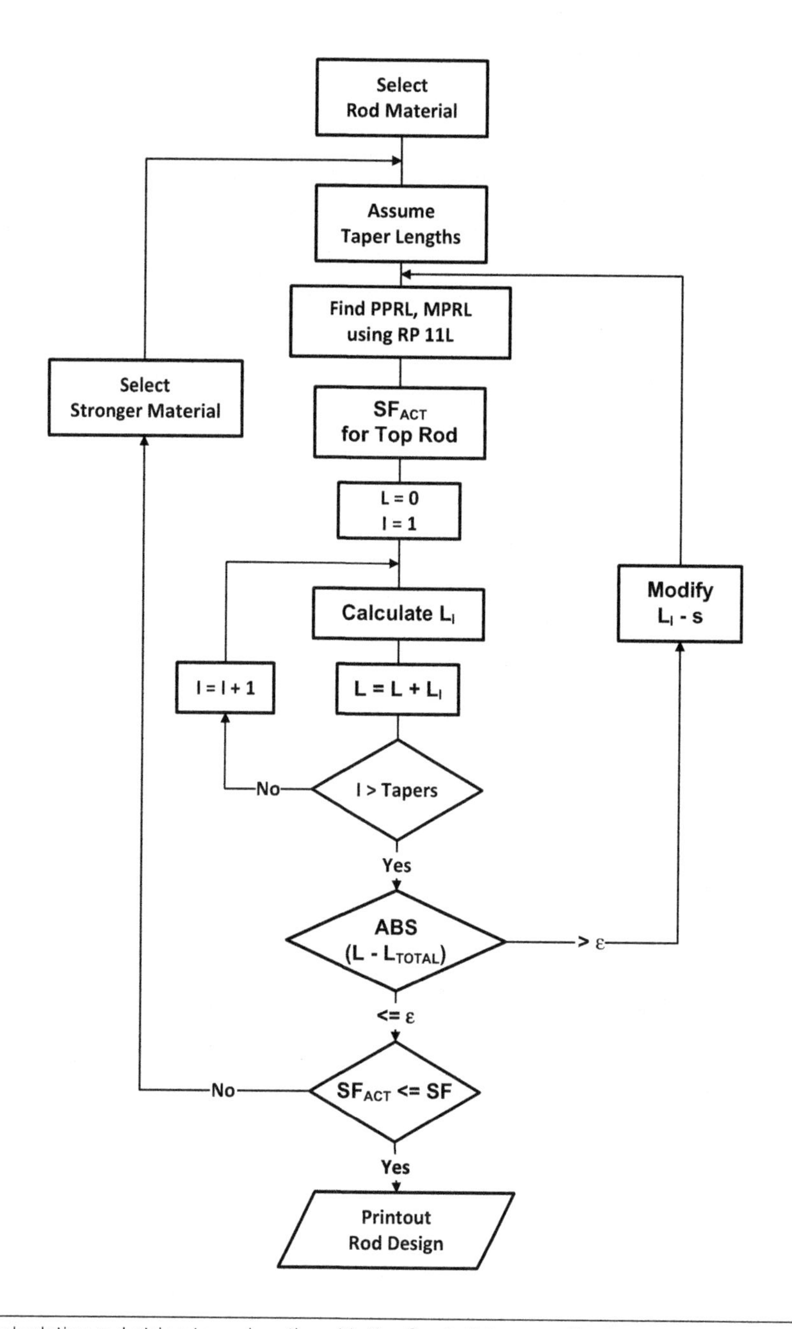

FIGURE 3.75

Flowchart of calculating rod string taper lengths with the Gault–Takacs method.

Comparison of the equal service factor method and the API taper percentages has shown [93] that the API percentages, for most operating conditions, give the top rods in the string a **larger** fraction than good design practice would dictate. This results in a **poorer** service factor and makes the top rods the most likely to **fail**. In addition, a greater percentage of the larger rods in the string adversely affects the peak polished rod load and increases the energy requirements. All these problems are **eliminated** if the equal service factor method is used to design the rod string.

EXAMPLE 3.14: USING THE DATA OF THE PREVIOUS EXAMPLES, DESIGN A ROD STRING WITH THE GAULT–TAKACS METHOD

Solution

Exhibit 3.6 presents detailed calculation results.

R O D   S T R I N G   D E S I G N   R E S U L T S

Rod String Weight: 13674 [lbs]   Avg. Rod Weight: 2.27 [lb/ft]

Size	Taper Lengths		Rod Stresses			SF
			Min.	Max.	Allow.	
[in]	[%]	[ft]	[psi]	[psi]	[psi]	[-]
1	33.1	1987	12230	28981	32067	0.8134
7/8	37.3	2239	9192	27559	30528	0.8125
3/4	29.6	1774	4536	25379	28171	0.8108

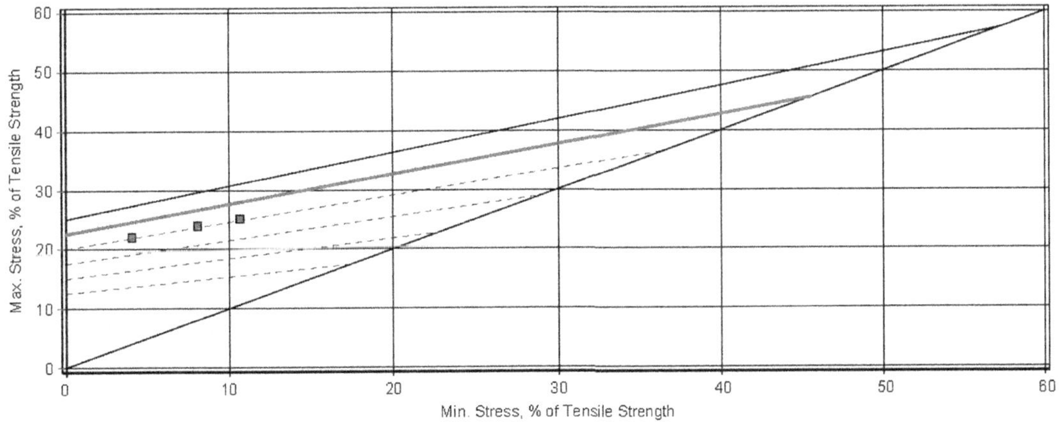

EXHIBIT 3.6

Sucker-rod string design results for Example 3.14.

3.5.5.5.5 The RodStar model

The commercial computer program package **RodStar** [94] includes the design of the rod string while performing a predictive analysis of the rod-pumping system. The design procedure is based on predicted rod loads that are calculated from the solution of the damped wave equation, so the design does not rely on approximate calculations like the procedures discussed so far. Taper lengths are selected using an iterative procedure so that their loading is identical at the top of each taper; loading is defined as follows:

$$Loading = \frac{S_{max1} - S_{min1}}{S_{a1} - S_{min1}} = \frac{S_{max2} - S_{min2}}{S_{a2} - S_{min2}} = \frac{S_{max3} - S_{min3}}{S_{a3} - S_{min3}} = \ldots = C \quad (3.46)$$

where:

$S_{max\ i}$, $S_{min\ i}$ = maximum and minimum rod stresses in the ith taper, psi,
$S_{a\ i}$ = allowed rod stress in the ith taper, found from Eq. (3.21), psi.

3.5.5.6 Comparison of design methods

3.5.5.6.1 Main features

The main features of available rod string design procedures are collected in Table 3.17, which contains information on the different ways authors estimate rod loads and also presents each model's basic design goal.

Investigation of the design goals of the different models reveals huge differences in basic principles. The **Bethlehem** model designs strings with the **same** maximum stresses at the top of each taper. This means that the $S_{min} - S_{max}$ points plotted on the modified Goodman diagram for the different taper sections will fall on a horizontal line that must inevitably cross several SF = const. lines. Lower tapers (with lower minimum stresses) have a higher service factor and a consequently reduced safety

Table 3.17 Basic Features of Available Sucker-Rod String Design Methods

Model	Year	Min. Load	Max. Load	Dynamic Loads	Design Goal
Bethlehem	1953	–	Fluid load plus rod weight in air	–	Equal max. stresses
West	1973	Rod weight in air	Fluid load plus rod weight in air plus dynamic loads	Mills acceleration factor	SF = const.
Neely	1976	Buoyant rod weight	Fluid load plus buoyant rod weight plus dynamic loads	Special formula	Equal modified stresses
Gault–Takacs	1990	Buoyant rod weight	Fluid load plus buoyant rod weight plus dynamic loads	From RP 11L	SF = const.

than the tapers higher up the string. The fatigue loading on lower tapers, therefore, is higher and these sections are more likely to experience premature failure.

Setting the service factors (*SF*s) equal in each taper, as done by **West** and **Gault and Takacs**, ensures the same amount of **safety** for every taper section. Rod strings designed this way have the same safety factor included in every taper and do not have any weak points.

Finally, the **Neely** design sets the modified stresses equal in each taper, meaning that $S_{min} - S_{max}$ points belonging to the different tapers, when plotted on the MGD, will lie on a **parallel** to the $SF = 1$ line. This line, however, **inevitably** crosses the lines corresponding to any service factor other than 1.0. This is because the modified stress line has a **tangent** of 0.5625, but the others are less steep, having tangents equal to 0.5625 *SF*. Therefore, the design generates different safety factors for every taper; upper tapers are relatively more loaded than lower ones. This situation is just the opposite of early design methods where usually the lower tapers were underdesigned.

3.5.5.6.2 Sample cases

String designs for the case given in Example 3.11 were performed using the procedures discussed so far; calculated strings are contained in Table 3.18, including the **Bethlehem** and the **RodStar** designs.

Table 3.18 Sample Rod String Designs for Example 3.11

Model	3/4 in taper	7/8 in taper	1 in taper
Bethlehem	2,364 ft	1,853 ft	1,783 ft
West	1,469 ft	2,400 ft	2,131 ft
Neely	1,543 ft	2,254 ft	2,203 ft
Gault–Takacs	1,774 ft	2,239 ft	1,987 ft
RODSTAR	1,825 ft	2,200 ft	1,975 ft

Comparison of the four conventional rod string designs is done on the modified Goodman diagram in Fig. 3.76. The stresses plotted are calculated according to each model's basic assumptions; for each taper and each design procedure the points $S_{min} - S_{max}$ are plotted on the nondimensional form of the MGD.

As seen, all four design methods produce strings that can safely handle the estimated well loads, as indicated by the fact that all maximum stresses are below the fatigue endurance limits of the rod material for $SF = 0.9$. Their behavior in relation to the fatigue loading of the individual tapers, however, is different. Two design models (**West** and **Gault–Takacs**) have their three points belonging to the three tapers located on $SF =$ const. lines; this is in accordance with their design principles. These strings are, therefore, uniformly loaded and have tapers designed with the same safety against fatigue failure.

The strings designed by **Neely's** and the **Bethlehem** procedure are not uniformly loaded because the lines connecting the three points corresponding to the three tapers are clearly seen to intersect $SF =$ const. lines. These methods, therefore, result in having different service factors in the design of each taper. Strings designed using the **Neely** method will fail in the top taper, whereas those designed with the **Bethlehem** model fail in the bottom taper.

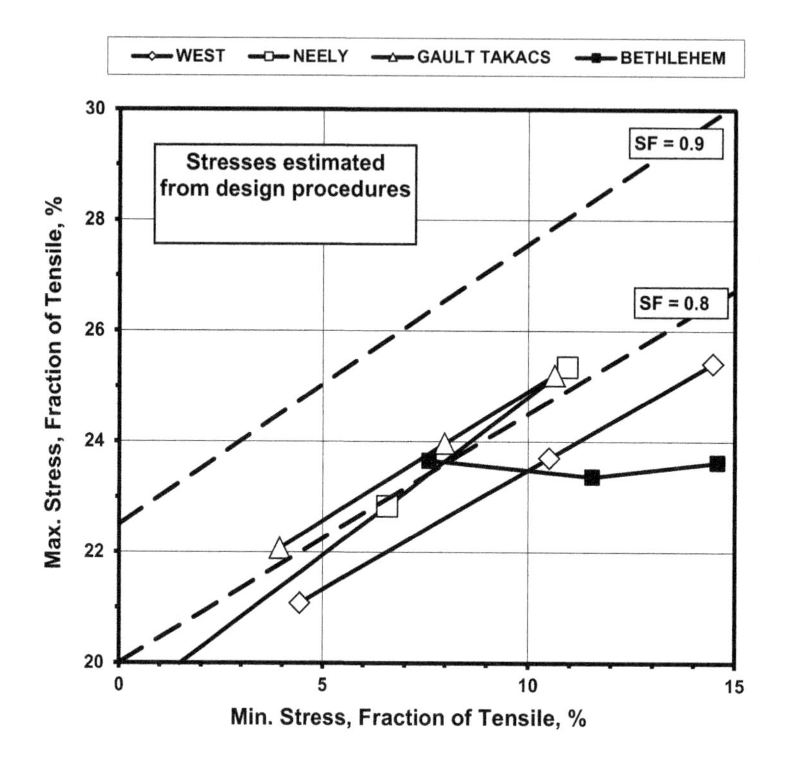

FIGURE 3.76

Comparison of sucker-rod string designs using estimated stresses.

The loads plotted in Fig. 3.76 can be considered as estimates only because of the approximate formulas used by each design method. In order to properly compare the features of the different rod string designs, one would have to actually measure the loads occurring in the different strings just designed. Since this would be close to impossible to do, this is, of course, not a viable approach. The best possible solution to the problem involves the calculation of "true" rod loads from the solution of the damped wave equation written on the rod string, as detailed in a later chapter of this book.

Figure 3.77 contains a comparison of the four design procedures based on the **predicted** rod stresses and plotted on the modified Goodman diagram. Comparison of Figs 3.76 and 3.77 makes it clear that a meaningful evaluation of the rod string design procedures must not be based on each model's calculation procedure because design loads are very different from predicted ones. The use of predicted loads received from the solution of the damped wave equation, on the other hand, ensures the highest possible approximation of measured loads and can thus give a reliable comparison of individual designs. A detailed study [95] based on predicted rod stresses proved that, except for the **Bethlehem** method, other design procedures result in taper percentages that **approximate** the ideal behavior, i.e., provide almost the same safety in each taper.

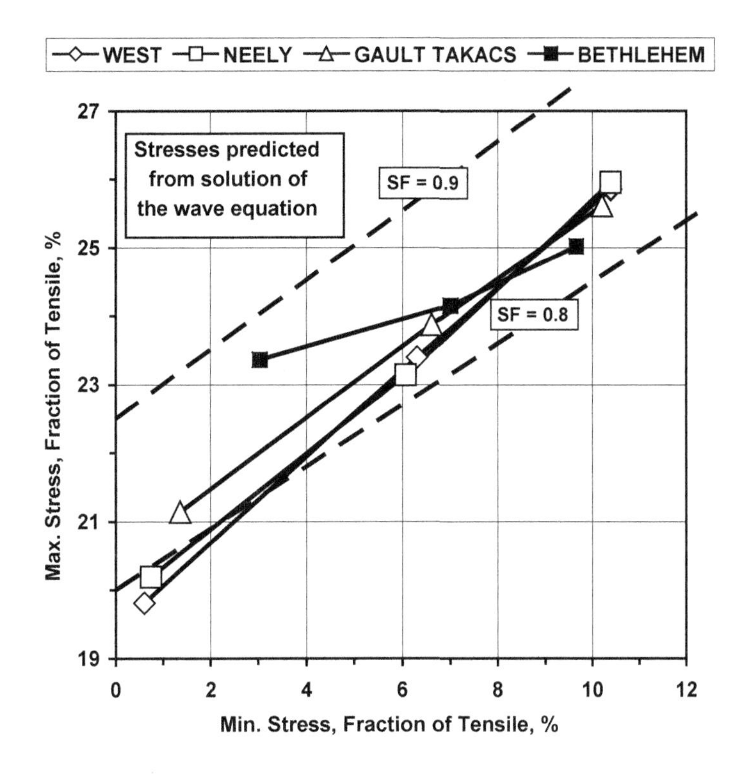

FIGURE 3.77

Comparison of sucker-rod string designs using predicted stresses.

3.5.5.6.3 Evaluation of the RodStar model

As already discussed, from the design models available, only the RodStar model uses realistic loads found from the solution of the wave equation for the calculation of the taper percentages. In order to investigate the model's virtues, let us substitute the allowed stress $S_{a\,i}$ from Eq. (3.21) into Eq. (3.46), which represents the goal of the design, and express the maximum stress in the ith taper section:

$$S_{\max\,i} = C\,SF\frac{T_a}{4} + S_{\min\,i}(C\,SF\,0.5625 - C + 1) \tag{3.47}$$

where:

$S_{\max\,i}$, $S_{\min\,i}$ = maximum and minimum rod stresses in the ith taper, psi,
C = uniform loading of the tapers, –,
SF = service factor, –,
T_a = minimum tensile strength of the rod material, psi.

In the formula C denotes the identical loading of the tapers and can attain values between 1 (fully loaded) and 0 (no load). In the case of $C = 1$ (fully loaded tapers) Eq. (3.47) is identical to Eq. (3.21)

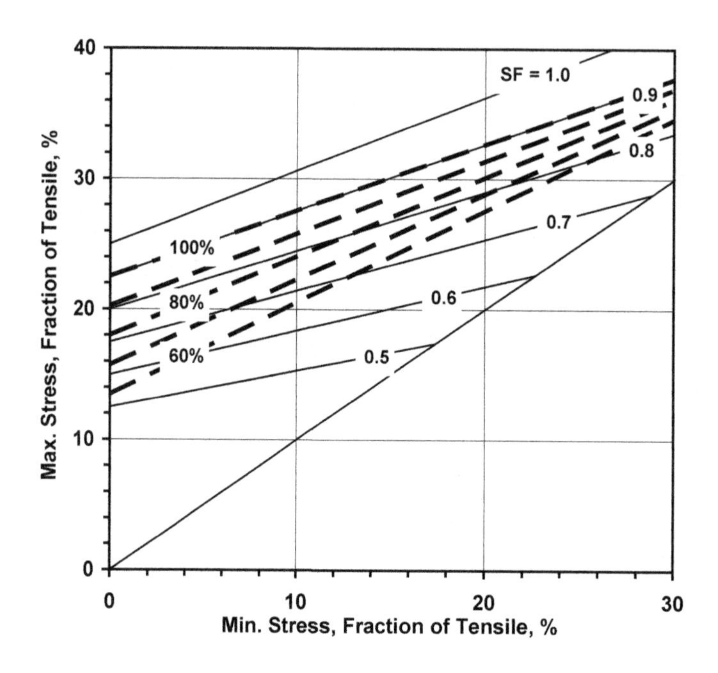

FIGURE 3.78

Illustration of the design principle used in the computer package RodStar.

and the string design is ideal because the safety against fatigue failure is the same in each rod taper. This situation, however, changes if the tapers are not fully loaded when the results deviate from the ideal case. This is clearly seen on the MGD presented in Fig. 3.78, where $C = \text{const.}$ lines calculated from Eq. (3.47) (shown in dashed lines) are plotted along with $SF = \text{const.}$ lines for a case when $SF = 0.9$ is used during the design process. As shown, the RodStar model gives a perfect design only if the string is fully (100%) loaded, but at any other loading the points belonging to the stresses in the different tapers will fall on lines intersecting the $SF = \text{const.}$ lines. In conclusion, the string design based on equal loadings does not meet the criteria for an ideal design because the safety against fatigue failure is different for each taper section.

3.5.5.6.4 Conclusions

Comparison of the available sucker-rod string design procedures resulted in the following general conclusions.

- The proper assessment of rod string designs, in lieu of measurements, must be based on predicted stresses calculated from the solution of the damped wave equation because loads and stresses estimated by the different procedures are not a true measure of the actual conditions.
- Optimum rod string design must provide the **same** safety against fatigue failures in each rod taper section; this is ensured by having the same service factor (*SF*) at the top of each taper.

- The early (**Bethlehem**) design principle of setting equal the maximum stresses at the top of each taper produces rod tapers with different safety; bottom tapers are usually more loaded than upper ones.
- The taper percentages proposed in **API RP 11L** can be misleading because actual well conditions are not properly taken into account.
- The **West**, **Neely**, and **Gault–Takacs** design methods behave properly and provide equal safety in each taper in most of the ranges.
- The design principle utilized in the RodStar program package does not meet the criteria of an ideal string design.

3.5.6 ROD STRING FAILURES

3.5.6.1 The nature of failures

All rod string failures can be classified into **two** easily recognizable groups: **tensile** and **fatigue** failures.

Tensile failures are **rare** and are caused by the **overstressing** of rods. When a great pull force is applied to the string (e.g., when trying to **unseat** a **stuck** pump), rod stresses can exceed the material's **tensile** strength, causing a tensile break. Usually, the great majority of such failures occur in the rod **bodies**, since these are the places along the string where the metal cross-section is the thinnest. Tensile breaks are **easily** identified by the permanent **stretch** of the rod at the point of failure and by the coarse, **granular** break faces. The rod is **necked down** (i.e., its diameter is considerably reduced) and the eventual break occurs at the middle of the necked-down area; see Fig. 3.79. The reduced cross-

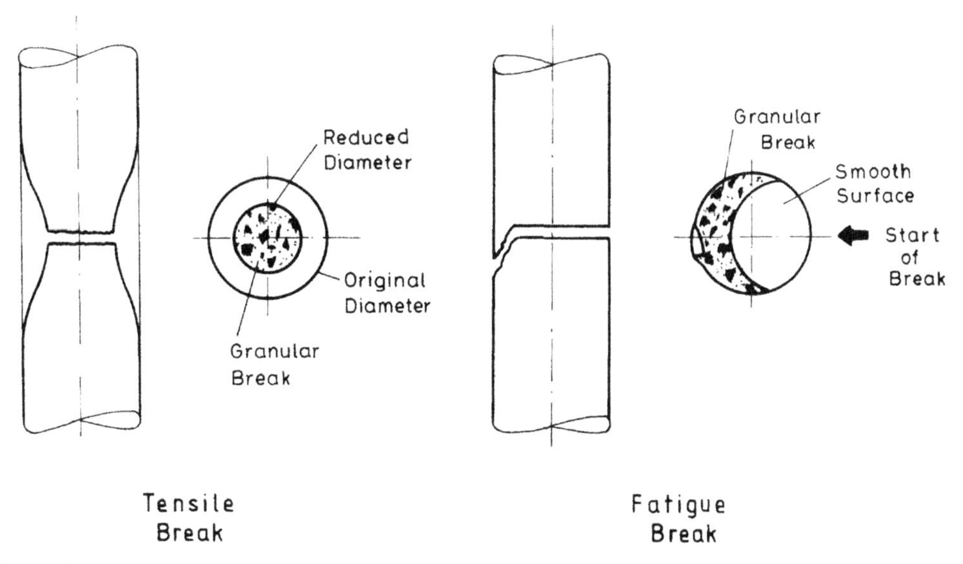

FIGURE 3.79

Typical tensile and fatigue breaks in the body of a sucker rod.

sectional area fails in a typical **tensile** break due to local overstressing with the two break faces showing uniform coarse appearance.

Tensile breaks can be **avoided** by limiting the **pull** applied to the string to about 90% of the **yield** strength of the material. Rods with an **apparent** permanent stretch should not be run, because they inevitably will fail during pumping. A **stretched** rod is **permanently** damaged, since its diameter is reduced in proportion to the amount of stretch. The reduction of the diameter gives rise to higher local stresses, which can cause the rod to part. In addition, stretched rods usually develop tiny **ruptures** on their surfaces that act as **stress raisers** and are the **starting** points of fatigue breaks. These are the reasons why stretched rods should be **discarded**. A general rule of thumb, used to recognize permanently stretched rods, is that their length should not exceed the nominal length by more than a coupling's length.

Most rod string failures are **fatigue-type breaks** that amount to 99% of the total failures. To describe the mechanism of fatigue failures, a **typical** rod break due to fatigue loading will be used in the following. As previously discussed, this type of failure is inherently associated with the **cyclic** loading of the rod string and occurs at much **lower** stresses than the yield strength of the material. It starts on the **surface** of the rod at some **stress raiser** (a nick, dent, corrosion pit, etc.), as a small **crack**. This initial crack reduces the **metal** cross-section and causes a local stress **concentration**. The increased stress induces an **overload** in the material and the crack **progresses** at right angles to the stress at an even rate across the material. The crack rate progressively increases as the metal cross-section is further reduced. The cyclic loading of the rod string causes the crack faces to periodically **separate** and rub against each other. After a great number of load reversals, the **remaining** metal area no longer can support the load and the rod **fails** in a tensile break.

A fatigue break is most easily identified by studying the **break face**. Such breaks do not involve rod **stretch** and a decrease in diameter, as seen in Fig. 3.79. The break face is divided into two **distinctive** areas: a **smooth**, polished, crescent-shaped pattern and a **coarse**, granular break surface typical for tensile breaks. The portion of the cross-section that failed due to tension exhibits a degree of final stretch that occurred just before the rod parted. The smooth surface is formed by the **polishing** action of the break faces rubbing together during the pumping cycles. The initial crack always starts at a **stress raiser** that is found on the surface of the rod, at the **middle** of the smooth portion of the break face. A close inspection of the rod surface around this area can usually reveal the **original** cause of the break.

As has been shown, the cause of fatigue **breaks** is identical to the cause of the stress raisers. **Stress raisers** are produced by mechanical damage, corrosive action, rod wear, or a bend in the rod. If none of these effects is present, the material can be expected to tolerate all rod stresses below its fatigue endurance limit, which is found from the **modified Goodman diagram**. The prime considerations for preventing rod failures, therefore, are:

- To keep rod stresses within the safe **limits** of the modified Goodman diagram.
- An equally important consideration should be the reduction of the **number** and **severity** of stress raisers by:
- Proper rod **handling** that prevents mechanical damage, and
- An effective **corrosion inhibition** program that reduces or eliminates corrosion problems.

A detailed analysis of sucker-rod failures including photographs of typical breaks is available from a leading manufacturer [96].

3.5.6.2 Rod body failures

Breaks in the rod body are almost **exclusively** fatigue failures caused by mechanical or corrosive damages, or a combination of both. Mechanical damage is classified as **surface** damage, **wear**, **flexing**, and damage occurring in **bent** rods. Corrosion problems are classified according to the type of corrosion, such as **sweet**, **sour**, **galvanic**, or **bacteria** corrosion.

The most common type of mechanical damage occurs on the surface of the rods and can take the form of nicks, cuts, or dents caused by tools or other steel parts. Rod elevators and wrenches, if not properly maintained, also can damage rods. Proper **handling** practices, as suggested in **API RP 11BR** [52], should be followed to reduce failures caused by these mechanical problems. In crooked wells, or in unanchored tubing, rod body **wear** can be excessive and can also accentuate rod failures.

The use of **bent** or **corkscrewed** sucker rods should not be allowed, especially in deep wells where higher stresses prevail. The basic rule is that a bent sucker rod is **permanently damaged** and will eventually fail. The convex area of a bent sucker rod (i.e., the outside of the bend) is permanently stretched. When run into the well and straightened by the well load, this portion of the cross-section carries no load. At the same time, the stress is concentrated in the **concave** side of the bend, causing an inevitable local **overload**. This is an ideal stress raiser that usually initiates a fatigue **crack** and leads to an ultimate rod part. Break faces are not **perpendicular** to the axis of the rod and bending fatigue failures are identified by an angle other than 90°. Sucker rods, therefore, should be **checked** for straightness before running in the well, and bent or kinked rods should be **discarded**. Under lighter loads, bent rods can also be used after a proper reduction in allowable stress is made, as shown in [97,98].

Rod **flexing** during the pumping cycle can also be a source of mechanical body failures. The rod string is supposed to move in a vertical direction only, and any **lateral** movement induces a **flexing** action that can increase the danger of fatigue failures. Flexing is most severe just above the rod **upset**, because this is the point where the lateral movement is transmitted by the rigid joint to the more elastic rod body. The high local stresses result in small fatigue **cracks** along the rod body that can lead to a final fatigue break. Fluid or gas **pound** is the most common causes of rod flexing, because both generate shock waves during the downstroke when the plunger approaches the fluid surface. By eliminating pounding conditions, shock loads are also eliminated and the flexing of the string is reduced.

Corrosion damage works on chemical or electrochemical principles and **removes** part of the metal from the surface of the rod. In lightly loaded rod strings, corrosion damage leads to an eventual **tensile break**, which is the result of the continuous reduction in rod metal area. When the remaining metal cross-section can no longer withstand the mechanical stress caused by the pumping load, the rod body fails in a simple tensile break. In deeper wells where greater rod loads are present, **stress corrosion fatigue** takes place, which is a **combined** action of corrosion and material fatigue. Stress corrosion fatigue starts with corrosion damage that leaves a corrosion **pit** on the surface of the rod body. The corrosion pit, under the high cyclic loads, acts as a stress raiser and is the **starting** point of a fatigue break.

Well fluids may contain different corrosive components and the nature of corrosive damage varies according to the environment present. The basic types of corrosion are **sweet** (CO_2), **sour** (H_2S), **galvanic,** and **bacteria**, which all result in corrosion pits of different shapes, sizes, and colors. An identification of the actual causes of corrosion damage can be done by comparing break specimens to the color photographs presented by **Moore** [99,100]. The reduction of corrosion problems requires the implementation of an efficient **inhibition** program, using the proper chemicals with the proper methods, but these procedures are outside the scope of the present book.

To identify fatigue failures in sucker-rod bodies, the parted rod sections and the break **faces** have to be examined. The cause of the failure is usually found on the spot where the break started. If a corrosion pit is found, the type of corrosion is identified from the characteristics of the pits. If mechanical damage is found, its cause can also be disclosed by a closer visual examination.

3.5.6.3 Joint failures

As already discussed in Section 3.5.2.2, the sucker-rod joint is a **friction-loaded**, **shouldered** connection between the pin and the coupling. In order to provide a firm connection, the pin shoulder face should always stay in **contact** with the coupling face. As long as this contact is maintained, the joint will not fail, in spite of the periodic loading imposed on it by the pumping movement. The proper **makeup** of the joint ensures a high initial stress in the mating parts, which should be sufficient to maintain the **tightness** in the threaded connection under the operating conditions. Most joint failures arise from **loss of tightness** and the corresponding **separation** of the pin shoulder face from the coupling. In this situation, one of the joining elements, the rod pin or the coupling, eventually will fail in fatigue. These failures usually occur with the same frequency in either the pin or the coupling.

The number of sucker-rod joint failures can effectively be reduced by the application of several **preventive** measures.

- The most important factor is to ensure that the joints are **properly** made up when they are run in the well.
- The widespread use of **rolled** threads on rod pins and couplings has substantially decreased the severity of fatigue failures by eliminating the sharp cuts at the thread roots.
- If allowed by the tubing size, **full-size** couplings should be used, since they can substantially increase the service life of the sucker-rod joint.
- All these preventive actions should be augmented by an effective **corrosion inhibition** program to reduce the effects of stress corrosion.

3.5.6.3.1 Pin failures

Practically all pin failures are **fatigue** failures, the most common cause of pin breaks being stress corrosion fatigue. All these breaks can be attributed to the **loss of tightness** in the joint due to improper **makeup**. When joint tightness is lost under the pumping load, the pin shoulder is no longer in contact with the coupling face, and the pin starts to move laterally inside the coupling. While the threaded section of the pin is held rigid by the coupling threads, the undercut section is periodically **bent** with the vibrations in the rod string. The **bending** moment increases the loading of the pin and causes a small fatigue **crack** to appear at the root of one of the threads. The small crack slowly **progresses** into the metal area, and the pin will break when the remaining metal area is insufficient to carry the well load.

The **break face** of a pin failure due to fatigue looks identical to that observed in a rod body failure. It is at right angles to the axis of the pin, and the smooth portion characterizing fatigue breaks can be identified clearly. The point of origin of the crack can be found with the same procedure as in the case of rod body breaks and is always at the midpoint of the smooth area's outside edge.

Pin breaks can occur at various places in the pin, depending on the grade of **looseness** in the joint (Fig. 3.80). Most frequently, the pin shoulder face and the coupling are still in contact during pumping, and the break occurs at the **first thread root** above the undercut. If the pin shoulder and the

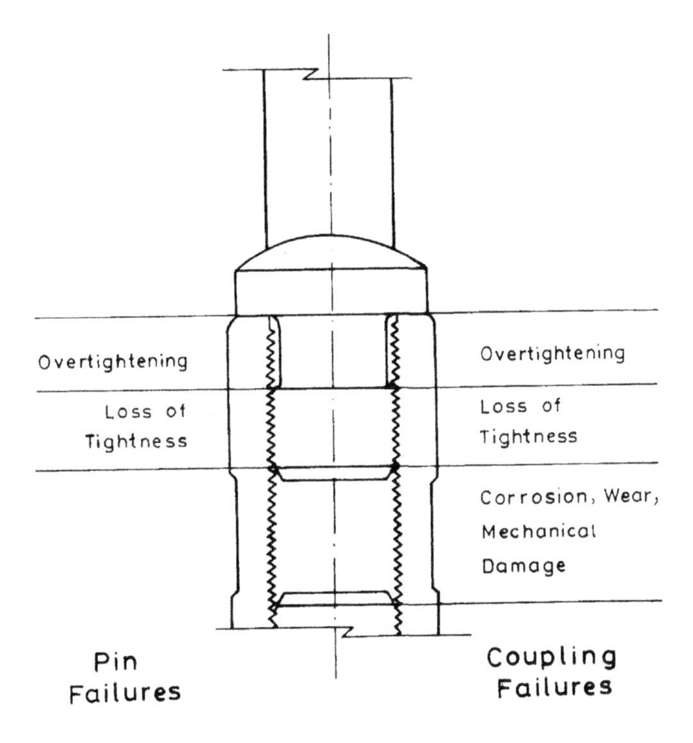

FIGURE 3.80

The distribution of pin and coupling failures in the sucker-rod joint.

coupling are further **separated** due to loss of makeup, then the pin fails at a thread root **higher** up the pin. The distance of the break from the first thread indicates the distance the shoulder and the coupling were separated at the time of failure. The pin may also fail in the undercut section due to fatigue caused by bending forces and the presence of stress raisers.

Tensile failures also are possible in the sucker-rod pin. These breaks always occur in the undercut area (see Fig. 3.80) and are characterized by a **necked-down** appearance of the pin and a coarse break face. Such failures are **rare** and are usually caused by overtightening of the joint (applying too much torque during rod makeup). They are easily identified by the reduced diameter in the undercut and the tensile-type break face situated at the middle of the elongated section.

The last kind of pin failure is thread **galling**, which occurs when the joint cross-threads during the makeup procedure, resulting in **flattened** or broken-off threads. This is a **rare** problem and is usually caused by damaged or dirty threads. Careful **cleaning** and lubrication of the threads and the following of standard makeup procedures can eliminate this problem.

3.5.6.3.2 Coupling failures

The main causes of coupling failures are the same as in the case of pin failures: fatigue, excessive tensile stress, and galling. Material **fatigue** is the exclusive cause of coupling breaks, and the failure most often starts from the inside of the coupling.

Coupling breaks are mainly attributed to the **loss of tightness** in the sucker-rod joint and the **bending** forces associated with it. The failure starts at a stress raiser and grows at right angles to the axis of the coupling, resulting in a characteristic fatigue-type break face described earlier. The place of the eventual break is usually opposite the **first full thread** on the end of the pin (see Fig. 3.80). If the pin shoulder–coupling contact is totally lost due to the loss of makeup (i.e., the shoulder face and the coupling are separated by a greater distance), the coupling breaks at a point closer to its end. Coupling failures starting from the inside due to corrosion at the threads between the pins are rare.

Fatigue failures starting from the outside usually occur in a section of the coupling between the pins and are initiated by **stress raisers** like corrosion pits, wear, or other mechanical damage. An examination of the break face can reveal whether the crack started from the inside or from the outside of the coupling. Figure 3.81 illustrates the patterns of the usual break faces in the two possible cases. As seen, the smooth pattern produced by the rubbing action of the crack diminishes into two sharp points. The fatigue crack started on the side of the coupling where these points are located. The position of the initial stress raiser is found at the **midpoint** of the polished crescent-like pattern, measured on the coupling side just determined [101]. In most cases, a simple visual inspection of this area can reveal the nature of the stress raiser: a nick, a corrosion pit, or the like. A typical mishandling example that causes surface damage on couplings is the "warming up" (hammering) to **loosen** sucker-rod joints.

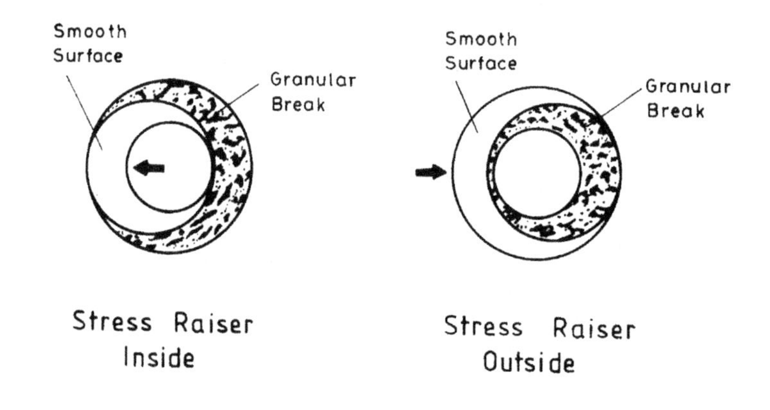

FIGURE 3.81

Typical break faces of coupling failures due to material fatigue.

Sucker-rod couplings also can be damaged by **overtightening** of the joint. During makeup, the coupling face is forced against the pin shoulder face owing to the high torque applied to the joint. Although a substantially high prestressing of the joint is an absolute necessity, if too much torque is applied the yield strength of the material can be reached and **permanent** deformation will result. The outside diameter of the coupling increases, the coupling **flares** out, and it can even **split** at the ends. This is tensile-type damage and always occurs at the two ends of the coupling (Fig. 3.80). Slimhole couplings are especially prone to this type of handling damage. Couplings found in a **flared-out** condition must be **discarded** and should not be run because they prevent the proper initial makeup and will lose their original tightness during pumping. The loss of tightness in the joint due to damaged couplings gives rise to fatigue failures in the pin and the coupling, as discussed before.

Finally, thread **galling** should be mentioned as the last and very rare kind of coupling failure. Proper cleaning, the use of the right lubricant, and proper makeup procedures can prevent galling of sucker-rod threads.

3.5.6.4 Fiberglass rod failures

Early fiberglass rods usually failed owing to manufacturing **defects**. Most rod body breaks were initiated by **knots** or **loops** in some of the glass fiber **rovings** making up the rod. Such rods showed no outside defects and their ultimate tensile strength also was not affected. But when run in the well and subjected to the cyclic loading produced by pumping forces, the small loops tended to **straighten** inside the rod body and eventually **cracked** the rod. Another source of rod failures was the **pulling** out of the rod body from the steel end connector due to insufficient bonding [71].

Improvements in fiberglass rod manufacturing processes **eliminated** most of these failure types. Today's third-generation rods have a sufficient service record to evaluate their long-term performance characteristics. Experience gained by a major user shows [102] that most failures occur either in the steel end fitting or in the rod body at the rod–connector bond. The end connector fails due to material **fatigue** of the metal at the pin or somewhere along its length. All other breaks start in the contact area of the rod body and the end connector. Manufacturing defects or overloading of the rods result in fatigue failures of the rod body inside or immediately above the connector. In addition, fiberglass rod bodies are more prone to **abrasion** than steel ones, and the use of rod **guides** is recommended. In general, fiberglass rods offer a shorter design **life** than steel ones, but they are resistant to **corrosion** and have a substantially lower **weight**. These advantages can, in some cases, offset the greater initial cost of a fiberglass rod string.

3.5.7 ANCILLARY EQUIPMENT

3.5.7.1 Polished rods

The polished rod is the top and **strongest** part of the sucker-rod string and it connects the rods to the pumping unit. Its main functions are:

- to transfer the pumping **loads** to the surface pumping unit, and
- to provide a **seal** in the installation's **stuffing box** to prevent well fluids from entering the atmosphere.

In order to fulfill its functions, the polished rod must be strong enough to carry the full load during the pumping cycle; therefore polished rods typically have diameters 1/4 in larger than the top rod section (see Table 3.19). To provide the necessary seal, the surface of the polished rod is machined to

Table 3.19 Recommended Polished Rod Sizes

Top rod size, in	PR size, in
5/8	1 1/8
3/4	1 1/8
7/8	1 1/4
1	1 1/2

close tolerances and has an extremely **smooth** surface, frequently provided by a **sprayed** metal coating. Abrasive action and corrosion by well fluids can damage the high-quality finish and can increase fluid leakage. In such cases, a roughened polished rod is repaired by installing a **liner** around the damaged rod.

Polished rods are standardized in **API Spec 11B** [49] and come in different sizes with outside diameters of 1 1/8 in, 1 1/4 in, and 1 1/2 in. Available standard lengths are between 8 ft and 36 ft; their **proper length** is calculated so as to equal the sum of the following measurements:

1. the maximum anticipated **stroke length** of the pumping unit,
2. the distance above the stuffing box to the lowest position of the **carrier bar**,
3. some extra lengths (safety distances) for accommodating (a) the polished rod **clamp** and (b) extra length below the stuffing box.

Basically, polished rods are simple solid **rods** with a constant outside diameter having **pin** (male) connections without undercuts on each end; see Fig. 3.82 (left side). Normally both pins have special polished rod (PR) pin threads **different** from the threads used on sucker rods. The corresponding polished rod coupling can be perfectly made up on the pin of the top rod, but the use of sucker rod couplings on the polished rod is **unreliable** and should be **avoided**. Figure 3.83 shows a comparison of polished rod and sucker rod pin threads; the upper one is a polished rod pin and the lower one is standard sucker-rod pin. As seen, polished rod pins do not have the undercut, stress-relieving neck.

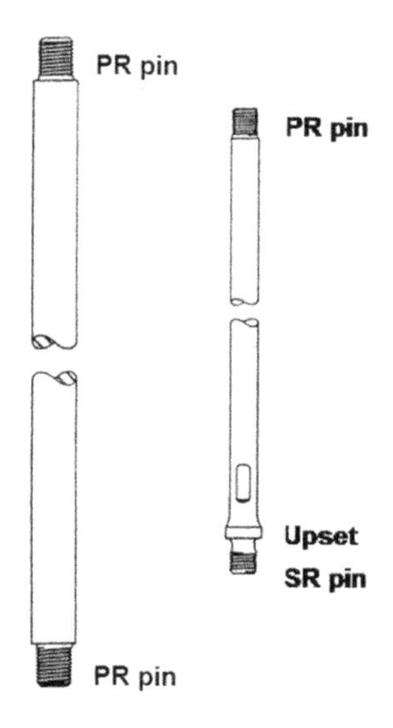

FIGURE 3.82

The two types of polished rods.

The other version of polished rods has an **upset** at one end, furnished with a conventional **undercut** sucker-rod (SR) pin; see Fig. 3.82 (right side). This allows the use of a standard rod coupling to connect the polished rod to the top sucker rod. Care should be taken to properly choose the coupling to be used.

Practically all failures in polished rods are caused by material **fatigue** due to cyclic loading, usually **bending**. Bending of the polished rod is unavoidable if the casinghead, tubing head, pumping

FIGURE 3.83

Comparison of sucker-rod and polished rod pin threads.

tee, and stuffing box are not on a **vertical**; a deviation not more than 1.5 in in a length of 20 ft is recommended. In order to maximize the working life of the polished rod, the following rules must be followed:

- The polished rod, at any position of the stroke, must be properly **aligned** with the wellhead,
- The **carrier bar** must have a flat and smooth top surface and must be level so that both sides of the **wireline hanger** carry equal loads, and
- Polished rod **clamps** must have a flat bottom and be placed on the uncoated part of the polished rod, not on the sprayed metal surface.

Polished rod **liners** provide a substitute for sprayed metal surfaces on uncoated polished rods and extend the life of those. Liners have an economic **benefit** because replacing them is less expensive than replacing the polished rod. They fit the polished rod closely and have an outside diameter 1/4 in larger than the polished rod; they are very thin, making their installation difficult. As shown in Fig. 3.84, the liner is fixed to the polished rod with screws (**A**) below the polished rod clamp and it has a packing element (**C**) to seal against well fluids.

Polished rod liners must not be used for wellhead pressures above about 500 psi because their packing could be a source of **leaking** well fluids to the surface.

3.5.7.2 Rod guides and scrapers

Rod-on-tubing friction and wear can be excessive in several cases: in **crooked** wells, in tubing or rod **buckling** situations, and in cases of fluid or gas **pounding**. To minimize the effects of wear on the

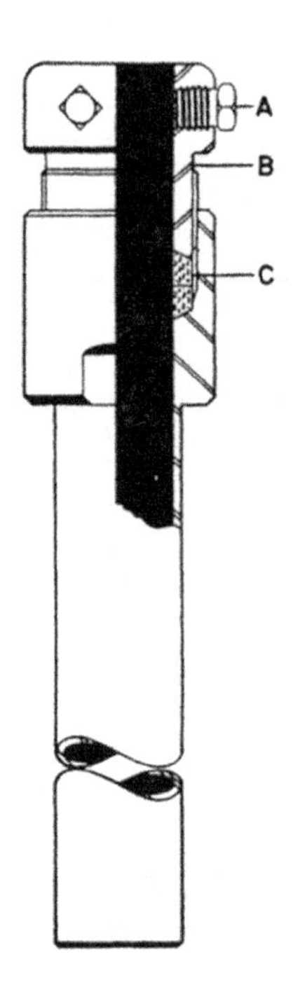

FIGURE 3.84

Construction of a polished rod liner: fixing screws (A), retainer (B), packing element (C).

metal parts and to decrease rod loads due to mechanical friction, the use of **rod guides** has been a common field practice for a long time. These guides, usually made of a **plastic** material, are attached to the rods at several depths, depending on well curvature and previous occurrences of high tubing wear. The application of rod guides can have the following general benefits:

- The number of rod and tubing failures can be drastically reduced by minimizing the **wear** of metal parts.
- The rod string in the tubing is **centralized** and the lateral (bending) forces increasing the danger of material fatigue are reduced.
- Rod guides or special scrapers remove **paraffin** depositions from the inside of the tubing.
- In composite rod strings rod guides on the **fiberglass** section protect fiberglass rods from wear on the tubing.

The protection of metal surfaces from wear depends on the volume available for **erosion** from the rod guide during its life. As shown in Fig. 3.85, the erodible wear volume (EWV) is the **amount** of material outside of the OD of the rod coupling, because as soon as this volume is eroded, the coupling will be in contact with the tubing inside wall, causing detrimental metal-on-metal wear. It must be clear that normally only one side of the guide is in contact with the tubing, causing a limited period of protection. **Rotating** the rod string, however, evenly **distributes** the wear on rod guides and substantially increases their operating life. The application of rod guides with rod rotators is a good solution for paraffin scraping and removal, too.

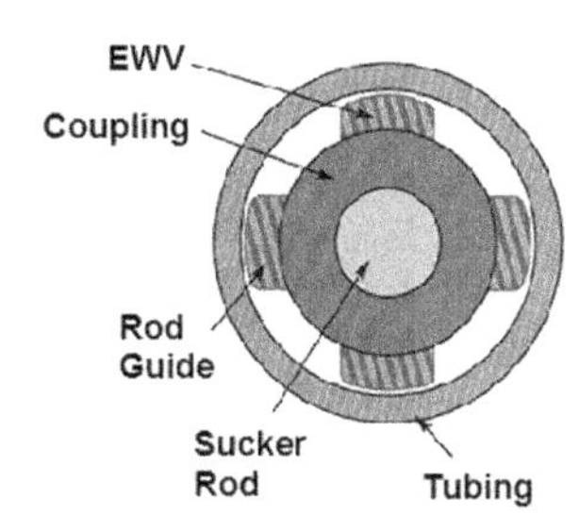

FIGURE 3.85

Definition of erodible wear volume (EWV) for sucker-rod guides.

In early practice, metal rod guides were welded to the rods, or short **subs** with molded-on rubber or plastic shapes were used between the rods. Recently most types of rod guides are **injection-molded** on the properly cleaned surface of the rods under factory conditions using some kind of polymer material, including nylon. The proper selection of guide materials is important and compatibility with well fluids must be checked [103].

The shape of factory-mounted, **molded-on** rod guides can vary; Fig. 3.86 shows a slanted vane design on the left that is an excellent paraffin control device. The guide on the right has straight vanes and tapered ends and is used to reduce fluid **turbulence** above and below the guide position. This is advantageous in cases where turbulence may remove the corrosion inhibitor film from the rod, causing increased corrosion damage around the rod guides.

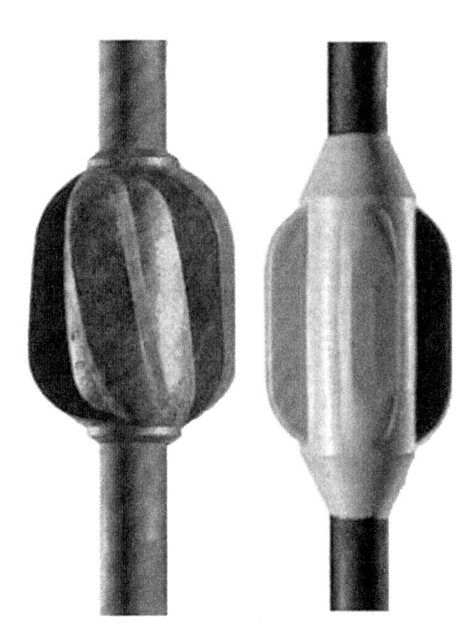

FIGURE 3.86

Molded-on rod guides.

Field-installable rod guides are usually of the **snap-on** type (Fig. 3.87) that are held in place by **friction** only and must tolerate high axial loads without slipping. They eliminate the costs related to factory mounting and shorten the time of returning to normal operations. Mostly used in low-volume, low-temperature wells with minor wear problems, snap-on guides offer an economic way to extend the service life of well equipment.

FIGURE 3.87

Snap-on type rod guide.

Among the materials available for rod guides are low-cost special **rubbers** that offer low friction; these can be used when sand abrasion is a problem. Higher temperatures require **nylon** to be used, and the ultimate material for high-temperature and corrosive service is glass-reinforced polyphenylene sulfide [104].

Finally, **wheeled guides** (Fig. 3.88) have several wheels placed vertically in a special coupling. The wheels, set at 45° to each other, **roll** on the inside surface of the tubing and provide a **centralized**, low-friction movement of the rod string.

3.5.7.3 Rod and tubing rotators

Wear on the tubing and rod strings can be made more **even** and the operating life of the tubing and rod strings can be extended if either a rod or a tubing **rotator** is used. **Rod rotators** are installed between the carrier bar and the polished rod clamp and provide the necessary torque to rotate the rod string. They have an actuator lever, the end of which, via an attached cable, is fixed to a stationary part of the pumping unit; see Fig. 3.89. As the carrier bar moves up and down, the actuator lever's movement is transmitted through a worm drive to turn the cover cap, on top of which the polished rod clamp rests.

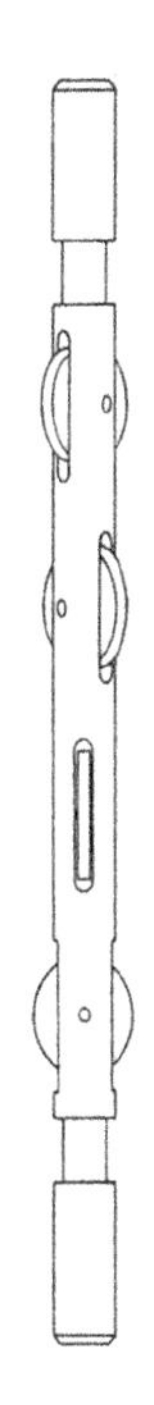

FIGURE 3.88

Wheeled rod guide.

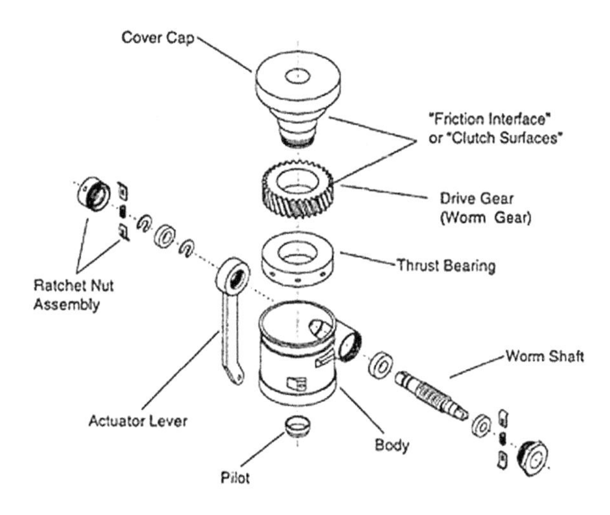

FIGURE 3.89

Main parts of a rod rotator.

Depending on the frictional forces between the carrier bar and the cover cap and between the drive gear and the cover cap, the rod string is **rotated** a small fraction of a full turn for every **stroke**. Rod rotation can happen during the up- or the downstroke, depending on the model. As experimentally verified [105], the typical torque developed by rod rotators never reaches the critical allowed torque on steel rods but, at the same time, may damage fiberglass rods.

Tubing rotators work on similar principles as rod rotators; the slow rotation they impart on the tubing string brings about similar advantages: **even** distribution of wear around the internal circumference of the tubing string. There are manual, mechanical, or electrical models available.

3.5.7.4 Sinker bars

Sinker bars are heavy solid steel **rods**, usually with standard rod **pins** on both ends, and are run as part of the rod string, usually located immediately **above** the pump. The running of sinker bars at the bottom of the rod string serves several purposes:

- To prevent **compression** and to reduce or completely eliminate **buckling** in lower rod sections during the downstroke.
- To help rods fall on the downstroke, especially if a viscous fluid is pumped.
- To keep fiberglass rods in tension.
- To increase the **overtravel** of the pump plunger, thereby increasing the pumping rate.

As discussed in Section 3.5.5.2.2, **buckling** of sucker rods should be avoided because of its many undesirable effects like rod and tubing wear, rod failures, etc. The root cause of rod buckling lies in the **slenderness** of standard sucker rods so it is understandable that the use of heavier rods, i.e., sinker bars with less tendency to buckling is advantageous. Since rod compression during the downstroke typically affects the bottom taper, sinker bars are used at the most dangerous place in the string: immediately above the pump. Their relatively high weight keeps the rods in tension, while they offer a greater resistance to buckling because of their lower **slenderness** ratio.

Adding sinker bars to the bottom of the rod string can increase the **overtravel** of the pump plunger during the downstroke due to the inertial effect of the **concentrated** mass situated above the plunger. This results in a longer plunger stroke length and increased liquid production of the pump.

In case of **composite** (built from steel and fiberglass sections) rod strings, the use of sinker bars is an absolute **necessity** because of the low weight of fiberglass rods and their high vulnerability to **buckling**. With the use of sinker bars, rod compression in fiberglass rods is eliminated, and the string is under tension during the whole pumping cycle. By running the sinker bars immediately above the pump, rod weight is **concentrated** at this point. This concentrated weight, combined with the high **elasticity** of the fiberglass section, can greatly increase plunger **overtravel** and the production capacity of the sucker-rod pump.

Specifications for sinker bars are covered in **API Spec. 11B** [49]; they have **pin** connections as well as wrench flats on both ends, as shown in Fig. 3.90. Elevator and fishing **necks** situated at only one end are **optional** and facilitate normal and emergency handling of the sinker bars. When calculating mechanical stresses in the sinker bar the **elevator neck** diameter must be used to calculate the cross-sectional area of the bar.

API materials for sinker bars are designated as **Grade 1** (carbon steel with a minimum tensile strength of 65,000 psi) and **Grade 2** (alloy steel, 90,000 psi tensile strength). The main physical parameters of sinker bars offered by a major manufacturer are given in Table 3.20. The sizing of sinker bar sections to avoid rod buckling was covered in Section 3.5.5.2.3.

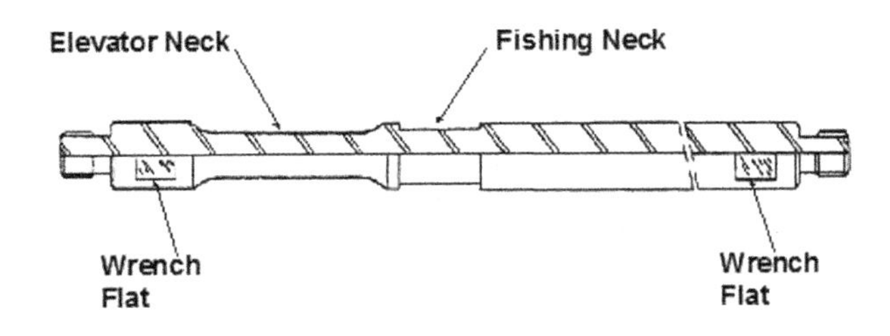

FIGURE 3.90

Construction of sinker bars.

Table 3.20 Main Physical Parameters of Sinker Bars Offered by a Major Manufacturer

SB size, in	Weight in Air, lb/ft	Metal Area, sq in	Corrected Area. sq in	Neck Diameter, in	Rod Length, ft	API Pin Size, in	Min. Tubing, in
1 1/4	4.172	1.227	1.231	0.875	25	5/8	2 3/8
1 1/4	4.172	1.227	1.231	0.875	25	3/4	2 3/8
1 3/8	5.049	1.485	1.489	1.000	25	5/8	2 3/8
1 3/8	5.049	1.485	1.489	1.000	25	3/4	2 3/8
1 1/2	6.008	1.767	1.772	1.000	25	3/4	2 3/8
1 1/2	6.008	1.767	1.772	1.000	25	7/8	2 3/8
1 5/8	7.051	2.074	2.080	1.000	25	7/8	2 7/8
1 3/4	8.178	2.405	2.413	1.000	25	7/8	2 7/8
2	10.680	3.142	3.151	1.000	25	1	2 7/8

3.5.7.5 On-and-off tools

The purpose of the **on-and-off tool** is to provide a means of **connecting** and **disconnecting** the rod string from the downhole pump. This feature can be advantageous in several cases:

- When an **oversize** pump is run below a smaller size of tubing,
- When parted rods are **fished** and the pump need not be pulled, and
- To eliminate **stripping** jobs, i.e., pulling the tubing and rod strings at the same time.

The tool (see Fig. 3.91) consists of two parts; the upper one (**A**) is connected to the end of the rod string and the lower one (**E**) is fixed to the pump. To connect the rod string to the pump already in place, the rods are run in the well with the **on-and-off tool** at the bottom. After the tool is engaged and lock **E** contacts follower **D**, the weight of the rods compresses spring **B** inside the tool body **C**, and the tool is locked in position by **rotating** the string by 90°. During normal pumping the same spring keeps the tool in the engaged position. Disconnecting the rod string from the plunger is also simple by rotating and picking up the rods. The use of this tool allows running or retrieving of the rod string independently from the tubing string.

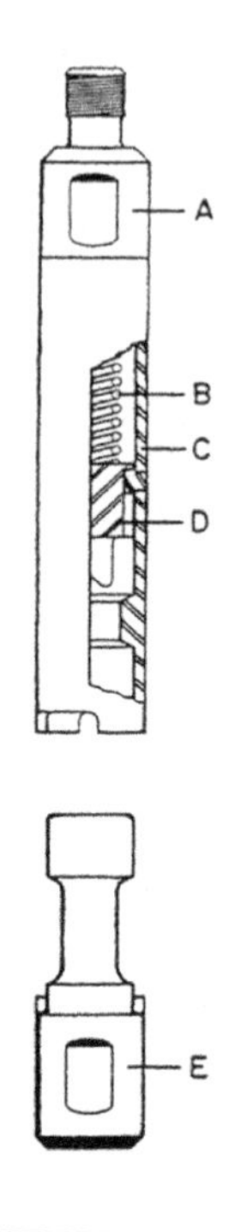

FIGURE 3.91

Construction of an on-and-off tool.

The main use of the **on-and-off tool** is with **oversize** tubing pumps that have plungers of a greater size than the inside diameter of the tubing string above the pump. These pumps are required in cases when the plunger size that fits the existing tubing size **restricts** the attainable liquid rate and use of a bigger size pump is desired. The disadvantage of oversize pumps is that inspection or change of any pump part is only possible if the rod and the tubing string are pulled **simultaneously** when the use of a **tubing drain** is recommended to avoid pulling a wet tubing string. These problems are eliminated with the use of the **on-and-off tool** that allows the pulling of the rod string separately from the tubing string.

The **on-and-off tool** is used to advantage also when the plunger of a tubing pump or the hold-down of a rod pump becomes **stuck**, usually because of scale or sand deposition. In those cases the rods cannot be pulled and a stripping job, i.e., simultaneous pulling of the rod and the tubing strings, is needed. With an **on-and-off tool** installed above the plunger, however, the rod string can be unlatched from the tool and retrieved easily.

3.5.7.6 Tubing drains

Tubing drains allow the fluid to be drained from the tubing string to avoid **wet trips**, i.e., pulling the tubing full with well fluid. They are primarily used in **oversize** tubing pump installations when the rod string is detached from the pump with the help of an **on-and-off tool** and the tubing is pulled separately.

The **tubing drain** contains a **sliding sleeve** that, when operated, uncovers large **ports** in the tubing wall and lets the fluid flow into the annulus. The drain, having a sufficiently large inside diameter that allows undisturbed movement of the rod string during normal operations, is run in the tubing string above the pump and below the **on-and-off tool**. Before pulling the tubing from the well, the rod string is **lifted** until the plunger contacts the tubing drain and lifts the sliding sleeve that, in turn, opens the ports on the tubing wall. When the fluid has drained from the tubing the rod string is lowered, the **on-and-off tool** is disengaged, and the rods can be pulled. The tubing, being empty of the rods and the liquid, can now be pulled.

3.6 WELLHEAD EQUIPMENT

3.6.1 INTRODUCTION

The wellhead arrangement of a typical sucker rod-pumped well is shown in Fig. 3.92. The **polished rod**, the uppermost part of the rod string, reciprocates with the movement of the walking beam, transmitted to the rods through the **carrier bar** by the wireline hanger. The polished rod moves inside the **tubing head**, on top of which a **pumping tee** is installed, which leads the fluids produced by the pump into the

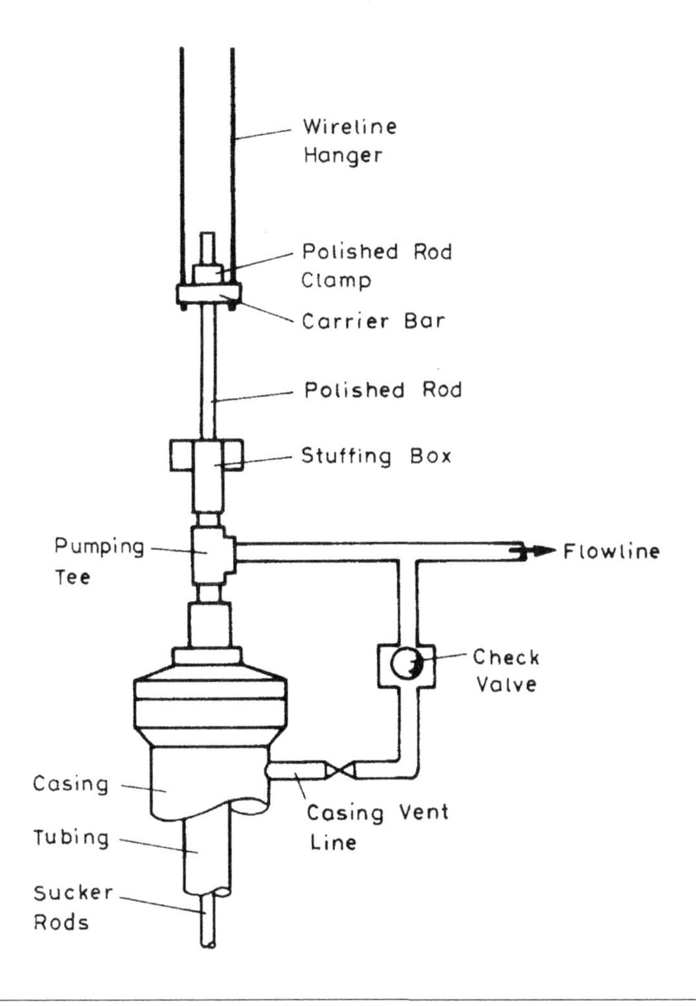

FIGURE 3.92

Wellhead arrangement of a typical sucker rod-pumped well.

flowline. Usually, the flowline and the casing vent line are connected with a short pipe section, enabling the gas that separates in the casing–tubing annulus to be led to the flowline. A **check valve** is installed on this line to prevent the fluids already produced from flowing back to the well. Above the pumping tee, a **stuffing box** is installed to eliminate leaking of well fluids into the atmosphere.

3.6.2 POLISHED ROD CLAMPS

The function of a polished rod clamp is to allow the carrier bar to **lift** the polished rod. The clamp is secured at the top of the polished rod, away from the spray metal coating. Two basic constructions exist, both being a **hinged** two-part construction with a maximum of three bolts to fix the clamp on the polished rod:

- **Friction** types have inside diameters **identical** to the outside diameter of the polished rod and rely on **friction** forces to stay in place, and
- **Indention** types have **smaller** inside diameters than the OD of the polished rod and contact the polished rod at **four points** along its perimeter; see Figs 3.93 and 3.94.

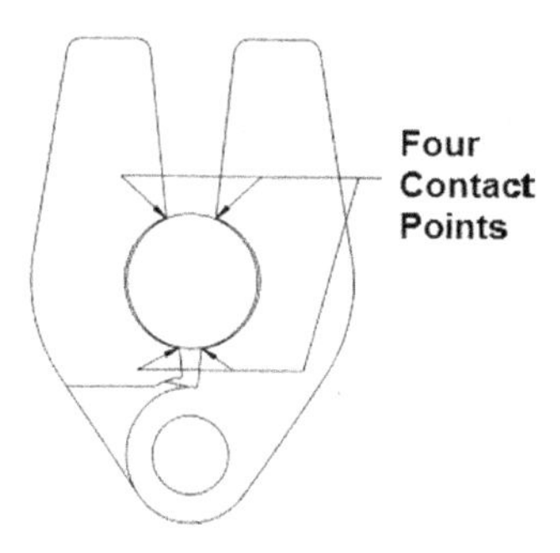

FIGURE 3.93

Operation of an indention-type polished rod clamp.

FIGURE 3.94

Illustration of a polished rod clamp.

Both types of clamps induce stress **concentrations** and thus reduce the **fatigue life** of the polished rod; this is the reason why most polished rod breaks occur at the **bottom** of the clamp. An extensive experimental study [106] concluded that despite the smaller contact area, indention-type polished rod clamps have the following **advantages** over the friction-type clamps:

- They cause **lower** stress concentrations on the polished rod.
- Even if their bolts are overtightened, the stresses in the polished rod do not increase.
- With the same torque applied to their bolts, they can support **higher** polished rod loads than friction-type clamps.

The use of indention-type clamps, therefore, is recommended and ensures a longer fatigue life of the polished rod.

3.6.3 POLISHED ROD STUFFING BOXES

The **stuffing box** is installed just above the **pumping tee**, its purpose being to prevent the **leakage** of well fluids into the atmosphere around the polished rod. A common type of stuffing box is illustrated in Fig. 3.95. Its operation is simple: by means of turning the handle on the cap, the

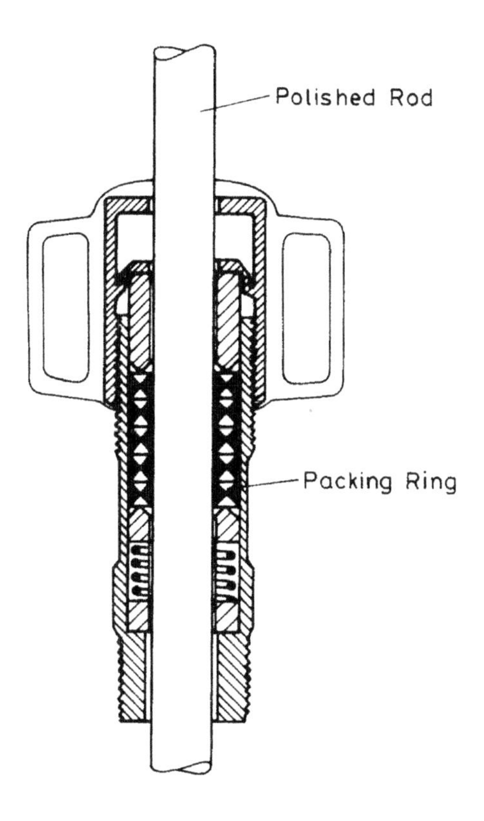

FIGURE 3.95

A common type of stuffing box.

resilient packing **rings** are compressed and squeezed against the polished rod. The packing rings are usually made of **rubber** or Teflon to offer a low friction while providing the required sealing action. It is important to periodically **adjust** the tightness of the packing rings to prevent leakage. At the same time, it is equally important not to **overtighten** them, in order to minimize the friction forces that arise in the polished rod. Too-high friction can also be caused by the drying out of the packing elements.

Normally, oil produced in the well stream **lubricates** the sealing surfaces, but intermittent pumping or a heading fluid production can result in the **drying** out of the packing. A dry packing can easily **burn**, and the stuffing box will **leak** profusely. A special **lubricator** with an oil reservoir (mounted above the stuffing box) provides a continuous lubrication on the polished rod in such situations.

Recent stuffing boxes use, instead of the compression-type packings described previously, **split-cone** packing rings made of various plastic materials [107]. The split construction allows **installation** or **removal** of the packing with the polished rod in place. The alignment of the splits in the cones should be staggered so as to prevent leakage of well fluids. Cones are usually inserted in the housing of the stuffing box pointing **upward**, and tightening the bolts squeezes them against the polished rod for an effective seal. A longer-lasting and more uniform seal is achieved by **inverted cone** packing rings that do not need excessive tightening and are inserted pointing **downward**, as shown in Fig. 3.96. Inverted cone packings have the advantage of being activated by fluid **pressure** and they automatically adjust the sealing force with changing flowline pressures.

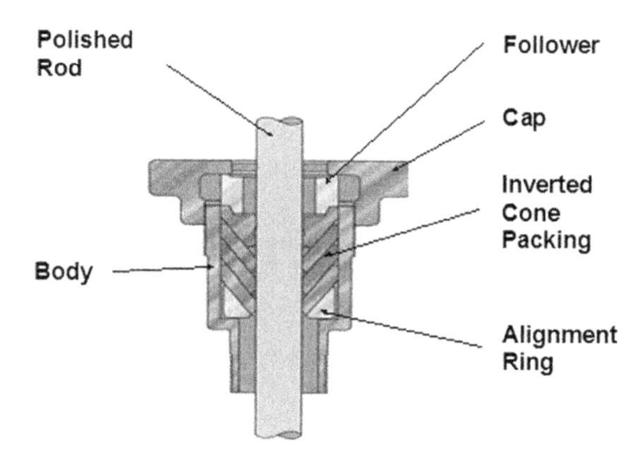

FIGURE 3.96

Stuffing box with inverted cone packings.

The stuffing box in Fig. 3.97 allows the change of the primary packing rings (inverted cones) under pressure. To do this the compression bolts are tightened, thereby forcing the lower packings against the polished rod to provide a positive seal against the wellhead pressure when the upper packing rings can safely be changed.

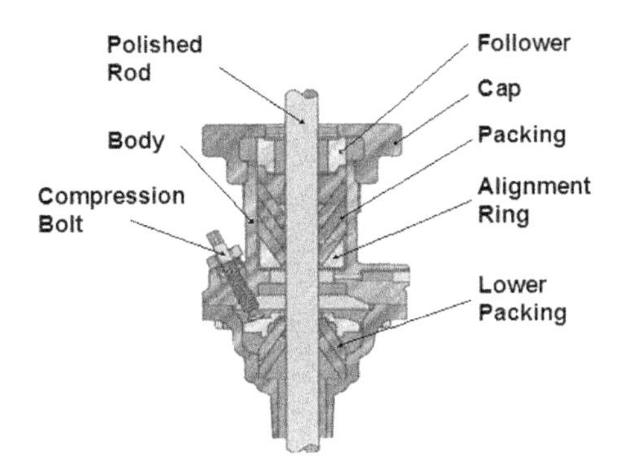

FIGURE 3.97

Stuffing box for greater wellhead pressures.

3.7 PUMPING UNITS

3.7.1 INTRODUCTION

The pumping unit is the mechanism that converts the rotary motion of the prime mover into the reciprocating vertical movement required at the polished rod. Most types of pumping units utilize a **walking beam**, inherited from the days of the **cable-tool** drilling rigs. Beam-type sucker rod pumping units are basically a four-bar **mechanical linkage**, the main elements of which are:

1. The **crank** arm, which rotates with the slow-speed shaft of the gear reducer,
2. The **pitman**, which connects the crank arm to the walking beam,
3. The portion of the walking **beam** from the equalizer bearing to the center bearing, and
4. The fixed distance between the saddle bearing and the crankshaft.

The operation of the above linkage ensures that the **rotary** motion input to the system by the **prime mover** is converted into a vertical **reciprocating** movement, output at the horsehead. The sucker rods, attached to the horsehead, follow this movement and drive the bottomhole pump. Although there are different arrangements of pumping units available, all employ the same basic component parts detailed above.

Since the vast majority of pumping units used worldwide belong to the **beam-type** class, the following sections will primarily deal with the structural and operational features of the several versions of these units. The ratings and other technical specifications of beam-pumping units are covered by the **American Petroleum Institute** in its publications. Non-API pumping units are also available and are mainly used for **long-stroke** sucker-rod pumping or for special conditions.

3.7.2 STRUCTURAL PARTS

The main structural parts of a common conventional pumping unit are shown in Fig. 3.98. The whole structure is built over a rigid steel **base**, which ensures the proper alignment of the components and is usually set on a concrete foundation. The **Samson post** may have three or four legs and is the strongest member of the unit, since it carries the greatest loads. On top of it is the center or **saddle bearing**,

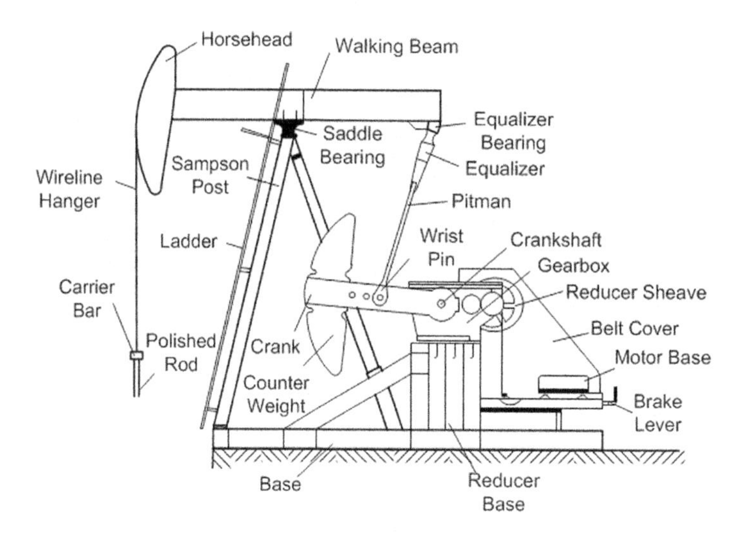

FIGURE 3.98

The structural parts of a common conventional pumping unit.

which is the pivot point for the walking beam. The **walking beam** is a heavy steel beam placed over the saddle bearing, with a sufficiently great metal cross-section to withstand the bending loads caused by the well load and the driving force of the pitmans.

The well side of the walking beam ends in the **horsehead**, which, through the **wireline hanger** (bridle), moves the polished rod. The horsehead has a curvature to ensure that the polished rod is moved in a **vertical** direction only; otherwise the resulting **bending** forces would quickly break the polished rod. In conventional units, the other end of the walking beam carries the **equalizer bearing**. To this the equalizer is connected, a short section of a lighter beam set across the walking beam, transmitting polished rod loads from the walking beam evenly to the two pitmans. The **pitmans** are steel rods that connect at their lower ends to the crank arms with the **wrist pins**. These pins are mounted on the wrist pin bearings, which allow the required rotary movement between the parts. There are several wrist pin bearings available in the two crank arms, and by connecting the pitmans to different bearings the **stroke** length of the polished rod can be changed. The **cranks** are situated on both sides of the gear reducer and are driven by the slow-speed shaft (**crankshaft**) of the gear reducer. The **counterweights** of the conventional unit are attached to the crank arms, allowing for adjustment along the crank arm axis; this arrangement was invented by **Trout** [108] back in 1926.

The proper operation of the pumping unit requires that friction losses in the structural bearings be minimal. In older units, the use of **sliding** (sleeve) bearings, usually made of bronze, was common. These bearings can tolerate very severe conditions with little damage but require regular lubrication. Today's pumping units, however, are almost exclusively equipped with antifriction **roller** bearings. These bearings are grease-lubricated and sealed, needing less maintenance than simple bronze bearings.

The remaining parts of the pumping unit are the speed **reducer** (gear reducer or **gearbox**), which reduces the high speed of the prime mover to the speed required for pumping; a **brake** assembly that can stop the unit at any point; and a **V-belt** drive and its cover. The unit is usually powered by an electric motor, but internal combustion engines are also used.

3.7.3 PUMPING UNIT GEOMETRIES

The different types of beam-pumping units, although all have the same basic components, can have various geometrical **arrangements** of their component parts. They are usually classified according to the criterion whether the walking beam operates as a **double-arm lever** (Class I) or as a **single-arm lever** (Class III). In the following, the basic features of pumping units are detailed with the usual letter designations of the component parts given in the next figures, as set forth in API specifications.

When describing the geometry and the kinematic parameters of pumping units, the following general **conventions** are used in the industry:

- The pumping unit's crank arm can be rotated in two **directions**: clockwise (**CW**) or counterclockwise (**CCW**). In the following and in all API publications, these directions are defined when looking at the pumping unit from the side with the wellhead to the observer's **right**.
- The crank angle, θ, is defined differently for the different pumping unit geometries; its **zero** is usually on a vertical line crossing the crankshaft while it is **measured** in the clockwise or counterclockwise direction. The proper positive direction and the zero angle are indicated in the following figures for each geometry of pumping units.

3.7.3.1 Conventional

The conventional unit is the **oldest** and most commonly used beam-type pumping unit and works on the same principle as the original cable-tool **drilling rig**. A schematic arrangement is shown in Fig. 3.99, from which the basic features of this pumping unit can be found:

- The walking beam acts as a double-arm **lever**, being driven at its rear end and driving the polished rod at its front end (Class I). This is also called a **pull-up** leverage system.

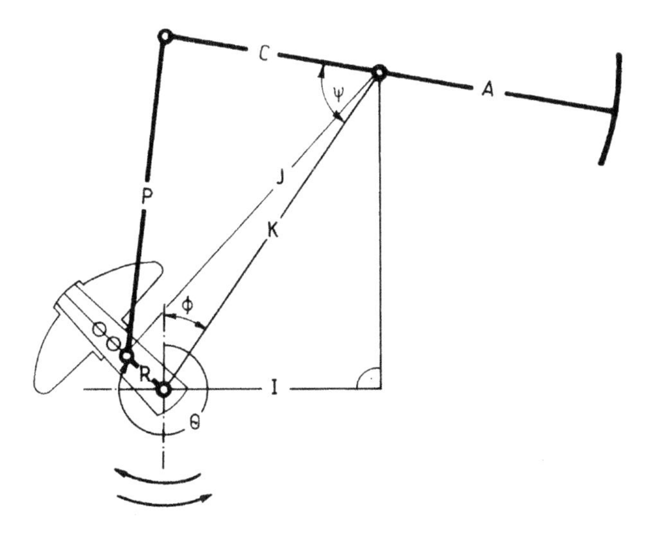

FIGURE 3.99

Schematic arrangement of a conventional-geometry pumping unit.

- When the beam is in the horizontal position, the equalizer bearing and the crankshaft are approximately on the same **vertical** line. Using the dimensions given in the figure, $C = I$.
- Counterweights are positioned either on the rear end of the beam (**beam-balanced** units) or on the crank arm (**crank-balanced** units).
- The unit can be driven in **both** (CW or CCW) directions.

3.7.3.2 Air balanced

These units were developed in the 1920s from the conventional unit by placing the horsehead on the driven part of the walking beam and by replacing the heavy counterweights with an air cylinder. They are about 40% lighter and 35% shorter than conventional units, making them ideal for portability and well testing. As seen in Fig. 3.100, the main features of this geometrical arrange-ment are:

- The beam works as a **single-arm lever** (**push-up** leverage or Class III system) since the horsehead and the pitman are on the same side of the beam.
- Dimensions I and C are approximately equal, just like in conventional units.
- Counterbalancing is ensured by the pressure force of compressed **air** contained in a cylinder, which acts on a piston connected to the beam.
- The unit can be driven in **both** (CW or CCW) directions.

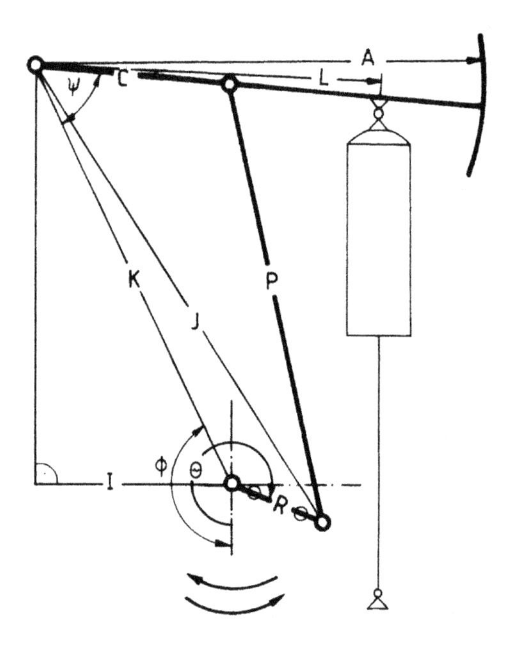

FIGURE 3.100

Schematic arrangement of an air-balanced pumping unit.

3.7.3.3 Mark II

The Mark II unit, a Class III lever system, was invented by **J. P. Byrd** in the late 1950s [109]. The main objectives of its development were to decrease the **torque** and **power** requirements of conventional units [110]. The schematic arrangement of a Mark II pumping unit is shown in Fig. 3.101; its basic features are the following:

- It utilizes a **push-up** leverage system.
- The equalizer bearing is located on the beam very close to the horsehead, making dimension *C* be greater than dimension *I*. This is one of the unique characteristics of Mark II units that improves performance beyond that of previous geometries.
- The rotary counterweights are placed on a separate counterbalance **arm** that is directed opposite to the crank arm and is **phased** by an angle, τ (usually about 24°). This unique feature ensures a more **uniform** net torque variation throughout the complete pumping cycle.
- This is a **unidirectional** pumping unit and must always be driven in the specified (CCW) direction shown in the figure.

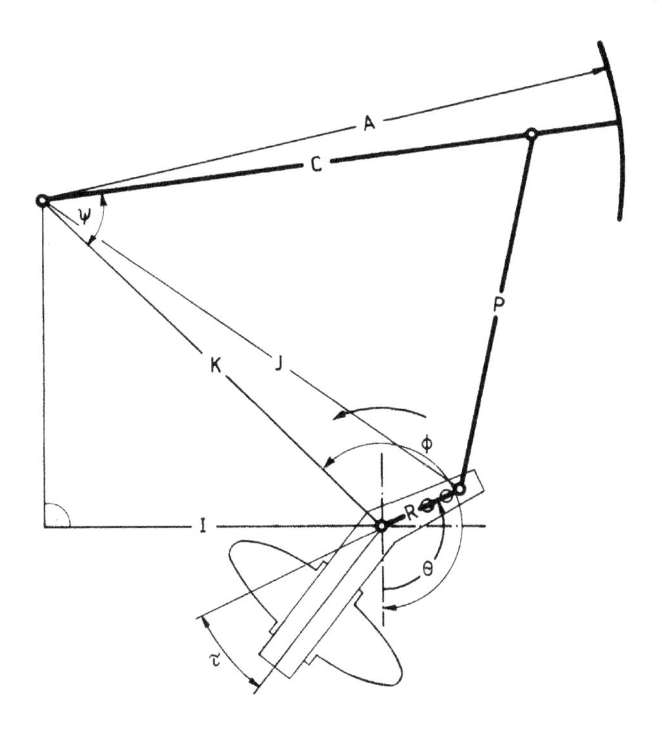

FIGURE 3.101

Schematic arrangement of a Mark II pumping unit.

3.7.3.4 TorqMaster or Reverse Mark

Units originally designed by **R. Gault**, who analyzed the advantages and disadvantages of previous pumping units by computer and tried to combine their best features in the new geometry, were marketed under the trade name **TorqMaster** in the 1980s; today they are known as **Reverse Mark** units, as offered by one of the major manufacturers.

The characteristic features of this unit, shown in Fig. 3.102, are:

- A **pull-up** geometrical arrangement,
- The gear reducer is set farther away from the Samson post, dimension *I* thus being greater than dimension *C*.
- The rotary counterweights are placed on the **crank** arm but are **phased** by an angle τ so that the crank arm leads the counterweights. The counterbalance phase angle varies with the size of the unit but is usually in the range of 8–15°.
- Crank arm rotation is **fixed** (CW), as shown in the figure.

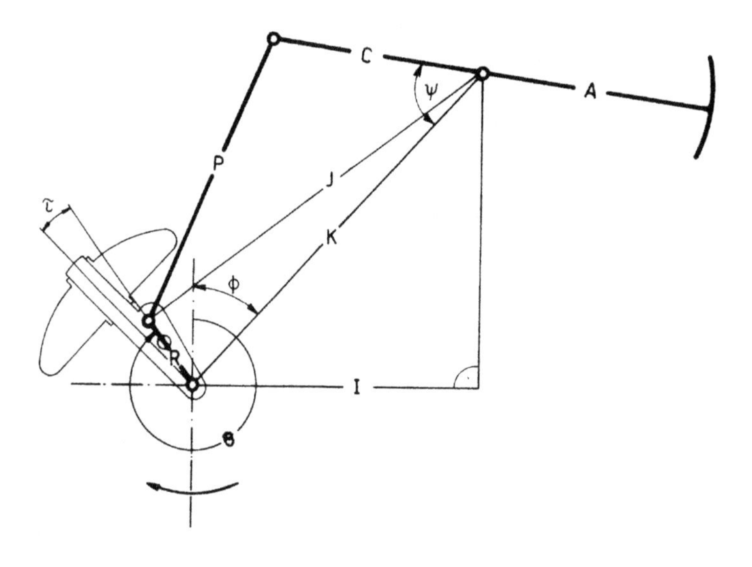

FIGURE 3.102

Schematic arrangement of a Torqmaster (Reverse Mark) pumping unit.

3.7.4 DESIGNATION OF PUMPING UNITS

In order to identify a given pumping unit, several **structural** and **operational** parameters have to be known. The most important parameter is the type of the unit's **geometrical** arrangement, detailed before. The unit's **structural capacity** (i.e., the maximum polished rod load allowed by the structure) as well as the maximum **torque capacity** of the gear reducer must also be known. Finally, the largest polished rod stroke **length** available in the given unit must be specified. All of these parameters are included in the unit's designation **code**, which is used by most manufacturers and operators. This code is quite easily interpreted; its first letters designate the geometry of the pumping unit:

B	for beam balanced conventional,
C	for crank balanced conventional,
A	for air balanced,
M	for Mark II,
RM	for Reverse Mark (TorqMaster) geometry classes

The next group of the code contains the **torque** rating of the gear reducer in thousands of in-lb, and a letter indicating the number of gear **reductions**, usually a letter "D" for **double** reduction. The following characters define the maximum polished rod **load** allowed on the pumping unit structure, in hundreds of pounds. The last group of the designation stands for the maximum polished rod stroke **length** available in the given unit.

Dimensional data of sucker-rod pumping units of a major manufacturer are presented in **Appendices B–E** for conventional, air-balanced, Mark II, and Reverse Mark geometries, respectively.

The API published standard combinations of the basic parameters of pumping units as **size ratings** in **API Spec. 11E** [111]. The standard peak torque capacities range from 6,400 to 3,648,000 in-lb, structural capacities range from 3,200 to 47,000 lb, and the greatest standard stroke length is 300 in.

EXAMPLE 3.15: INTERPRET THE FOLLOWING PUMPING UNIT DESIGNATION: M-320D-213-120

Solution

The unit is of the Mark II geometry and has a double-reduction gearbox with a 320,000 in-lb maximum capacity. The peak polished rod load allowed on the unit is 21,300 lb, and the available maximum stroke length equals 120 in.

3.7.5 PUMPING UNIT KINEMATICS

The pumping unit is essentially a **driving mechanism** that performs its useful work at the polished rod. The loading of this mechanism during the pumping cycle is rather **irregular**, because a heavy load (the weight of the rods plus the fluid load) is lifted on the **upstroke**, while no power is needed on the **downstroke**. In order to calculate power and torque requirements throughout the cycle, the **kinematic** parameters (velocity and acceleration) of the polished rod's movement must be known. Accordingly, several methods were developed to evaluate the motion of the polished rod. It will be shown in the following that early investigators utilized approximate solutions and developed formulae of limited validity. The use of exact kinematic models, on the other hand, ensures the calculation of the required parameters with a high degree of accuracy.

3.7.5.1 Early models

The simplest and oldest approach to describe the motion of the polished rod is the assumption of a **simple harmonic motion**. According to this model, the polished rod's movement is approximated with the vertical component of a point's uniform movement on the circumference of a circle. The polished rod displacement, expressed as a function of the crank angle, is:

$$s(\theta) = \frac{S}{2}(1 - \cos\theta) \tag{3.48}$$

where:

$s(\theta)$ = polished rod displacement, in,
S = polished rod stroke length, in,
θ = crank angle, rad.

The velocity of the polished rod is found by **differentiating** the above function with respect to time:

$$v(\theta) = \frac{ds(\theta)}{d\theta}\frac{d\theta}{dt} = \frac{S}{2}\omega \sin\theta \tag{3.49}$$

where:

$\omega = \frac{N\pi}{30}$ = angular frequency, 1/s.

After substituting the stroke length S in inches, the formula becomes:

$$v(\theta) = \frac{S\,N}{229}\sin\theta \tag{3.50}$$

Polished rod acceleration can similarly be derived:

$$a(\theta) = \frac{S\,N^2}{2{,}189}\cos\theta \tag{3.51}$$

where:

$v(\theta)$ = polished rod velocity, ft/s
$a(\theta)$ = polished rod acceleration, ft/s^2
N = pumping speed, SPM
S = stroke length, in
θ = crank angle, rad.

As seen, the kinematic parameters of the polished rod's movement are simple **harmonic** functions of the crank **angle**; hence the name of this model. An investigation of these functions allows the following conclusions to be drawn:

- The crank angle ranges for the upstroke and the downstroke are **equal**, i.e., 180°, and
- The maximum and minimum accelerations occur at the **start** of the upstroke and the downstroke, respectively. Their absolute values are equal, as given below:

$$|a_{max}| = |a_{min}| = \frac{S\,N^2}{2{,}189} \tag{3.52}$$

This formula was first given by **Mills** [112] and has long been used to approximate peak and minimum polished rod loads. It will be shown later that this kinematic model has a limited applicability, mainly due to its incorrect basic assumptions.

A more accurate approach for describing the polished rod's motion was the assumption of the **crank and pitman** motion. Figure 3.103 shows a schematic drawing of this driving mechanism. The crank arm (R) rotates at a constant angular velocity and moves the equalizer bearing (E) through the pitman (P). As shown, the equalizer bearing is assumed to move on a **vertical** line, although actually it travels along an **arc** drawn around the center bearing. Also, the crankshaft and the equalizer bearing fall on the same vertical line, which holds for conventional geometry units only. Polished rod position and velocity versus crank angle are found from trigonometric calculations:

$$s(\theta) = \frac{A}{C}\left[P + R(1 - \cos\theta) - \sqrt{P^2 - R^2 \sin^2\theta}\right] \tag{3.53}$$

$$v(\theta) = \frac{A\,R}{12\,C}\omega \sin\theta \left[1 + \frac{R\cos\theta}{\sqrt{P^2 - R^2 \sin^2\theta}}\right] \tag{3.54}$$

where:

A, C, P, R = dimensions defined in Figs 3.99–3.102, in
ω = angular frequency, 1/s
θ = crank angle, rad.

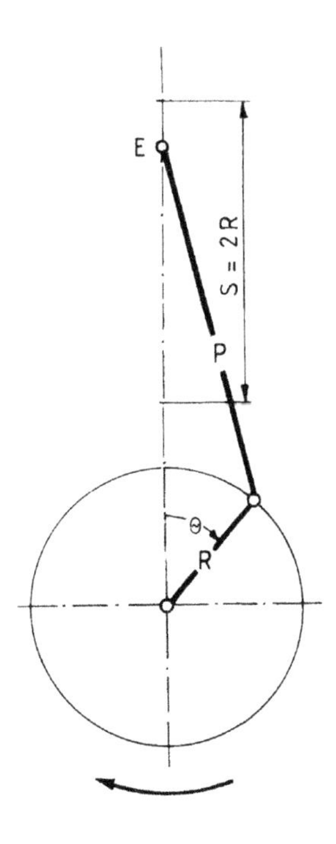

FIGURE 3.103

Schematic drawing of a "crank and pitman" driving mechanism.

In this model, the maximum and minimum accelerations are of different magnitude:

$$a_{max} = \frac{A}{C} \frac{S N^2}{2{,}189} \left(1 + \frac{R}{P}\right) \tag{3.55}$$

$$a_{min} = -\frac{A}{C} \frac{S N^2}{2{,}189} \left(1 - \frac{R}{P}\right) \tag{3.56}$$

where:

A, C, P, R = dimensions defined in Figs 3.99–3.102, in
S = polished rod stroke length, in
N = pumping speed, SPM

The application of the crank and pitman model to the kinematic analysis of pumping units leads to the following conclusions:

- The up- and downstroke crank angle **ranges** are equal, and
- Maximum acceleration occurs at the **start** of the upstroke and is greater than the minimum acceleration that occurs at the start of the downstroke.

3.7.5.2 Exact methods

The kinematic behavior of actual pumping units widely differs from those of the simplified models previously discussed. Although this fact has long been known, it was only in the 1960s that **Gray** [113] came up with an exact kinematic analysis method. He utilized the analogy between the pumping unit and a four-bar **mechanical linkage** used in driving mechanisms. The elements of this **mechanism** are the crank arm, the pitman, and the walking beam. The connection points of the parts are the crankshaft, the crank pin bearing, the equalizer bearing, and the center (saddle) bearing. These driving members are easily recognized in all pumping unit geometries shown in Figs 3.99–3.102.

Gray's calculation model, with slight modifications, was later adopted by the **API** and is included in the appendices of **API Spec. 11E** [111]. The calculation model developed by **Svinos** [114] is also based on a sound theoretical basis and complements the API procedures by presenting formulae for the calculation of polished rod velocity and acceleration. In the following, these exact kinematic analysis procedures are discussed in detail.

3.7.5.2.1 The API model

Polished Rod Position

The derivation of the polished rod **position** versus crank angle function is illustrated on a conventional geometry unit shown in Fig. 3.99. If a clockwise crank rotation is assumed, the variable angle ψ will have its maximum value at the start of the upstroke and its minimum value at the beginning of the downstroke. Designating these angles by ψ_b and ψ_t, respectively, while ψ is an intermediate angle, polished rod **position** at any crank angle is found from the following formula:

$$s(\theta) = A(\psi_b - \psi) \tag{3.57}$$

where:

$s(\theta)$ = polished rod position (displacement), in,
A = dimension defined in Figs 3.99–3.102, in,
ψ, ψ_b = angles defined above.

Polished rod stroke **length** can also be expressed using the two extremes of the angle ψ:

$$S = A\,(\psi_b - \psi_t) \tag{3.58}$$

where:

S = polished rod stroke length, in.

Dividing Eq. (3.57) by the above expression gives the position of the polished rod in a **dimensionless** from, which is denoted as *PR* (position of rods):

$$PR(\theta) = \frac{\psi_b - \psi}{\psi_b - \psi_t} \tag{3.59}$$

where:

$PR(\theta)$ = position of rods, –,
ψ, ψ_b, ψ_t = angles defined above.

This formula is valid for all pumping unit geometries, and the variable angles on the right-hand side are found as functions of the crank angle. The relevant equations to be used for calculating these angles are published in the Appendices of **API Spec. 11E** [111].

Figure 3.104 gives a comparison of the polished rod motions calculated with the simple **harmonic** and the **exact** kinematic models for an **M-228D-213-86** pumping unit. As seen, large errors can occur if the approximate calculations are used.

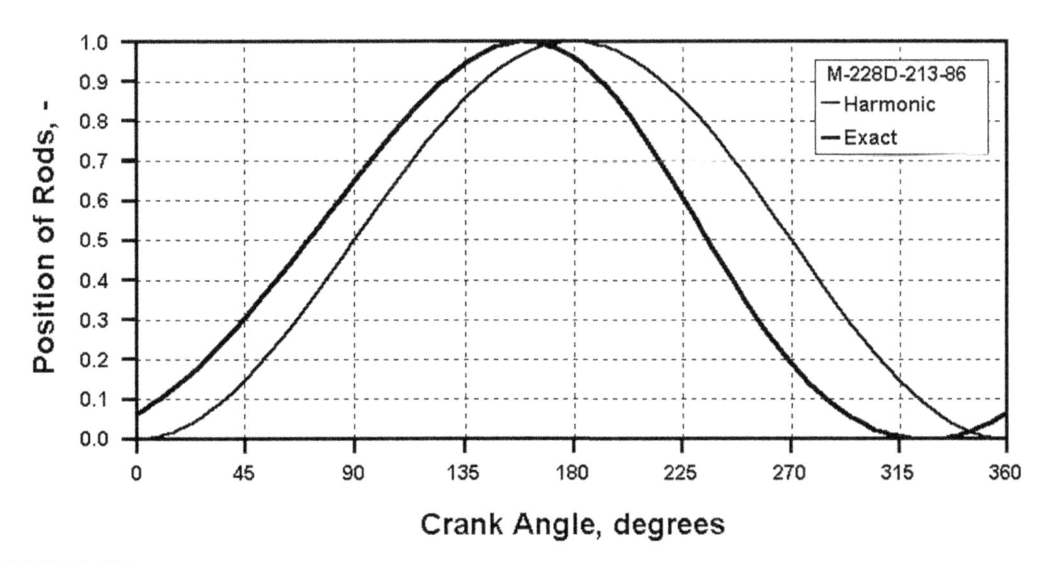

FIGURE 3.104

Comparison of the simple harmonic and the exact polished rod motion for an M-228D-213-86 pumping unit.

In general, the exact kinematic models, in contrast to the harmonic or the crank and pitman assumptions, **correctly** describe the motion of the polished rod. It can be shown that the **start** of the up- and downstroke do not occur at crank angles of 0 and 180°, as predicted by the previous models. Thus the crank angle ranges for the up- and downstroke also will be different. Table 3.21 presents, for the various geometries, the expressions for the angles θ_u and θ_d, which represent the crank angles at the start of the up- and downstroke, respectively. The variables given in the table refer to the geometrical data of the pumping units as defined in Figs 3.99–3.102.

Table 3.21 Crank Angles at the Top and the Bottom of the Polished Rod Stroke for Different-Geometry Pumping Units

Geometry	Θ_u	Θ_d
Conventional or Torqmaster	$\Phi - \varepsilon_1$	$\Phi + \pi - \varepsilon_4$
Mark II	$\Phi + \pi - \varepsilon_2$	$\Phi - \varepsilon_3$
Air balanced	$\Phi - \pi + \varepsilon_2$	$\Phi + \varepsilon_3$
The definition of angles $\boldsymbol{\varepsilon}_i$ is given below		
$\varepsilon_1 = arc\ \sin\frac{C\ \sin\Psi_b}{P+R}$	$\varepsilon_2 = arc\ \sin\frac{C\ \sin\Psi_b}{P-R}$	
$\varepsilon_3 = arc\ \sin\frac{C\ \sin\Psi_t}{P+R}$	$\varepsilon_4 = arc\ \sin\frac{C\ \sin\Psi_t}{P-R}$	

EXAMPLE 3.16: FIND THE CRANK ANGLE RANGES OF THE UP- AND DOWNSTROKE FOR A C-228D-213-86 AND AN M-228D-213-86 PUMPING UNIT. DIMENSIONAL DATA (IN INCHES) ARE GIVEN BELOW

	C-228D-213-86	M-228D-213-86
C	96.05	186.00
I	96.00	124.00
K	151.34	169.37
P	114.00	135.75
R	37.00	31.50

Solution

Consider the conventional unit first.

To find θ_u and θ_d, the parameters ϕ, ε_1, ε_4, must be known. These are found from formulae in the Appendices of **API Spec. 11E** [111].

$$\phi = \sin^{-1}(96.0/151.34) = \sin^{-1}0.6343 = 39.4°$$

$$\psi_b = \cos^{-1}\left[151.34^2 + 96.05^2 - (37+114)^2\right] / [2 \times 151.34 \times 96.05] = \cos^{-1}0.3208 = 71.28°$$

$$\psi_t = \cos^{-1}\left[151.34^2 + 96.05^2 - (114 - 37)^2\right] / [2 \times 151.34 \times 96.05] = \cos^{-1}0.9012 = 25.68°$$

$$\varepsilon_1 = \sin^{-1}[96.05 \sin 71.28]/[114 + 37] = \sin^{-1}0.6024 = 37.05°$$

$$\varepsilon_4 = \sin^{-1}[96.05 \sin 25.68]/[114 - 37] = \sin^{-1}0.5406 = 32.72°$$

Finally, using Table 3.21:

$$\theta_u = 39.4 - 37.05 = 2.4°$$

$$\theta_d = 39.4 + 180 - 32.76 = 186.6°$$

The up- and downstroke crank angle ranges are thus:
upstroke = 186.6 − 2.4 = 184.2°,
downstroke = 360 − 184.2 = 175.8°.
The same calculations for the Mark II unit follow:

$$\phi = \sin^{-1}(124/169.37) + 180 = \sin^{-1}0.7321 + 180 = 47.06 + 180 = 227.06°$$

$$\psi_b = \cos^{-1}\left[169.37^2 + 186^2 - (135.75 - 31.5)^2\right] / [2 \times 169.37 \times 186] = \cos^{-1}0.8319 = 33.7°$$

$$\psi_t = \cos^{-1}\left[169.37^2 + 186^2 - (31.5 + 135.75)^2\right] / [2 \times 169.37 \times 186] = \cos^{-1}0.5604 = 55.9°$$

$$\varepsilon_2 = \sin^{-1}[186 \sin 33.7]/[135.75 - 31.5] = \sin^{-1}0.9899 = 81.86°$$

$$\varepsilon_3 = \sin^{-1}[186 \sin 55.9]/[135.75 + 31.5] = \sin^{-1}0.9209 = 67.06°$$

Finally, using Table 3.21:

$$\theta_u = 227.06 + 180 - 81.86 = 325.2 = -34.8°$$

$$\theta_d = 227.06 - 67.06 = 160°$$

The up- and downstroke crank angle ranges are thus:
upstroke = 160 + 34.8 = 194.8°,
downstroke = 360 − 194.8 = 165.2°.

It is noted that both the harmonic motion and the crank and pitman motion models resulted in the same 180° for the upstroke and downstroke crank angle ranges, which considerably differ from the accurate values calculated here.

Polished Rod Velocity, Torque Factors

Polished rod **velocity** is an important parameter of the pumping motion because it provides a direct way to calculate the **power** requirements of pumping. If instantaneous polished rod loads and velocities are known, the product of these values gives the instantaneous **power** needed at the polished rod. Polished rod velocity can be found by **differentiating** the displacement with respect to time:

$$v(\theta) = \frac{ds(\theta)}{d\theta}\frac{d\theta}{dt} = \frac{ds(\theta)}{d\theta}\omega \tag{3.60}$$

where:

$v(\theta)$ = polished rod velocity, in/s,
ω = angular velocity of the crank arm, 1/s,
$ds(\theta)/d\theta$ = derivative of polished rod displacement function, in.

Polished rod velocity has great importance when calculating the torque loading on the gearbox and the following derivation is provided to show its effect on torque calculations. For this, let's neglect friction in the structural bearings of the pumping unit; then the power **input** at the crankshaft should be equal to the work **done** at the polished rod. The energy **balance**, written in consistent units, is the following:

$$T\, d\theta = F\, v(\theta) dt \tag{3.61}$$

where:

T = torque on the crankshaft, in-lb,
F = polished rod load, lb,
$v(\theta)$ = polished rod velocity, in/s,

This equation can be solved for the torque T required at the crankshaft:

$$T = F\, v(\theta)\, \frac{dt}{d\theta} = F\, \frac{v(\theta)}{\omega} \tag{3.62}$$

Upon substitution of Eq. (3.60) into the above formula, we get:

$$T = F\, \frac{ds(\theta)}{d\theta} \tag{3.63}$$

The term $ds(\theta)/d\theta$ on the right-hand side is the **derivative** of the polished rod displacement function with respect to crank angle and is closely bound together with the polished rod velocity $v(\theta)$ by Eq. (3.60). Knowing that torque is normally found as force times lever arm, this term can be defined as the **imaginary** lever arm which, when multiplied by the polished rod load, gives the **torque** at the crankshaft resulting from that load. Therefore, it is called the **torque factor** and is usually designated by the symbol *TF*. Torque factors, like polished rod loads, **vary** with the crank angle, and instantaneous torques at any crank angle are calculated from the final equation given here:

$$T(\theta) = TF(\theta) F \tag{3.64}$$

where:

$T(\theta)$ = crankshaft torque at crank angle θ, in-lb
$TF(\theta)$ = torque factor at crank angle θ, in
F = polished rod load at crank angle θ, lb.

Torque factors, in principle, can be calculated from **measured** values of polished rod velocity by using Eq. (3.60), but the procedure is tedious and impractical. The proper way to derive them utilizes the **geometrical** data of the pumping unit; the case of a conventional unit is considered whose schematic drawing is given in Fig. 3.105. Since crankshaft torque is found by multiplying the tangential force, F_t, by the crank arm length, R, this tangential force must be determined first. A momentum **balance** is written up on the two arms of the beam, and using the notations in the figure, we get:

$$A\, F = C\, F_p\, \sin\beta \tag{3.65}$$

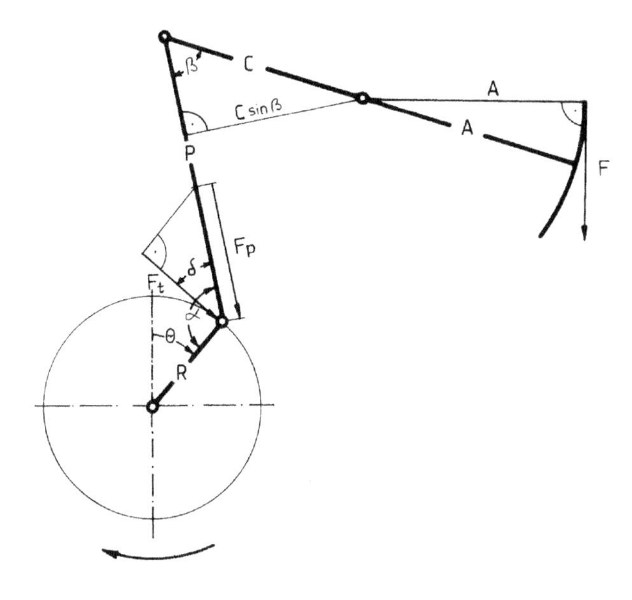

FIGURE 3.105

Derivation of torque factors from geometrical data of a conventional pumping unit.

This equation is solved for the force arising in the pitman:

$$F_p = \frac{A}{C \sin\beta} F \tag{3.66}$$

The tangential force sought is the orthogonal component of this pitman force and can be written as:

$$F_t = F_p \cos\left(\alpha - \frac{\pi}{2}\right) = F_p \sin\alpha \tag{3.67}$$

Substituting F_p from Eq. (3.66) and expressing the torque on the crankshaft:

$$T = R\,F_t = \frac{R\,A}{C} \frac{\sin\alpha}{\sin\beta} F \tag{3.68}$$

This equation is **analogous** to Eq. (3.64), where the concept of torque factor was introduced. Therefore, the multiplier of the polished rod load in the above formula is **identical** to the torque factor. Thus the final formula for calculating torque factor (*TF*) values, based on the geometrical data of a given pumping unit, is:

$$TF(\theta) = \frac{R\,A}{C} \frac{\sin\alpha}{\sin\beta} \tag{3.69}$$

where:

$TF(\theta)$ = torque factor at crank angle θ, in
A, C, R = dimensions of the pumping unit defined in Figs 3.99–3.102, in
α, β = variable angles defined in the Appendices of API Spec. 11E [111], degrees.

The torque factor versus crank angle relationship is a characteristic parameter of the different pumping units. This is why manufacturers **should** supply, in addition to *PR* (position of rods) values, the values of torque factor as well. **API Spec. 11E** suggests that these functions be given for every **15 degrees** of crank rotation, which is sufficient for hand calculations only. Accurate calculations require a closer **spacing** or the use of computer programs.

A sample comparison of the torque factors calculated with the simple harmonic and the exact kinematic models is presented in Fig. 3.106, which shows *TF* values versus crank angle for an **M-228D-213-86** pumping unit. The inaccuracies of assuming a simple harmonic motion are relevant; the magnitude of the peak torque factors as well as the length of the upstroke and downstroke are different from those calculated with the approximate formulae. This holds for other pumping unit geometries as well, leading to the final conclusion that accurate torque calculations require the use of **exact** kinematic models.

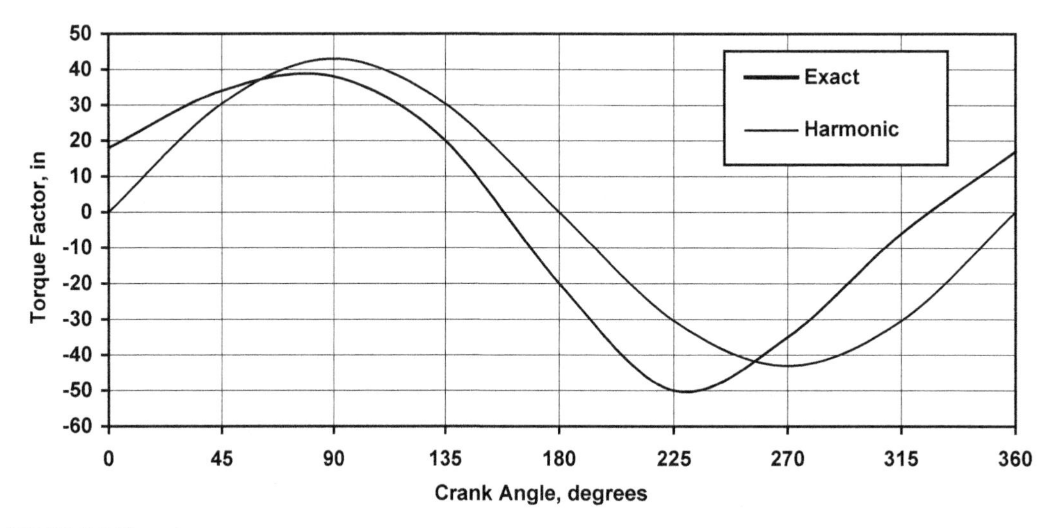

FIGURE 3.106

Comparison of torque factor values calculated with the simple harmonic motion and the exact kinematic analysis.

Polished Rod Acceleration

Although the **API Spec. 11E** model does not provide direct calculation of the polished rod's acceleration, it can be found by differentiating the polished rod velocity as detailed in the following:

$$a(\theta) = \frac{\mathrm{d}v(\theta)}{\mathrm{d}\theta}\frac{\mathrm{d}\theta}{\mathrm{d}t} = \frac{\mathrm{d}TF(\theta)}{\mathrm{d}\theta}\frac{\mathrm{d}^2\theta}{\mathrm{d}t^2} \tag{3.70}$$

where:

$a(\theta)$ = polished rod acceleration, in/s^2
$TF(\theta)$ = torque factor at crank angle θ, in
$\mathrm{d}^2\theta/\mathrm{d}t^2$ = angular acceleration of the crank, 1/s^2.

In the above formula, Eq. (3.60) was used to express polished rod velocity with torque factors.

The importance of the exact knowledge of polished rod acceleration history is explained by the fact that during the pumping cycle the polished rod moves large **masses**: the rod string and the fluid column above the pump. The **inertia** of the moving masses produces inertial forces, which are proportional to the **acceleration** that occurs at the polished rod. In order to calculate these forces, it is usually assumed that the rod string is an **inelastic** concentrated mass, accelerated and decelerated by the polished rod during the pumping cycle. In addition to this assumption, the frictional forces, which can influence polished rod loads, are neglected. In this **idealized** case, Newton's second law can be applied to the sucker-rod string's movement: it states that the sum of forces acting at the polished rod is equal to the mass being moved multiplied by the acceleration occurring. In consistent units it can be written as:

$$\sum F = \frac{W_r}{g} a \tag{3.71}$$

where:

ΣF = sum of forces acting on the polished rod,
W_r = rod string weight,
g = acceleration of gravity,
a = polished rod acceleration.

The sum of forces acting at the polished rod is found by subtracting the weight of the rod string from the load measured at the polished rod, because they are opposing forces:

$$\sum F = PRL - W_r \tag{3.72}$$

Upon substitution of this equation into Eq. (3.71), and solving for the polished rod load:

$$PRL = W_r\left(1 + \frac{a}{g}\right) = W_r(1+\delta) \tag{3.73}$$

where:

PRL = polished rod load, lb
W_r = weight of the rod string, lb
δ = acceleration factor, –

The term $(1+\delta)$ is commonly referred to as the **impulse factor** and was widely used to **approximate** the total (static plus inertial) polished rod loads. The value of the acceleration factor is directly related to the acceleration valid at the polished rod. For approximate calculations, it is customary to use the simple harmonic motion, and the maximum and minimum acceleration values are found from the **Mills formula** [112], given in Eq. (3.52). Upon substitution, the following equation is reached:

$$\delta = \pm \frac{S\,N^2}{2{,}189\,g} = \pm \frac{S\,N^2}{70{,}500} \tag{3.74}$$

where:

δ = acceleration factor, –
S = polished rod stroke length, in
N = pumping speed, SPM

The validity of the above formulae is limited mainly to **shallow** wells, because of the rough approximation of the polished rod's motion, and also due to the fact that the rod string is far from being an inelastic and **concentrated** mass.

3.7.5.2.2 The Svinos model

The elegant procedure proposed by **J. Svinos** [114] for the description of the kinematic behavior of pumping units provides a simple and **exact** calculation model for finding the kinematic parameters: displacement, velocity, and acceleration of the polished rod. He used two-dimensional **vectors** to simulate the movement of the pumping unit's main **components**: the crank, the pitman, and the walking beam. This approach allows one to find the position, velocity, and acceleration of **any** component during the pumping cycle. Compared to the **API Spec. 11E** [111] model, velocities and accelerations of the pumping unit's component are **directly** calculated and provide much higher **accuracy** than the numerical differentiation required in the API procedure. One of the most important features of the **Svinos** model is that by providing accurate acceleration data, **inertial** torques acting on the gear reducer are easy to calculate, as described in a later section.

The calculations of polished rod position, *PR*, and torque factor, *TF*, values are identical to those detailed previously for the API model (see Eqs 3.59 and 3.69); that is the reason why in the following only the procedures for calculating polished rod velocity and acceleration are given.

Instead of the crank angle, θ, generally used when calculating the kinematic parameters of different pumping unit geometries (see Figs 3.99–3.102) **Svinos** chose a different angle as the **independent** variable for his kinematic calculations; this angle, θ_2, is defined as:

$$\theta_2 = 2\pi - \theta + \arcsin\left(\frac{I}{K}\right) \quad \text{for } \textbf{Conventional} \text{ and Reverse Mark units} \tag{3.75}$$

$$\theta_2 = \pi - \theta + \arcsin\left(\frac{I}{K}\right) \quad \text{for } \textbf{Mark-II} \text{ and } \textbf{Air Balanced} \text{ units} \tag{3.76}$$

where:

I = vertical distance between the center bearing and the crankshaft, in,
K = distance between the center bearing and the crankshaft, in, and
θ = crank angle, rad.

Normally the variation of crank angle, θ, with time is calculated from data received from rod-pumping analysis tools or pump controller units. Based on this, angular **velocity** and **acceleration** of the crank's movement can also be found. After differentiating any of the previous two formulas we can find how the angular velocity and acceleration of Svinos' independent variable, θ_2, is related to the kinematic parameters of the crank's motion:

$$\frac{d\theta_2}{dt} = -\frac{d\theta}{dt} \tag{3.77}$$

$$\frac{d^2\theta_2}{dt^2} = -\frac{d^2\theta}{dt^2} \tag{3.78}$$

where:

$d\theta/dt$ = instantaneous angular velocity of the crank, 1/s, and
$d^2\theta/dt^2$ = instantaneous angular acceleration of the crank, $1/s^2$.

Svinos defined a variable angle, θ_b, between the centerline of the walking beam and the line connecting the crankshaft to the center (saddle) bearing. Polished rod **velocity** and **acceleration** can be expressed by using the angular velocity and acceleration of this angle, as given here:

$$v(\theta) = A\,\frac{d\theta_b}{dt} \tag{3.79}$$

$$a(\theta) = A\,\frac{d^2\theta_b}{dt^2} \tag{3.80}$$

where:

$v(\theta)$ = polished rod velocity, in/s,
$a(\theta)$ = polished rod acceleration, in/s^2, and
A = distance from the center bearing to the horsehead, in.

In order to find velocity and acceleration values from the previous final formulas, the time derivatives of the variation of angle θ_b must be evaluated from the following formulas:

$$\frac{d\theta_b}{dt} = \frac{R}{C}\,\frac{d\theta_2}{dt}\,\frac{\sin(\theta_3 - \theta_2)}{\sin(\theta_3 - \theta_b)} \tag{3.81}$$

$$\frac{d^2\theta_b}{dt^2} = \frac{d\theta_b}{dt}\left[\frac{\frac{d^2\theta_2}{dt^2}}{\frac{d\theta_2}{dt}} - \left(\frac{d\theta_3}{dt} - \frac{d\theta_b}{dt}\right)\cot(\theta_3 - \theta_b) + \left(\frac{d\theta_2}{dt} - \frac{d\theta_3}{dt}\right)\cot(\theta_2 - \theta_3)\right] \tag{3.82}$$

where:

C = distance from the center bearing to the equalizer bearing, in, and
R = crank length, in.

Equation (3.82) is used, in addition to the determination of the polished rod acceleration from Eqn. (3.80), to calculate the angular acceleration of the **beam** and can be used to calculate the **articulating** inertial torques resulting from the inertia of the beam, horsehead, equalizer, etc., as will be shown in Chapter 4 of this book.

The angles not yet defined in the previous formulas are the following:

$$\theta_3 = \arccos\left(\frac{P^2 + L^2 - C^2}{2\,P\,L}\right) - \beta \tag{3.83}$$

$$\theta_b = \arccos\left(\frac{P^2 - C^2 - L^2}{2\,C\,L}\right) - \beta \tag{3.84}$$

where:

$P =$ length of the pitman, in.

The definitions of the two variables L and β featuring in the previous formulas are given here:

$$L = \sqrt{K^2 + R^2 - 2\,K\,R\,\cos\theta_2} \tag{3.85}$$

$$\beta = \mathrm{j}\,\arccos\left(\frac{L^2 + K^2 - R^2}{2\,K\,L}\right) \tag{3.86}$$

where:

$K =$ distance between the crankshaft and the center bearing, in.

The sign j of variable β varies with angle θ_2 as follows:

$$j = 1 \quad \text{for } 0 < \theta_2 < \pi$$
$$j = -1 \quad \text{for } \pi < \theta_2 < 2\pi$$

The only remaining term not defined yet is:

$$\frac{\mathrm{d}\theta_3}{\mathrm{d}t} = \frac{R}{P}\,\frac{\mathrm{d}\theta_2}{\mathrm{d}t}\,\frac{\sin(\theta_b - \theta_2)}{\sin(\theta_3 - \theta_b)} \tag{3.87}$$

where:

$P =$ length of the pitman, in, and
$R =$ crank length, in.

Following the calculation model of **Svinos** allows one to include the effects of variable crankshaft angular velocities, which was not possible when using the **API Spec. 11E** [111] procedure. In most of the cases the crank turns at a relatively **constant** speed (especially when a low-slip electric motor is used) and the angular velocity of the crank is **constant** while its acceleration is zero. Expressing this condition with the new angle notation, we get the following formulas that must be used in Eqs (3.81) and (3.82) for the calculation of polished rod velocities and accelerations if the unit turns at a constant speed:

$$\frac{\mathrm{d}\theta_2}{\mathrm{d}t} = -\frac{N\,\pi}{30} \tag{3.88}$$

$$\frac{\mathrm{d}^2\theta_2}{\mathrm{d}t^2} = 0 \tag{3.89}$$

where:

$N =$ pumping speed, SPM.

In cases when the crank's angular velocity is **fluctuating**, e.g., when high-slip electric motors are used to drive the pumping unit or the unit is poorly balanced, then **instantaneous** pumping speeds during the pumping cycle must be known. Either a recording tachometer is used to measure pumping speeds or a digital dynamometer system is used to calculate the crank's instantaneous angular velocity;

crank acceleration is found from differentiation of this data. The calculation procedure is then performed by using the appropriate historical data of the velocity and acceleration of Svinos' independent variable, θ_2, as defined in Eqs (3.77) and (3.78).

Svinos's procedure is straightforward and results in accurate values for the polished rod's velocity and acceleration. As already mentioned, the accurate calculation of the accelerations of the different parts of the pumping unit provides a reliable way to find the **inertial** torques on the gearbox resulting from both the acceleration of the walking beam and the crankshaft.

EXAMPLE 3.17: CALCULATE THE KINEMATIC PARAMETERS OF A C-228D-213-86 PUMPING UNIT, IF THE UNIT'S ROTATION IS CLOCKWISE AND THE PUMPING SPEED IS CONSTANT AT 10 SPM. THE UNIT'S DIMENSIONAL DATA ARE GIVEN HERE IN INCHES

$K = 151.34$	$C = 96.0$
$P = 114.0$	$A = 111.0$
$I = 96.0$	$R = 37.0$

Solution

Let's detail the calculation of polished rod velocity and acceleration values for the crank angle of $\theta = 60°$.

First find the independent variable angle θ_2 from Eq. (3.75):

$$\theta_2 = 2\pi - 60\pi/180 + \arcsin\,(96/151.34) = 5.92 \text{ rad.}$$

Since the unit rotates at a constant speed, angular velocity and acceleration of angle θ_2 are found from Eqs (3.88) and (3.89):

$$\mathrm{d}\theta_2/\mathrm{d}t = -10\pi/30 = -1.047 \;\; 1/\text{sec.}$$

$$\mathrm{d}^2\theta_2/\mathrm{d}t^2 = 0.$$

Now find variables L and β from Eqs (3.85) and (3.86):

$$L = \left[151.34^2 + 37^2 - 2 \times 151.34 \times 37 \cos(5.92)\right]^{0.5} = 117.44 \text{ in.}$$

$$\beta = -1 \arccos\left[\left(117.44^2 + 151.34^2 - 37^2\right)/(2 \times 151.34 \times 117.44)\right] = -0.111 \text{ rad.}$$

The angles θ_3 and θ_b can now be calculated from Eqs (3.83) and (3.84):

$$\theta_3 = \arccos\left[\left(114^2 + 117.44^2 - 96^2\right)/(2 \times 114 \times 117.44)\right] + 0.111 = 0.966 \text{ rad.}$$

$$\theta_b = \arccos\left[\left(114^2 - 96^2 - 117.44^2\right)/(2 \times 96 \times 117.44)\right] + 0.111 = 2.142 \text{ rad.}$$

The derivatives of these are evaluated from Eqs (3.81) and (3.87):

$$\mathrm{d}\theta_3/\mathrm{d}t = -37/114 \times 1.047 \sin(2.142 - 5.92)/\sin(0.966 - 2.142) = 0.2197 \;\; 1/\text{sec.}$$

$$\mathrm{d}\theta_b/\mathrm{d}t = -37/96 \times 1.047 \sin(0.966 - 5.92)/\sin(0.966 - 2.142) = 0.4242 \;\; 1/\text{sec.}$$

Polished rod velocity at $\theta = 60°$ can now be found from Eq. (3.79):

$$v = 111 \times 0.4242 = 47.09 \text{ in/sec.}$$

To find the acceleration of the walking beam, the second derivative of θ_b is found from Eq. (3.82):

$$\mathrm{d}^2\theta_b/\mathrm{d}t^2 = 0.4242[0 - (0.2197 - 0.4242)\cot(0.966 - 2.142) + (-1.047 - 0.2197)\cot(5.92 - 0.966)]$$
$$= 0.098 \;\; 1/\text{sec}^2.$$

Finally, polished rod acceleration is evaluated from Eq. (3.80):

$$a = 111 \times 0.098 = 10.87 \text{ in/sec}^2.$$

The detailed results of the complete kinematic calculation are given in Exhibit 3.7.

CALCULATION OF KINEMATIC PARAMETERS OF PUMPING UNITS

API Size = C-228D-213-86
Geometry Type = Conventional
Unit Rotation = Clockwise
Linkage Dimensions (in):

K =	151.34	C =	96.00
P =	114.00	A =	111.00
I =	96.00		

Crank Length -in-	Calc. Stroke -in-	Start of UP -deg-	Start of DWN -deg-	Stroke Ranges UP -deg-	Stroke Ranges DWN -deg-
37.00	88.35	2.3	186.6	184.3	175.7

Crank Angle -deg-	Position of Rods	Torque Factor -in-	Pol. Rod Velocity -in/sec-	Pol. Rod Acceleration -in/sec²-	Beam Acceleration -1/sec²-
0	0.001	-2.4	-2.5	64.5	0.581
15	0.017	13.4	14.1	66.5	0.599
30	0.079	28.4	29.7	56.4	0.509
45	0.181	39.5	41.3	35.3	0.318
60	0.308	45.0	47.1	10.8	0.098
75	0.442	45.1	47.2	-8.4	-0.075
90	0.571	41.6	43.6	-19.4	-0.175
105	0.687	36.3	38.1	-23.9	-0.216
120	0.786	30.4	31.9	-25.2	-0.227
135	0.867	24.4	25.5	-25.6	-0.231
150	0.931	18.2	19.0	-26.8	-0.242
165	0.975	11.4	12.0	-29.8	-0.268
180	0.997	3.8	4.0	-34.6	-0.312
195	0.996	-5.2	-5.4	-40.3	-0.364
210	0.966	-15.3	-16.0	-43.8	-0.395
225	0.905	-25.6	-26.8	-41.6	-0.375
240	0.815	-34.6	-36.3	-33.0	-0.298
255	0.702	-41.1	-43.0	-20.7	-0.187
270	0.575	-44.5	-46.6	-7.8	-0.070
285	0.442	-44.8	-46.9	4.7	0.042
300	0.312	-42.3	-44.3	16.8	0.151
315	0.194	-36.8	-38.5	29.1	0.262
330	0.097	-28.3	-29.6	42.1	0.379
345	0.029	-16.7	-17.5	54.7	0.493
360	0.001	-2.4	-2.5	64.5	0.581

EXHIBIT 3.7

Kinematic parameters of a pumping unit for Example 3.17.

3.7.5.3 Comparison of pumping units

The selection of the **proper** pumping unit to be used on a given well is a complex problem that requires a detailed analysis of the available **geometrical** types and sizes. Although many factors must be considered, the **kinematic** behavior of the unit can play an important role in the selection process. This is why a clear understanding of the relative advantages and drawbacks of the different pumping unit geometries must be acquired before a decision is made. In the following, a short **comparison** of the kinematic properties of different pumping unit geometries is given. The calculations required for an actual comparison can be based on the exact models detailed before, or the model of **Laine et al.** [115] can be followed, which gives accurate approximates of *PR* (position of rods) and *TF* (torque factor) values and needs modest calculation time. This was achieved by fitting the exact polished rod movement with harmonic functions using Fourier series.

The requirements for an **ideal** kinematic behavior of pumping units can be summed up as follows:

- A long, **slow** upstroke is desirable. A crank rotation of substantially **more** than 180° for the upstroke provides sufficient time for the pump barrel to fill up with fluids. In addition, a **reduced** peak torque can be expected on the gearbox, since the total work necessary for the lifting of the fluid is done over a **longer** period of time.
- Downstroke can be **faster**, the only constraint being the **free-fall** velocity of the rods in the well fluids [116].
- Low torque factors on the upstroke are a **desirable** feature in any pumping unit [117]. Since polished rod loads are **greatest** during the upstroke, low torque factors result in **less** torque required on the crankshaft. Power consumption is thus **reduced** and the increase in pumping efficiency leads to lower operating costs.
- The acceleration pattern of the unit should exhibit a **reduced** acceleration at the **bottom** of the stroke. At the start of the upstroke, the fluid load is transferred to the rod string, and this large load combined with a great acceleration would give **excessive** inertial forces at the polished rod. Lower polished rod accelerations at the start of the upstroke, therefore, result in lower peak loads.

After the discussion of the desirable features of pumping units, the kinematic parameters of the different geometries will be examined. To better illustrate the features of each geometry, pumping units of the same API size (**320D-256-120**) will be used in the following comparisons.

Figure 3.107 presents the variation of torque factors with crank angle for the four major geometrical types of pumping units of the same size. **Conventional units** exhibit upstrokes and downstrokes that are approximately **equal** in crank angle ranges. Although they may be driven in **either** of the two possible directions, the small differences between the upstroke and downstroke crank angle ranges can mean greater differences in operating conditions. Therefore, the **preferred** direction of rotation should be found and followed. The **air-balanced** unit's operation is also **symmetrical**, i.e., the up- and downstroke crank angle ranges are approximately equal; this is the reason why this unit can also be driven in both directions (CW or CCW). The **Mark II** unit, however, must be driven in the counter-clockwise direction when it exhibits a substantially longer and **slower** upstroke (up to 10% longer) than the downstroke. Turning it in the wrong direction, therefore, eradicates all the benefits of a longer upstroke. The **Reverse Mark** geometry behaves very similarly to the Mark II and also has a **longer** upstroke but should always be driven in the clockwise direction.

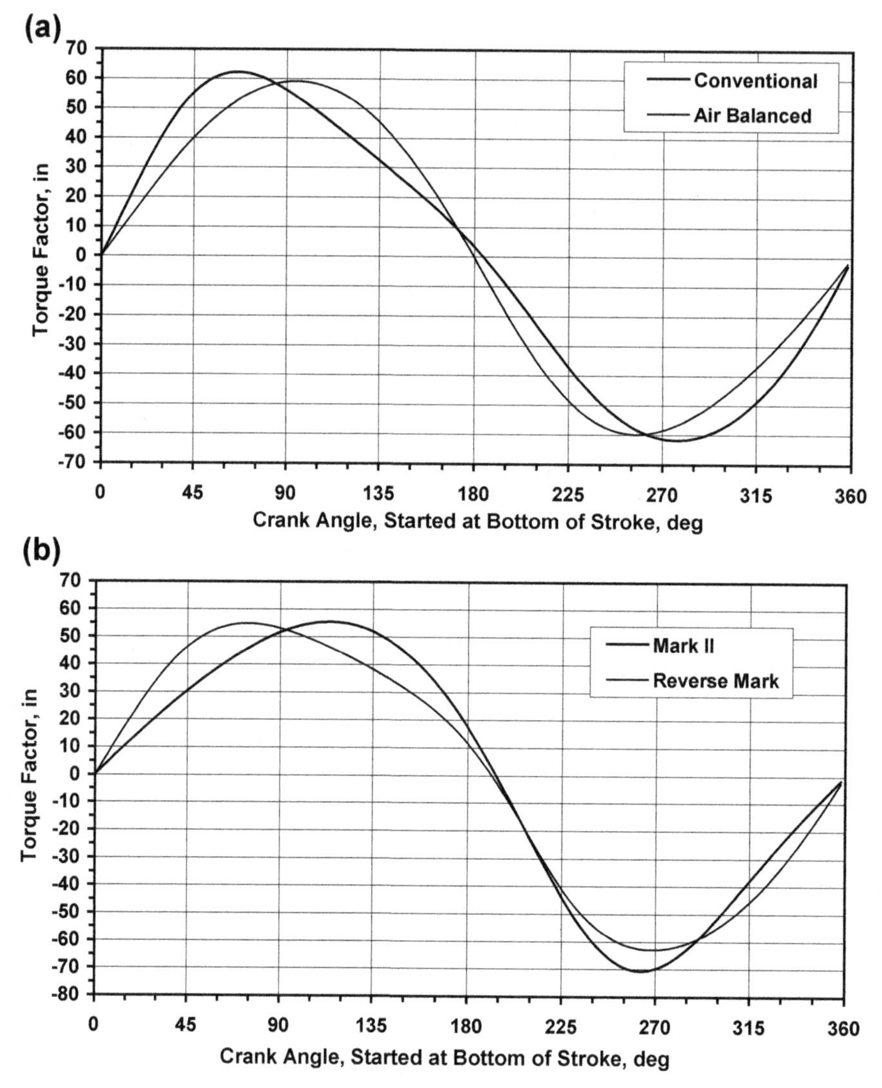

FIGURE 3.107

Torque factor versus crank angle functions for different geometry (a = Conventional and Air Balanced, b = Mark II and Reverse Mark) pumping units of API 320D-256-120 size.

Comparison of the **torque factor** patterns of the four units shows that **conventional** and **air-balanced** types behave very similarly and their maximum upstroke torque factors occur quite **early** in the upstroke. This feature leads to high crankshaft torques on the upstroke because of the high polished rod loads during the upstroke. As seen in Fig. 3.107, the **Reverse Mark** unit has lower torque factors in the upstroke than the previous units, but the lowest possible values are provided by the **Mark II**. The consequence of the lower torque factors combined with the fact that their maximum occurs much later in the upstroke is that under the same well conditions the **Mark II** unit needs the **least** gearbox torque. It also requires the lowest prime mover **power**, making the unit ideal for deep high-volume applications.

The polished rod accelerations that occur in different types of pumping units of the same API size (**320D-256-120**) at a pumping speed of 10 SPM are presented in Fig. 3.108. The **conventional-**

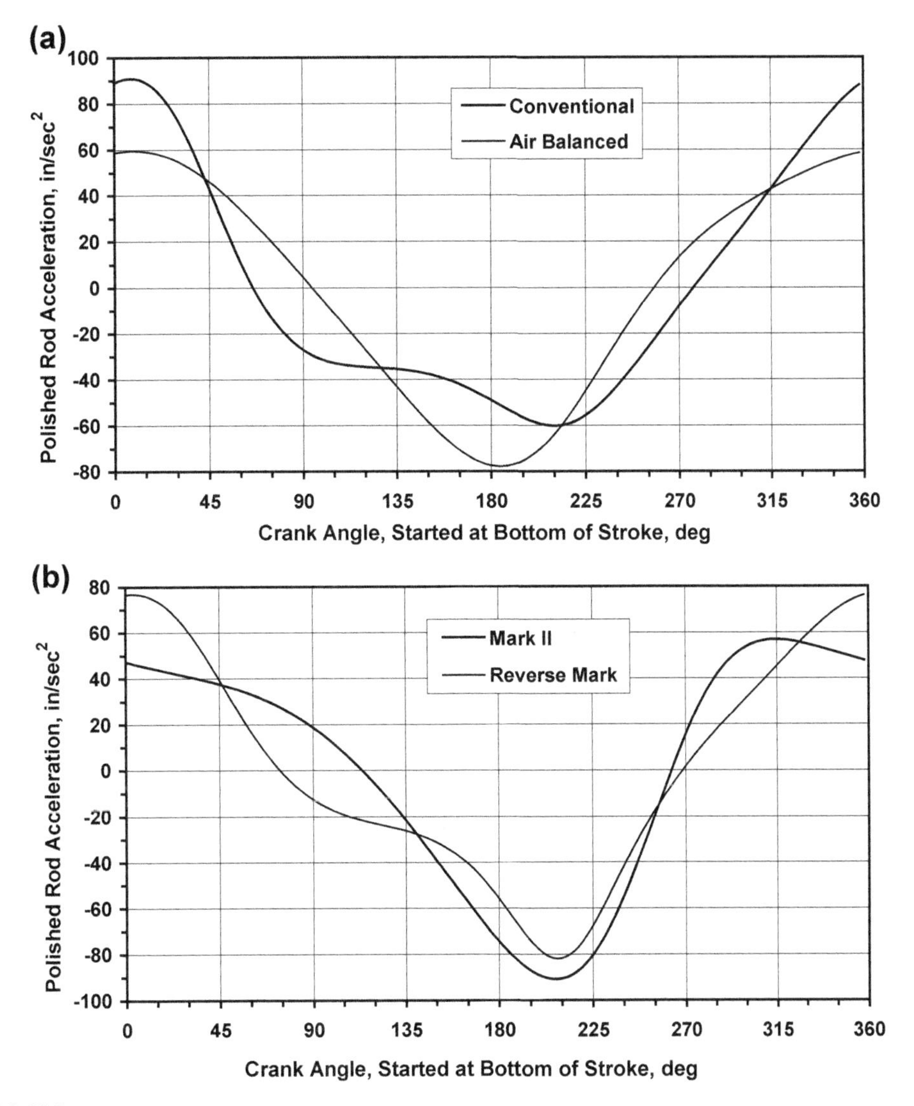

FIGURE 3.108

Polished rod acceleration versus crank angle functions for different geometry (a = Conventional and Air Balanced, b = Mark II and Reverse Mark) pumping units of API 320D-256-120 size.

geometry unit's polished rod acceleration reaches its maximum immediately after the **start** of the upstroke, a very **undesirable** feature. The combined effects of the high acceleration and the high polished rod loads valid at this time result in exceptionally high torques that may overload the speed reducer. The **air-balanced** unit, in comparison to the conventional geometry, has lower upstroke accelerations while its minimum acceleration is higher. The **Reverse Mark** geometry has an acceleration pattern similar to that of the conventional unit but with lower acceleration values. The lowest polished rod accelerations are provided by the **Mark II** geometry; another beneficial feature is that peak acceleration occurs **before** the start of the upstroke; thus, a reduction in peak loads can be expected. An added positive feature of this geometry is that usually the range of rod stresses decreases, extending the service life of the rod string [118].

3.7.6 SPECIAL PUMPING UNITS

In addition to the beam-type pumping units discussed so far, other units of widely different constructions are also available. Long-stroke pumping units are covered in Chapter 7 of this book, and the following short discussion covers selected developments only and is not intended as a complete treatment of this ever-changing topic.

3.7.6.1 Low-profile units

Low-profile pumping units were developed to provide a limited vertical dimension, needed in areas with traveling **irrigation** systems or in urban areas. Figure 3.109 shows a unit manufactured by **Lufkin Industries** that utilizes a walking **head** instead of a walking beam. It is a unidirectional pumping unit producing a longer-lasting upstroke than downstroke. Another low-profile unit (Fig. 3.110) uses a two-bar mechanical linkage. The pitmans are connected to wirelines running on pulleys. With fewer moving parts than a conventional unit, power savings can be realized.

The pumping units mentioned above all have the usual gearboxes to reduce the speed of the prime mover. Some new designs entirely eliminate the gearbox and have **belt drives** instead [119]. This type

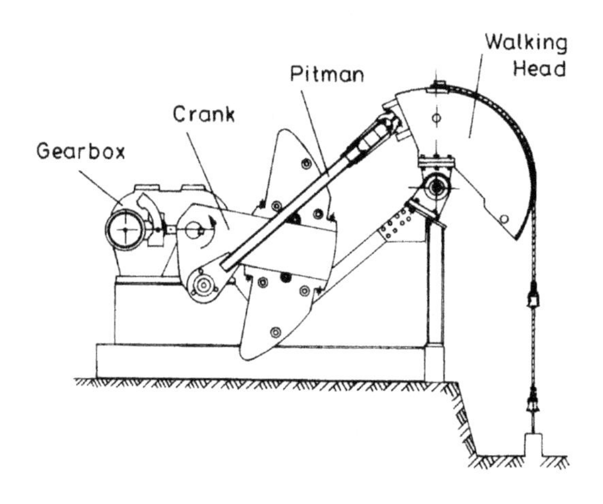

FIGURE 3.109

Low-profile pumping unit with walking head.

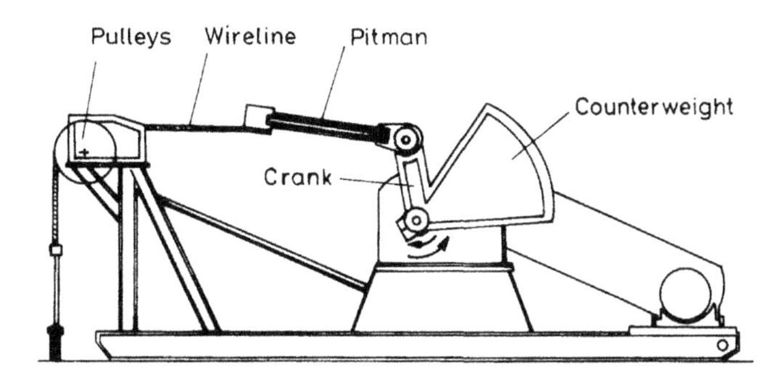

FIGURE 3.110

Low-profile pumping unit.

of drive permits a more efficient power transfer than conventional geared reducers. Unit geometry is also unique, allowing for a slow, long upstroke and a faster, short downstroke, making these units the ideal choice for installations with fiberglass rod strings.

3.7.6.2 The Linear Rod Pump

Instead of the beam-pumping unit, the **Linear Rod Pump** (LRP), available from Unico, Inc [120,121], utilizes a linear actuator to move the polished rod. As shown in Fig. 3.111, the unit has a **rack-and-pinion** mechanism driven by a **reversible** electric motor. The motor and the gearbox are fixed to a

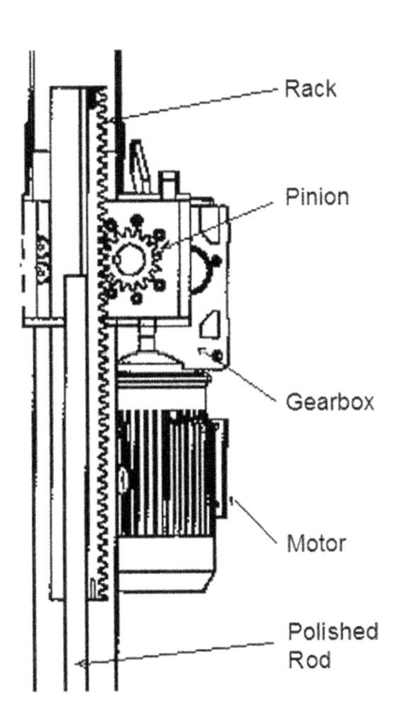

FIGURE 3.111

Schematic drawing of the Linear Rod Pump (LRP).

vertical tube mounted directly to the wellhead. Inside the tube the rack gear is raised and lowered by the pinion (driven by the gearbox), thereby moving the polished rod fixed to the rack. Since the motor is connected to a variable-speed drive (VSD), proper programming of the VSD unit allows a full control of the polished rod's kinematic behavior during the complete pumping cycle.

Programming of the VSD unit that drives the Linear Rod Pump allows the operator to select the variation of most of the kinematic parameters of the polished rod's movement during the pumping cycle. Polished rod velocities for long **portions** of both the upstroke and the downstroke are **constant** and can be selected independently of each other; acceleration/deceleration at stroke reversals is also selectable. These features greatly decrease dynamic forces and reduce **energy** requirements. The nearly instantaneous adjustment of polished rod velocity is facilitated by the low inertia of the system.

LRP units are available in smaller sizes with a maximum polished rod stroke length of 144 in and a maximum load capacity of 30,000 lb [121]. Their principal advantages over beam-pumping units are [122]:

- Low investment and installation costs related to their small size and low weight.
- Lower energy requirements than those of beam units because of the simpler design and computer control.
- Minimum pumping speed can be as low as 1 SPM without any operational problems; the limit for beam-pumping units is 4–5 SPM.
- Much smaller footprint and low profile.

3.8 GEAR REDUCERS

Speed or, more commonly, **gear reducers** (gearboxes in short) are the heart of the pumping unit and represent about 50% of the investment cost of any pumping unit. Their main function is to **reduce** the high rotational speed of the prime mover to the pumping speed required and, at the same time, to **increase** the output torque to meet well loads. The usual speed reduction ratio of pumping unit gearboxes is about 30 to 1, the maximum output speed is around 20 SPM. Speed reducer sizes are standardized by the **API** in **Spec. 11E** [111]; the rating relates to the maximum mechanical torque allowed on the reducer. The standard peak torque range is from 6,400 to 3,648,000 in-lb, and the corresponding gearboxes are designated as sizes 6.4 and 3648. These ratings are valid at different nominal pumping speeds; the speeds used are a function of unit size. Gearboxes of size 2,560 and greater are tested at 11 SPM, smaller sizes at proportionally greater speeds; gearbox sizes 320 and smaller at 20 SPM. Two kinds of speed reducers are used: **geared** and **chain** reducers.

Gear reducers utilize double- or triple-reduction **gearing**; Fig. 3.112 shows a schematic of the most popular double-reduction unit. It contains three **shafts**: the high-speed **input** shaft, an **intermediate** shaft, and a **slow-speed** shaft. The high-speed shaft is driven by the prime mover through a **V-belt sheave**, and the slow-speed shaft drives the crank arms of the pumping unit. Since torque is **increased** at every shaft as speed is reduced, shaft diameters are also **increased** with the decrease of speed. They are designed to withstand the high torsional and bending loads and ensure that the gear faces are in full contact even at the rated torque of the reducer. The shafts run in **bearings** mounted in the reducer housing. **Sleeve** bearings (usually bronze) are used at the slow-speed shaft and straight roller bearings on the other shafts; preferably, all shafts are equipped

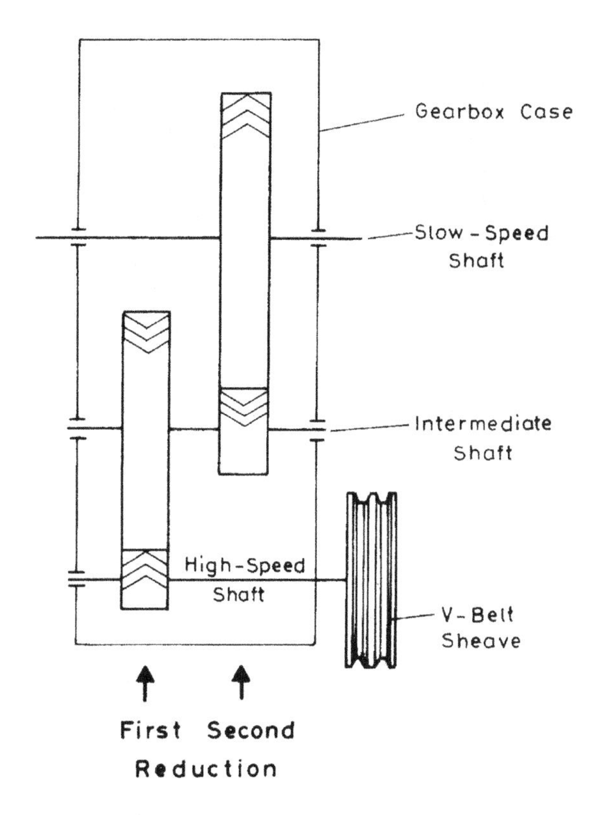

FIGURE 3.112

Schematic drawing of a double-reduction gear reducer.

with antifriction **roller** bearings. Figure 3.113 shows the arrangement of the gears and bearings of a typical gearbox.

The tooth forms most frequently used on the gears are the **herringbone** (see Fig. 3.113) or the double helical tooth, which provides **uniform** loading and quiet operation. Herringbone gears are less sensitive to misalignment and resist torque reversals better than other types of gears. The gears are precision machined to fine tolerances and heat-treated to achieve the hardness required.

The speed reduction ratio of the gearbox is found from the diameters of its gears, as given in the following formula; typical values are about 30:1, i.e., the high-speed (input) shaft turns at a speed about 30 times higher than the slow-speed (output) shaft.

$$Z = \frac{D_{sl}}{D_{hi}} \frac{D_{ii}}{D_{io}} \tag{3.90}$$

where:

$Z =$ the gearbox's speed reduction ratio, –,
D_{sl}, $D_{hi} =$ pitch diameter of slow-speed and high-speed gears, in,
D_{ii}, $D_{io} =$ pitch diameter of intermediate input and output gears, in.

FIGURE 3.113

Double-reduction gearbox with open housing.

The proper operation and the life of the gear reducer mainly depend on the proper **lubrication** of the moving parts. Lubrication oil of a suitable viscosity is contained in the bottom of the **housing** to provide an oil bath for the gears, which carry the oil upward as they turn. Oil brought to the top of the gears is taken off by **wipers;** excess oil is directed to oil channels that lead the lubricant to every bearing. This works only with sufficiently high pumping speeds above 5 SPM when enough oil is carried upward by the gears. At speeds less than 5 SPM, an additional wiper has to be added to the high-speed gear to ensure proper lubrication. Regular inspection and change of the oil is a prime requirement for trouble-free operation.

The high-speed shaft of gearboxes may be turned by the motor in any of the two possible directions; gearboxes operate properly **independent** of the direction of rotation. During a normal pumping cycle, however, the actual direction of rotation may change several times because the prime mover will change its sense of rotation depending on its torque loading. Under positive loads the prime mover **drives** the gearbox while negative torques cause (1) the gearbox to become the **driving** member of the system, and (2) the motor to become an electrical generator. The two operating modes and the reversal of loading is indicated by the typical **backlash** sound emitted from the gearbox due to the transfer of load from one side of the gear teeth to the other. If coupled with heavy loads this kind of load reversal can eventually break the teeth of the gears, especially under special conditions like fluid pounding. The operating **life** of the gearbox is mostly determined by its loading; overloaded reducers fail much earlier than those properly loaded, as given in Fig. 3.114. This shows the grave importance of monitoring the torque loading on gear reducers and the need for maintaining optimum counterbalance conditions.

Typical technical data of different sizes of pumping unit gearboxes are given in Table 3.22. The table lists the gearboxes' overall speed reduction ratio, the diameter of the output (slow speed) shaft, the lubricating oil capacity, and the range of sheave sizes that can be fitted on the crankshaft.

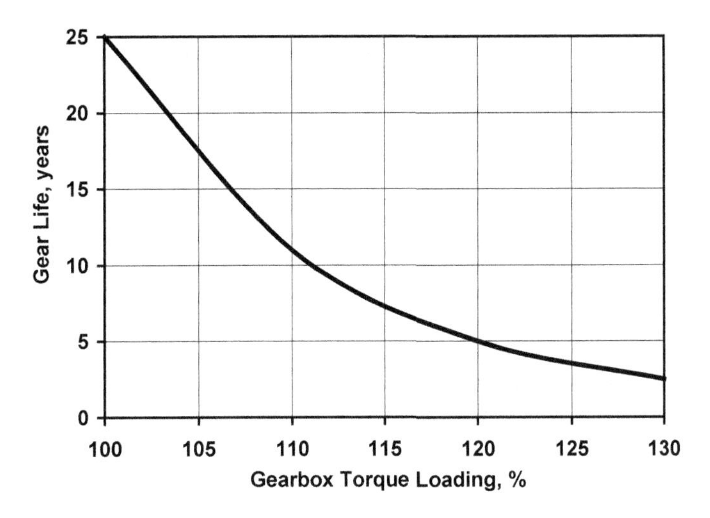

FIGURE 3.114

Estimated operating life of gearboxes.

Table 3.22 Typical Technical Data of Pumping Unit Gearboxes

		Shaft Diameter		Oil	Sheave Size	
Gearbox Size-	Gear Ratio-	Mark II, in	Other, in	Capacity, gal	Max. in	Min. in
2560D	34.53 to 1		11.7500	235	68.0	55
1824D	28.33 to 1	10.5000	9.0000	165	68.0	55
1280D	28.05 to 1	10.5000	8.5000	120	68.0	35
912D	28.72 to 1	9.0000	7.0000	107	55.2	33
640D	28.60 to 1	9.0000	7.0000	70	55.6	22
456D	29.04 to 1	9.0000	7.0000	55	50.0	22
320D	30.12 to 1	8.5000	6.4375	50	47.0	24
228D	28.45 to 1	7.0000	6.0000	34	41.0	24
160D	28.67 to 1	7.0000	5.4375	22	38.0	20
114D	29.40 to 1	6.4375	4.4375	17	33.6	20
80D	29.15 to 1		4.4375	17	30.0	20
57D	29.32 to 1		4.0000	13	27.6	20
40D	29.20 to 1		4.0000	7	24.0	20
25D	28.90 to 1		3.0000	5	18.4	
16D	35.70 to 1		2.5000	5	15.3	

Chain reducers use sprockets and chains for speed reduction and are available in double- or triple-reduction configurations. The chains used are double or (more frequently) triple **antifriction** roller chains. The use of chain reducers is not very common; most pumping units are equipped with geared speed reducers, called gearboxes. **API Spec. 11E** [111] covers both types of speed reducers and gives detailed design parameters with which manufacturers should comply.

3.9 V-BELT DRIVES

3.9.1 CONSTRUCTION

The pumping unit's gearbox is connected to the prime mover through a sheave and belt drive, or simply called **V-belt drive**. The purpose of this drive is to further **reduce** the speed of the prime mover's relatively high rotational speed. In case of a NEMA D electric motor the average motor speed is about 1,170 *RPM*; this speed is reduced by a typical gearbox having a reduction ratio of 30:1 to a pumping speed of 1,170/30 = 39 SPM. Since this speed would be too fast for any sucker-rod pumping unit, the V-belt drive's further reduction is needed to reach practical pumping speeds.

Other purposes of the V-belt drive are:

- It provides the means to **change** pumping speeds,
- Its use allows **mounting** of the prime mover away from the pumping unit's revolving cranks.

The V-belt drive consists of the following **components**, as shown in Fig. 3.115:

- Gearbox (or unit) **sheave** that is delivered with the pumping unit; manufacturers usually supply the largest size that can physically fit on the gearbox.
- Prime mover **sheave** that is of a smaller size; this sheave is usually changed in the field to attain the required pumping speed.
- **V-belts** (usually C section type); the proper number of belts to be used depends on the power to be transmitted and the sizes of the sheaves.

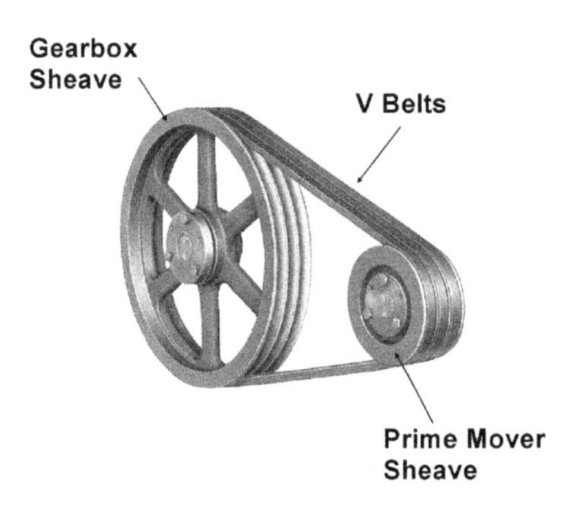

FIGURE 3.115

Components of the V-belt drive used on pumping units.

Mounting of the sheaves on the shafts of the gearbox and the prime mover can be accomplished by following the two procedures described here:

1. The conventional method needs a tight fit between the shaft and the bore of the sheave and uses a tapered **key** to secure the two parts together. Removal of the sheave may become difficult when the parts are corroded.
2. The up-to-date solution includes an intermediate component, called **QD bushing**, placed between the shaft and the sheave. As shown in Fig. 3.116, the bushing is saw cut through one side; this feature facilitates an easy mounting on the shaft. Since the outside surface of the bushing is **tapered**, the sheave with a tapered **bore** can be easily fitted. By tightening the cap screws a firm and reliable connection between the sheave, the bushing, and the shaft is achieved that can be removed without difficulty.

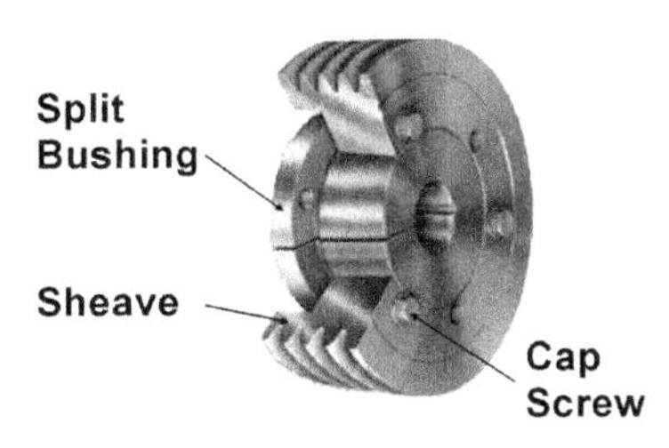

FIGURE 3.116

Use of a "QD bushing" to attach V-belt sheaves to shafts.

Sheaves usually have several **grooves** where V-belts can be fitted. If there are more grooves than V-belts, then use the grooves close to the motor to reduce the loading of motor bearings. The **number** of V-belts to be used has to be selected properly: too many cause excessive load on the motor and its bearings, but too few cause slipping and power waste.

3.9.2 OPERATING CONDITIONS

The **overall** speed reduction ratio of the drive train is composed of the gearbox's and the V-belt drive's reduction ratios, and assuming that the V-belts do not slip we get:

$$Z_d = Z\frac{D}{d} \tag{3.91}$$

where:

Z_d = overall speed reduction ratio of the drive train, –,
Z = speed reduction ratio of the gearbox, –,
D = gearbox sheave pitch diameter, in,
d = prime mover sheave pitch diameter, in.

The pumping **speed** of an existing rod-pumping installation is found from the speed of the prime mover and the overall speed reduction:

$$N = \frac{RPM}{Z_d} = \frac{RPM}{Z}\frac{d}{D} \tag{3.92}$$

where:

N = pumping speed, 1/min,
RPM = prime mover speed, 1/min,
d = prime mover sheave pitch diameter, in
D = gearbox sheave pitch diameter, in
Z = speed reduction ratio of the gearbox, –.

The length, BL, of the V-belts to be used on the given drive is calculated from:

$$BL = 2\,CD + \frac{(D+d)\pi}{2} + \frac{(D-d)^2}{4\,CD} \tag{3.93}$$

where:

BL = required length of the V-belt, in,
d = prime mover sheave pitch diameter, in,
D = gearbox sheave pitch diameter, in,
CD = center distance between the gearbox and the prime mover shafts, in.

In case belts with the calculated BL length are not available, then the prime mover is **moved** on its base to change CD to accommodate a **standard**-size belt. This feature is available on most pumping units.

The operating conditions of V-belts are affected by their **velocity,** which must be kept between 2,000 and 5,000 ft/min. Velocities less than 2,000 ft/min result in short belt life, while velocities above 5,000 ft/min need special, balanced sheaves. Belt velocity, BV, in ft/min units is calculated from the following formula:

$$BV = \frac{RPM\; d\; \pi}{12} \tag{3.94}$$

where:

BV = belt velocity, ft/min,
d = prime mover sheave pitch diameter, in,
RPM = average rotational speed of the prime mover, 1/min.

As already discussed, the pumping speed of a given sucker-rod pumping installation is usually changed by changing the prime mover sheave size, d. The proper sheave size to provide a desired pumping speed can be found from solving Eq. (3.92), as follows:

$$d = \frac{N\,Z\,D}{RPM} \tag{3.95}$$

where:

d = required prime mover sheave size, in,
N = desired pumping speed, 1/min,

D = gearbox sheave pitch diameter, in,
Z = speed reduction ratio of the gearbox, –,
RPM = average rotational speed of prime mover, 1/min.

Since sheaves are available in **standard** sizes only, the next standard sheave must be selected.

EXAMPLE 3.18: FIND THE NECESSARY PRIME MOVER SHEAVE SIZE AND THE REQUIRED BELT LENGTH IF A PUMPING SPEED OF 10 SPM IS DESIRED. THE 320D GEARBOX HAS A SPEED REDUCTION RATIO OF $Z = 30.12$, THE GEARBOX SHEAVE DIAMETER IS $D = 47$ IN, THE ELECTRIC MOTOR'S AVERAGE SPEED IS $RPM = 1{,}170$ AND THE PRIME MOVER AND GEARBOX CENTER DISTANCE IS $CD = 75$ IN. ALSO CHECK WHETHER BELT VELOCITY IS IN THE RECOMMENDED RANGE

Solution

The required prime mover sheave diameter is found from Eq. (3.95):

$$d = 10 \times 30.12 \times 47/1{,}170 = 12.1 \text{ in.}$$

A standard size of 12 in is selected and Eq. (3.93) is used to calculate the required belt length:

$$BL = 2 \times 75 + (47 + 12)\pi/2 + (47 - 12)^2/(4 \times 75) = 246.8 \text{ in.}$$

In practice, the next standard belt is selected and the centerline distance is adjusted accordingly. Belt velocity is checked by Eq. (3.94):

$$BV = 1{,}170 \times 12\pi/12 = 3{,}676 \text{ ft/min; this is in the recommended range.}$$

3.10 PRIME MOVERS

3.10.1 INTRODUCTION

Early pumping units were powered by **steam** engines, and then the slow-speed **gas** engines became standard. The use of **electric motors** gained wide acceptance during the late 1940s, and nowadays the majority of pumping units are run by electricity. Originally, the main advantages of electric power were the low **cost** of electric power, lower investment costs due to low price of electric motors, and the easy adaptation of motors to intermittent pumping. Some of these advantages still exist, although the cost of electricity has substantially increased over the years.

The choice between **electric** and **gas** power is based on several factors. The availability of gas or electricity at the well site has prime importance, but a proper decision cannot be reached without an analysis of the operating costs involved. The investment cost of a gas **engine** is much higher than that of an electric motor, but on the other hand, gas engines have a much **longer** service life. The energy costs when using electric motors have steadily increased over the last years due to increased power costs. Gas, if available, can turn out to be more economical, and can even cost **nothing** if it previously was vented. In summary, to decide on the type of prime mover to be used in a given installation, a comparison of the anticipated operating costs is required.

3.10.2 INTERNAL COMBUSTION ENGINES

Internal combustion engines usually run on gas, which can be dry separator gas or, more frequently, **wet** wellhead gas led from the casinghead. A **gas scrubber** is installed at the wellhead to knock out oil and water from the gas before it enters the carburetor. Engines can run on any wellhead gas except sour gases containing H_2S. The available units are classed into two broad categories: the slow-speed and the high-speed engines.

Slow-speed engines have operational speeds between 200 and 800 rpm and can be two-cycle or four-cycle single-cylinder or multicylinder engines. The most popular type in this category is the horizontal single-cylinder two-cycle engine with a large **flywheel**. The inertia of the flywheel is used to smooth out the speed variations during operation. The performance curves of a typical engine are given in Fig. 3.117, where engine torque and power (BHP = brake horsepower) are plotted against speed. The torque is fairly **constant** over a large range of speeds, ensuring proper operation under the **cyclic** loads of pumping. Engines are available in several sizes ranging from ten to several hundred horsepower. Their main advantages are the relatively large torque developed, few component parts, and long life.

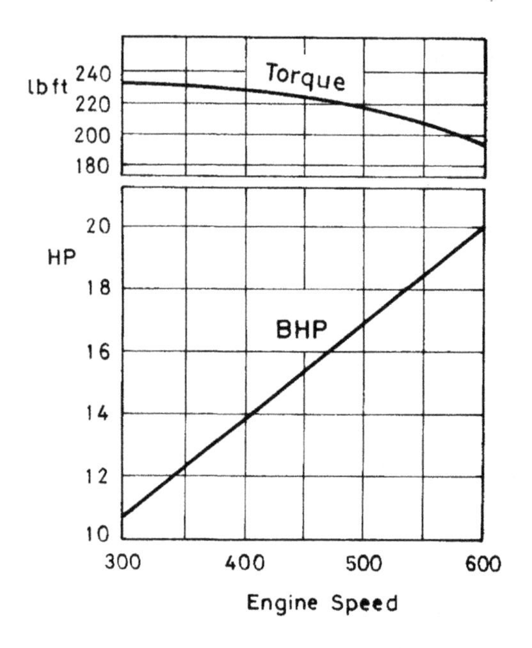

FIGURE 3.117

Performance curves of a typical internal combustion slow-speed engine.

High-speed engines run at 750–2,000 rpm and are usually four-cycle or diesel **engines**. They develop less torque than slow-speed engines and are more complex; they also have more parts. Although they are less **expensive** than slow-speed engines, their service life is **reduced** due to the higher operational speeds.

Rating and testing specifications of engines are given in **API Spec. 7B-11C** [123]; details on installation, starting, and control can be found in [124]. A compete treatment of engines used in sucker-rod pumping service is provided by **Bommer and Podio** [125].

3.10.3 ELECTRIC MOTORS

3.10.3.1 Introduction

Most sucker-rod pumping units are driven by electric motors connected to a three-phase, 60 Hz AC power supply at 480 V, although motors with 460 V nameplate rating are typically used. The reasons for the popularity of electric motors are the relatively low **cost**, easy control, and adaptability to automatic operation. Generally, three-phase squirrel-cage **induction** motors with six poles are used; these have the simplest construction among electric motors. They are also the **most reliable** motors due to the fact that their rotor is not connected to the electric supply. At the same time, they are the **most efficient** electric motors available, which explains their popularity in the oilfield.

Induction motors work on the principle of electromagnetic induction, which states that an electric current is induced in any conductor moving inside a magnetic field. The magnetic field is generated in the **stator**, the standing part of the motor containing coils for the each phase. This field rotates as the direction of the AC current changes because electromagnets change their magnetic poles twice for every cycle of the AC current. Inside the stator is, attached to the shaft of the motor, the **squirrel cage** type **rotor**, consisting of short-circuited copper bars not connected to the power source. The rotating magnetic field maintained by the stator windings induces the flow of an AC current in the rotor; thus the rotor becomes a set of electromagnets. The magnetic poles of the rotor's field are attracted and repelled by the unlike and like poles, respectively, of the stator, and the rotor maintains the continuous rotation of the motor shaft.

The rotational speed of the stator's magnetic field is the motor's **synchronous speed,** which depends on the frequency of the AC current and the number of poles the stator has:

$$N_{synch} = \frac{120\,f}{p} \tag{3.96}$$

where:

N_{synch} = the motor's synchronous speed, *RPM*,
f = frequency of the AC power, Hz,
p = number of poles in the stator, –.

As seen from Eq. (3.96), the speed of the motor depends mainly on the **frequency** of the AC power. Since most motors in sucker-rod pumping service have six poles, their synchronous speed at 60 Hz operation is 1,200 *RPM*. If driving of the sucker-rod pump at different speeds is required, the change of the driving power's frequency is the most feasible solution; this is used in installations with **variable-speed drives**.

3.10.3.2 Motor slip

Since the operation of the motor depends on electromagnetic induction, requiring a **relative** movement between the stator's (primary) magnetic field and the rotor's wires, it must be clear that the rotor must always revolve at a slower speed than the motor's synchronous speed. The difference of the synchronous speed and the speed under fully loaded conditions is an important parameter of the motor. It is usually expressed as a percentage of the synchronous speed and is called the **slip** of the motor, defined below:

$$slip = \frac{N_{synch} - N}{N_{synch}} 100 \tag{3.97}$$

where:

$slip$ = the slip of the motor, %,
N_{synch} = the motor's synchronous speed, RPM, and
N = nominal (full-load) motor speed, RPM.

Motors with different slip values are available, and they behave differently under the cyclic loading conditions of pumping. A **low-slip motor** has a small **range** of speed changes and draws **high** currents when more heavily loaded to keep its full-load speed. Motors with higher slip can **slow** down under higher loads and require **less** current than low-slip motors. It was proved that high motor speed variation has several beneficial effects on the performance of the pumping system. The **torque** required to move the rod string is **reduced** due to the lower acceleration of the polished rod. Another important effect is the utilization of the **inertial energy** of the rotating counterweights to decrease the peak torque values during the pumping cycle. When the motor slows down, inertial energy is **released** from the counterweights; when the speed increases then energy is stored; and the cycle repeats.

In order to fully utilize the benefits of high-slip motors, special motors with **ultrahigh**-slip (UHS) characteristics (slips as high as 50%) were developed for the specific conditions of rod pumping. These motors more efficiently utilize the inertia of the rotating parts of the pumping system and can thus experience lower power peaks than motors with normal slip. The high **investment** cost and relatively low efficiency of UHS motors, however, greatly limits their use for normal applications; their use is usually justified in installations where high mechanical loads cannot be reduced by changing the surface conditions like pumping speed, stroke length, and counterbalancing.

3.10.3.3 Motor loading

The nature of the pumping load, as seen in previous sections, is **cyclic**: power demand is **high** during the upstroke, when the rod string and the fluid load is lifted, and practically no power input is required during the downstroke, when the rods **fall** due to their own weight. As no prime mover would operate efficiently under such conditions, various means are employed to **counterbalance** the unit and to **smooth** out the cyclic effects of the pumping loads. But mechanical counterbalancing, although efficient, cannot completely eliminate the **fluctuations** in power demand. All this means that the electric motor is heavily loaded during part of the pumping cycle when it is **driving** the pumping unit, while it acts as an electric **generator** during other parts of the cycle when it is **driven** above its synchronous speed by the falling rod string. No other industry uses induction motors under such operating conditions and these conditions will never allow the optimum performance of the electric motor [126]. The proper selection of the motor type and size should, therefore, take into account the special conditions under which a pumping unit prime mover must operate [127,128].

Electric motors are designed for a permissible temperature rise. The heating effect of an electric current is a function of the **root mean-square** (rms) of the instantaneous current values. For a constant loading situation, where motors are designed to work, the **rms** and the average currents are equal to each other. It follows from the definition of the **rms** current that for **fluctuating** currents, the **rms** value is always **higher** than the simple average current. Thus the temperature rise occurring in an electric motor subjected to cyclic loading will be inherently **higher** than it would be if the same but **steady** average load were applied. In order to limit this temperature rise, the motor should be **derated** (oversized), i.e., a motor with a higher power capacity should be chosen [129]. The increase in required prime mover power is indicated by the cyclic load factor (*CLF*); the derating factor is the reciprocal of the *CLF* value defined here:

$$CLF = \frac{rms\ \ current}{avg\ \ current} \tag{3.98}$$

where:

CLF = cyclic load factor, –,
$rms\ current$ = rms or thermal motor current, amperes, and
$avg\ current$ = average motor current, amperes.

The *CLF* value is a measure of the **evenness** of the current drawn by the motor and should be calculated from the **instantaneous** currents. As already mentioned, for constant loads the **rms** and the average currents are equal and $CLF = 1$. Under cyclic loading conditions, however, the cyclic load factor is always **greater** than unity. Typical values suggested by **Lufkin Industries** are listed in Table 3.23 [83].

Table 3.23 Typical Cyclic Load Factor Values, According to Brown [83]

Pumping Unit Geometry	Typical CLF Values	
	NEMA C or L-S Engine	NEMA D or H-S Engine
Conventional, air balanced	1.897	1.375
Mark II	1.517	1.100

These considerations are valid only for the **continuous** operation of an electric motor. At the **start-up** of pumping, however, the whole pumping system should be set into motion from a **standstill**, and this requires very high **instantaneous** torques at the shaft of the motor. This is why motors with a **breakaway torque** (the torque valid at zero speed) of at least 2–3 times higher than the full-load torque are required for starting the pumping unit. The electric power taken by the motor during start-up (that takes less than 1 s) is at least 3 times as much as the nameplate power.

3.10.3.4 Motor types, motor performance

Electric motors used for pumping service are designated by the National Electrical Manufacturers Association (**NEMA**) as **B**, **C** and **D** motors [130]. The torque-vs-speed characteristics of these motors, when connected "across the line," are shown in Fig. 3.118. Their main features are:

NEMA B: normal slip (below 3%), full load efficiency above 92%, and breakaway torque (at zero speed) 100–175% of full-load torque.

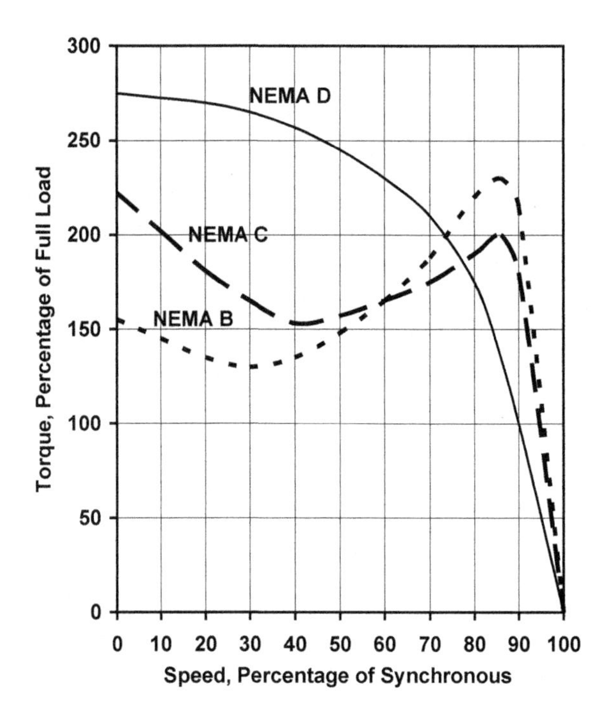

FIGURE 3.118

Torque–speed characteristics of different types of NEMA electric motors.

NEMA C: normal slip below 5%, full load efficiency above 90%, and breakaway torque 200–250% of full-load torque.

NEMA D: high slip of 5–8%, full load efficiency above 88%, and breakaway torque 275% of full-load torque.

Motors are available with the following **enclosure** types, the selection of which is based on environmental considerations:

- **Open drip-proof** (ODP) is the standard type used on pumping units; this enclosure prevents any particles (like raindrops, sand, etc.) to enter the motor from an angle less than 15° from the vertical.
- **Weather-protected** (WP) enclosures protect the motor against horizontal rain.
- **Totally enclosed fan-cooled** (TEFC) motors offer the highest protection against the most severe weather conditions, such as dust or wet weather.

The most popular electric motor in the oilfield is the **NEMA D**; the main advantageous features of this motor type in sucker-rod pumping applications are:

- It has the highest breakaway torque, facilitating easy **starting** of the pumping unit from a standstill.
- Its peak output torque is close to the locked motor torque.

- The higher slip provided by the NEMA D motors has the following advantages:
  - The higher motor speed variation increases inertial effects in the rotating components of the pumping unit.
  - Gearbox torques and motor currents decrease due to the beneficial effects of higher inertial torques.

Typical qualitative performance curves for **NEMA D** motors are shown in Fig. 3.119, where motor torque, current drawn, and motor efficiency are plotted as a function of motor **speed,** assuming a constant line voltage at the motor terminals. The most important parameter is **torque** output, which, at zero speed (locked rotor), must be sufficient to start the pumping unit; this value is called the **breakaway** torque, which is the greatest torque developed by the motor. As speed increases, the motor torque decreases to the rated torque indicated by the small circle at the motor's rated speed. Around the rated speed torque changes linearly with speed and diminishes to zero at the **synchronous** speed. When the motor is driven by the well load past its synchronous speed, it develops negative torques and the motor becomes a mechanical **brake** and an electrical **generator** at the same time. In the braking (regenerative) mode the motor returns electrical power to the distribution system; credit for this power is usually prevented by power companies by installing **detent** meters on well sites.

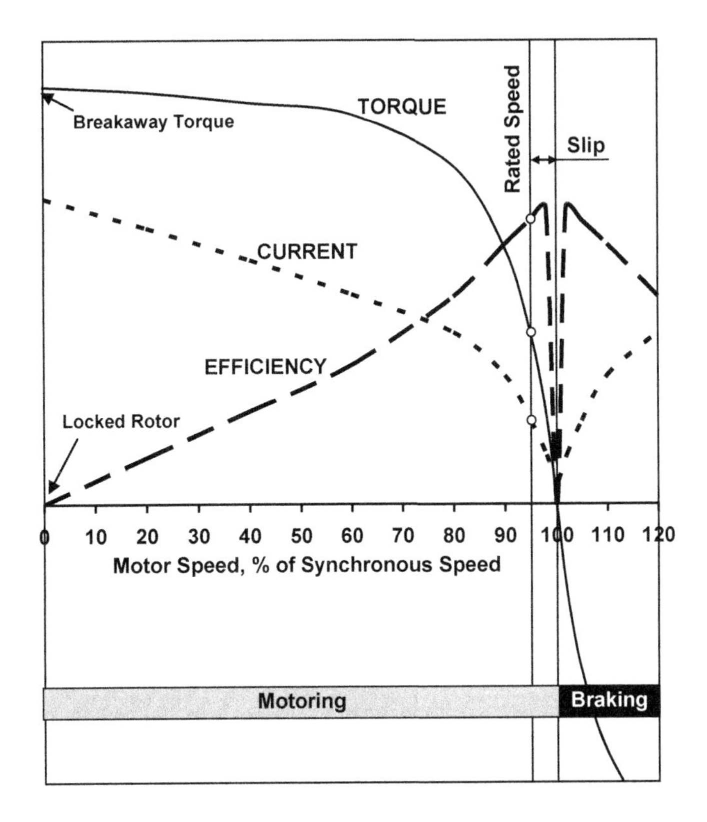

FIGURE 3.119

Qualitative performance curves of NEMA D electric motors.

At zero speed (locked rotor) when the motor starts from a standstill and must overcome the inertia of the system, motor **current** is at a maximum, which is several times greater than the rated current. Current continuously decreases to the rated current of the motor, indicated by a small circle in Fig. 3.119, as speed increases. Minimum current is reached at the synchronous speed, which is usually about one third of the motor's rated current and equals the **magnetizing** current, which provides the rotating magnetic field in the motor's stator. If the motor is driven by the well load at speeds above the synchronous speed, current generated in the motor increases and mirrors the motoring curve in the braking mode.

Motor **efficiency** defines the effectiveness of the motor to convert electrical power to mechanical power and is found as the ratio of the motor's mechanical output power to the input electrical power. At locked rotor conditions (zero speed) efficiency is zero because the motor does not turn. Efficiency is zero also at the motor's synchronous speed because no output torque (i.e., mechanical power) is created. Motor efficiency has a maximum between zero speed (locked rotor) and the synchronous speed near the motor's rated speed. At speeds close to synchronous, motor efficiency sharply drops and becomes zero at the synchronous speed. In the braking (regenerative) mode, efficiency mirrors its variation below the synchronous speed. This behavior explains why **oversized** motors running close to their synchronous speeds as well as **undersized** motors running at speeds much lower than synchronous both have low efficiencies.

The performance curves discussed so far are experimentally attained by applying different constant loads on the shaft of the electric motor tested while electric current, shaft speed, and other necessary parameters are measured at **steady-state** conditions. These curves should also be available for the **generating** region, where motor speed is above the synchronous speed and the motor acts as a generator. Although valid under steady-state conditions, the curves thus obtained are normally utilized to describe motor performance under the **dynamic** conditions of the sucker-rod pumping cycle because of the very **rapid** response of the motor to any changes in loads. Compared to the quick response of the motor, changes in motor loads during the pumping cycle occur relatively slowly. This is why steady-state performance curves combined with the time variation of mechanical loads on the motor can be used to predict the performance of electric motors driving sucker-rod pumping units. Since manufacturers usually do not publish detailed performance data, the necessary data can be requested by using a form published in **API Bulletin 11L5** [131].

As discussed in Section 3.10.3.2, motors with ultrahigh-slip properties (UHS motors) have several advantages in rod pumping. Application of ultrahigh-slip motors started many years ago [132]; their main benefits include the reduction of pumping unit structural loads, peak torques, and power consumption [133]. These motors have **special** stator and rotor constructions and can be **switched** to different operation modes. Usually, four modes are available that offer different values of breakaway torques and motor slip values. Typical performance curves for an **ultrahigh-slip motor** are given in Fig. 3.120, showing the torque–speed relationship for the four available modes. Motor slip and breakaway torque are functions of the torque mode; the usual ranges of these parameters for different motor sizes are given below:

Torque Mode	Slip	Breakaway Torque
High	23–36%	410%
Medium	29–39%	320%
Medium-low	33–51%	260%
Low	42–63%	225%

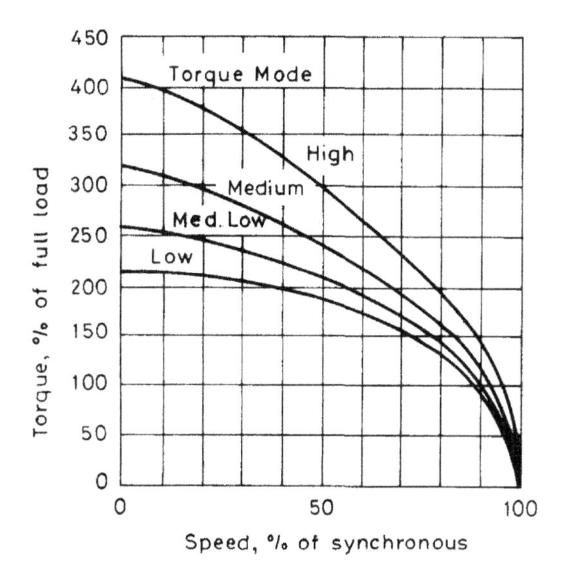

FIGURE 3.120

Typical performance curves of an ultrahigh-slip electric motor.

3.10.3.5 Motor power

When calculating the electric power taken by an electric motor, the phase angle between the voltage and the current must be properly taken into account. Because of this phase angle the **line current** (as measured by an ammeter) can be broken into two components, one in the direction of the voltage (real current) and one at right angles to it (magnetizing current). Since electric power is defined as voltage multiplied by current, one can calculate two kinds of powers depending on the current component used: (1) **apparent power** is found from the line current, and (2) **real power** from the real current.

With identical currents in each phase and equal voltages between any pairs of wires, the motor's **apparent power** (often designated as *KVA* and measured in thousands of volt amperes) is calculated from the formula:

$$KVA = \sqrt{3}\,\frac{U\,I}{1{,}000} = 1.732\ 10^{-3}\ U\ I \tag{3.99}$$

where:

KVA = apparent motor power, 1,000 VA,
U = motor voltage, V, and
I = motor amperage, A.

The **real electric power** input to the motor (often designated as **KW** and measured in kW units) is found by considering that real current equals $I_{real} = I\cos\varphi$ and:

$$KW = \sqrt{3}\,\frac{U\ I}{1{,}000}\cos\varphi = 1.732\ 10^{-3}\ U\ I\cos\varphi \tag{3.100}$$

where:

KW = real motor power, kW.

The term $\cos\varphi$ is called the **power factor** (often designated as PF) and is, by definition, the ratio of the real power to the apparent power, real power being equal to the power measured by a wattmeter and apparent power being calculated from ammeter and voltmeter readings. The actual value of power factor is a measure of the **power efficiency** of the electric motor: the greater it is, the more efficient the power consumption is in the motor. If voltage and current are in phase, then the phase angle is zero and $PF = \cos(0) = 1$; this never happens in a motor. At the synchronous speed the motor's power factor equals zero because the phase angle is 90°; at any other speed it is greater than zero but less than unity.

REFERENCES

[1] API Spec. 11AX Specification for subsurface sucker rod pumps and fittings. 12th ed. Dallas (Texas): American Petroleum Institute; 2006.

[2] Manual M-65 Pump and rod engineering manual. Longview (Texas): Axelson Inc.; 1983.

[3] Subsurface pumps: selection and application. Garland (Texas): Oilwell Division of US Steel; 1978.

[4] Matheny SL. Longest rod pump goes to work. OGJ September 22, 1980:83–4.

[5] Simon DJ. Are variables in the design details of subsurface rod pumps causing high lifting costs and reduced performance?. Proc. 28th annual southwestern petroleum short course. 1981. p. 342–347.

[6] www.flotekind.com/catalog/artificial-lift.

[7] Jayroe B. The petrovalve plus: an innovation in sucker rod pump valve technology. Proc. 43rd southwestern petroleum short course. 1995. p. 115–123.

[8] Ivey RK. Valve cage for a rod drawn positive displacement pump, US Patent 5,593,292; 1997.

[9] Ivey RK. HIVAC (High volume and compression) cages for subsurface sucker-rod pumps. Proc. 45th southwestern petroleum short course. 1998. p. 51–66.

[10] Juch AH, Watson RJ. New concepts in sucker-rod pump design. JPT March 1969:342–54.

[11] Johnson LE, Botts L. New solutions to old problems in pumping gaseous wells. Proc. 17th annual west Texas oil lifting short course. 1970. p. 115–119.

[12] Williams LR. A way to eliminate fluid and gas pound with the new two-stage charger valve. Proc. 23rd annual southwestern petroleum short course. 1976. p. 279–282.

[13] Cox BR, Williams BJ. Methods to improve the efficiency of rod-drawn subsurface pumps. Paper SPE 18828 presented at the SPE production operations symposium, Oklahoma City, Oklahoma. March 13–14, 1989.

[14] Muth GM. Pump systems and methods, US Patent 6,543,543; 2003.

[15] Muth GM, Walker TM. Extending downhole pump life using new technology. Paper SPE 68859 presented at the western regional meeting held in Bakersfield. March 26–30, 2001.

[16] www.muthpump.com.

[17] Fischer Ch K, Williams BJ. Downhole pump with bypass around plunger, US Patent 6,273,690; 2001.

[18] Parker RM, Wacker J, Watson B, Dimock J. The panacea pump tool. Proc. 49th annual southwestern petroleum short course. 2002. p. 88–95.

[19] Williams B. Sand-Pro sucker rod pump for fluid with sand production conditions in down-hole sucker rod pumps. Proc. 54th annual southwestern petroleum short course. 2007. 172–174.

[20] Sands R, Castro P. Sand-Pro Conger FMT presentation. Proc. 56th annual southwestern petroleum short course. 2009. p. 141–142.

[21] Skillman M. Downhole sucker rod pump and method of use, US Patent 6,497,561; 2002.

[22] Pickett A. Technologies optimize artificial lift. The American Oil & Gas Reporter; June 2010.

[23] Petterson JC, Curfew JV, Bucaram SM. Minimizing equipment failures in rod pumped wells. Proc. 40th southwestern petroleum short course, Lubbock, Texas. 1993. p. 96–105.

[24] Hein NW, Loudermilk MD. Review of new API pump setting depth recommendations. Paper SPE 24836 presented at the 67th annual technical conference and exhibition held in Washington, D.C. October 4–7, 1992.

[25] API RP 11AR Recommended practice for care and use of subsurface pumps. 4th ed. Dallas (Texas): American Petroleum Institute; June 2000.

[26] McCafferty JF. Importance of compression ratio calculations in designing sucker-rod pump installations. Paper SPE 25418 presented at the production operations Symposium held in Oklahoma City. March 21-23, 1993.

[27] Keelan RF. How to select downhole rod pumping equipment. WO August 1, 1984:29–34.

[28] Lubinski A, Blenkarn KA. Buckling of tubing in pumping wells, its effects and means for controlling it. Trans AIME 1957;210:33–48.

[29] Kent RA. How and why tubing anchors reduce operating costs of rod pumped wells. Proc. 5th annual west Texas oil lifting short course. 1958. p. 87–106.

[30] Logan JL. Tubing movement and tubing anchor payout in pumping wells. Proc. 11th annual southwestern petroleum short course. 1964. p. 115–120.

[31] Clegg JD. Multi-chamber gas anchor, US Patent 4,515,608; 1985.

[32] Clegg JD. Another look at gas anchors. Proc. 36th annual southwestern petroleum short course. 1989. p. 293–307.

[33] McCoy JN, Rowlan OL, Becker D, Podio AL. Optimizing downhole packer-type separators. Proc. 60th annual southwestern petroleum short course. 2013. p. 91–113.

[34] Harris JW. Development of rod pumps and subsurface accessories for pumping gaseous wells. Proc. 11th annual southwestern petroleum short course. 1964. p. 111–114.

[35] Carruth DV. Downhole gas/liquid separator and method, US Patent 7,104,321; 2006.

[36] Samayamantula JS. An innovative design for downhole gas separation. Proc. 57th annual southwestern petroleum short course. 2010. p. 169–177.

[37] McCoy JN, Rowlan OL, Becker D, Podio AL. Downhole diverter gas separator. Proc. 59th annual southwestern petroleum short course. 2012. p. 103–114.

[38] McCoy JN, Podio AL, Rowlan OL, Becker D. Evaluation and performance of packer-type downhole gas separators. Paper SPE 164510 presented at the production and operations symposium held in Oklahoma City. March 23–26, 2013.

[39] Podio AL, McCoy JN, Wood MD. Decentralized, continuous-flow gas anchor. Paper SPE 29537 presented at the production operations symposium held in Oklahoma City. April 2–4, 1995.

[40] McCoy JN, Podio AL, Wood MD, Nygaard H, Drake B. Downhole Gas separator increases beam-pumped well production. Pet Eng Int May 1995:27–34.

[41] McCoy JN, Podio AL, Woods M, Guillory C. Field tests of a decentralized downhole gas separator. Paper SPE 36599 presented at the annual technical conference and exhibition held in Denver, Colorado. October 6–9, 1996.

[42] Echometer improved collar-size gas separator, Brochure available at www.echometer.com.

[43] Campbell JH, Brimhall RM. An engineering approach to gas anchor design. Paper SPE 18826 presented at the SPE production operations symposium in oklahoma City, Oklahoma. March 13–14, 1989.

[44] McCoy JN, Rowlan OL, Podio AL. Collar size separator performance and animation. Proc. 60th annual southwestern petroleum short course. 2013. p. 115–128.

[45] McCoy JN, Podio AL, Lisigurski O, Patterson JC, Rowlan OL. A laboratory study with field data of downhole gas separators. SPE Prod Oper February 2007:20–40.

[46] McCoy JN, Patterson J, Podio AL. Downhole gas separators – a laboratory and field study. J Can Petrol Technol May 2007:48–55.

[47] Bohorquez RR, Ananaba OV, Alabi OA, Podio AL, Lisigurski O, Guzman M. Laboratory testing of downhole gas separators. SPE Prod Oper November 2009:499–509.

[48] [Chapter 9]. In: Bradley HB, editor. Petroleum engineering handbook. Richardson (Texas): Society of Petroleum Engineers; 1987.

[49] API Spec. 11B Specification for sucker rods. 26th ed. Dallas (Texas): American Petroleum Institute; 1998.

[50] Hardy AA. Correcting sucker rod troubles as seen by a manufacturer. Proc. 3rd west Texas oil lifting short course. 1956. p. 29–34.

[51] Hardy AA. Sucker-rod joint failures. API Drill Prod Pract 1952:214–25.

[52] API RP 11BR Recommended practice for care and handling of sucker rods. 9th ed. Dallas (Texas): American Petroleum Institute; 2008.

[53] Hardy AA. Why sucker rods fail. Part 2. OGJ August 12, 1963:109–12.

[54] Lozano A, Boeri J, Zalazar B, Olmos DE, Ameglio A. Development and field experience of a new concept in sucker rods technology. J Can Pet Technol December 2005:1–6.

[55] Carstensen K. Connectable rod system for driving downhole pumps for oil field installations, US Patent 7,108,063; 1996.

[56] Carstensen K. Newly designed API-modified sucker rod connection & precision makeup method. Paper SPE 86920 presented at the international thermal operations and heavy oil symposium and western regional meeting held in Bakersfield. March 16–18, 2004.

[57] PRO/KC sucker rod connection and makeup services. Weatherford brochure; 2008.

[58] Telli FD, Muse D, Fernandez E, Pereya M, Toscano R. New sucker-rod connection designed for high-load applications. Paper SPE 120627 presented at the production and operations symposium held in Oklahoma City. April 4–8, 2009.

[59] Telli FD. Increasing sucker rods' working capacities. J Pet Technol January 2010:26–9.

[60] Metters EW. Super hi-strength sucker rods. Proc. 20th southwestern petroleum short course. 1973. p. 1–6.

[61] Patton LD. Continuous rod design. PE August 1970:72–8.

[62] COROD continuous rod and well services. Weatherford brochure; 2012.

[63] COROD brochure. Highland AL systems. 1994.

[64] Coiled rod solutions. PRO-ROD brochure; 2006.

[65] Hood LE. The flexible sucker rod - an innovation in pumping. API Drill Prod Pract 1968:250–60.

[66] Rice RJ. Mobil tests new flexible sucker-rod system in Hugoton field. OGJ April 8, 1974:76–85.

[67] Zaba J. Modern oil well pumping. Tulsa (Oklahoma): The Petroleum Publishing Company; 1962.

[68] McDannold GR. Pumping through Macaroni (hollow) sucker rods. Proc. 7th west Texas oil lifting short course. 1960. p. 51–54.

[69] Hicks AW. Fiberglass sucker rods - an historical over-view. Proc. 32nd annual southwestern petroleum short course. 1985. p. 379–392.

[70] Watkins DL, Haarsma J. Fiberglass sucker rods in beam-pumped oil wells. JPT May 1978:731–6.

[71] Saul HE, Detterick JA. Utilization of fiberglass sucker rods. JPT August 1980:1339–44.

[72] API Spec. 11C Specification for reinforced plastic sucker rods. 1st ed. Dallas (Texas): American Petroleum Institute; 1986. now part of API Spec 11B.

[73] Hallden DF. Enhancement of fiberglass sucker-rod design is offered. OGJ September 30, 1985:58–61.

[74] Lea JF, Pattillo PD, Studenmund WR. Interpretation of calculated forces on sucker rods. SPE PF February 1995:41–5.

[75] Cox JC, Lea J, Nickens H. Beam pump rod buckling and leakage considerations. Proc. 51st annual southwestern petroleum short course. 2004. p. 67–84.

[76] Does buoyancy cause buckling of drill collars and drill pipe? Oil Gas J September 13, 1965:108–10.

[77] Lubinski A, Blenkarn KA. Buckling of tubing in pumping wells, its effects and means for controlling it. Pet Trans AIME 1967;210:73–88.

[78] Nickens H, Lea JF, Bhagavatula R, Garg D. Downhole beam pump operation: slippage and buckling forces transmitted to the rod string. JCPT May 2005:56–61.

[79] Bishop D, Long SW. Minimize rod buckling to reduce tubing failures. Proc. 57th annual southwestern petroleum short course. 2010. p. 23–33.

[80] Young WC, Budynas RG. ROARK'S formulas for stress and strain. 7th ed. New York (NY): McGraw-Hill; 2002.

[81] Hardy AA. Polished rod loads and their range of stress. Proc. 5th west Texas oil lifting short course. 1958. p. 51–54.

[82] Hardy AA. Sucker-rod string design and the Goodman diagram. Paper 64-Pet-2 presented at the petroleum mechanical conference of the ASME, Los Angeles. September 20–23, 1964.

[83] Brown KE [Chapter 2]. The technology of artificial lift methods, vol. 2a. Tulsa (Oklahoma): Petroleum Publishing Company; 1980.

[84] Morrow F. Using the API specification 11C and the stress range diagram. Proc. 35th southwestern petroleum short course. 1988. p. 272–277.

[85] Sucker rod handbook. Bethlehem (Pennsylvania): Bethlehem Steel Company; 1953.

[86] Eickmeier JR. Diagnostic analysis of dynamometer cards. JPT January 1967:97–106.

[87] West PA. Improved method of sucker rod string design. Proc. 20th southwestern petroleum short course. 1973. p. 157–163.

[88] West PA. Improving sucker rod string design. PE July 1973:68–77.

[89] Neely AB. Sucker rod string design. PE March 1976:58–66.

[90] API RP 11L Recommended practice for design calculations for sucker rod pumping systems (conventional units). 4th ed. Washington, D.C: American Petroleum Institute; 1988.

[91] Neely AB. Personal Communication, July 1979.

[92] Gault RH. Rod stresses from RP 11L calculations. Proc. 37th southwestern petroleum short course. 1990. p. 292–301.

[93] Gault RH, Takacs G. Improved rod string taper design. Paper SPE 20676 presented at the 65th annual technical conference and exhibition of the SPE in New Orleans. Louisiana; September 23–26, 1990.

[94] User Manual RodStar-D/V. La Habra (CA): Theta Oilfield Services; 2006.

[95] Takacs G, Gajda M. Critical evaluation of sucker rod string design procedures. Proc. 60th annual southwestern petroleum short course. 2013. p. 213–222.

[96] A Special Report from Norris Sucker rod failure analysis. NORRIS A Dover Company; 2007.

[97] Salama MM, Mehdizadeh P. Reduction in allowable stress for bent sucker rods set. OGJ August 14, 1978: 95–8.

[98] Bellow DG, Kumar A. Stress analysis of bent sucker rods. J Can Pet Technol 1978;(3):76–81.

[99] Moore KH. Learn to identify and remedy sucker-rod failures. OGJ April 9, 1973:73–6.

[100] Moore KH. Stop sucker rod failures to save money. PEI July 1981:27–42.

[101] Steward WB. Sucker rod failures. OGJ April 9, 1973:54–68.

[102] Jacobs GH. Cost-effective methods for designing and operating fiberglass sucker rod strings. Paper SPE 15427 presented at the 61st annual technical conference and exhibition of the SPE in New Orleans, Louisiana. October 5–8, 1986.

[103] Tubing wear prevention solutions. Willis (Texas): R&M Energy Systems; 2004.

[104] Murtha TP, Pirtle J, Beaulieu WB, Waldrop JR, Wood HV. New high-performance field-installed sucker rod guides. Paper SPE 16921 presented at the 62nd annual technical conference and exhibition of SPE, Dallas, Texas. September 27–30, 1987.

[105] Smith D. Rod rotator torque in rod strings. Willis, Texas: R&M Energy Systems; 2008.

[106] Angelo L. Effects of polished rods clamps on polished rod fatigue life. R&M Energy Systems; January 1995.
[107] Hercules stuffing boxes. Borger (Texas): R&M Energy Systems.
[108] Trout WC. Counterbalance for crank shafts, US Patent 1,588,784; 1926.
[109] Byrd JP. Pumping device, US Patent 3,029,650; 1958.
[110] Doyle D. Performance characteristics of the mark II improved geometry pumping unit. Proc. 52nd annual southwestern petroleum short course. 2005. p. 21–32.
[111] API Spec. 11E Specification for pumping units. 18th ed. Washington, DC: American Petroleum Institute; 2008.
[112] Mills KN. Factors influencing well loads combined in a new formula. Pet Eng April 1939. p. 37.
[113] Gray HE. Kinematics of oil-well pumping units. API Drill Prod Pract 1963:156–65.
[114] Svinos JG. Exact kinematic analysis of pumping units. Paper SPE 12201 presented at the 58th annual technical conference and exhibition of the SPE, San Francisco, California. October 5–8, 1983.
[115] Laine RE, Cole DG, Jennings JW. Production technology: harmonic polished rod motion. Paper SPE 19724 presented at the 64th annual technical conference and exhibition of the SPE, San Antonio, Texas. October 8–11, 1989.
[116] Byrd JP. A faster downstroke in a beam and sucker rod pumping unit - is it good or bad?. Proc. 27th annual southwestern petroleum short course. 1980. p. 201–223.
[117] Watson J. Comparing class I and class III varying pumping unit geometries. Proc. 30th annual southwestern petroleum short course. 1983. p. 373–384.
[118] Byrd JP. History, background, and rationale of the mark II, beam type, oil field pumping unit. Proc. 37th annual southwestern petroleum short course. 1990. p. 272–291.
[119] Lea JF. What's new in artificial lift. WO May 1985:39–46.
[120] Beck Th L. et al. Linear rod pump apparatus and method, US Patent 8,152,492; 2012.
[121] Linear rod pump. Franksville, WI: Brochure available from UNICO; 2011.
[122] Beck Th, Peterson R. A comparison of the performance of linear actuator versus walking beam pumping systems. Proc. 56th annual southwestern petroleum short course. 2009. p. 143–165.
[123] Specifications for internal-combustion reciprocating engines for oil field service. API Spec. 7B-11C. 8th ed. Dallas (Texas): American Petroleum Institute; 1981.
[124] [Chapter 10]. In: Bradley HB, editor. Petroleum engineering handbook. Richardson (Texas): Society of Petroleum Engineers; 1987.
[125] Bommer PM, Podio AL. The beam lift handbook. 1st ed. University of Texas at Austin; 2012.
[126] Audas BE, DaCunha JJ, Walker RO. Behavior and analysis of rod pumped wells: computing mechanical and electrical loading for induction motors. Proc. 59th annual southwestern petroleum short course. 2012. p. 3–16.
[127] Durham MO, Lockerd CR. Beam pump motors: the effect of cyclical loading on optimal sizing. Paper SPE 18186 presented at the 63rd annual technical conference and exhibition of the SPE, Houston, Texas. October 2–5, 1988.
[128] McCoy JN, Podio AL, Jennings J, Drake B. Motor power, current and torque analysis to improve efficiency of beam pumps. Proc. 40th annual southwestern petroleum short course. 1993. p. 44–58.
[129] Howell JK, Hogwood EE. Electrified oil production. Petroleum Publishing Co.; 1962.
[130] NEMA standard publications MG 10-2001. Rosslyn, VI: National Electrical Manufacturers Association; 2001.
[131] API BUL 11L5 Bulletin on electric motor performance data request form. 1st ed. Washington DC: American Petroleum Institute; 1993.
[132] Chastain J. How to pump more for less with extrahigh-slip motors. OGJ March 4, 1968:62–8.
[133] Lea JF, Durham MO. Study of the cyclical performance of beam pump motors. Paper SPE 18827 presented at the production operations symposium held in Oklahoma city. March 13–14, 1989.

CHAPTER

CALCULATION OF OPERATIONAL PARAMETERS

4

CHAPTER OUTLINE

Sucker-Rod Pumping Handbook. http://dx.doi.org/10.1016/B978-0-12-417204-3.00004-2

4.1 INTRODUCTION

This chapter deals with the different **calculation** models that enable the determination of the most important **operating** parameters of sucker-rod pumping. The available calculation procedures will be presented in the order of their introduction, starting with the earliest and simplest formulae of limited accuracy and ending with the up-to-date complex simulation models. The accurate prediction of the operating conditions of a rod-pumping system is of vital importance for the design of new installations and also for the analysis and optimization of existing installations. There are some very basic operational parameters; most of the additional data required for design or analysis can be derived from:

- the polished rod **loads** occurring during pumping,
- the **downhole** stroke length of the plunger, and
- the **torques** required at the speed reducer.

Due to the importance of the above parameters, several approximate formulae and calculation methods have been developed in the past to find their values. Of these, discussion includes early procedures that give fairly good estimations for shallow wells when pumping light fluid loads. Under such circumstances, the rod string can be treated as a **concentrated** mass, and this assumption leads to quite simple physical and mathematical models. As well depths increase, however, the assumptions of these conventional predictions are no longer valid, and the calculation accuracies attained rapidly deteriorate.

Section 4.3 presents a detailed treatment of the **API RP 11L** procedure, which ensures a much higher degree of accuracy in the calculation of pumping parameters than the simple formulae used previously. This calculation model is more **general**, can be used under widely varying conditions, and has been considered the standard method to determine the operational conditions of sucker-rod pumping installations for the last 40 years. Its main assumptions and limitations are detailed, and a step-by-step discussion of its use is presented.

The most accurate way to find the operating conditions of a sucker-rod pumping installation involves the solution of the one-dimensional damped **wave equation** that can be written to describe the **behavior** of the sucker-rod string. The solution of this partial differential equation is one of the engineer's most powerful **tools** to predict the performance of the rod-pumping system and to design or analyze future installations. Although most authors start from the same differential equation, several methods of solution are available. These approaches are covered, as well as details on the use of these procedures for the analysis of pumping installations.

The last two topics covered in this chapter are the determination of the torsional loads at the speed reducer and the calculation of the power requirements of pumping. The procedures for the analysis of gearbox torques occurring during the pumping cycle are presented, along with the calculations required to find the size of the prime mover. All methods are illustrated by example problems that guide the reader through the calculation procedures presented.

4.2 EARLY CALCULATION MODELS

A common feature of the simple predictions available for the determination of pumping parameters is that they treat the **elastic** behavior of the rod string using **simplified** mechanical models. The reason for the need to use simple models lies in the complexity of describing the actual behavior of the pumping system. Most of the approximate formulae were derived from the assumption that the rod string is a **concentrated** mass that is moved by the polished rod. With this approach, the performance of the pumping system is simulated by its analogy to a **spring** moving a concentrated mass. Such models usually allow for easy mathematical solutions and result in simple formulae for the calculation of the main parameters of pumping.

In addition to the crude description of the rod string's elastic behavior, the approximate calculations employ further simplifying assumptions. Conventional pumping unit geometry is usually assumed, the kinematics of the polished rod's movement is approximated by a simple harmonic motion, etc. Although all of these assumptions introduce some errors in the calculations, such simple predictions were used for many years to find the operating characteristics of sucker-rod pumping. The next sections give detailed coverage of the most widely used predictive methods.

4.2.1 POLISHED ROD LOADS

The **components** of the polished rod load, in general, are the following:

- the **weight** of the rod string,
- a **buoyant** force that decreases the rod weight,
- mechanical and fluid **friction** forces along the rod string,
- **dynamic** forces occurring in the string, and
- the **fluid** load on the pump plunger.

The sum of the rod string weight and the buoyant force is usually expressed by the "**wet weight**" of the rods, which is quite simple to calculate. The effects of the friction forces are **not** included in most calculation procedures because they are difficult or impossible to predict.

Dynamic forces at the polished rod are also easy to find if the **concentrated** mass model is used. As previously discussed in **Chapter 3**, the inertial forces are calculated by multiplying the mass being moved with the acceleration at the polished rod. It is customary to utilize the "**acceleration factor**" formula of **Mills** [1], as derived in Eq. 3.74 and reproduced below:

$$\delta = \pm \frac{S\,N^2}{70,500} \tag{4.1}$$

where:

δ = acceleration factor, –,
S = polished rod stroke length, in, and
N = pumping speed, SPM.

The dynamic forces stem from the **inertia** of the moving masses: the rod string and the fluid column. They are additive to the static loads during the upstroke and must be subtracted from the static rod weight on the downstroke. It should be noted at this point that the dynamic loads calculated by the above procedure do not include the effects of the stress **waves** occurring in the rod string. They only represent the forces required to accelerate the rods and the fluid column, which are assumed to be concentrated, inelastic masses. Therefore, the simple formulae to be presented below can yield sufficient calculation accuracy as long as this assumption is valid, i.e., in shallow- to medium-depth wells with light pumping loads.

An expression to **approximate** the peak polished rod load (*PPRL*) can now be written as the sum of the fluid load on the plunger and the static plus dynamic loads. In the familiar **Mills formula**, given below, the buoyancy of the rods is **neglected** to account for the friction forces:

$$PPRL = F_o + W_r(1 + \delta) \tag{4.2}$$

where:

$PPRL$ = peak polished rod load, lb,
F_o = fluid load on the plunger, lb,
W_r = total rod string weight in air, lb, and
δ = acceleration factor, –.

Fluid load on the plunger is found from:

$$F_o = 0.433\, H\, A_p\, SpGr \tag{4.3}$$

where:

F_o = fluid load on plunger, lb,
H = depth of the dynamic fluid level, ft,
A_p = plunger area, sq in, and
$SpGr$ = specific gravity of the produced fluid, –.

During the **downstroke**, the buoyant weight of the rod string must be decreased by the dynamic force to find the **minimum** polished rod load, because they act in opposite directions:

$$MPRL = W_{rf} - W_r\, \delta \tag{4.4}$$

The **buoyant rod weight**, W_{rf}, can be expressed with the rod weight in air, as follows:

$$W_{rf} = W_r \frac{\rho_r - \rho_l}{\rho_r} = W_r\left(1 - \frac{62.4\, SpGr}{487.5}\right) = W_r(1 - 0.128\, SpGr) \tag{4.5}$$

Substituting the above expression in Eq. (4.4), and by collecting like terms, we get the final formula for the minimum polished rod load:

$$MPRL = W_r[1 - 0.128\, SpGr - \delta] \tag{4.6}$$

where:

$MPRL$ = minimum polished rod load, lb,
W_r = total rod string weight in air, lb,
$SpGr$ = specific gravity of the produced fluid, –, and
δ = acceleration factor, –.

There are several **variants** of the polished rod load formulae just derived, as published by different authors and manufacturers. A detailed review of these methods is not considered necessary here; for further information, the reader can consult **Szilas** [2] and **Brown** [3]. The main differences between the approximate prediction formulae can usually be attributed to one of the following reasons:

- the dynamic loads can also be calculated to take into account the acceleration of the **fluid** column,
- the acceleration factor can be **modified** to describe the crank and pitman motion,
- the fluid load is sometimes calculated on the **net** plunger area, and
- the formulae can be modified for other pumping unit **geometries**.

EXAMPLE 4.1: FIND THE MAXIMUM AND MINIMUM POLISHED ROD LOADS FOR THE FOLLOWING CASE. THE ROD STRING IS COMPOSED OF 42.3% OF 3/4 IN, 40.4% OF 5/8 IN, AND 17.3% OF 1/2 IN RODS IN A 5,000 FT WELL. PLUNGER SIZE IS 1.5 IN, THE FLUID LEVEL IS AT 4,800 FT. PUMPING SPEED IS 10 SPM, STROKE LENGTH IS 120 IN, AND THE SPECIFIC GRAVITY OF THE FLUID PUMPED IS 0.95

Solution

First calculate the total rod weight, using the specific rod weights found in Table 3.9:

$$W_r = 5,000(0.423 \times 1.634 + 0.404 \times 1.135 + 0.173 \times 0.726) = 5,000 \times 1.275 = 6,375 \text{ lb.}$$

The value of the acceleration factor is calculated with the use of Eq. (4.1):

$$\delta = 120 \times 10^2/70,500 = 0.17.$$

The fluid load on the plunger, using Eq. (4.3):

$$F_o = 0.433 \times 4,800 \times 1.52 \times 3.14/4 \times 0.95 = 3,487 \text{ lb.}$$

The peak polished rod load can now be calculated with Eq. (4.2):

$$PPRL = 3,487 + 6,375(1 + 0.17) = 3,487 + 7,459 = 10,946 \text{ lb.}$$

Minimum polished rod load is found from Eq. (4.6):

$$MPRL = 6,375(1 - 0.128 \times 0.95 - 0.17) = 6,375 \times 0.708 = 4,513 \text{ lb.}$$

4.2.2 PEAK NET TORQUE

The **net crankshaft torque** on the speed reducer (gearbox) of a pumping unit is the sum of the torques required to move the polished rod and the counterweights. Thus, actual torque loading heavily depends on the **counterbalancing** of the unit. The approximate calculation models are all based on the assumptions that:

- the unit is **perfectly** counterbalanced, and
- maximum and minimum polished rod loads occur at crank angles where the torque factor is at a **maximum**.

An approximate **ideal counterbalance effect** (i.e., the force required at the polished rod to **perfectly** counterbalance the unit) can be found as the rod string weight plus half the fluid load. This can also be expressed with the use of the polished rod loads:

$$CBE = \frac{PPRL + MPRL}{2} \tag{4.7}$$

where:

CBE = counterbalance effect, lb,
$PPRL$ = peak polished rod load, lb, and
$MPRL$ = minimum polished rod load, lb.

The net torque at the crankshaft is calculated using the concept of **torque factors** by multiplying the net polished rod loads with the torque factor values. If a conventional geometry pumping unit and a simple harmonic motion at the polished rod is assumed, then it can be shown (see Eq. 3.49) that the maximum value of the torque factor equals half of the polished rod stroke length. An approximate formula for the peak net torque is derived accordingly:

$$PT = (PPRL - CBE)\frac{S}{2} = (PPRL - MPRL)\frac{S}{4} \tag{4.8}$$

where:

PT = peak net torque, in-lb,
$PPRL$ = peak polished rod load, lb,
$MPRL$ = minimum polished rod load, lb, and
S = polished rod stroke length, in.

EXAMPLE 4.2: CALCULATE THE APPROXIMATE PEAK GEARBOX TORQUE FOR THE CASE PRESENTED IN THE PREVIOUS EXAMPLE

Solution

Use Eq. (4.8):

$$PT = (10{,}946 - 4{,}513) \times 120/4 = 192{,}990 \text{ in lb.}$$

4.2.3 PLUNGER STROKE LENGTH

The **surface** stroke length, as measured at the polished rod, and the plunger's **downhole** stroke length can differ **considerably**, mainly due to the elastic behavior of the rod string. **Plunger stroke** is defined as the travel of the pump plunger **relative** to the working barrel, because this relative movement causes the displacement of the produced fluids in the pump. After they are run in the well, the tubing and the rod strings are both **loaded** by their own weight. Accordingly, the rod and tubing strings are permanently **stretched**, but their elongations do not influence the downhole plunger stroke, since these loads are always present during the pumping cycle. The load that causes a **variable** stretch in both the tubing and the rod strings is the **fluid load**, which alternately acts on the plunger and on the tubing. To find the downhole plunger stroke, therefore, the prime consideration should be to accurately calculate the elongations of the tubing and rod strings during the pumping cycle.

Assume that a sucker-rod pump is driven with a sufficiently **slow** pumping speed to make the dynamic loads in the rod string **negligible**. Consider an **unanchored** tubing string, as shown in Fig. 4.1. At the start of the polished rod's upstroke, the fluid load is **transferred** to the plunger when the traveling valve closes, causing the rods to stretch. Thus the plunger, in relation to the working barrel, is also **moved** downward by an amount equivalent to the rod stretch. At the same time, the tubing string is **unloaded** and returns to its unstretched position, since the standing valve no longer carries the fluid load. The pump barrel, therefore, moves upward, and this movement has the same effect as if the plunger moved in the downward direction. As a result, the plunger's movement at the start of the upstroke is directed downward and amounts to the sum of the **elongations** of the tubing and the rod strings. But the rod string is simultaneously lifted by the polished rod, which has to cover a distance equivalent to the plunger's movement before the plunger can start its upstroke. From this moment on, the polished rod and the plunger move **together** until the top of the upstroke is reached. This elementary description of the pumping motion shows that, for **low** pumping speeds, the difference between polished rod stroke and downhole plunger stroke equals the sum of tubing and rod string elongations that occur due to the fluid load.

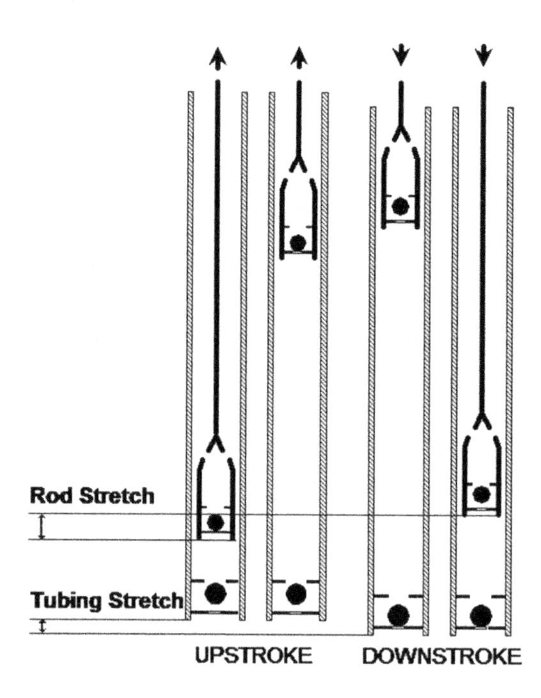

FIGURE 4.1

Approximate determination of plunger stroke length.

Thus, for **slow** pumping speeds, **plunger** stroke length for an unanchored tubing string is calculated as:

$$S_p = S - (e_r + e_t) \tag{4.9}$$

where:

S_p = plunger stroke length, in,
S = polished rod stroke length, in,
e_r = rod string stretch, in, and
e_t = tubing stretch, in.

The elongations of the tubing and rod strings are found by using **Hooke's law**, which is given in consistent units below:

$$e = \frac{F\,L}{A\,E}$$

In the case of a single rod section, and using the fluid load F_o as the force elongating the rod string, rod **stretch** (in field units) is found from the formula:

$$e_r = \frac{12\,F_o\,L}{A_r\,E} \tag{4.10}$$

The term $12/A_r/E$ is called the **elastic constant** of the given rod and can be found from Table 3.9 for steel rods or from Table 3.12 for fiberglass rods. Introducing the elastic constant in Eq. (4.10), the final formula for rod stretch is:

$$e_r = F_o\,E_r\,L \tag{4.11}$$

where:

e_r = rod stretch, in,
E_r = rod elastic constant, in/lb/ft,
L = length of the rod section, ft, and
F_o = fluid load on plunger, lb.

For **tapered sucker-rod strings**, Eq. (4.11) is modified to take into account the different elastic properties of the rod sizes making up the string. The total rod stretch of a tapered sucker-rod string is calculated with the following formula:

$$e_r = F_o \sum_{i=1}^{N} E_{\mathrm{ri}} L_i \tag{4.12}$$

where:

e_r = rod stretch, in,
F_o = fluid load on plunger, lb,
E_{ri} = elastic constant of the ith rod section, in/lb/ft,
L_i = length of the ith rod section, ft, and
N = number of tapers in the string.

For an **anchored tubing string**, no tubing elongation is allowed by the tubing anchor. For unanchored tubing strings, the stretch of the tubing is derived analogously to rod stretch calculations:

$$e_t = F_o \, E_t \, L_t \tag{4.13}$$

where:

e_t = tubing stretch, in,
L_t = tubing length, ft,
E_t = elastic constant of the tubing string, in/lb/ft, and
F_o = fluid load on plunger, lb.

For illustration, basic properties of the some common tubing sizes are listed in Table 4.1, based on data from a major manufacturer. The elastic constant of the tubing, E_t (in/lb/ft), represents the stretch in inches of a 1-ft-long tubing section under a load of 1 lb. Its values were calculated from $E_t = 12/A_t/E_t$; where $E_t = 3.0E7$ psi is the Young's modulus for steel.

Table 4.1 Basic Mechanical Data of Common Tubing Sizes

Tubing Size (in)	Weight (lb/ft)	Metal Area (sq in)	Elastic Constant (in/(lb-ft))
2 3/8	4.00	1.158	3.454E-07
	4.60	1.304	3.067E-07
	5.95	1.692	2.364E-07
	7.45	2.152	1.859E-07
2 7/8	6.40	1.812	2.208E-07
	8.60	2.484	1.610E-07
	9.45	2.708	1.477E-07
3 1/2	9.20	2.590	1.544E-07
	10.20	2.915	1.372E-07
	12.70	3.682	1.086E-07

Up to this point a **slow** pumping speed was assumed and dynamic loads were **disregarded** in the calculation of plunger stroke length. If the **concentrated** mass principle is still used and **actual** pumping speeds are considered, it is easy to see that due to the acceleration pattern of the rod string, **inertial forces** will arise during the pumping cycle. These inertial forces are maximal at the two endpoints of the stroke where maximum accelerations occur. According to **Hooke's law**, the dynamic forces cause a further **elongation** of the rod string that will increase the plunger stroke, since the inertial forces try to move the plunger further in the direction of its original movement. This phenomenon was first described by **Marsh** [4] and the dynamic component of the plunger stroke was called "**overtravel.**" This term indicates that the plunger's stroke is longer in dynamic conditions than for slow pumping speeds.

The amount of plunger **overtravel** is quite easy to find if the rod string is assumed to be a **concentrated** mass. Using the simple harmonic motion to approximate the polished rod's acceleration pattern, the dynamic force is calculated by the **Mills equation**. The rod stretch caused by this load equals the overtravel of the plunger and is calculated from **Hooke's law**. It was **Coberly** [5] who used these principles to develop the following formula for calculating plunger overtravel:

$$e_o = 1.36\text{E-}6\, L^2\, \delta \tag{4.14}$$

where:

e_o = plunger overtravel, in,
L = rod string length, ft, and
δ = Mills acceleration factor, –.

In summary, rod and tubing **elongation** due to the fluid load tends to **decrease** plunger stroke, whereas **dynamic** effects **increase** it. Combining Eqs (4.9) and (4.14), the final expression for the approximation of plunger stroke length, when using an unanchored tubing string, is given below. (For an anchored tubing, the tubing stretch term e_t is set to zero.)

$$S_p = S - (e_r + e_t) + e_o \tag{4.15}$$

where:

S_p = downhole plunger stroke length, in,
S = surface polished rod stroke length, in,
e_r = rod stretch, found from Eq. (4.12), in,
e_t = tubing stretch, found from Eq. (4.13), in, and
e_o = plunger overtravel, found from Eq. (4.14), in.

As with the other approximate formulae for the calculation of pumping parameters, several **versions** of this equation were proposed by different investigators. All have the common feature of assuming a **concentrated** mass moved by the polished rod and disregarding the dynamic forces occurring in the rod string due to the elastic stress waves. In spite of their limitations, such formulae are widely used to estimate downhole plunger stroke length. For increased calculation accuracy, however, more sophisticated mechanical and mathematical modeling of the rod string's elastic behavior is required.

EXAMPLE 4.3: CALCULATE THE PLUNGER STROKE LENGTH FOR THE PREVIOUS EXAMPLES IF THE SIZE OF THE UNANCHORED TUBING IN THE WELL IS 2 7/8 IN

Solution

Rod stretch is calculated from Eq. (4.12), using the elastic constants for the different rod sizes found in Table 3.9, and the fluid load value calculated in Example 4.1:

$$e_r = 3,487 \times 5,000(8.762 \times 10^{-7} \times 0.423 + 1.262 \times 10^{-6} \times 0.404 + 1.99 \times 10^{-6} \times 0.173)$$
$$= 17,435,000 \times 1.22 \times 10^{-6} = 21.3 \text{ in.}$$

The elongation of the unanchored tubing is calculated from Eq. (4.13), using the tubing elastic constant from Table 4.1:

$$e_t = 2.208 \times 10^{-7} \times 5,000 \times 3,487 = 3.9 \text{ in.}$$

Plunger overtravel is calculated from Eq. (4.14), and using the value of the acceleration factor already found in Example 4.1 we get:

$$e_o = 1.36 \times 10^{-6} \times 5,000^2 \times 0.17 = 5.8 \text{ in.}$$

Using Eq. (4.15), the plunger stroke is finally calculated:

$$S_p = 120 - (21.3 + 3.9) + 5.8 = 100.6 \text{ in.}$$

4.3 THE API RP 11L METHOD

4.3.1 INTRODUCTION

Because of the inaccuracies generally experienced when using the early approximate procedures for calculating the operational parameters of rod pumping, the need arose to develop highly **reliable**, exact prediction methods. The conventional analysis methods, however, if applied to the pumping system, do not give satisfactory results, mainly due to the great number of influencing parameters. Therefore, the only efficient way to achieve the goal is the **modeling** of the total pumping system. This approach was used by **Sucker Rod Pumping Research Inc.**, founded in 1954 and sponsored by several US manufacturing and operating companies. This firm's only task was to **investigate** the many complex problems of sucker-rod pumping and to **develop** accurate calculation procedures. In the first phase of the research project, a **mechanical** model was created to simulate the operation of the pumping system. This proved to be too complicated to manage and control, and an **electric analog model** had to be developed instead. The electric analog was much simpler to operate and had the added advantage of providing outputs that were easy to process. Details of the development of the analog model and the calculation procedure are described by **Gibbs** [6].

During the construction of the electric analog model, great effort was done to include the effects of all parameters influencing the operation of the pumping system [7]. The most important improvement was the exact description of the propagation of longitudinal **stress waves** that travel in the rod string during pumping. This feature of the mathematical model allowed the calculation of the forces and displacements at any depth **along** the rod string [8]. Thus the **dynamic** behavior of the rod string could be described exactly; this has been completely neglected in the previous calculation procedures that assumed a concentrated inelastic rod string.

The basic **assumptions** used for the electric analog model (which limit the use of the results) are:

- A **conventional** geometry pumping unit drives the polished rod.
- The pump completely **fills** with liquid in every cycle; the effect of free gas at the pump suction is not considered.
- The pumping unit is perfectly **counterbalanced**.
- The tubing string is assumed to be vertical and **anchored** at the pump setting depth.
- **Steel** rods are used.
- Prime mover **slip** is low, equivalent to the use of a NEMA D electric motor.
- Normal downhole **friction** due to viscous damping is assumed. Rod–tubing mechanical friction is not considered.

Thousands of experimental runs were made on the analog model representing broad ranges of the independent variables (stroke length, pumping speed, plunger size, etc.). The analog computer was capable of creating synthetic dynamometer cards and making predictions of pump capacity, power requirements, and the loading of the pumping unit and the rod string. The output parameters obtained, due to their great number and the complex interactions between them, could be correlated only by using **nondimensional** parameter groups. The results of the final correlations were plotted on several charts that were first published by the **American Petroleum Institute** (API) in 1967. The calculation procedure based on these analog studies is contained in the original publication **API RP 11L**, which has been regularly updated since that time [9] and has become the standard method of determining the operational parameters of sucker-rod pumping. Since its first publication this procedure has proven very accurate and even today compares very favorably with the results obtained from the solution of the wave equation, but at a much reduced computing effort.

4.3.2 DYNAMIC BEHAVIOR OF ROD STRINGS

As already pointed out, the **RP 11L** procedure correctly accounts for the **elastic** behavior of the rod string. To achieve this, the dynamic behavior of the string, i.e., the propagation of the force **waves** and their effects, had to be studied first [10]. During pumping, the rod string is **excited** at both ends, at the top by the polished rod and at the bottom by the downhole pump's action. The characteristics of the rod string's surface motion are determined by the **kinematic** parameters of the pumping unit that drives the polished rod. This motion is transmitted by the **elastic** rod string in the form of elastic **force waves** or vibrations along its length. The force waves are met by similar waves generated by the pump at the bottom of the string. The interactions and reflections of the force waves greatly influence the forces and displacements that occur at different points along the rod string.

In order to properly describe the vibration characteristics of rod strings, their **resonant** behavior must be known. It has been shown [10] that the undamped **natural frequency** (synchronous speed) of a straight uniform sucker-rod string, expressed in strokes/min, can be found from the formula provided because one fourth of a complete wave occurs in the rod:

$$N_o = \frac{v_s}{4\,L} 60 = \frac{15\,v_s}{L} \tag{4.16}$$

where:

N_o = undamped natural frequency, strokes/min,
v_s = sound velocity in rod material, ft/s, and
L = length of the rod string, ft.

The velocity of force wave propagation equals the **sound velocity** valid in the given rod material. A typical value for steel rods is about 16,300 ft/s, after also taking into account the effects of the rod couplings. Substituting this value into this equation, the formula suggested in **RP 11L** is arrived at:

$$N_o = \frac{245{,}000}{L} \tag{4.17}$$

where:

N_o = undamped natural frequency, strokes/min, and
L = length of the rod string, ft.

As seen, the **fundamental** natural frequency, also called the **synchronous pumping speed**, of a straight sucker-rod string is the sole function of its **length** and does not depend on the size of the rod used. For tapered rod strings, however, the natural frequency differs considerably from the value suggested by this equation. It can be shown that the natural frequency of a tapered string made of **steel** is always **higher** than that of a straight rod string of the same length. In order to find synchronous speeds for tapered strings, the **RP 11L** introduces the concept of the **frequency factor**, F_c in the following formula:

$$N'_o = F_c\, N_o \tag{4.18}$$

where:

N'_o = natural frequency of a tapered string, strokes/min,
F_c = frequency factor, –, and
N_o = natural frequency of a straight rod string of the same length, strokes/min.

For steel rods, the value of the frequency factor is always **greater** than unity and is a function of the relative lengths and cross-sectional areas of the individual taper sections. The **RP 11L** [9] publication contains charts for two particular rod taper combinations that can be used to find frequency factors.

EXAMPLE 4.4: FIND THE SYNCHRONOUS PUMPING SPEED OF A THREE-TAPER ROD STRING 6,000 FT LONG AND COMPOSED OF 2,268 FT OF 7/8 IN, 2,220 FT OF 3/4 IN, AND 1,512 FT OF 5/8 IN ROD SECTIONS

Solution

First the natural frequency for a straight string with the same total length is calculated using Eq. (4.17):

$$N_o = 245{,}000/6{,}000 = 40.8 \text{ strokes/min.}$$

To find the frequency factor, first the taper percentages are calculated:

$$\text{percentage } 7/8 \text{ in} = 2{,}268/6{,}000 \times 100 = 37.8\%.$$

$$\text{percentage } 5/8 \text{ in} = 1{,}512/6{,}000 \times 100 = 25.2\%.$$

The frequency factor is read off from the appropriate chart in **RP 11L** [9] as $F_c = 1.168$, and the synchronous speed of the tapered string is thus:

$$N'_o = 1.168 \times 40.8 = 47.7 \text{ strokes/min.}$$

4.3.3 CALCULATION PROCEDURE

4.3.3.1 Independent variables

As already mentioned, the results of the simulation runs on the analog model were correlated using **dimensionless** (nondimensional) parameters. These parameters were selected after a thorough dimensional analysis of the pumping system's performance. It was found that the following **dimensionless groups** can be used reliably as independent (input) variables for correlating the conditions of widely different pumping cases:

N/N_o = dimensionless pumping speed, –,
N/N_o' = dimensionless pumping speed, –,
$F_o/S/k_r$ = dimensionless rod stretch due to fluid load, –, and
$W_{rf}/S/k_r$ = dimensionless rod stretch due to buoyant rod string weight, –.

where:

N = actual pumping speed, strokes/min,
N_o = synchronous pumping speed for a straight rod string, strokes/min,
N_o' = synchronous pumping speed for the tapered rod string, strokes/min,
F_o = fluid load on the plunger, lb,
S = polished rod stroke length, in,
k_r = spring constant of the rod string, lb/in, and
W_{rf} = buoyant rod string weight, lb.

The first two parameters are the ratio of the actual pumping speed N to the rod string's natural frequencies. They only differ from each other in their denominators, where the natural frequency of a straight rod string or the tapered string's actual frequency is used. The next two parameters need some clarification. Both of them contain the factor k_r, which is the **spring constant** of the rod string, i.e., the force in pounds required to stretch the total rod string by a length of 1 in. Since the individual rod taper sections are connected in **series**, the net spring constant of the total string is calculated by:

$$\frac{1}{k_r} = \sum_{i=1}^{N} L_i\, E_{ri} \tag{4.19}$$

where:

k_r = spring constant of the rod string, lb/in,
L_i = length of the ith taper section, ft, and
E_{ri} = elastic constant of the ith rod taper section, found in Table 3.9, in/lb/ft.

Fluid load on the plunger is calculated on the gross area of the plunger:

$$F_o = 0.34\, H\, d^2\, SpGr \tag{4.20}$$

where:

F_o = fluid load on plunger, lb,
H = dynamic fluid level, ft,

d = plunger diameter, in, and
$SpGr$ = specific gravity of the produced fluid, –.

The buoyant rod string weight is found from:

$$W_{rf} = W_r(1 - 0.128\ SpGr) \quad (4.21)$$

where:

W_{rf} = rod string weight including buoyancy, lb,
W_r = total rod string weight in air, lb, and
$SpGr$ = specific gravity of the produced fluid, –.

Based on these formulas, the dimensionless independent parameters can be easily calculated.

4.3.3.2 Calculated parameters

Just as was done with the **independent** input variables, the **dependent** parameters (i.e., the outputs of the analog model) were also processed with the introduction of dimensionless groups. The resultant correlations are plotted in **API RP 11L** (now **API TL 11L** [9]) on several **charts** as a function of the independent variables discussed above. The calculation of the operational parameters of pumping is based on the use of these charts; the procedure to be followed is detailed in the following.

Plunger Stroke Length

For an **anchored** tubing string, the plunger stroke is calculated from:

$$S_p = S\frac{S_p}{S} \quad (4.22)$$

where:

S_p = length of the plunger's downhole stroke length, in,
S_p/S = dependent variable read from Fig. 4.1 in **API TL 11L** [9], –, and
S = polished rod stroke length, in.

In case the tubing is **not** anchored, the elongation of the tubing due to the fluid load decreases the plunger's stroke length, as given below:

$$S_p = S\frac{S_p}{S} - \frac{F_o}{k_t} \quad (4.23)$$

where the parameters not defined above are:

F_o = fluid load on the plunger (Eq. 4.20), lb, and
k_t = spring constant of the tubing string, lb/in.

The unanchored tubing's **spring** constant is calculated by the following expression, where the elastic constant of the tubing string E_t is found from Table 4.1:

$$k_t = \frac{1}{E_t\ L} \quad (4.24)$$

where:

k_t = spring constant of the tubing string, lb/in,
E_t = elastic constant of the tubing string, in/lb/ft, and
L = pump setting depth, ft.

Plunger Displacement

Based on a 100% filling efficiency (i.e., assuming the pump barrel to fill up **completely** with liquid during every stroke), the daily volume **displaced** by the pump can be calculated based on the plunger stroke length:

$$PD = 0.1166\, d^2\, S_p\, N \tag{4.25}$$

where:

PD = pump displacement, bpd,
S_p = plunger stroke length, in,
N = pumping speed, strokes/min, and
d = plunger size, in.

Polished Rod Loads

The peak and the minimum **loads** on the polished rod during the pumping cycle are calculated as follows:

$$PPRL = W_{rf} + \frac{F_1}{Sk_r}\, Sk_r \tag{4.26}$$

$$MPRL = W_{rf} - \frac{F_2}{Sk_r}\, Sk_r \tag{4.27}$$

where:

$PPRL$ = peak polished rod load, lb,
$MPRL$ = minimum polished rod load, lb,
W_{rf} = buoyant rod string weight (Eq. 4.21), lb,
$F_1/S/k_r$ = dependent variable read from Fig. 4.2 in **API TL 11L** [9], –,
$F_2/S/k_r$ = dependent variable read from Fig. 4.3 in **API TL 11L** [9], ,
S = polished rod stroke length, in, and
k_r = spring constant of the rod string (Eq. 4.19), lb/in.

Polished Rod Power

The **power** required to drive the polished rod is given by:

$$PRHP = 2.53\text{E-}6\, \frac{F_3}{Sk_r}\, S^2\, N\, k_r \tag{4.28}$$

where:

$PRHP$ = polished rod horsepower, HP,
$F_3/S/k_r$ = dependent variable read from Fig. 4.5 in **API TL 11L** [9], –,
S = polished rod stroke length, in,

N = pumping speed, strokes/min, and
k_r = spring constant of the rod string (Eq. 4.19), lb/in.

Peak Net Torque

The peak net torque on the speed reducer is calculated assuming a **perfectly** balanced pumping unit. Thus, the actual **counterbalance effect** at the polished rod is assumed to be equal to its ideal value, given below:

$$CBE = 1.06(W_{rf} + 0.5\,F_o) \tag{4.29}$$

where:

CBE = ideal counterbalance effect at the polished rod, lb,
W_{rf} = buoyant rod string weight (Eq. 4.21), lb, and
F_o = fluid load on the plunger (Eq. 4.20), lb.

Under these conditions, the **peak** net torque is calculated by:

$$PT = \frac{2T}{S^2 k_r}\,\frac{S^2}{2}\,k_r\left[1 + 10\left(\frac{W_{rf}}{Sk_r} - 0.3\right)T_a\right] \tag{4.30}$$

where:

PT = peak net torque on speed reducer, in-lb,
$2T/S^2/k_r$ = dependent variable read from Fig. 4.4 in **API TL 11L** [9], –,
S = polished rod stroke length, in,
k_r = spring constant of the rod string (Eq. 4.19), lb/in,
$W_{rf}/S/k_r$ = dimensionless independent variable, –, and
T_a = torque adjustment factor, read from Fig. 4.6 in **API TL 11L** [9], –.

In summary, the **RP 11L** calculations require the following main steps to be performed:

1. Collection of **input** data. These can come from an existing installation, or the predicted data of a future installation can be used.
2. Calculation of the **independent** dimensionless parameters, based on the input data.
3. Using the **design** charts the **dependent** dimensionless parameters are found.
4. From the dependent variables, the **operational** parameters of pumping are determined.

EXAMPLE 4.5: CALCULATE THE MAIN PARAMETERS OF THE PUMPING SYSTEM GIVEN IN EXAMPLE 4.1, USING THE RP 11L PROCEDURE

Solution

The calculations are given below, according to the four steps detailed above.

Step 1. Input data collection.

Well data were given in Example 4.1 and are reproduced here.

Pump setting depth: L = 5,000 ft.

Dynamic liquid level: H = 4,800 ft.

Plunger size: d = 1.5 in.

Tubing size: d_t = 2 7/8 in.

Pumping speed: N = 10 strokes/min.

Stroke length: S = 120 in.

Liquid specific gravity: $SpGr = 0.95$.
Rod string: 42.3% of 3/4 in, 40.4% of 5/8 in, 17.3% of 1/2 in rods.

Step 2. Calculation of independent variables.
The N_o synchronous speed is calculated from Eq. (4.17) as:

$$N_o = 245{,}000/5{,}000 = 49 \text{ strokes/min.}$$

The frequency factor is found as $F_c = 1.17$ from the appropriate RP 11L chart. Using this value, N_o' is found from Eq. (4.18):

$$N_o' = 1.17 \times 49 = 57.3 \text{ strokes/min.}$$

The fluid load and the total rod string weight were calculated in Example 4.1:

$$F_o = 3{,}487 \text{ lb.}$$
$$W_r = 6{,}375 \text{ lb.}$$

The buoyant rod weight is found from Eq. (4.21):

$$W_{\text{rf}} = 6{,}375(1 - 0.128 \times 0.95) = 5{,}600 \text{ lb.}$$

The spring constant of the rod string is calculated from Eq. (4.19), using the elastic constants of the rods taken from Table 3.9:

$$1/k_r = 5{,}000\left(0.423 \times 8.762 \times 10^{-7} + 0.404 \times 1.262 \times 10^{-6} + 0.173 \times 1.99 \times 10^{-6}\right) = 5{,}000 \times 1.224 \times 10^{-6}$$
$$= 6.123 \times 10^{-3} \text{ in/lb.}$$
$$k_r = 1/6.123 \times 10^{-3} = 163.3 \text{ lb/in.}$$

Now the dimensionless independent variables can be calculated:

$$N/N_o = 10/49 = 0.2.$$
$$N/N_o' = 10/57.3 = 0.17.$$
$$F_o/S/k_r = 3{,}487/120/163.3 = 0.18.$$
$$W_{\text{rf}}/S/k_r = 5{,}600/120/163.3 = 0.29.$$

Step 3. The dependent dimensionless parameters are determined from **RP 11L** charts as follows:

$$S_p/S = 0.88.$$
$$F_1/S/k_r = 0.31.$$
$$F_2/S/k_r = 0.11.$$
$$F_3/S/k_r = 0.20.$$
$$2T/S^2/k_r = 0.241.$$
$$T_a = 0.037.$$

Step 4. The operational parameters are calculated.
The tubing is unanchored, and Eq. (4.23) is used to find the plunger stroke length; the elastic constant of the tubing is $E_t = 2.208 \times 10^{-7}$, as read from Table 4.1.

$$S_p = 120 \times 0.88 - 3{,}487 \times 2.208 \times 10^{-7} \times 5{,}000 = 101.8 \text{ in.}$$

Equation (4.25) is used to calculate pump displacement:

$$PD = 0.116 \times 101.8 \times 10 \times 1.5^2 = 265.7 \text{ bpd.}$$

Polished rod loads are evaluated with Eqs (4.26) and (4.27):

$$PPRL = 5{,}600 + 0.31 \times 120 \times 163.4 = 11{,}678 \text{ lb.}$$
$$MPRL = 5{,}600 - 0.11 \times 120 \times 163.4 = 3{,}443 \text{ lb.}$$

The power required at the polished rod is calculated with Eq. (4.28):

$$PRHP = 2.53 \times 10^{-6} \times 0.20 \times 120^2 \times 10 \times 163.4 = 11.9 \text{ HP.}$$

The counterbalance effect needed to perfectly counterbalance the unit is found by the use of Eq. (4.29):

$$CBE = 1.06(5,600 + 0.5 \times 3,487) = 7,784 \text{ lb.}$$

The peak net torque on the speed reducer is determined by using Eq. (4.30):

$$PT = 0.241 \times 120^2/2 \times 163.4[1 + 10(0.29 - 0.3)0.037] = 282,483 \text{ in lb.}$$

4.3.4 COMPUTER CALCULATIONS

Although the **RP 11L** procedure is quite straightforward, and a form is also provided by the API to guide the user, it still requires hand calculations and **visual** read-off, which can influence the **accuracy** obtained. Computerized solutions, therefore, not only can speed up calculation time but also can reduce **errors**. Several programs were developed for hand-held calculators [11,12], but the use of personal computers [13,14] has become standard practice. A common problem of these programs is the handling of the **charts** from which the dependent variables are read. In most cases, charts published in **RP 11L** are digitized and the data stored in a tabular form, and a lookup procedure is used to find the required values. Some authors also include minor enhancements to the original **RP 11L** procedure, to be detailed later.

The calculation of the **frequency factor**, F_c, poses a special problem because the relevant charts in **RP 11L** cannot easily be described. One **reliable** mathematical solution is proposed in [15], based on the theory behind the concept of frequency factor. It was shown in the original study on sucker-rod string vibrations [10] that **resonance** occurs when the following condition is met:

$$\tan\left(\frac{\pi}{2}x_1F_c\right)\tan\left(\frac{\pi}{2}x_2F_c\right) - \frac{A_1}{A_2} = 0 \tag{4.31}$$

where:

$A_1, A_2 =$ cross-sectional areas of rod sections, sq in,
$x_1, x_2 =$ taper lengths as fractions of the total string length, –, and
$F_c =$ frequency factor, –.

The above equation holds for a two-taper rod string, but similar expressions can be derived for strings with more tapers. The calculation of frequency factor values is thus reduced to finding the root of this equation. Using the **Newton–Raphson method**, the following iterative formula is proposed, which yields sufficient accuracy with a moderate computational effort:

$$F_c^{n+1} = F_c^n - \frac{f(F_c)}{f'(F_c)} \tag{4.32}$$

where:

F_c^{n+1} and $F_c^n =$ the successive approximations of the frequency factor,
$f(F_c) =$ the left-hand side of Eq. (4.31), and
$f'(F_c) =$ the derivative of $f(F_c)$.

EXAMPLE 4.6: FIND THE PUMPING PARAMETERS FOR THE FOLLOWING CASE. THE ROD STRING IS COMPOSED OF 2,850 FT OF 3/4 IN, 4,650 FT OF 5/8 IN GRADE D STEEL RODS IN A 7,500 FT WELL WITH AN ANCHORED TUBING STRING. PLUNGER SIZE IS 1.25 IN, THE FLUID LEVEL IS AT 7,000 FT, PUMPING SPEED IS 10 SPM, STROKE LENGTH IS 120 IN, AND THE SPECIFIC GRAVITY OF THE FLUID PUMPED IS 1.0. THE WELLHEAD PRESSURE IS 100 PSI

Solution

Exhibit 4.1 presents the output of the calculations.

```
                     PUMPING SYSTEM DESIGN CALCULATIONS
                              USING API RP 11L

              W E L L    D A T A  of Well       New Well #1

Pump Setting Depth                           :        7500.0  [ft]
Liquid Spec. Gravity                         :         1.000  [ - ]
Dynamic Liquid Level                         :        7000.0  [ft]
Wellhead Pressure                            :         100.0  [psi]
Estimated Volumetric Efficiency of the Pump  :        100.00  [%]

       I N P U T    D A T A  for  D E S I G N    C A L C U L A T I O N S

Plunger Size    :      1 1/4  [in]      Tubing is anchored
Stroke Length   :      120.0  [in]
Pumping Speed   :      10.00  [SPM]

    Taper          Rod          Size         Length        Weight        Modulus
                  Grade
                                [in]          [ft]         [lb/ft]        [psi]
    Steel         API D          3/4         2850.0         1.63        3.100E+07
    Steel         API D          5/8         4650.0         1.14        3.100E+07

  C A L C U L A T E D     N O N - D I M E N S I O N A L     V A R I A B L E S

                Fo/Skr  = 0.268                    N/No    = 0.306
                Wrf/Skr = 0.609                    N/No'   = 0.273

                 ** If Fo/Skr > 0.6 then program uses 0.6 **
             ** If N/No or N/No' is > 0.6 then program uses 0.6 **

   C A L C U L A T E D    O P E R A T I N G    C H A R A C T E R I S T I C S

Plunger Stroke Length :      98.0   [in]     Hydraulic Power      :      9.2   [HP]
Plunger Displacement  :     178.5   [bpd]    Polished Rod Power :       12.8   [HP]
                                             Lifting Efficiency :       71.9   [%]
                             CONVENTIONAL     AIR BALANCED       MARK II
Peak Polished Rod Load:          15661           15202             14895   [lbs]
Min. Polished Rod Load:           5834            5375              5068   [lbs]
Counterbalance Effect :          11314           10905             11040   [lbs]
Peak Net Torque       :         313716          301168            233129   [in-lb]
Appr. HS Motor Power  :           23.9            23.9              19.1   [HP]
Appr. NS Motor Power  :           29.8            29.8              23.8   [HP]
```

EXHIBIT 4.1

Detailed results of **API RP 11L** calculations for Example 4.6.

4.3.5 CALCULATION ACCURACY

After its first publication in 1967, the **RP 11L** became the **accepted** practice in the industry for sizing and analyzing sucker-rod pumping units. Its main **advantage** over previous calculations is that the resonant behavior of the sucker-rod string is properly accounted for. When no special problems (low pump efficiency, fluid or gas pound, excessive friction, etc.) are present, the accuracies obtainable are much higher than with the use of the early approximate methods.

The **RP 11L**'s calculation accuracy was first evaluated by **Griffin** [16–18], who utilized dynamometer cards taken under controlled conditions as a basis to compare the capabilities of the conventional and the **RP 11L** calculations. His results for conventional-geometry pumping units are summarized below:

	Avg. Errors in *PPRL*		**Avg. Errors in *PT***	
	RP 11L	**Conventional Calculations**	**RP 11L**	**Conventional Calculations**
77 wells [16]	1.41%	−3.43%	7.26%	−18.8%
124 wells [17,18]	1.90%	−12.6%	8.50%	−28.4%

Takacs et al. [15] evaluated the **RP 11L** using data of 25 wells of moderate depth and obtained the following average errors:

Parameter	**Error**
PPRL	1.20%
MPRL	2.55%
PT	5.02%
PRHP	1.58%

4.3.6 ENHANCEMENTS OF THE RP 11L MODEL

During the many years of widespread use, several problems arose that required modifications or improvements to be included in the **RP 11L** calculation procedure. In the following, those proposals are discussed that improve calculation accuracy and increase the application ranges of the original procedure.

4.3.6.1 Minor improvements

Originally, the effect of the **wellhead pressure** was neglected in the development of the **RP 11L** procedure, and this assumption could introduce some errors, especially in shallow wells with higher wellhead pressures. The inclusion of the load due to wellhead pressure acting on the plunger area, therefore, can improve the accuracy of calculated rod loads and other parameters [13,15]. **Clegg** [13] also proposed to account for the weight of the polished rod and the pump plunger by adding 100 lb to the rod string weight in air.

A frequent problem occurs when pumping installations on **shallow wells** are analyzed or designed with the original procedure. In these cases the actual value of the independent variable $F_o/S/k_r$ can fall below 0.1, its minimum shown on the **RP 11L** charts, and no predictions can be made. To overcome the problems of exceeding the original parameter ranges, additional curves are added to the design charts.

4.3.6.2 Considerations for other geometries

Since the **RP 11L** was developed to simulate a **conventional**-geometry pumping unit, its application to other geometries can lead to serious errors. For estimation purposes, however, the following formulae can be used for pumping units with geometries other than conventional, proposed by **Lufkin Industries** [19]. In the equations below, the original nomenclature of **RP 11L** is used, as defined previously.

Air-Balanced Units

$$PPRL = W_{\mathrm{rf}} + F_o + 0.85\left(\frac{F_1}{Sk_r}\,Sk_r - F_o\right) \tag{4.33}$$

$$MPRL = PPRL - \left(\frac{F_1}{Sk_r} + \frac{F_2}{Sk_r}\right)Sk_r \tag{4.34}$$

$$CBE = 1.06\frac{PPRL + MPRL}{2} \tag{4.35}$$

$$PT = 0.96\,\frac{2T}{S^2k_r}\,\frac{S^2}{2}\,k_r\left[1 + \left(\frac{W_{\mathrm{rf}}}{Sk_r} - 0.3\right)\frac{T_a}{10}\right] \tag{4.36}$$

Mark II Units

$$PPRL = W_{\mathrm{rf}} + F_o + 0.75\left(\frac{F_1}{Sk_r}\,Sk_r - F_o\right) \tag{4.37}$$

$$MPRL = PPRL - \left(\frac{F_1}{Sk_r} + \frac{F_2}{Sk_r}\right)Sk_r \tag{4.38}$$

$$CBE = 1.04\frac{PPRL + 1.25\,MPRL}{2} \tag{4.39}$$

$$PT = \left(0.93\,F_{p\ \max} - 1.2\,F_{p\ \min}\right)\frac{S}{4} \tag{4.40}$$

In order to use these equations, only the last phase of the original **RP 11L** procedure must be changed when the operation characteristics are calculated. The determination of the independent dimensionless parameters and the use of the design charts is the same for all geometries. This approach provides a quick way to estimate the effects of using pumping units of different geometries on the same well and is widely used for this purpose.

EXAMPLE 4.7: CALCULATE THE OPERATIONAL PARAMETERS OF PUMPING IF AN AIR-BALANCED OR A MARK II UNIT IS USED ON THE WELL GIVEN IN EXAMPLE 4.5

Solution

Steps 1–3 are the same as in Example 4.5 and the results found there are used.

First, find the parameters for an air-balanced unit:

Polished rod loads are found using Eqs (4.33) and (4.34):

$$PPRL = 5,600 + 3,487 + 0.85(0.32 \times 120 \times 162.5 - 3,487) = 9,087 + 2,340 = 11,427 \text{ lb.}$$

$$MPRL = 11,427 - (0.32 + 0.11) \times 120 \times 162.5 = 11,427 - 8,385 = 3,042 \text{ lb.}$$

The counterbalance effect to perfectly counterbalance the unit is calculated from Eq. (4.35):

$$CBE = 1.06(11,427 + 3,042)/2 = 7,669 \text{ lb.}$$

The peak net torque, by Eq. (4.36) is:

$$PT = 0.96 \times 0.245 \times 120^2/2 \times 162.5[1 + (0.29 - 0.3)4/10] = 274,083 \text{ in lb.}$$

The same calculations for the Mark II unit, with peak and minimum loads found from Eqs (4.37) and (4.38):

$$PPRL = 5,600 + 3,487 + 0.75(0.32 \times 120 \times 162.5 - 3,487) = 9,087 + 2,065 = 11,152 \text{ lb.}$$

$$MPRL = 11,152 - (0.32 + 0.11) \times 120 \times 162.5 = 7,022 - 8,385 = 11,152 - 8,385 = 2,767 \text{ lb.}$$

The counterbalance effect and peak net torque are calculated from Eqs (4.39) and (4.40), respectively:

$$CBE = 1.04(11,152 + 1.25 \times 2,767)/2 = 7,598 \text{ lb.}$$

$$PT = (0.93 \times 11,152 - 1.2 \times 2,767) \times 120/4 = 211,529 \text{ in lb.}$$

4.3.6.3 Major revisions

Basically, there are two reliable methods available today to design or analyze sucker-rod installations: the **RP 11L procedure** and the mathematical solution of the one-dimensional **wave equation** that describes the longitudinal vibrations in the rod string. The latter (to be detailed in Section 4.4) gives more **accurate** results but involves complex calculations and requires the use of computers. The **RP 11L**, on the other hand, is much easier to apply and requires less effort. In order to combine the relative advantages of the two methods, the **RP 11L** calculation model is sometimes utilized with charts developed from the results of simulation runs using the solution of the wave equation. Such procedures retain the familiar construction of the **RP 11L** calculations (and thus require little computational effort) but ensure a much higher calculation accuracy than the original method would allow.

The first calculation model using the above approach was described by **Clegg** [20]. Its aim is to account for the presence of free gas in the pump barrel at suction conditions that can greatly reduce the deliverability of the sucker-rod pump. The RP 11L charts are replotted using data obtained from the solution of the wave equation. Different charts for 75%, 50%, and 25% liquid fillage are developed and a conventional pumping unit is assumed. In another case, also developed at **Shell Oil Co.**, design charts for three different pumping unit geometries were constructed: conventional, Mark II, and Torqmaster [21]. These charts are used in lookup tables and a calculation procedure is presented that closely follows the steps of the **RP 11L**. The advantage of the above models is the relative ease of application and the increased calculation accuracy as compared to the original procedure. Their widespread use, however, is made impossible by the fact that only a part of the required charts is published.

4.3.7 COMPOSITE ROD STRINGS

The **API RP 11L** model was developed for rod strings made up from rods of a **single** material, steel. Therefore, to simulate the behavior of **composite rod strings**, i.e., strings composed of steel and fiberglass rod sections, some modifications of the original calculation procedure are required. In the following, the suggestions of **Jennings and Laine** [22] are detailed, in which the applicability of the **RP 11L** procedure is extended to sucker-rod pumping installations utilizing composite rod strings.

All the required modifications relate to the determination of the composite rod string's **elastic** behavior. The natural frequency of any rod string, expressed in strokes/min, is found from Eq. (4.16), which states:

$$N_o = \frac{15\ v_s}{L} \tag{4.41}$$

where:

N_o = undamped natural frequency, strokes/min,
v_s = sound velocity in rod material, ft/s, and
L = length of the rod string, ft.

The equivalent sound velocity, v_s, for a uniform composite rod string can be derived on the basic assumption that rod sections behave like **springs** connected in series. From this assumption follows:

$$v_s = \sqrt{\frac{32.2}{\sum_{i=1}^{N} \frac{x_i}{E_i\ A_i} \sum_{i=1}^{N} w_i\ x_i}} \tag{4.42}$$

where:

v_s = sound velocity in rod material, ft/s,
x_i = length fraction of the ith taper section, –,
E_i = Young's elastic modulus for the material of the ith taper section, psi,
A_i = cross-sectional area of the ith taper section, sq in,
w_i = specific weight of the ith taper section, lb/ft, and
N = total number of rod tapers in the string, –.

By substituting the sound velocity, v_s, calculated from the above expression into Eq. (4.41), the natural frequency for an equivalent, **uniform** composite rod string is found. Just like for all-steel strings, the natural frequency of a tapered composite string differs from the value of N_o thus calculated. To find the tapered string's natural frequency, the frequency factor concept, introduced in the **RP 11L**, is still used. A basic difference in the behavior of steel and composite strings is that the frequency factor, F_c, is always **less** than unity for composite strings, while it is greater than unity for all-steel strings.

The procedure for calculating the frequency factor, F_c, is based on the original study of the sucker-rod string's resonant behavior [10]. It is shown by **Jennings and Laine** [23] that the basic equations developed for steel rods must be modified to reflect the effects of the different materials of the taper sections. For a two-taper composite rod string the equation (to be solved for F_c values) is given below. (Compare with Eq. (4.31), valid for the all-steel case.)

$$\tan\left(\frac{\pi}{2}\, x_1 \frac{v_s}{v_{s1}}\, F_c\right) \tan\left(\frac{\pi}{2}\, x_2 \frac{v_s}{v_{s2}}\, F_c\right) - \frac{v_{s2}}{E_2 A_2} \frac{E_1 A_1}{v_{s1}} = 0 \tag{4.43}$$

where:

x_i = length fraction of the ith taper section, –,
E_i = Young's elastic modulus for the material of the ith taper section, psi,
A_i = cross-sectional area of the ith taper section, sq in, and
v_s = equivalent sound velocity of a uniform composite rod string, ft/s.

The actual sound velocities of the individual tapers, v_{s1} and v_{s2}, are found from the fundamental expression describing the force wave propagation speed in solid materials:

$$v_{si} = \sqrt{\frac{32.2\, E_i A_i}{w_i}} \tag{4.44}$$

where:

v_{si} = sound velocity in the ith taper section, ft/s,
w_i = specific weight of the ith taper section, lb/ft,
E_i = Young's elastic modulus of the ith taper section, psi, and
A_i = cross-sectional area of the ith taper section, sq in.

To solve Eq. (4.43) for F_c values, the **Newton–Raphson** iteration method can be used, as previously detailed.

The remaining steps of the **RP 11L** calculation procedure need not be modified and the original charts can be used to arrive at the operating parameters of pumping. This approach provides an easy way to estimate the operational conditions of pumping installations utilizing composite rod strings. In case a higher degree of accuracy is desired, a more complex simulation of the pumping system's behavior involving the solution of the wave equation is necessary.

EXAMPLE 4.8: CALCULATE THE MAIN OPERATING PARAMETERS FOR THE WELL GIVEN IN EXAMPLE 4.6 IF A COMPOSITE ROD STRING IS USED. THE TOP TAPER IS 7,000 FT OF 1 IN FIBERGLASS RODS WITH A 500-FT-LONG 2 IN API GRADE 1 SINKER BAR SECTION BELOW

Solution

Exhibit 4.2 contains the calculated data.

```
               PUMPING SYSTEM DESIGN CALCULATIONS
                       USING API RP 11L

        W E L L   D A T A  of Well      New Well #1

Pump Setting Depth                           :    7500.0  [ft]
Liquid Spec. Gravity                         :     1.000  [ - ]
Dynamic Liquid Level                         :    7000.0  [ft]
Wellhead Pressure                            :     100.0  [psi]
Estimated Volumetric Efficiency of the Pump  :    100.00  [%]

     I N P U T   D A T A  for  D E S I G N   C A L C U L A T I O N S

Plunger Size    :    1 1/4  [in]      Tubing is anchored
Stroke Length   :    120.0  [in]
Pumping Speed   :    10.00  [SPM]

    Taper          Rod          Size        Length       Weight       Modulus
                  Grade
                                [in]         [ft]        [lb/ft]       [psi]
  Fiberglass      Norris         1          7000.0        0.94      7.100E+06
  Sinker Bar   API Grade 1       2           500.0       10.70      3.100E+07

  C A L C U L A T E D    N O N - D I M E N S I O N A L    V A R I A B L E S

            Fo/Skr  = 0.485                  N/No    = 0.456
            Wrf/Skr = 1.104                  N/No'   = 0.573

              ** If Fo/Skr > 0.6 then program uses 0.6 **
          ** If N/No or N/No' is > 0.6 then program uses 0.6 **

   C A L C U L A T E D   O P E R A T I N G   C H A R A C T E R I S T I C S

Plunger Stroke Length :     145.4  [in]      Hydraulic Power     :    13.7  [HP]
Plunger Displacement  :     264.7  [bpd]     Polished Rod Power  :    14.2  [HP]
                                             Lifting Efficiency  :    96.0  [%]
                          CONVENTIONAL       AIR BALANCED        MARK II
Peak Polished Rod Load:       15319             14911             14639   [lbs]
Min. Polished Rod Load:        6398              5990              5718   [lbs]
Counterbalance Effect :       11316             11077             11329   [lbs]
Peak Net Torque       :      296619            284754            207633   [in-lb]
Appr. HS Motor Power  :        35.5              35.5              28.4   [HP]
Appr. NS Motor Power  :        44.1              44.1              35.3   [HP]
```

EXHIBIT 4.2

Detailed results of **API RP 11L** calculations for Example 4.8.

Comparison of the results of Examples 4.6 and 4.8 indicate the advantage of using composite rod strings in sucker-rod pumping. Due to the extreme elasticity of the fiberglass rods, downhole pump stroke has increased from 98 to 145.4 in, as compared to the all-steel string. Plunger displacement has accordingly increased from 179 to 265 bpd. At the same time, downhole energy losses have decreased, as shown by the increase in lifting efficiency from 72% to 96%.

4.4 SIMULATION OF THE ROD STRING'S BEHAVIOR

4.4.1 INTRODUCTION

It was recognized early by rod-pumping experts that the **key** to the proper description of the pumping system is the exact **simulation** of the rod string's **behavior**. Only this can provide the necessary accuracy in calculating the operational parameters valid at the surface and at downhole conditions.

The sucker-rod string's most important feature is its **elasticity**, which is responsible for the difficulty with which the downhole conditions can be calculated from surface data. Due to the rod string's highly elastic nature, all **impulses** generated by the pumping unit's motion at the surface are instantly **transmitted** downhole. The operation of the subsurface pump also sends similar signals to the surface. All these impulses take the form of elastic force or **stress waves** that travel along the string with the speed of the sound. The **interferences** and reflections of these waves have a drastic effect on the **displacements** and **loads** that can be observed at different points **along** the string.

The rod string fully satisfies the physical criteria for an **ideal slender bar**, making the propagation of stress waves in it a **one-dimensional** phenomenon. Several early attempts with different simplifying assumptions were tried to simulate the behavior of such stress waves [24,25]. As was discussed before, the **API RP 11L** procedure is also the result of such an approach [7,8]. Although the basic principles were seen clearly, a long time passed before **Gibbs** [26,27] published the first reliable universal method for solving the one-dimensional damped wave equation. With the advent of powerful personal computers, however, the time has come when the required complex calculations are now easily accomplished, even in the oilfield [28,29].

4.4.2 BASIC FORMS OF THE WAVE EQUATION

4.4.2.1 The Gibbs model

Figure 4.2 shows a rod string section with a uniform cross-sectional area, A, and length, L. The co-ordinate axes x and u are directed **downward** and represent axial **distance** and **displacement** of the rod along the string, respectively. In order to find the governing equation for the movement of the string, a **force balance** must be written for an **incremental** element of the rod. As shown in Fig. 4.2, the following forces act on the rod element:

- W = the buoyant weight of the rod element,
- F_x = a tension force that represents the pull from above on the rod element,
- $F_{x+\Delta x}$ = another tension force representing the downward pull on the rod element, and
- F_d = a damping force opposing the movement of the rod element, which is the result of fluid and mechanical friction on the rod element's surface.

Using **Newton's second law**, the mass of the element times its acceleration should be equal the sum of the forces acting on the element:

$$m\frac{\partial^2 u}{\partial t^2} = F_x - F_{x+\Delta x} + W - F_d \tag{4.45}$$

The weight of the rod element, W, is a **static** force that is constant during the pumping cycle. Therefore, it can be dropped from the equation, and its effect can later be superimposed on the solution

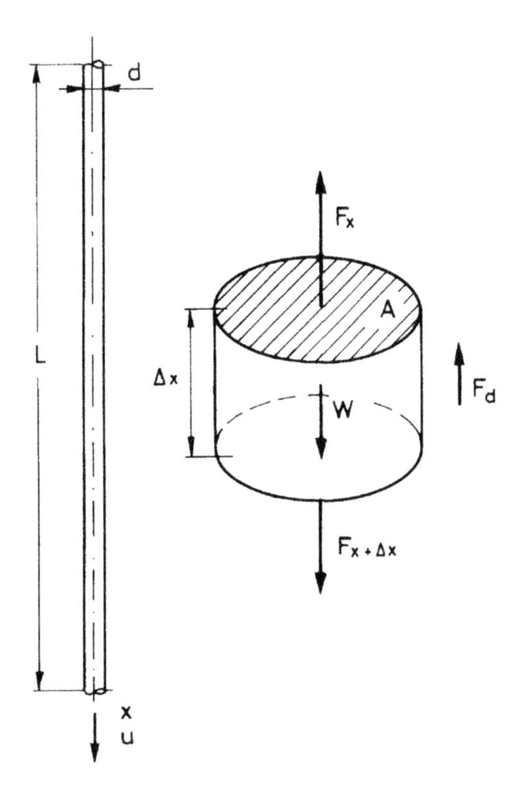

FIGURE 4.2

Illustration of the forces acting on an incremental element of the rod string.

of the wave equation. The tension forces F_x and $F_{x+\Delta x}$ can be expressed with the mechanical **stresses** present in the rod sections at the axial distances x and $(x+\Delta x)$:

$$F_x = S_x A \tag{4.46}$$

$$F_{x+\Delta x} = S_{x+\Delta x} A \tag{4.47}$$

where:

F_x and $F_{x+\Delta x}$ = rod loads in sections x and $x + \Delta x$, psi,
S_x and $S_{x+\Delta x}$ = rod stresses in sections x and $x + \Delta x$, psi, and
A = cross-sectional area of the rod, sq in.

Substituting the above expressions into Eq. (4.45):

$$m \frac{\partial^2 u}{\partial t^2} = (S_{x+\Delta x} - S_x) A - F_d \tag{4.48}$$

Since sucker rods in normal operation undergo an **elastic** deformation, **Hooke's law** can be applied, which states that the stress at any cross-section is proportional to the deformation of the actual rod element:

$$S = E \frac{\partial u}{\partial x} \tag{4.49}$$

where:

S = mechanical stress, psi,
E = Young's modulus of elasticity for the rod material, psi, and
$\partial u/\partial x$ = rod strain, i.e., the change of rod displacement over rod length, –.

Using the above definition for rod stress and substituting the appropriate terms in Eq. (4.48), we get:

$$m \frac{\partial^2 u}{\partial t^2} = E\,A \left(\left.\frac{\partial u}{\partial x}\right|_{x+\Delta x} - \left.\frac{\partial u}{\partial x}\right|_{x} \right) - F_d \tag{4.50}$$

The multiplier of the term *EA* on the right-hand side can be expressed with the second derivative of displacement, *u*, with respect to distance, *x*. Introducing this and expressing the mass, *m*, with the volume and density of the rod element, we arrive at the following equation:

$$\frac{\Delta x\, A\, \rho}{144\, g_c} \frac{\partial^2 u}{\partial t^2} = E\, A\, \Delta x \frac{\partial^2 u}{\partial x^2} - F_d \tag{4.51}$$

where:

ρ = density of rod material, lb/ft^3, and
g_c = 32.2 gravitational constant.

In order to develop the final form of the wave equation, the **damping force**, F_d, remains to be determined. This covers a complex mixture of forces always acting in a direction that **opposes** the rod displacement and includes:

- **Fluid** friction forces on the rods, couplings, and the tubing, and
- **Mechanical** (Coulomb) friction between
  - the stuffing box and the polished rod,
  - the tubing and the rods and couplings along the rod string, and
  - the plunger and the barrel of the downhole pump.

The effect of the damping forces is that they **remove** energy from the system along the rod string. If an energy balance between the two endpoints of the rod string is made, then the **difference** between the power introduced at the polished rod and the power available at the pump is equal to the power **lost** due to the damping forces. It follows from this that a perfect description of the downhole pump's operating conditions is only possible if damping (friction) forces are perfectly simulated, as proved by **DaCunha-Gibbs** [30].

The proper simulation of damping forces is complicated by the absolutely different **nature** of the two kinds of friction. Fluid friction is the result of **viscous** forces arising in the annular space around the whole length of the rod string and it is proportional to the relative **velocity**, i.e., the shear velocity between the fluid and the rods. Coulomb friction, on the other hand, occurs at points where stationary and moving system components are in **contact**; its magnitude is a function of the normal force pushing them against each other and the friction factor specific for the surfaces in contact. The **net** friction force, therefore, cannot be described by a **single** formula because of the different behavior of its two main components. Consequently, if Coulomb (mechanical) friction is substantial (especially in deviated wells), the wave equation must be modified to include two damping terms.

The problem caused by the two kinds of damping forces is easily solved if a near-vertical well is assumed where Coulomb friction is normally negligible because its greatest component (i.e., friction between the rods and the tubing) is zero. The other parts of Coulomb friction that occur in the stuffing box and the downhole pump are usually low and constant over the pumping cycle. For such cases practically all investigators **approximate** the sum of all damping forces with some kind of a **viscous** force. In the following, the widely used approach of **Gibbs** [26] will be detailed. This approach provides a perfect simulation of the rod string's elastic behavior as long as Coulomb friction is negligible in the well. The version of the damped wave equation derived this way can be considered the **conventional** wave equation. For wells with significant Coulomb friction, especially those with high deviations from the vertical, a different form of the wave equation must be derived and used, as will be shown later.

Assuming only viscous damping, energy is continuously **lost** along the rod string because the well fluids impart a viscous force on the outer surface of the rods. This **viscous damping force**, at the low pumping velocities involved, is proportional to the relative **velocity**, i.e., the shear velocity $\partial u/\partial t$ between the fluid and the rods. **Gibbs** also assumed the damping force to be proportional to rod **mass** and gave the following semi-empirical formula:

$$F_d = c\,\frac{\Delta x}{144}\,\frac{\rho\,A}{g_c}\,\frac{\partial u}{\partial t} \tag{4.52}$$

where:

F_d = damping force, lb,
$c = \frac{\pi\, v_s\, \nu}{2\, L}$ = damping coefficient, 1/s,
ν = dimensionless damping factor, –,
v_s = sound velocity in the rod material, ft/s,
L = total rod length, ft,
Δx = rod length increment, ft,
ρ = density of rod material, lb/ft^3,
g_c = 32.2 gravitational constant,
A = cross-sectional area of the rod, sq in, and
$\partial u/\partial t$ = shear velocity between the fluid and the rods, ft/s.

Upon substitution of the formula into Eq. (4.51) and dividing both sides by Δx:

$$\frac{\rho\,A}{144\,g_c}\,\frac{\partial^2 u}{\partial t^2} = E\,A\,\frac{\partial^2 u}{\partial x^2} - c\,\frac{\rho A}{144 g_c}\,\frac{\partial u}{\partial t} \tag{4.53}$$

This equation is the final form of the conventional **one-dimensional wave equation** describing the propagation of force **waves** in the sucker-rod string. In this form, it is valid for variable rod diameters, i.e., for tapered rod strings. From it, the more familiar equation for a uniform rod section is found by a simple mathematical operation; the formula given here indicates that rod displacement, u, is a function of both the position, x, and the time, t:

$$\frac{\partial^2 u(x,t)}{\partial t^2} = v_s^2\,\frac{\partial^2 u(x,t)}{\partial x^2} - c\,\frac{\partial u(x,t)}{\partial t} \tag{4.54}$$

where:

$u(x,t)$ = rod displacement, ft,
c = viscous damping factor, 1/s,
$v_s = \sqrt{\frac{144\ g_c\ E}{\rho}}$ = sound velocity in the rod material, ft/s,
x = position of rods, ft, and
t = time, s.

This is the most widely used form of the one-dimensional wave equation, which is strictly valid for cases with viscous damping only. It is a **linear**, second-order **hyperbolic** partial differential equation, the different solution methods of which will be detailed later. For a complete derivation of the wave equation, consult **Gibbs**' recent book [31].

4.4.2.2 Other models

Second-order partial differential equations are more complicated to solve than first-order differential equations, which led several investigators to propose **modified** versions of the wave equation. Instead of the usual second-order differential equation, **Norton** [25] and, later, **Bastian et al.** [32] developed a coupled set of **first-order** differential equations. This set is simpler to solve and can decrease the required computational effort. The derivation of the governing equations is started with **Newton's second law**, written on the rod element as before:

$$F_x - F_{x+\Delta x} + W - F_d = m\frac{\partial v}{\partial t} \tag{4.55}$$

The **damping** force is calculated with the formula proposed by **Gibbs** (Eq. 4.52), where $\partial u/\partial t$ is substituted with rod velocity v:

$$F_d = c\,\frac{\Delta x\ \rho\ A}{144\ g_c}\ v \tag{4.56}$$

Substituting this, and also the expression for rod mass, into Eq. (4.55):

$$F_{x+\Delta x} - F_x - c\frac{\Delta x\ \rho\ A}{144\ g_c}v = \frac{\rho\ A}{144\ g_c}\frac{\partial v}{\partial t} \tag{4.57}$$

After dividing both sides by Δx, the first term on the left-hand side is equal to the first forward difference of the tension force F. This is a close approximation for the derivative of the same force with respect to axial distance x. Upon substitution, the **first** basic equation of the model is found:

$$\frac{\partial F}{\partial x} - c\rho A v = A\ \rho\frac{\partial v}{\partial t} \tag{4.58}$$

The second basic equation comes from the application of **Hooke's law** to the tension force:

$$F = E\,A\frac{\partial u}{\partial x} \tag{4.59}$$

Differentiating this formula with respect to time, we arrive at the **second** basic equation of the model:

$$\frac{\partial F}{\partial t} = E\,A\frac{\partial^2 u}{\partial x\ \partial t} = E\,A\frac{\partial v}{\partial x} \tag{4.60}$$

Equations (4.58) and (4.60) constitute a **coupled** set of first-order differential equations that describes the behavior of the rod string. Their **simultaneous** solution allows a direct calculation of forces and velocities along the rod string. Rod displacements at different well depths are found by integration of the velocities valid at the given axial distance. **Norton** [25] proposed an analytical solution of the basic equations, whereas **Bastian et al.** [32] utilized numerical methods involving finite differences. The latter authors claim that this approach provides a faster and more accurate solution to the damped wave equation than the models using the second-order differential equation.

4.4.3 SOLUTIONS OF THE WAVE EQUATION

Solutions of the damped wave equation allow the calculation of the displacement of the rods, u, at any axial distance, x, and at any time, t. Thus the movement of any rod element is a function of both the vertical **distance** and the **time**. There are basically two ways to find displacements and forces along the rod string: either starting at the surface and proceeding downward or vice versa. The two main uses of the wave equation can be classified according to this criterion and are:

- The **diagnostic** analysis, which involves calculating the downhole displacements and forces based on surface measurements, i.e., the construction of pump cards.
- The **predictive** analysis, which, as its name implies, predicts the surface conditions based on the description of the sucker-rod pump's operation.

Since the conventional wave equation, Eq. (4.54), contains second-order derivatives in both time and space, its solution requires, in general, two **initial** and two **boundary** conditions. Initial conditions are chosen to represent the rod string at rest when at any axial distance, x, rod displacements and velocities are zero. These conditions are only used in the predictive analysis, because the diagnostic analysis seeks steady-state solutions only. **Boundary** conditions, on the other hand, must be used in both the diagnostic and the predictive model. They must represent the physical phenomena at either end of the rod string and are normally provided by:

- The time histories of polished rod **load** and polished rod **displacement** in the **diagnostic** model.
- The time histories of the polished rod **displacement** and of the pump **displacement** in the **predictive** model.

Both the above analyses can be accomplished by the use of different solutions applied to the one-dimensional wave equation. The solution methods most frequently used for these purposes are described in the following.

4.4.3.1 The analytical solution

Gibbs proposed an analytical procedure for the solution of the wave equation [27,33]. He used the classical method of the **separation** of variables, which involves splitting of the original partial differential equation into two ordinary differential equations. These are solved, and the individual results obtained are combined into a product solution that satisfies the wave equation with the given boundary conditions. In order to facilitate the inclusion of the **boundary** conditions into the final solution, the polished rod load and polished rod displacement versus time functions are described by **Fourier series** approximations. The relevant harmonic functions are as follows:

$$D(t) = F(t) - W_{rf} = \frac{\sigma_0}{2} + \sum_{n=1}^{N} [\sigma_n \cos(n\,\omega t) + \tau_n \sin(n\,\omega t)] \tag{4.61}$$

$$u(t) = \frac{\nu_0}{2} + \sum_{n=1}^{M} [\nu_n \cos(n\,\omega t) + \delta_n \sin(n\,\omega t)] \tag{4.62}$$

where:

$D(t)$ = dynamic load at the polished rod versus time,
$F(t)$ = polished rod load versus time,
W_{rf} = buoyant weight of the rod string,
σ_n, τ_n = Fourier coefficients of the dynamic load function,
ν_n, δ_n = Fourier coefficients of the displacement function,
ω = angular frequency of pumping,
N = number of dynamic load coefficients, and
M = number of displacement coefficients.

The above Fourier series are **truncated** at N and M and the adequate number of Fourier coefficients is difficult to determine. Optimum values are found by considering the degree of fit to the original functions and the computation time required. The **displacement**-versus-time function is **smooth**, so a lesser number of coefficients is needed than for the dynamic load function. Increasing the number of Fourier coefficients over a certain level does not give further gains in accuracy but, at the same time, it considerably increases the computational effort required. In general, 10–15 coefficients are adequate for most situations [34].

The values of the coefficients in Eqs (4.61) and (4.62) are evaluated with the classical integration formulae, e.g.:

$$\sigma_n = \frac{\omega}{\pi} \int_{t=0}^{2\pi} D(t) \cos(n\,\omega t) dt \tag{4.63}$$

where:

$n = 0, 1, \ldots, N.$

In practice, however, these integrals cannot be evaluated analytically because the load and displacement functions are known at some discrete points only. Therefore, a **numerical** integration procedure must be used. **Gibbs** proposed the use of the **trapezoidal rule**, but other methods are also available. As regards the number of polished rod load and displacement values, at least 50 pairs equally spaced in time are required to provide sufficient accuracy [34].

Leaving out derivations, only Gibbs' final formulae are given below, which describe the rod displacement and the dynamic force in the functions of the time, t, and the distance from the surface, x:

$$u(x,t) = \frac{\sigma_0}{2\,E\,A} + \frac{\nu_0}{2} + \sum_{n=1}^{M} [O_n \cos(n\,\omega t) + P_n \sin(n\,\omega t)] \tag{4.64}$$

$$D(x,t) = E\,A \left\{ \frac{\sigma_0}{2\,E\,A} + \sum_{n=1}^{N} \left[O_n' \cos(n\,\omega t) + P_n' \sin(n\,\omega t) \right] \right\} \tag{4.65}$$

The Fourier coefficients in the equations above are determined from the coefficients of the surface load and displacement functions (Eqs (4.61) and (4.62)) using a procedure developed by **Gibbs** and not reproduced here. With the knowledge of these coefficients, the displacements and forces for different assumed times are determined easily at the bottom of a uniform rod section, enabling a downhole dynamometer card (dynagraph) to be plotted.

In summary, the **analytical solution** of the wave equation involves the following main steps:

1. The polished rod load and polished rod displacement versus time functions are determined. Today this is predominantly achieved by using a **digital** dynamometer, but **digitizing** a conventional dynamometer card is also possible.
2. The buoyant rod string weight, W_{rf}, is **subtracted** from the loads to get the dynamic loads, $D(t)$.
3. The displacement and dynamic load functions are **approximated** by Fourier series and the coefficients for each are found.
4. Using a calculated or estimated damping factor and the procedure of **Gibbs**, the Fourier coefficients valid at the lower end of the rod section are found. The effects of the buoyant rod weight are superimposed on the calculated displacement and load functions.
5. In the case of a tapered string, the loads and displacements just calculated are utilized for the top of the next-lower rod taper. Then Gibbs' recursive formulae can be used to find displacements and forces at the bottom of the actual rod section.
6. The forces and displacements at the bottom of the string are identical to the forces and displacements at the pump and a downhole (pump) dynamometer card can be constructed. The operational conditions of the sucker-rod pump are analyzed from the shape and the operational parameters read from the pump card.

EXAMPLE 4.9: CALCULATE THE DOWNHOLE DYNAMOMETER CARDS AND THE OPERATIONAL PARAMETERS OF PUMPING FOR THE WELL PRESENTED IN EXHIBIT 4.3

Solution

Exhibit 4.3 contains the calculated results for the given case. The calculated downhole cards are plotted in Fig. 4.3, where the surface dynamometer and the pump cards are in solid line. The diagram in dashed line represents the rod loads versus displacement at the top of the second (5/8 in) taper. From this figure, the rod loads in any taper section are easily determined and the sucker-rod pump's actual stroke length also can be found. The numerical values of these and other calculated parameters are contained in Exhibit 4.3.

4.4.3.2 Numerical solutions

As is the case with every differential equation, the damped wave equation also can be recast in a **difference** form. The transformation involves replacing the **derivatives** with **difference** quotients, an approximation frequently used in the numerical solution of differential equations. The difference quotients include **finite** differences; hence the name of this approach: **finite difference solution**. One possible set of difference quotients is given below, which can be used to approximate the following terms in the wave equation:

$$\frac{\partial u}{\partial t} \approx \frac{u(x, t + \Delta t) - u(x, t)}{\Delta t} \tag{4.66}$$

```
                DIAGNOSTIC ANALYSIS OF DYNAMOMETER CARDS
             WITH THE ANALITICAL SOLUTION OF THE WAVE EQUATION


              I N P U T   D A T A   from   FILE  : demo1.CRD

      Name of Field : Example               Well ID : Ex. 4.9 & 10

      Pump Setting Depth :     3000.0 ft    Measured Prod. Rate :   198.0 bpd
      Dynamic Liquid Level :   3000.0 ft    Liquid Spec. Gravity :  1.000

           Plunger Size :        2.50 in       Tubing Anchored :  YES
           Stroke Length :      50.00 in
           Pumping Speed :       9.00 SPM

                     P U M P I N G   U N I T   D A T A

      Manufacturer :  NN Manufacturing      API Size :  C-228-213-120

                       R O D   S T R I N G   D A T A

                      Size      Taper    Material    Young's
                                Length               Modulus
                       in        ft                    psi

                      3/4      1956.0     Steel      0.31E+08
                      5/8      1044.0     Steel      0.31E+08

                 D Y N A M O M E T E R   C A R D   D A T A

           Date Card Taken :                  Time Card Taken :
                      Number of Digitized Points :  145

                     A S S U M E D   C O N D I T I O N S

           Number of Fourier Coefficients :   20  for Force
                                              10  for Displacement
           Dimensionless Damping Factor :  0.100


              C A L C U L A T I O N   R E S U L T S

  Buoyant Rod Weight :      3831 lbs    Fluid Load on Pump :      6386 lbs
  Plunger Stroke     :     33.49 in     Plunger Displ.     :     219.6 bpd
  Polished Rod Power :       4.9 HP     Hydraulic Power    :       4.9 HP
  Loss in String     :       0.3 HP     Pump Power         :       4.5 HP
                     Lifting Efficiency :   99.9 %

           C A L C U L A T E D   R O D   L O A D I N G

      Taper Size :  3/4  in      Material :   Steel

      Max. Load :      11189 lbs          Min. Load :      2482 lbs
      Max. Stress :    25310 psi          Min. Stress :    5615 psi

                    S.F.     Percent Loading, %
                         Grade K  Grade C  Grade D
                    1.00   103.7     98.7     79.4
                    0.80   129.6    123.4     99.3
                    0.60   172.8    164.5    132.3

      Taper Size :  5/8  in      Material :   Steel

      Max. Load :       7700 lbs          Min. Load :       616 lbs
      Max. Stress :    25075 psi          Min. Stress :    2005 psi

                    S.F.     Percent Loading, %
                         Grade K  Grade C  Grade D
                    1.00   112.1    106.2     84.0
                    0.80   140.1    132.8    105.0
                    0.60   186.8    177.0    140.0
```

EXHIBIT 4.3

Results of the diagnostic analysis of the dynamometer card presented in Fig. 4.3 using **Gibbs'** analytical solution.

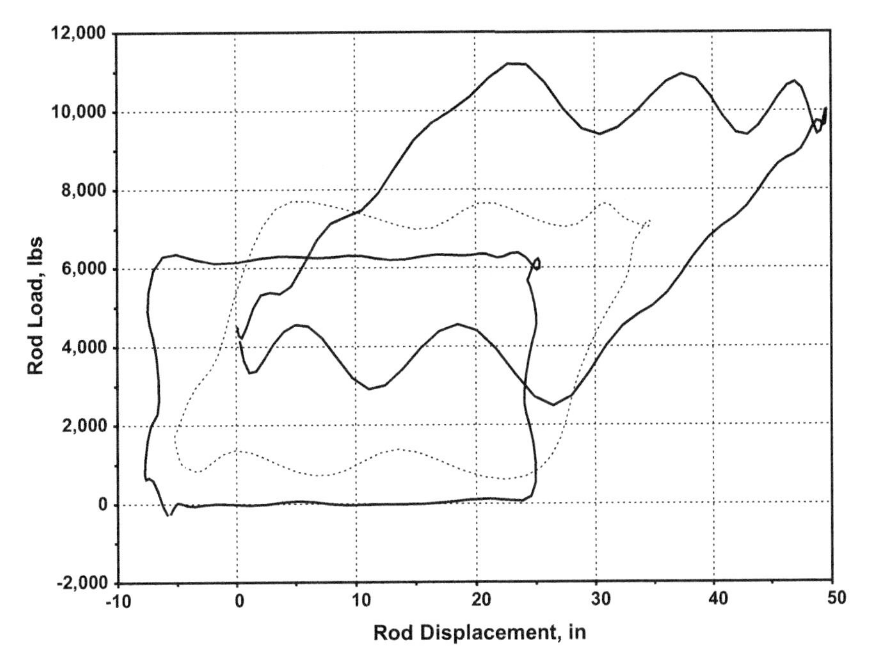

FIGURE 4.3

Calculated downhole cards for Example 4.9.

$$\frac{\partial^2 u}{\partial x^2} \approx \frac{u(x+\Delta x, t) - 2u(x,t) + u(x-\Delta x, t)}{\Delta x^2} \tag{4.67}$$

$$\frac{\partial^2 u}{\partial t^2} \approx \frac{u(x, t+\Delta t) - 2u(x,t) + u(x, t-\Delta t)}{\Delta t^2} \tag{4.68}$$

The above formulae are **Taylor series** approximations neglecting the higher-order expansions. If they are inserted into any version of the wave equation, the resultant finite difference equation can be solved numerically [26]. There are two possibilities: such equations can be solved for $u(x + \Delta x, t)$, (i.e., for the displacement at the same time but at the next distance); or they can be solved for $u(x, t + \Delta t)$ (i.e., for the rod displacement at the same place but ahead in time). These two approaches conform to the **diagnostic** and the **predictive** analysis methods of simulating the behavior of the sucker-rod string.

For illustration only, a possible form of a finite difference solution formula [34] is given below for the **predictive** case:

$$u(x, t+\Delta t) = \frac{\left\{\frac{\Delta t^2 v_s^2}{\Delta x^2}\left[u(x+\Delta x, t) - 2u(x,t) + u(x-\Delta x, t)\right] + 2u(x,t) - u(x, t-\Delta t) + c\,\Delta t\, u(x,t)\right\}}{1 + c\,\Delta t} \tag{4.69}$$

4.4.3.2.1 Diagnostic analysis

The **diagnostic** analysis involves the calculation of displacements, u, along the length of the rod string, x, for the same values of the time, t. Since only steady-state solutions are sought in this type of analysis, no initial conditions are required. Thus the derivatives of force and displacement with respect to time do not change from one pumping cycle to another. The boundary conditions that must be used are provided by the surface **dynamometer** card, which gives the time **histories** of the dynamic force and the polished rod movement at place $x = 0$, i.e., the functions $D(0, t)$ and $u(0, t)$.

Application of the finite difference method involves dividing the rod string into a number of Δx segments to facilitate a step-wise solution. For general work, the Δx should be from 500 to 750 ft [34], but smaller values are used for increased accuracy. A time increment, Δt, must also be assumed, which is usually defined by the number of points read from the surface dynamometer diagram. The values of the two increments are **interrelated** and the following **stability** criterion applies [35]:

$$\Delta x \leq \Delta t \, v_s \tag{4.70}$$

where:

Δx = length increment, ft,
Δt = time increment, s, and
v_s = sound velocity in the rod material, ft/s.

The main calculation steps of the **diagnostic** model using finite differences can be summarized as follows:

1. The polished rod load and displacement-versus-time functions are determined, and their values are found at given Δt time intervals. The rod string length increment, Δx, is established based on the stability criterion, and the rod string is divided into the appropriate number of segments.
2. The initial displacements of the rod string at the surface, $u(0, t)$, are set to the polished rod positions at every time step.
3. The displacement at the next-lower **segment**, $u(x+\Delta x, t)$, is calculated with the finite difference formula. This is repeated for all time steps involved to cover the whole pumping cycle.
4. Step 3 is repeated for the next consecutive rod string elements until a junction of the different taper sections is reached. At such points, rod displacements are corrected for the static rod stretch and dynamic forces are calculated with **Hooke's law**.
5. At the bottom of the string, after correction for buoyant rod weight, the calculated displacements and loads define the operating conditions of the sucker-rod pump and the pump card is constructed.

EXAMPLE 4.10: EVALUATE THE PUMPING SYSTEM PRESENTED IN THE PREVIOUS EXAMPLE WITH THE NUMERICAL SOLUTION. USE IDENTICAL DAMPING FACTORS OF 0.1 FOR BOTH THE UP- AND DOWNSTROKE

Solution

Exhibit 4.4 contains the calculated results and Fig. 4.4 presents the plots of the downhole cards. As seen, results obtained are very similar to those of the analytical solution shown in Fig. 4.3. Comparison of the two figures also reveals an important **shortcoming** of the analytical method: the Fourier series approximation cannot precisely describe the

CALCULATION RESULTS

Buoyant Rod Weight :	3831 lbs	Fluid Load on Pump :	6386 lbs
Plunger Stroke :	32.43 in	Plunger Displ. :	212.7 bpd
Polished Rod Power :	4.9 HP	Hydraulic Power :	4.7 HP
Loss in String :	0.4 HP	Pump Power :	4.5 HP

Lifting Efficiency : 96.5 %

CALCULATED ROD LOADING

Taper Size : 3/4 in Material : Steel

Max. Load :	11586 lbs	Min. Load :	2343 lbs
Max. Stress :	26209 psi	Min. Stress :	5299 psi

S.F.	Percent Loading, % Grade K	Grade C	Grade D
1.00	108.2	102.9	82.7
0.80	135.2	128.7	103.4
0.60	180.3	171.6	137.8

Taper Size : 5/8 in Material : Steel

Max. Load :	8001 lbs	Min. Load :	493 lbs
Max. Stress :	26056 psi	Min. Stress :	1606 psi

S.F.	Percent Loading, % Grade K	Grade C	Grade D
1.00	117.6	111.4	88.0
0.80	147.0	139.3	110.0
0.60	196.0	185.7	146.6

EXHIBIT 4.4

Results of the diagnostic analysis of the dynamometer card presented in Fig. 4.3 using a numerical solution.

sharp corners in the diagrams. In the case of the pump card, however, sharp corners are usually **inevitable**, mainly due to the sudden release and picking up of the fluid load by the plunger during the pumping cycle. The numerical solution can easily handle this problem, as seen in Fig. 4.4, but the analytical model tends to round the corners of the pump card; see Fig. 4.3.

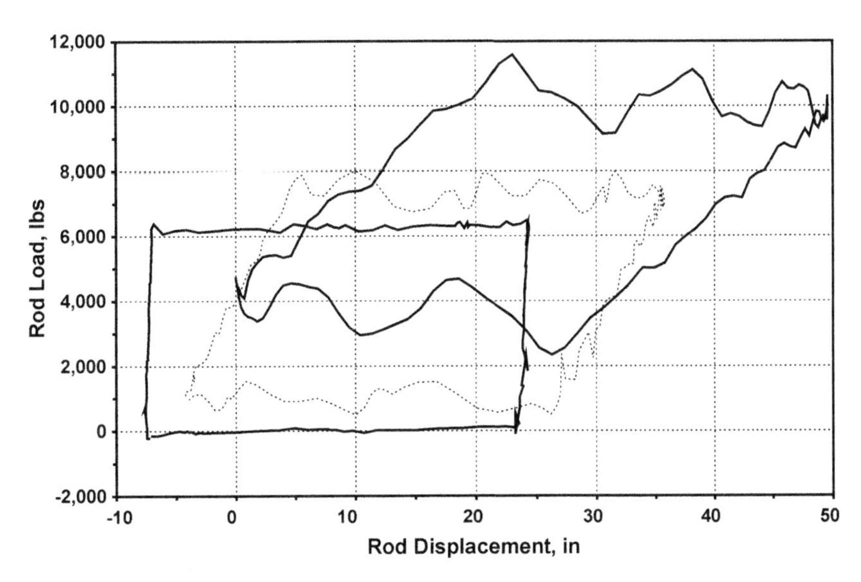

FIGURE 4.4

Calculated downhole cards for Example 4.10.

4.4.3.2.2 Predictive analysis

In comparison to the diagnostic model, the **predictive** analysis also considers time and uses a calculation formula that gives the rod displacements ahead in time, $u(x, t + \Delta t)$. Since **transient** solutions are also possible, a sufficient number of pumping cycles must be simulated before a **stable** solution is reached. The calculation procedure usually converges to a steady-state solution in about 3–5 pumping cycles [34]. The boundary conditions are provided by the motion of the polished rod versus time, $u(0, t)$, and the operation of the downhole pump. This model can be used to **predict** the surface loads and to construct the surface dynamometer diagram based on the motion of the pumping unit and the subsurface pump.

Just as in the diagnostic case, the rod string is divided into a number of **segments**, and the segment length is determined in a similar fashion as in the diagnostic case. The stability of the solution requires that the time increment, Δt, satisfy the condition given below [35] (note the difference from the criterion for the diagnostic case):

$$\Delta t \leq \frac{\Delta x}{v_s} \tag{4.71}$$

where:

Δx = length increment, ft,
Δt = time increment, s, and
v_s = sound velocity in the rod material, ft/s.

The greatest difficulty in the predictive analysis is the simulation of the downhole pump's **performance**. If a single-phase liquid is pumped, pump displacements and loads are relatively easy to determine; however, any downhole problems, especially the effects of **free** gas, make the description of the pump movement an extremely complex task. This involves the exact timing of the valve opening and closing, usually accomplished by testing the displacements at the bottom of the rod string with preselected criteria. The use of this technique allows the simulation of different downhole conditions encountered in practice [26].

In summary, the predictive analysis consists of the following main calculation steps:

1. The rod **length** increment, Δx, is defined and the time increment, Δt, that satisfies the stability criterion is found.
2. The boundary conditions, i.e., the values of the polished rod displacement versus time function, $u(0, t)$, are determined from the pumping unit's kinematic evaluation.
3. The displacement at the next-lower rod segment is determined using the finite difference formula. This procedure is repeated, taking into account the changes in rod size, until the bottom of the string is reached.
4. The action of the sucker-rod pump is taken into account as detailed above.
5. Steps 3 and 4 are repeated for all time steps.
6. The whole procedure is repeated for several pumping cycles to reach a steady-state solution without any transient effects.

7. The final rod displacements and loads valid at the end of the calculations represent the conditions at the polished rod and at the subsurface pump. Then the **surface** and **downhole** dynamometer diagrams (dynagraphs) can be plotted to analyze the operation of the pumping system.

4.4.3.3 Boundary and initial conditions

The solution of any variant of the **wave equation** requires that **boundary** and **initial** conditions at arbitrary point(s) along the string be known. Depending on the type of analysis desired, different input data are needed:

- For the **diagnostic** analysis, surface displacements and loads versus time supply the required boundary conditions.
- For the **predictive** analysis, the boundary conditions are provided by the polished rod's movement versus time and a description of the operating conditions at the subsurface pump.

The polished rod displacement and load-versus-time functions are not readily available if traditional (hydraulic) dynamometers are used because they register polished rod **load** as a function of polished rod **position** and contain no reference to time. This is why the implementation of Gibbs' original method necessitated the use of a **special** dynamometer arrangement [33]. The first special dynamometers recorded displacement and load data on **strip** charts for later evaluation [36–39], but up-to-date electronic models are directly linked to a computer for online data acquisition [28,40–42]. Such units are more expensive than ordinary dynamometers but, due to their many beneficial features, their use is common today.

At earlier times, only **conventional** dynamometers were available; surface dynamometer cards taken with those only **implicitly** contain the boundary conditions for the solution of the wave equation. As first proved by **Takacs-Papp** in 1978 [43], the displacement- and load-versus-time functions necessary for the solution of the wave equation can be determined from a **conventional** dynamometer card, provided the angular velocity of the crank is **constant** over the pumping cycle. In this case, the polished rod displacement-versus-time function is easily found from a **kinematic** analysis of the pumping unit (see Section 3.7.5). Using this relationship and taking the matching load and displacement pairs (read or digitized from the dynamometer card), the points of the polished rod load versus time function sought can be determined. In most of the cases, this technique allows performing diagnostic analyses for users having traditional dynamometers only [44], because normally the crank turns at a constant speed. This was proved by experiences gained by a major oil company on thousands of rod-pumped wells, when practically **identical** results were received to those attained with special dynamometers, except when the pumping unit was driven by a fully loaded high-slip motor [45].

One of the boundary conditions required for the **predictive** analysis of the pumping system is the time history of the polished rod's motion. This function is best described by a **kinematic** analysis of the pumping unit, which supplies exact position data. Since such a high precision is not usually required, it is customary to use the equation for a simple **harmonic** motion [46]. Thus the number of required input data is decreased and, at the same time, sufficient simulation accuracy is reached with decreased calculation effort [32]. **Laine et al.** [47] developed approximate relationships to calculate polished rod position values for different pumping unit geometries.

In addition to the data just described, the predictive analysis requires the description of the pump's performance. This can be a very difficult task, especially when downhole problems such as gas interference are present. Usually, the conditions at the bottom of the string are continuously monitored, and the actions of the pump valves are simulated according to different presumed conditions. This approach can be used to simulate the behavior of the rod string under the conditions usually found in the oilfield [26].

4.4.4 DETERMINATION OF THE DAMPING COEFFICIENT

As discussed before, the **damping** term in the wave equation stands for the **irreversible** energy losses that occur along the rod string during its movement. Although these losses originate from a wide variety of very complex phenomena, the **conventional** wave equation can model only those damping forces that are of **viscous** nature, as was discussed in conjunction with Eq. (4.52). Other than viscous friction effects like **Coulomb** friction between the tubing and the rod string cannot be properly described by the usual damping term and the conventional wave equation. This is the reason why, especially for deviated wells where heavy loads occur due to Coulomb friction, a different and more comprehensive basic equation must be defined and solved.

The **magnitude** of viscous damping forces is **different** during the upstroke and the downstroke of the rod string because the relation of fluid flow and rod movement differs in the two cases. During the upstroke the velocities of the fluid flow and the rod's movement point in the **same** direction: upward, whereas in the downstroke they point in **opposite** directions. It is easy to see that viscous drag between the rods and the fluid is **greater** during the downstroke when the fluid flows against the movement of the rods than during the upstroke. Therefore, at least in theory, **different** damping factors must be used for the two conditions, but many investigators use one single factor only to describe damping losses.

In the following, the available models for damping coefficient determination are discussed for use in the **conventional** wave equation. This means that the solutions discussed are valid for cases where mechanical friction (a.k.a. **unaccounted** friction) in the well is negligible, i.e., in vertical or nearly vertical wells producing a light oil without any paraffin or scale problems.

4.4.4.1 Approximate methods

The first method for finding damping losses was proposed by **Gibbs** [33], who presented the graph shown in Fig. 4.5 to calculate the dimensionless **damping factor**. This is a simple empirical correlation, which gives the damping factor as the **sole** function of the polished rod's average velocity.

Gibbs' second method [33] involves an energy **balance** written for the two ends of the rod string. The power **input** at the polished rod must overcome the sum of the useful work at the downhole pump and the energy consumed by the **losses** along the string. Since polished rod power is easily found from the area of the surface dynamometer **card**, and pump power equals the **useful** hydraulic power (which can also be calculated), it only remains to find the energy losses. In calculating these losses, Gibbs

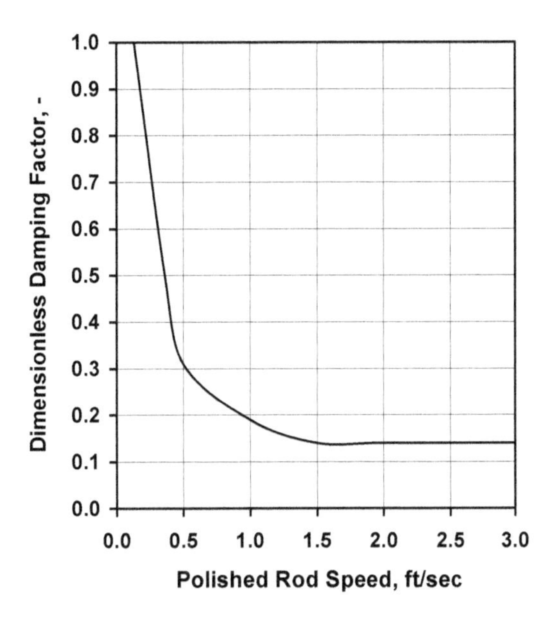

FIGURE 4.5

The empirical correlation of **Gibbs** [33] to determine the dimensionless damping factor.

assumed a simple **harmonic motion** of the rods and developed an equation for the energy losses due to damping. The energy balance then can be solved for the damping coefficient and the following formula is obtained:

$$c = \frac{550 \times 144\left(PRHP - P_{\text{hydr}}\right)T^2\, g_c}{\sqrt{2}\pi\, WS^2} \tag{4.72}$$

where:

c = damping coefficient, 1/s,
g_c = 32.2, gravitational constant,
$PRHP$ = polished rod power, HP,
P_{hydr} = hydraulic power used for fluid lifting, HP,
T = period of the pumping cycle, s,
W = total weight of the rod string, lb, and
S = polished rod stroke length, in.

Note that this equation is valid for **any** rod material and is a modification of Gibbs' original formula, proposed by **Everitt and Jennings** [35]. They pointed out that the use of Eq. (4.72) relies heavily on the knowledge of the hydraulic power, P_{hydr}, which in many cases can only be approximated.

4.4.4.2 Calculation models

The previous theory of determining the damping coefficient is theoretically sound, but its **practical** application is troublesome. It is evident that in order to find the proper value of the damping coefficient, the hydraulic power must be known. The hydraulic power represents the **useful** work expended to lift the given amount of liquid from the dynamic level to the surface:

$$P_{hydr} = 7.36\text{E-}6\ Q\ SpGr\ L_{dyn} \tag{4.73}$$

where:

P_{hydr} = hydraulic power, HP,
Q = liquid production rate, bpd,
$SpGr$ = specific gravity of the produced fluid, –, and
L_{dyn} = dynamic liquid level in the well, ft.

Equation (4.73) includes the pumping rate, Q, which depends on the pump's downhole stroke length. Pump stroke length, in turn, can only be obtained after the downhole pump card is calculated. However, the calculation of the downhole card requires the use of the proper **damping** coefficient, which is about to be determined. Consequently, the value of the damping coefficient can be found only by an **iterative** procedure, as pointed out by several investigators [32,35].

An iterative solution is presented by **Everitt and Jennings** [35] and the required calculations follow the flowchart given in Fig. 4.6. This procedure is based on the observation that a change in the damping coefficient has a much **greater** impact on the **area** of the pump card than on the net plunger stroke **length**. Therefore, first a converged value for the net plunger stroke, S_{net}, is found and is used in the rest of the procedure to calculate the hydraulic power, P_{hydr}. Hydraulic power being thus determined, the proper value of the damping coefficient is sought next, using an **iterative** scheme. This involves the calculation of the pump **card** and its area, the latter being proportional to the power output by the pump, P_{pump}. In cases where the **hydraulic** power equals the **pump** power, the proper value of the damping coefficient is found. Otherwise, the coefficient is adjusted accordingly, and the iterative procedure is repeated.

The big advantage of the **Everitt–Jennings** procedure is that no manual intervention is required in calculating downhole dynamometer cards because the damping factor is automatically determined during the calculation process. Modifications of the original theory include an altered iteration scheme and the use of different damping factors for the upstroke and the downstroke [48,49].

The same technique is proposed by **Bastian et al.** [32], with the only difference being that they used a more exact formula than Eq. (4.72) for the determination of damping coefficients. As pointed out in Section 4.4.2.2, these authors directly calculated the **velocities** at different points along the string and were thus not forced to use the **harmonic** motion to approximate the rod string's displacements. After the average velocity for a complete pumping cycle is found, Eq. (4.72) can be modified as:

$$c = \frac{550\ 144 \left(PRHP - P_{hydr}\right) T\ g_c}{2\ W\ S\ v_{avg}} \tag{4.74}$$

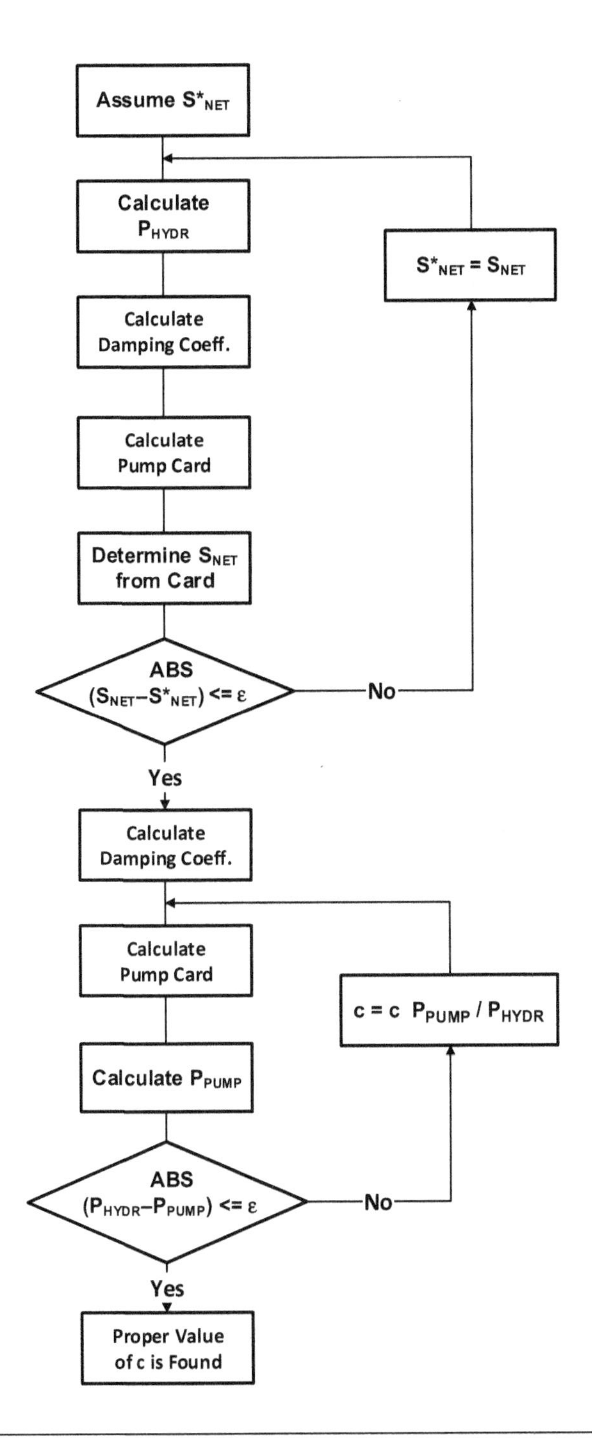

FIGURE 4.6

Flowchart of an iterative solution of the damping coefficient.

Proposed by ***Everitt and Jennings*** *[35].*

where:

c = damping coefficient, 1/s,
g_c = 32.2, gravitational constant,
$PRHP$ = polished rod power, HP,
P_{hydr} = hydraulic power used for fluid lifting, HP,
T = period of the pumping cycle, s,
W = total weight of the rod string, lb,
S = polished rod stroke length, in, and
v_{avg} = average velocity of the rod string, ft/s.

The previously discussed **Bastian et al.** simulation model, including the detailed way of determining the damping coefficient, is utilized to solve the example problem below.

EXAMPLE 4.11: FIND THE VALUE OF THE DAMPING COEFFICIENT AND CALCULATE THE DOWNHOLE CARDS FOR THE WELL GIVEN IN EXHIBIT 4.5

Solution

Exhibit 4.5 contains detailed calculation results for the example case; the surface and downhole cards are presented in Fig. 4.7. The tubing in this well is unanchored, as seen from the shape of the pump card. The gross plunger stroke length is 74.5 in, whereas the net stroke equals 55.7 in.

The proper value of the damping coefficient is calculated as $c = 0.589$ 1/s, and the use of this value resulted in a hydraulic power (8.6 HP) very closely approximating the power required at the pump (8.5 HP). The total power input at the polished rod (11.1 HP) covers the sum of the required pump power and the losses, which amount to 2.6 HP.

The merits of the above procedure, originally proposed by **Everitt and Jennings** [35] for finding the damping coefficient, are shown in Fig. 4.8. This figure shows the useful hydraulic and the pump powers plotted against the damping coefficient in the example case. As discussed before, the hydraulic power, which is directly proportional to the **net** plunger stroke, is fairly **constant** over a wide range of the damping coefficient. The power used by the **pump** (computed from the area of the pump card), however, drops **drastically** as the damping coefficient is increased, because more energy is wasted along the string. The proper value is found where the two curves **intersect**. The damping coefficient thus determined assures that the damping losses just **calculated** are equal to the actual amount of energy **removed** from the pumping system.

4.4.4.3 Practical methods

The practical methods of finding the damping factor, used in most of the wave equation program packages [50,51], are based on the **definition** of the damping losses. These losses occur along the rod string and their effect increases the surface energy requirement of pumping beyond the power needed to operate the downhole pump. In case of a diagnostic analysis, however, **downhole** conditions are estimated from measured **surface** data so damping losses must be **deducted** from the input power at the polished rod. Starting from the surface dynamometer card and assuming that the conventional wave equation **correctly** describes the behavior of the rod string, the use of the **proper** damping factor would yield a pump card of **ideal** shape and proportions because exactly the right amount of energy losses would be removed from the system. Therefore, the proper damping factor can be found by **trial and error**, i.e., by calculating the downhole card with different damping factor values and observing the changes of the pump cards.

```
              DIAGNOSTIC ANALYSIS OF DYNAMOMETER CARDS
           WITH THE NUMERICAL SOLUTION OF THE WAVE EQUATION
         INCLUDING THE DETERMINATION OF THE DAMPING COEFFICIENT


           I N P U T   D A T A   from   FILE  : demo2.CRD

Name of Field : Example                 Well ID : Ex. 4.11

Pump Setting Depth :     5000.0 ft     Measured Prod. Rate :    250.0 bpd
Dynamic Liquid Level :   5000.0 ft     Liquid Spec. Gravity :   0.900

      Plunger Size :       2.00 in        Tubing Anchored :  NO
      Stroke Length :     90.00 in        Tubing Size :      2 7/8 in
      Pumping Speed :     10.00 SPM

                    P U M P I N G   U N I T   D A T A

Manufacturer :  NN Manufacturing       API Size :  C-228D-213-100

                      R O D   S T R I N G   D A T A

                  Size      Taper     Material     Young's
                            Length                 Modulus
                   in        ft                      psi

                  7/8      3000.0      Steel       0.31E+08
                  3/4      2000.0      Steel       0.31E+08

                 D Y N A M O M E T E R   C A R D   D A T A

      Date Card Taken :                     Time Card Taken :
                    Number of Digitized Points :    75

                   C A L C U L A T I O N   R E S U L T S

               Calculated Damping Coefficient :  0.589 1/s

Buoyant Rod Weight :      8819 lbs     Fluid Load on Pump :       6131 lbs
Gr. Plunger Stroke :     74.53 in      Gr. Plunger Displ. :      347.5 bpd
Net Plunger Stroke :     55.74 in      Net Plunger Displ. :      259.9 bpd
Polished Rod Power :      11.1 HP      Hydraulic Power    :        8.6 HP
Loss in String     :       2.6 HP      Pump Power         :        8.5 HP

                    Lifting Efficiency :   77.9 %
```

EXHIBIT 4.5

Results of the diagnostic analysis of the dynamometer card presented in Fig. 4.7 using a numerical solution and the determination of the damping factor.

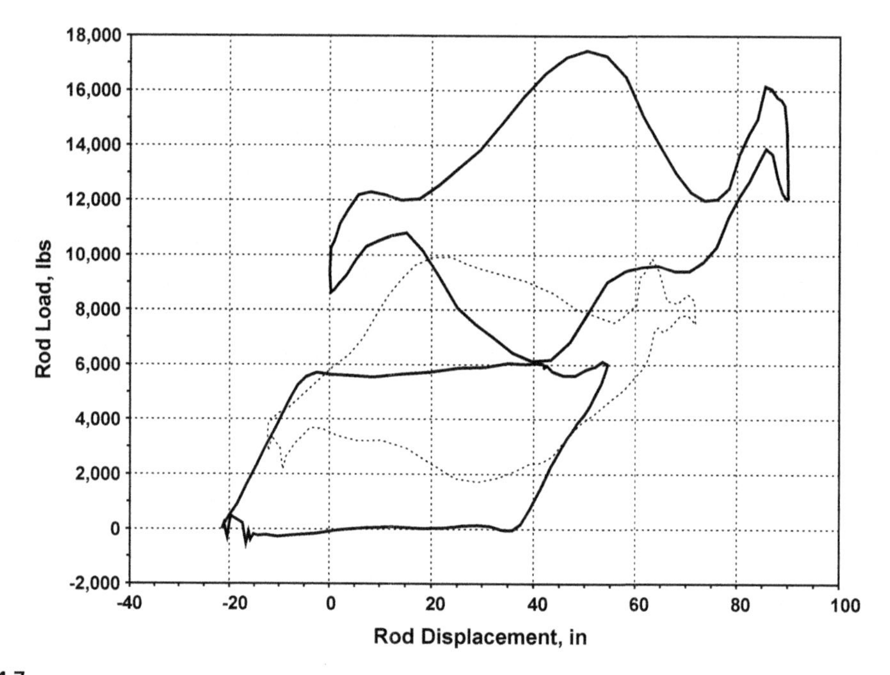

FIGURE 4.7

Calculated downhole cards for Example 4.11.

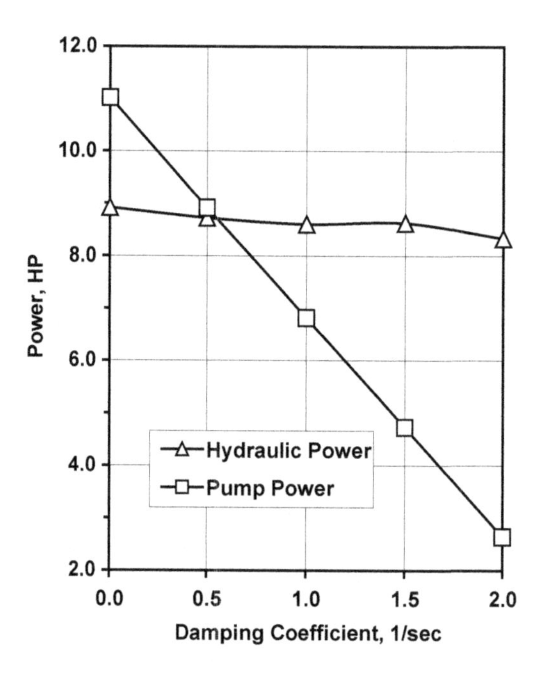

FIGURE 4.8

Graphical determination of the proper value of damping coefficient for Example 4.11.

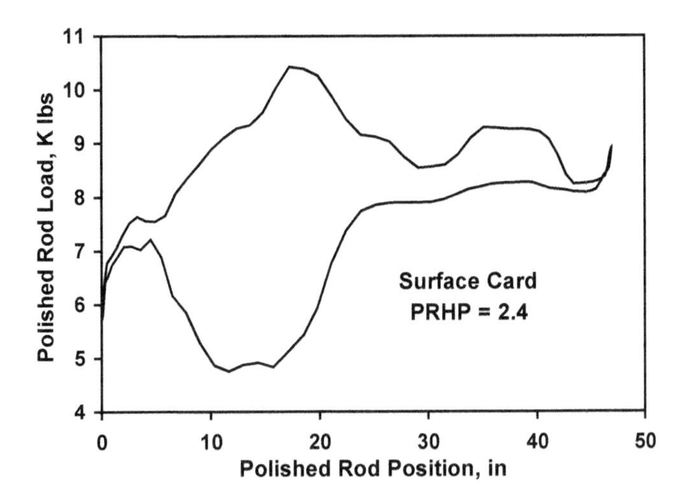

FIGURE 4.9

An example surface dynamometer card.

The effect of the damping factor on the calculated downhole pump cards is illustrated by presenting an example. Figure 4.9 shows a surface card taken on a well with negligible downhole mechanical friction and an unanchored tubing string. The pump cards calculated for different assumed damping factors are presented in Fig. 4.10. Using zero damping, i.e., assuming no energy losses along the rod string, results in a pump card with the same pump power as the polished rod horsepower, *PRHP*. As the value of the damping factor is **increased**, pump power **decreases** because more and more power is consumed by downhole damping **losses**; at the same time, the **shape** of the card changes as well. Investigation of the effects of the damping factor on the downhole pump card allows the following **conclusions** to be drawn:

- The plunger stroke **length** does not significantly change with the change of the damping factor.
- For low amounts of damping, the downhole cards are "**fat**" (convex) and have rounded upper and lower sides.
- Too much damping causes the pump card to be **distorted** and to present a concave shape.
- Proper damping is detected when the top and bottom sides of the pump card are near **horizontal**.

Based on these rules, the proper damping factor for the example case should lie between 0.05 and 0.07.

In conclusion, the time-proven practical way of finding the proper damping factor when using software packages is by **trial and error**. One has to change input values until the downhole pump card is of the required shape: characterized by nearly horizontal top and bottom sides. This can be achieved only if the pump is in good **mechanical** condition without any major malfunctions, and if negligible Coulomb friction (a.k.a. **unaccounted** friction) occurs downhole. The process requires some **experience**; pumped-off conditions, as presented in Fig. 4.10, are the easiest to tackle because the "**nose**" of the pump card is a good indicator of the proper damping level.

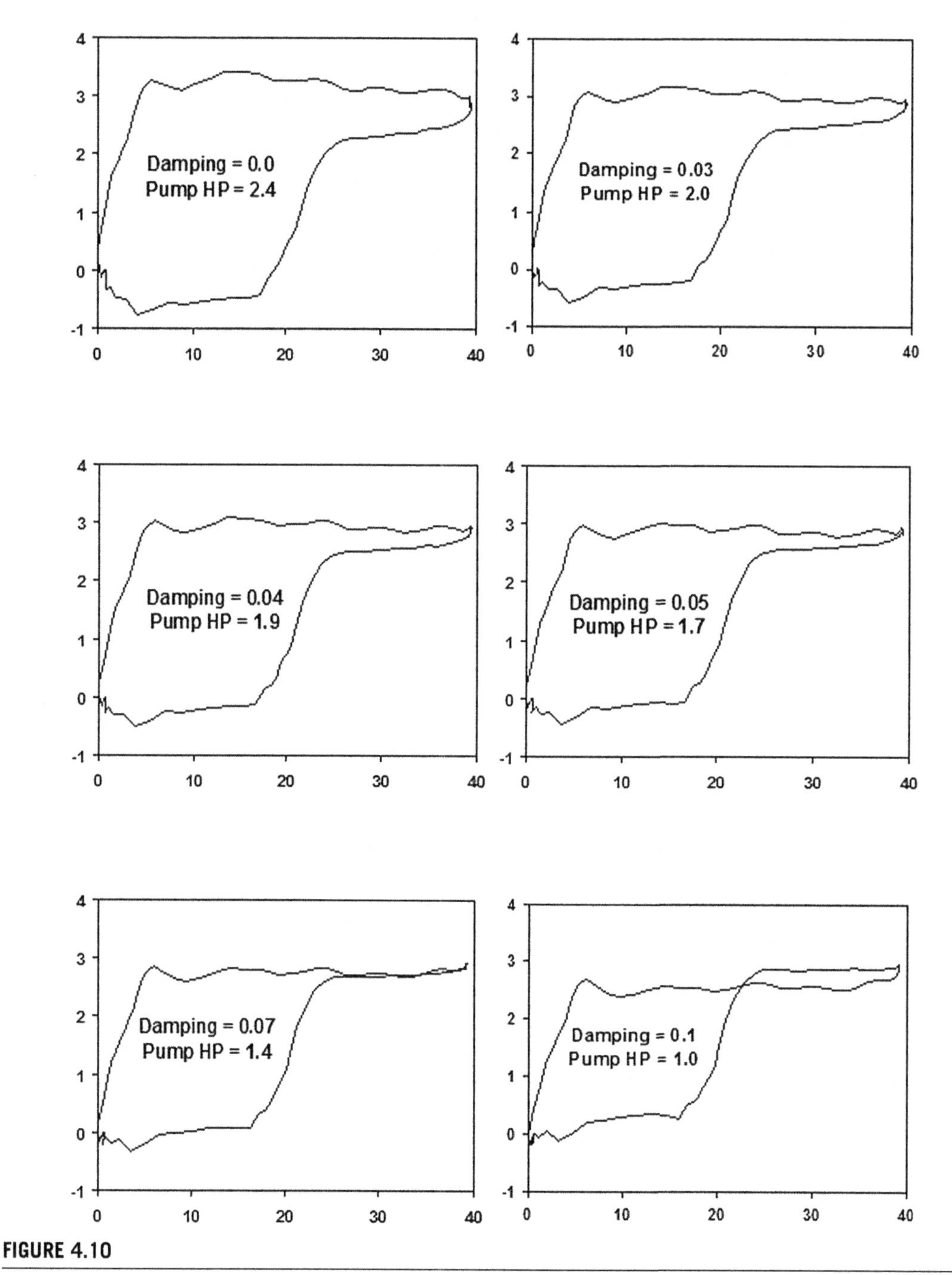

FIGURE 4.10

Calculated pump cards using different damping factors for the surface card in Fig. 4.9.

4.4.5 APPLICATION OF THE SIMULATION MODELS

The simulation techniques described in this chapter are widely used to **diagnose** and **predict** the behavior of sucker rod-pumped wells, and computer programs based on these procedures are the ultimate tools of the rod-pumping analyst. The benefits of applying these methods in diagnostic and predictive analysis are manifold. The most important feature of the **diagnostic** analysis is the ability to calculate **downhole** pump cards, which are much more easily analyzed than surface cards. This topic will be covered in Chapter 6, along with the other techniques of analyzing pumping installations. In addition to this very important application, surface **loads** and **torques**, as well as rod string **stresses**, are also made available by the diagnostic technique.

The main application of **predictive** analysis methods is to find the operational parameters of a planned **future** installation. Several pumping variants can be investigated before the final decision is made on the kind and size of equipment, as well as its operating conditions. Similarly, different operational **problems** encountered in practice can be simulated, and their effects on the downhole and surface conditions can be determined. The following is a list of the most important applications of the **predictive** analysis methods, as discussed by **Gibbs** [26]:

- Different volumetric **efficiencies** can be assumed at the sucker-rod pump. This enables the simulation of pumping a gassy mixture and the study of the effects on the operational parameters of free gas entering the pump.
- Pumping **malfunctions** can be simulated: gas-locked pump, improper valve action, mechanical pump problems, etc.
- The changes of the pumping **parameters** with the changes in well performance can be evaluated, e.g., a pumped-off condition is easily simulated.
- The use of different pumping **units** on the same well, and their relative merits, can be evaluated.
- The effects of prime mover speed **variation** can be included in the predictive model, and the performance of motors with different slips can be compared.
- By changing the taper percentages of the string, an **optimum** rod string taper design can be found.

Of the above possible applications, the evaluation of different pumping units is illustrated in the following example problem.

EXAMPLE 4.12: PREDICT THE SURFACE AND DOWNHOLE PUMP CARDS FOR THE WELL CONDITIONS GIVEN BELOW, IF A C-228D-213-86 CONVENTIONAL GEOMETRY UNIT OR AN M-228D-213-120 MARK II UNIT IS USED. THE TUBING STRING IS ANCHORED AND THE PUMP IS ASSUMED TO COMPLETELY FILL WITH LIQUID IN EVERY PUMPING CYCLE. USE A DIMENSIONLESS DAMPING FACTOR VALUE OF 0.2

Well depth = 6,000 ft	Pumping speed = 8 strokes/min
Dynamic liquid level = 6,000 ft	Stroke length = 88 in
Plunger size = 1.75 in	Liquid *SpGr* = 1.0

The rod string is made up of API Grade D steel rods with the following taper lengths: 2,268 ft of 7/8 in, 2,226 ft of 3/4 in, and 1,506 ft of 5/8 in rods.

Solution

Calculated dynamometer cards for the conventional and the Mark II units are presented in Figs 4.11 and 4.12, respectively.

Evaluation of the two cases shows very similar operating parameters. The gross pump stroke for the Mark II unit is 62 in, slightly greater than for the conventional unit. Peak polished rod load is greater for conventional (18,263 lb), as compared to 18,005 lb for the Mark II unit, but the opposite holds for the minimum polished rod load. Polished rod powers compare closely (about 9.7 HP for both units), as do hydraulic powers (7.7 HP).

An initial comparison of the above data can suggest that there are no benefits of using either of the units, but this is not the case. The evaluation cannot be considered complete without calculating the torques at the speed reducer for the two cases, which will be done in Section 4.6.5.

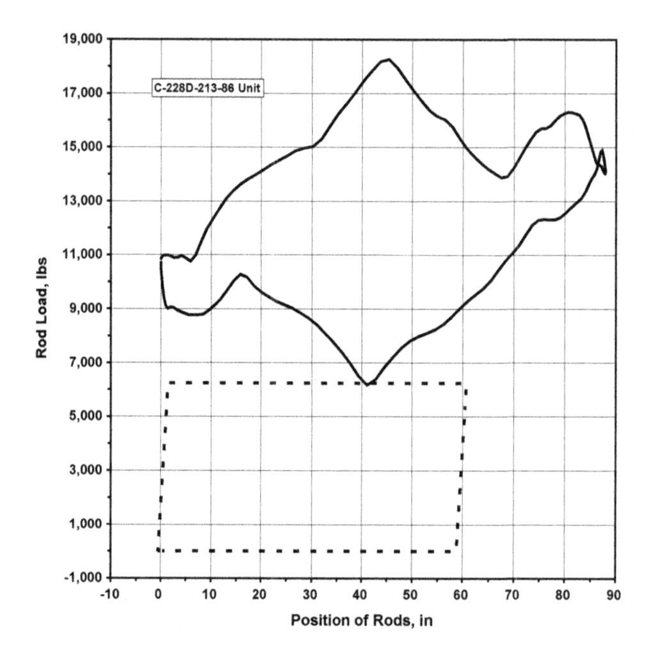

FIGURE 4.11

Predicted surface and downhole cards for Example 4.12, using a conventional pumping unit.

4.4.6 THE WAVE EQUATION IN SPECIAL CASES

The previous sections dealt with the description and use of the **conventional** wave equation, which considered energy losses along the rod string related to viscous damping only. Friction between the produced fluid and the rod string, which is proportional to the shear velocity between the fluid and the rod string, is properly handled by the wave equation's damping term. This kind of damping term and the conventional wave equation, however, cannot simulate those cases where, in addition to viscous damping, other factors also influence the behavior of the rod string. Two such special cases will be discussed in the following: the effects of (1) Coulomb friction that is especially important in deviated wells and (2) fluid inertia that dominates high-rate production from shallow wells.

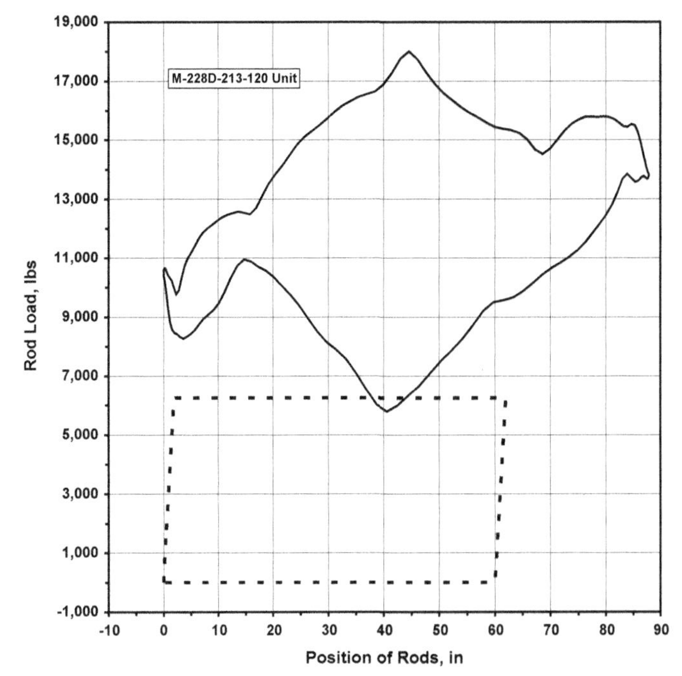

FIGURE 4.12

Predicted surface and downhole cards for Example 4.12, using a Mark II pumping unit.

4.4.6.1 Simulation of deviated wells

4.4.6.1.1 Well deviation basics

In sucker-rod pumping a vertical well provides **ideal** operating conditions and is preferred over a deviated or crooked one; in practice, however, wells are **never** fully vertical. Even when a vertical well trajectory is desired, the usual drilling technology results in an **unintentionally** crooked well. A great number of wells, at the same time, are **intentionally** drilled to reach targets that call for extremely deviated trajectories. Deviated wells, very popular in the offshore environment, are drilled from a central platform to reach different parts of the same reservoir and allow saving of huge investment costs. On land locations surface restrictions like urban areas may prevent the use of vertical boreholes. Although deviated or crooked wells are not the best candidates for sucker-rod pumping, present-day technology provides the means to produce them economically.

Well trajectory is controlled during the drilling process and the parameters of the borehole are usually known after the completion of the well; special **deviation surveys** can also be conducted to define the well path in three dimensions. Definition of the three-dimensional (3D) well trajectory is normally provided at several survey stations along the well path and includes the following measured parameters (see Fig. 4.13):

- **Measured depth** (MD): measured from the surface along the center line of the tubing string, feet,
- **Inclination** (α in the figure): the angle measured between the borehole direction and the vertical, degrees, and

- **Azimuth:** the angle of the borehole direction as projected to a horizontal plane and relative to due north, measured clockwise, degrees.

Figure 4.13 shows two stations (measurement points) only; further stations in a real well follow downhole with different combinations of the three basic parameters. Since these measurements are made normally at **discrete** points in the well, the complete well trajectory is defined, as a first approximation, by a series of **straight** line segments in the 3D space. At every survey station (at given MD or true vertical depth levels), the borehole's **inclination, azimuth**, or **both** may change, creating a **dogleg**, i.e., where the trajectory of the wellbore in the three-dimensional space changes rapidly.

Doglegs create several problems in drilling and production operations; in sucker-rod pumping they are especially harmful because of the mechanical **wear** on the rod and the tubing strings as they rub on each other. The magnitude of damage, of course, very much depends on the rate of **change** of wellbore direction at each station. To quantify trajectory changes in wells, the parameter **dogleg severity** (*DLS*) is introduced, which is a normalized estimate of the overall (3D) curvature of the wellbore between two consecutive directional survey stations. It is expressed in two-dimensional degrees per 100 ft of wellbore length with the following formula, based on the minimum curvature method:

$$DLS = 100 \frac{\arccos[\cos\alpha_1 \cos\alpha_2 + \sin\alpha_1 \sin\alpha_2 \cos\Delta Az]}{MD_2 - MD_1} \tag{4.75}$$

where:

DLS = dogleg severity, degrees/100 ft,
α_1, α_2 = well inclinations at two consecutive stations, degrees,
ΔAz = change of azimuth between two consecutive stations, degrees, and
MD_1, MD_2 = measured depths at two consecutive stations, ft.

The variation of **dogleg severity** with well depth is an excellent indicator of the changes in wellbore direction and thus provides valuable information for designing and analyzing sucker-rod

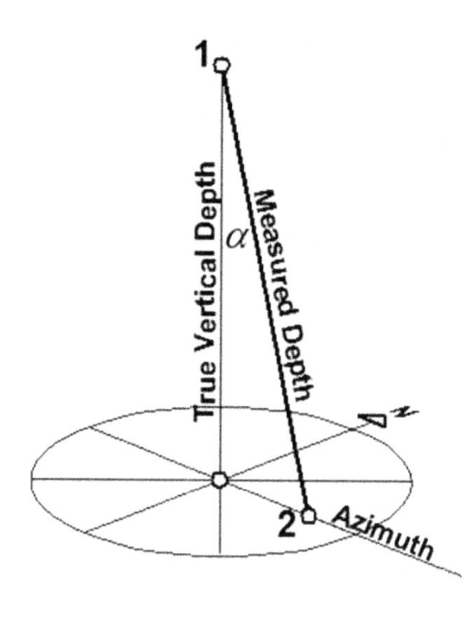

FIGURE 4.13

Definition of the parameters used in well inclination surveys.

pumping operations. This is the reason why, when working with deviated wells, the deviation survey must be available for the rod pump analyst.

4.4.6.1.2 Problems caused by well deviation

As discussed in Section 4.4.2.1, the conventional wave equation works well in wells without considerable deviation from the vertical because practically all damping forces along the rod string are caused by **fluid** friction only. In deviated or crooked wells, however, the magnitude of **mechanical** (Coulomb) friction forces occurring between the rod string and the tubing is significant and cannot be neglected. These two kinds of damping forces behave absolutely differently: fluid friction is characterized by **viscous** forces proportional to the shear velocity between the fluid and the rods, whereas Coulomb friction depends on the **normal** force and the **friction coefficient** between the rods and the tubing wall. As a result, the wave equation for use in deviated wells must include **two** damping terms of different nature. In order to understand the background of the universal wave equation (to be introduced later), let's first investigate some simplified cases where the various differences between vertical and deviated wells are highlighted.

The well path may contain shorter or longer inclined but **straight** sections where both the inclination and azimuth angles are constant. One significant difference as compared to the conditions in a vertical section is related to the weight of the rod element:

- Rod weight in a vertical well always acts **vertically** and it fully contributes to rod loading; this is the reason why it is not included in the conventional wave equation as discussed earlier.
- In an inclined well section, as shown in Fig. 4.14, only one component, W_α, of the rod weight, W, creates axial load in the rod string because the rods rest on the tubing wall. Since the axial component of weight depends on the **inclination** of the given section, the deviated wave equation must include a **gravity term** to take care of the variation of axial loads created by rod weight along the string.

Even more significant than the gravity force is the effect of **Coulomb** friction between the rod string and the tubing inside wall. In general, mechanical friction between solid surfaces is

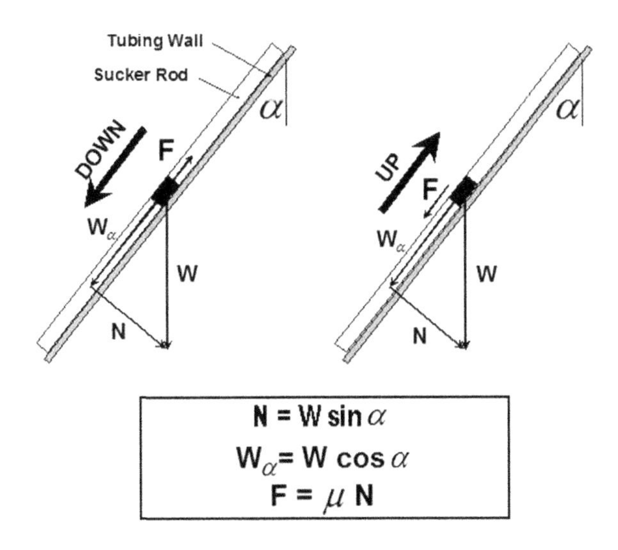

FIGURE 4.14

Graphical presentation of rod weight and mechanical friction force in a straight section of an inclined well.

proportional to the normal (perpendicular) **force** pressing the two surfaces against each other and the **friction coefficient** valid for the given materials. In a vertical well, friction along the rod string is zero because the rods and the tubing are not in contact. In a straight but inclined well, however, friction forces are found from the **normal** force, N, and the friction coefficient, μ (see Fig. 4.14):

$$F = \mu W \sin \alpha \tag{4.76}$$

where:

F = friction force, lb,
W = rod weight in fluid, lb,
α = well inclination, degrees, and
μ = friction coefficient, –.

Coulomb friction always acts **opposite** the direction of motion, and it (a) decreases axial loads during the downstroke of the string and (b) increases axial loads during the upstroke, as indicated in Fig. 4.14. In this simplified case of a straight inclined well section the magnitude of frictional forces, according to Eq. (4.76), is a function of rod **weight**, rod **inclination**, and the **friction coefficient**.

Treatment of friction forces is much more complicated in well sections where inclination and azimuth angles may change with well depth. To simplify the conditions and to present some basic concepts, take a dogleg where wellbore inclination changes only while the azimuth is constant; this section of the borehole trajectory is, therefore, two-dimensional only. The case is illustrated in Fig. 4.15, which depicts the inclination angles and forces contained in a vertical plane. As before, frictional force between the rods and the tubing is found from the normal (a.k.a. **side**) force, N; this, in turn, can be calculated from the axial loads acting on the rod element. As shown in Fig. 4.15, if axial upward and downward loads are assumed equal, then adding them as **vectors** at the point of inclination change creates a rhombus whose shorter diagonal is the required side force, N. Since diagonals of rhombuses are perpendicular and bisect opposite angles, the following relation holds:

$$N = 2L \sin(\Delta\alpha/2) \tag{4.77}$$

where:

N = side (normal) force, lb,
L = axial load in the rod, lb, and
$\Delta\alpha$ = change in inclination between two consecutive stations, degrees.

The change of inclination, $\Delta\alpha$, is easily found from the variation of dogleg severity along the well for the 2D case studied here. Since the change in azimuth, ΔAz, is zero, substituting $\Delta Az = 0$ in the formula for dogleg severity (Eq. 4.75) we get:

$$DLS = 100\frac{\arccos[\cos\alpha_1 \cos\alpha_2 + \sin\alpha_1 \sin\alpha_2]}{MD_2 - MD_1} = 100\frac{\arccos[\cos(\alpha_1 - \alpha_2)]}{MD_2 - MD_1} = 100\frac{\Delta\alpha}{MD_2 - MD_1} \tag{4.78}$$

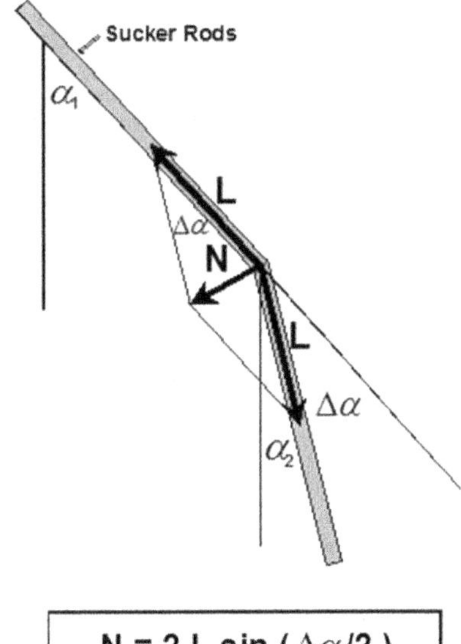

FIGURE 4.15

Approximate distribution of forces in a dogleg.

where:

DLS = dogleg severity, degrees/100 ft,
$\Delta\alpha$ = change in inclination between two consecutive stations, degrees, and
MD_1, MD_2 = measured depths at two consecutive stations, ft.

This means that the angle $\Delta\alpha$ needed to calculate the Coulomb friction force at any depth is inherently determined by the DLS–depth function, which is part of the well's deviation survey. Although this relationship is verified here for the 2D case, side forces in real 3D trajectories are determined very similarly.

The friction force is now easily calculated from the normal force, and using Eq. (4.77):

$$F = 2\,\mu\,L\sin(\Delta\alpha/2) \tag{4.79}$$

where:

F = friction force, lb,
L = axial load in the rod, lb,
$\Delta\alpha$ = change in inclination between two consecutive stations, degrees, and
μ = friction coefficient, –.

This formula allows drawing of the following **conclusions** on the behavior of Coulomb friction forces in two-dimensional doglegs:

- Frictional loads on the rod string at any dogleg are proportional to the **axial load**, L, valid at the given position and independent of rod weight (compare to the case of a straight rod section

discussed earlier). An important consequence is that the **same** $\Delta\alpha$ change in inclination, if situated **higher** up the well, causes **greater** Coulomb forces because of the higher axial loads due to rod string weight.
- Friction forces increase for greater **differences** in inclination, $\Delta\alpha$, i.e., at depths where well deviation changes more rapidly.

It must be noted that the previous discussions do not present an **exact** description of the conditions valid in a deviated well; they are meant only to **illustrate** some important differences between vertical and inclined rod behavior. In reality, well trajectory is usually represented as a succession of circular **arcs** and rod loads are much more complicated to determine because upward-pointing and downward-pointing axial rod loads at a dogleg are not equal. This is why the calculation of Coulomb friction forces necessitates an **iterative** procedure where the vector sum of all relevant forces (axial rod loads, side force, rod weight, and Coulomb force) at a given dogleg must add up to zero. Calculations are started at the bottom of the string and continue upward, taking into account the trajectory of the well; computer time requirements are at least 10 times higher than those needed for a vertical well.

4.4.6.1.3 The deviated wave equation

After the discussion of some specific problems caused by well deviation, let's extrapolate the conclusions drawn from the simplified cases to the real case of a **three-dimensional** well trajectory. In addition to the proper handling of **fluid damping** forces, the basic requirements for the development of a **deviated** wave equation can be summed up as follows:

- A **gravity term** must be included to represent the variation of axial loads created by the interaction of rod weight and well inclination.
- Frictional **Coulomb** rod loads must be treated differently in straight sections and doglegs:
  - In **straight** but inclined sections they are a function of rod weight.
  - At **doglegs** they mainly depend on the axial load present in the rods.

The first presentation of a deviated wave equation was made by **Gibbs** [52,53], who modified his original one-dimensional equation (see Eq. 4.54) to include the effects of well deviation. In contrast to the one-dimensional case, the differential equation is written for a 3D rod element, ds, and rod displacement, u, is a function of rod length measured along the well trajectory, s. The final partial differential equation is reproduced here in a schematic form:

$$\frac{\partial^2 u(s,t)}{\partial t^2} = v_s^2 \frac{\partial^2 u(s,t)}{\partial s^2} - c\frac{\partial u(s,t)}{\partial t} - C(s) + g(s) \tag{4.80}$$

where:

$u(s,t)$ = rod displacement, ft,
c = viscous damping factor, 1/s,
v_s = sound velocity in the rod material, ft/s,
$C(s)$ = term representing the effects of Coulomb friction, ft/s^2,
$g(s)$ = term to account for axial load from rod weight, ft/s^2,
s = rod length along the trajectory, ft, and
t = time, s.

Since the first two terms on the right-hand side are similar to those in the one-dimensional wave equation, the last two terms need explanation. The gravity term, $g(s)$, stands for the **component** of rod weight, which is parallel to well inclination; its variation with well depth is easily found from deviation survey data. The Coulomb friction term, $C(s)$, includes the **two** effects discussed before and requires the use of different formulas in **straight** and **curved** sections of the well. In a dogleg, rod element ds is considered part of a **circle** whose radius is found from the actual value of dogleg severity, as given here:

$$R = \frac{360}{2\pi}\frac{100}{DLS} = \frac{5,730}{DLS} \tag{4.81}$$

where:

R = radius of curvature, ft, and
DLS = dogleg severity, degrees/100 ft.

The side force points to the **center** of curvature; its magnitude is found from the axial load in the rod and geometrical considerations, similarly to the procedure discussed in conjunction with Fig. 4.15. The frictional force is then found after the selection of the proper friction coefficient.

The **friction coefficient**, μ, in general, is a function of the materials in contact, their roughnesses, and their lubrication. An average value of $\mu = 0.2$ is generally used for bare rods and tubing, but actual values vary with the **lubricity** of the well fluid. Oil-lubricated smooth surfaces are characterized by $\mu = 0.1$; rough steel surfaces with water lubrication result in $\mu = 0.3$ [54]. The effect of rod guides on friction coefficients, which are normally installed in deviated sections of rod-pumped wells, depends on the type of guides used. **Molded-on** guides are very unfavorable because they increase the friction coefficient by about 150% on average; **wheeled** guides, on the other hand, reduce it to about 10%. Using an original coefficient of 0.2, molded guides would give a final value of $\mu = 0.3$, whereas wheeled guides would reduce it to $\mu = 0.02$; the friction load on the rod string would change accordingly.

Actual friction coefficients can be **inferred** from valve tests of the sucker-rod pump [55] because Coulomb friction between the rods and the tubing distorts the loads measured on the surface. During a standing valve test (SV test) the load measured by a dynamometer on the surface is **less** than the buoyant weight of rod string, W_{rf}, because part of the rod weight is supported by friction between the rod string and the tubing. Similarly, a traveling valve test (TV test) gives measured surface loads **greater** than the buoyant weight of rods plus fluid load, $W_{rf} + F_o$, because in this case frictional loads add to the well load. The difference of the measured and the calculated loads permits the calculation of frictional loads and the friction coefficient valid in the given well.

The deviated wave equation is usually solved **numerically** using finite differences. Solution is much more complex than for vertical wells because changes in well deviation must be accurately followed; this requires dividing the rod string into at least **10** times more segments than in a vertical well. The required computational time is also increased because of the many iterations involved in the final solution. Most commercial computer program packages include the design of rod **guide** placement, which is a basic requirement of decreasing well service costs related to rod and tubing failures [55].

4.4.6.1.4 An illustrative example

To illustrate the use of the deviated wave equation, an example is presented with the following basic data and using the RodStar [56] program package.

Pump setting depth, ft	8,000	Tubing size, in	2 7/8
Pump size, in	2	Tubing anchor, ft	7,900
Polished rod stroke, in	193.3	Pumping speed, SPM	6.0
API rod taper code	86	Unit rotation	clockwise
C-1280D-365-120 unit			

The well trajectory is close to vertical down to a depth of about 4,800 ft, from where it starts to deviate; a 3D deviation survey plot is presented in Fig. 4.16.

Dogleg severity, which is calculated from the data of the deviation survey, is plotted versus measured depth in Fig. 4.17; calculated side loads follow the plot presented in Fig. 4.18. As seen, there is a tight correlation between the side load and the DLS at any given depth, as expected from previous discussions.

Assuming that the downhole pump fills completely at every stroke, the surface dynamometer card was constructed using the solution of the wave equation for two cases: by considering and by disregarding well deviation. The resultant cards are depicted in Fig. 4.19; comparison of the two cards shows that Coulomb friction along the rod string, in general, increases upstroke loads and decreases downstroke loads.

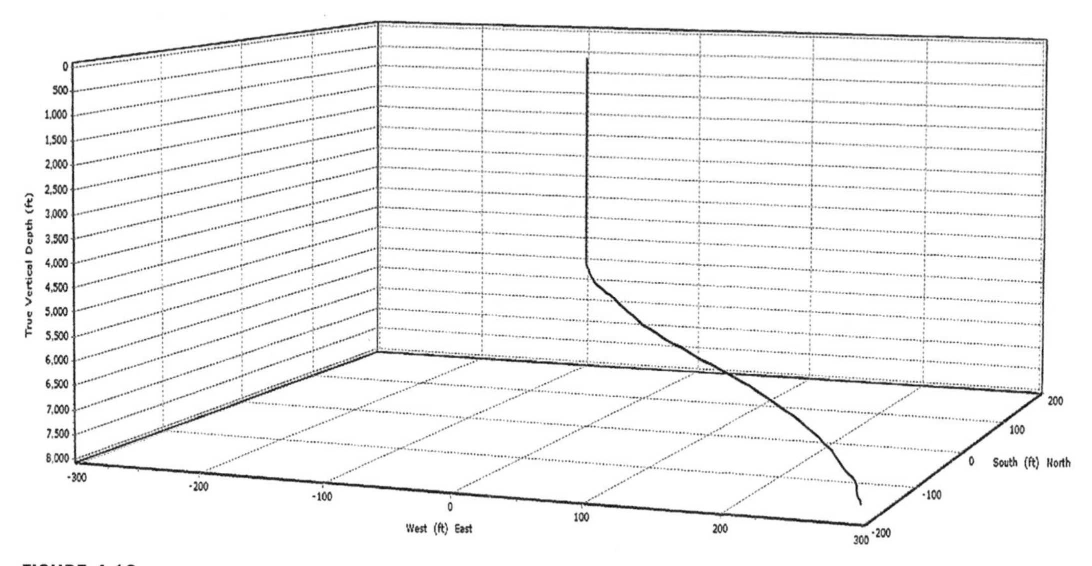

FIGURE 4.16

3D deviation survey plot of an example well.

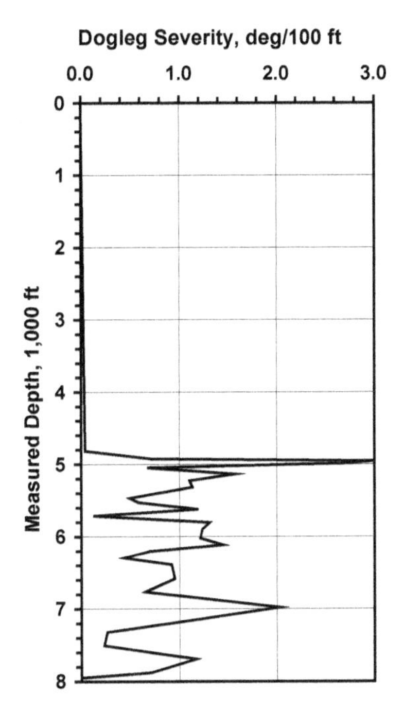

FIGURE 4.17

Dogleg severity versus well depth in the example well.

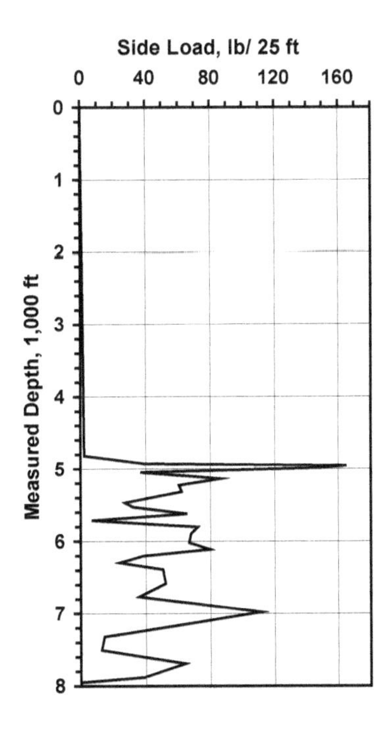

FIGURE 4.18

Side load versus well depth in the example well.

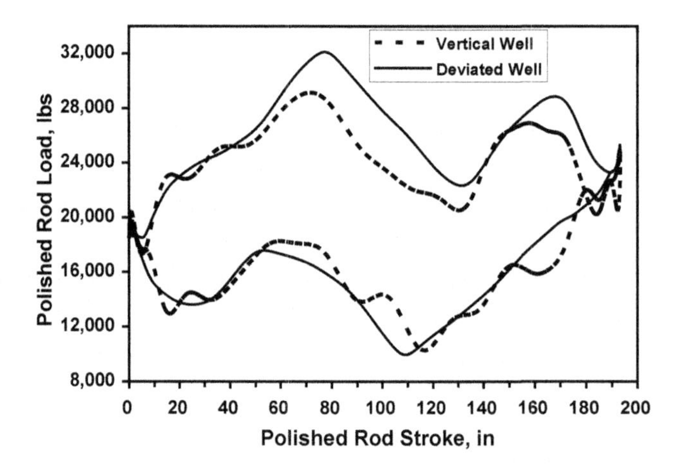

FIGURE 4.19

Comparison of surface cards calculated with and without consideration of well deviation for an example well.

The different loading conditions for the vertical and the deviated case have changed the energy conditions of the system and the loading of the pumping system's components very substantially. As detailed in the following, polished rod horsepower has increased, as well as the loading of the pumping unit structure and that of the gearbox; all these add up to a drop in overall system energy efficiency.

Operational Parameter	Vertical	Deviated
Polished rod horsepower, HP	24.9	29.7
Unit structural load, %	80.0	88.0
Gearbox load, %	81.6	92.3
System efficiency, %	56.0	42.9

One general conclusion can be distilled from the example presented that relates to the diagnostic analysis of deviated wells using the vertical model of the wave equation. In such cases the conventional wave equation does not remove all the energy losses that occur along the rod string because it considers losses due to fluid friction only. Calculated pump cards, therefore, have shapes typical for the "**unaccounted friction**" phenomenon with exaggerated areas. The apparently high calculated pump powers indicate clearly that the conventional wave equation cannot handle the effects of mechanical Coulomb friction in deviated wells.

4.4.6.2 Considerations for fluid inertia

4.4.6.2.1 Introduction

The simulation models described in the preceding sections have one basic assumption in common that has not been mentioned yet: they only include the **dynamics** of the rod **string** and disregard the dynamic effects arising in the **fluid** column. In these procedures, the fluid in the tubing is considered a **concentrated** mass that only imparts its hydrostatic pressure on the plunger as well as a viscous drag on the rod string. In reality, however, these assumptions are not universally valid, because in some cases the **inertia** of the fluid can greatly affect the loads and displacements along the string. The fluid contained in the tubing string, being compressible to some extent, behaves the same way as the rod string does: pressure waves induced by the plunger's movement travel along its length. These waves also are reflected from the two ends of the column, just like the stress waves in the solid rod string. Based on these considerations, the actual effects of the fluid column on the behavior of the pumping system can be classified as: (1) the effects of the dynamic (inertial) in addition to the static fluid **loads** on the plunger, and (2) the viscous **damping** along the rod string.

Fortunately, in the majority of pumping wells, dynamic fluid loads are negligible, and the assumptions of the traditional simulation models are valid. According to **Laine et al.** [57], **fluid inertia** is only significant in the following cases:

- In **shallow wells**, due to the smaller total rod weight and the smaller fluid column pressure, the relative importance of fluid inertia increases, and this can have a pronounced effect on rod loads.
- Pumping units operating relatively **large** pumps in high-producer wells move large liquid masses, causing the fluid inertial forces to increase.
- Greater pumping **speeds** also increase the inertial forces, both in the rod string and in the fluid column.

A practical classification of sucker-rod pumping systems based on the importance of fluid inertia effects was proposed by **Svinos** [58], as follows:

- **Group One** wells are **not affected** by fluid inertia, and the use of the conventional wave equation to simulate operating conditions is justified. Wells belonging here have:
  - Pump setting depths greater than 4,000 ft and any plunger size, or
  - Pump setting depths less than 4,000 ft with plunger sizes of less than or equal to 2 in.
- **Group Two** wells are considerably **affected** by fluid inertia and require special wave equations to simulate their downhole operating conditions. Wells with shallow depth and high rates go into this group with:
  - Pump setting depths less than 4,000 ft with plunger sizes of 2.25 in or larger.

4.4.6.2.2 Calculation models

As discussed previously, fluid inertial effects are significant in **shallow** wells produced by relatively **big** pumps. In such wells, an extra force is required (in addition to the buoyant rod weight) to accelerate the fluid column in the tubing and through the flowline toward the separator. This dynamic force is the result of a pressure surge in the fluid column and is very similar to the "**water hammer**" experienced in liquid pipelines. The effect is due to the **stiffness** of the short rod string that transmits a high acceleration to the fluid column, as well as due to the great fluid **mass** moved by the big pump; all these result in great inertial loads on the rod string during the upstroke. Fluid inertial effects are substantial when producing highly **incompressible** liquids like high water-cut oil, but decrease as fluid compressibility increases due to produced gas because gas bubbles very efficiently **damp** out the pressure waves in the fluid. Similarly, fluid inertia is negligible in deeper wells with longer and more flexible rod strings because rod stretch during the upstroke efficiently decreases fluid acceleration.

The first treatment on fluid inertia problems was presented by **Doty and Schmidt** [59], who developed a calculation procedure that included simulation of such effects. Their basic observation is that the rod string and the fluid column are lightly **coupled** at the traveling valve of the sucker-rod pump so the partial differential equations describing the movement of those two components can be solved separately. The behavior of the rod string is characterized by a **system** of partial differential equations consisting of (1) momentum balance and (2) Hooke's law. The dynamics of the fluid column are represented by (1) the equation of motion, (2) the continuity equation, and (3) the equation of state. The two systems of partial differential equations are solved with the method of characteristics. Other investigators, although using the same basic concepts, solve the problem by utilizing finite-difference methods [57,60].

A schematic pump card illustrating fluid inertia effects is presented in Fig. 4.20, where the first peak (hump) on the upstroke represents the plunger load due to the **initial** pressure wave generated by fluid inertia; the second peak shows the effect of the **reflected** wave. Fluid inertia causes the following basic changes in the pumping system's performance:

- The plunger **stroke** decreases,
- Maximum plunger **loads** and peak polished rod **loads** increase, and
- The **power** required at the polished rod increases.

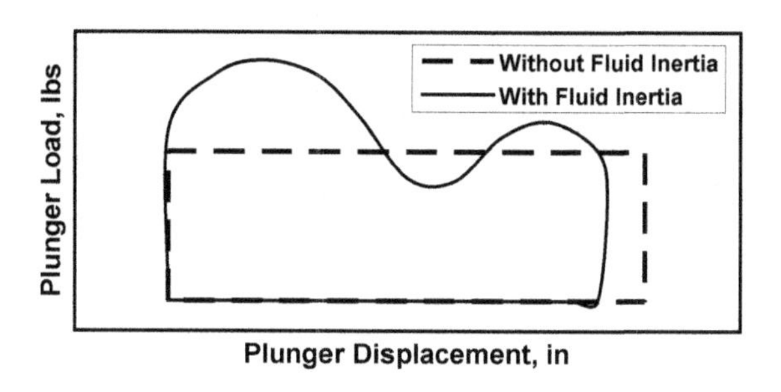

FIGURE 4.20

Schematic pump card illustrating fluid inertia effects.

A sample calculation is presented for a shallow well with an oversize sucker-rod pump and the following operational data:

Pump setting depth, ft	1,500	Tubing size, in	3 1/2
Pump size, in	3.75	Tubing anchor, ft	1,500
Polished rod stroke, in	100.7	Pumping speed, SPM	9.0
API rod taper code	77	Fluid compressibility, 1/psi	2.0E-6
C-320D-213-100 unit		Unit rotation	Counterclockwise

Figure 4.21 depicts surface dynamometer cards calculated by considering and disregarding fluid inertial effects; the two pressure waves are clearly recognizable on the card including inertia. Operating conditions are very different for the two cases; fluid inertia causes higher loading of the equipment and a lower overall energy efficiency of the system.

	Fluid Inertia	
Operational Parameter	**Included**	**Disregarded**
Polished rod horsepower, HP	19.5	18.2
Unit structural load, %	67.0	61.0
Gearbox load, %	99.4	80.9
System efficiency, %	56.0	60.0

The biggest operational problem caused by fluid inertia is the **overload** on the gearbox, which shortens equipment life and increases production costs. To eliminate the harmful effects of fluid inertia in shallow, high-volume California wells, **Svinos** [61] developed a pulsation dampener that, set just above the pump in the tubing string, effectively removed the peaks from the pump card and radically reduced the loading of the gearbox.

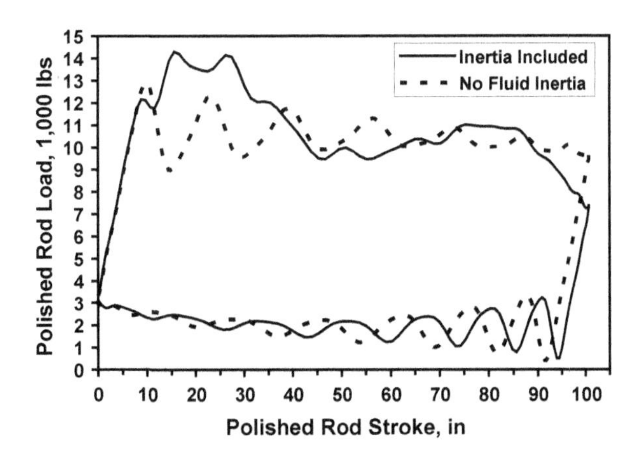

FIGURE 4.21

Surface cards calculated by considering and disregarding fluid inertia for a sample case.

4.5 PRODUCTION RATE CALCULATIONS

4.5.1 PUMP DISPLACEMENT

After the plunger's downhole **stroke** length, S_p, has been determined with some method (using either the **RP 11L** procedure or a simulation model), the daily volumetric displacement of the sucker-rod pump can be found. Assuming that the barrel **completely** fills up with **liquid** in every cycle for 24 h of operation per day, the liquid volume displaced by the pump is calculated as given here:

$$PD = 0.1166\, d^2\, S_p\, N \tag{4.82}$$

where:

PD = liquid volume displaced by the pump, bpd,
S_p = plunger stroke length, in,
N = pumping speed, strokes/min, and
d = plunger size, in.

Depending on the stroke length used in the equation, the production rate calculated can refer to the **gross** or the **net** plunger stroke. The net stroke length already reflects the effects of an unanchored tubing, if appropriate, as well as the stroke length reduction due to gas interference. If available, which is the case when a simulation model is used, the net plunger stroke length should be used to find a more realistic figure for the production rate.

The volumetric rate received from the above formula should be corrected for **liquid shrinkage** to find the stock tank **volume**. In the case of highly volatile crude oils, neglecting this effect can introduce some errors due to the relatively higher volume factors of such oils. The formula to account for the oil's shrinkage from bottomhole conditions to the stock tank is given in the following, where the volume factor, B_o, can be approximated using several different methods. Since B_o is always greater than unity, the oil volume measured in the stock tank is always **less** than the oil volume displaced by the downhole pump.

$$Q_{sc} = \frac{PD}{B_o} \tag{4.83}$$

where:

Q_{sc} = stock tank volumetric rate, STB/d,
PD = pump displacement, bpd, and
B_o = volume factor of the oil at pumping conditions, –.

4.5.2 PUMP LEAKAGE LOSSES

While lifted to the surface, liquids produced by the sucker-rod pump can leak at several places along their path to the surface and the amount of leakage decreases the production rate measured at the surface. Leakage in the pump itself is called **plunger slippage** and is the result of part of the produced liquid **slipping** back through the plunger–barrel **clearance**. This clearance, although very small and on the order of a few thousandths of an inch, is **essential** for the proper operation of the pump for two reasons:

- The first and most important one is the **lubrication** of the metal surfaces; a liquid film between the metal plunger and the metal barrel prevents **galling** of those two components. Produced fluid provides adequate lubrication if plunger slippage is about 2–5% of the pumping rate.
- The second reason is to eliminate the harmful effects of **solid** particles that pass through the plunger–barrel clearance.

Slippage in sucker-rod pumps takes two forms: static and dynamic slippage.

- **Static slippage** is the **dominant** factor and occurs only during the **upstroke** of the pump; it is caused by the pressure differential across the plunger–barrel fit. The high hydrostatic pressure present in the tubing string, acting on top of the plunger with the traveling valve closed, forces liquid to **slip** past the plunger into the pump chamber between the traveling and the standing valves.
- **Dynamic slippage**, on the other hand, takes place **both** on the up- and on the downstroke of the pump and is caused by the plunger's **movement**, its magnitude being proportional to the plunger **velocity**, i.e., the pumping speed used. The direction of liquid slippage is different for the up- and downstroke: during upstroke liquid **falls** below the traveling valve, while during the downstroke liquid flows **upward** and decreases the amount of liquid passing through the traveling valve.

Static slippage in sucker-rod pumps has been a heavily discussed topic and several correlations are available for estimating the liquid rate lost due to this effect. All theoretical and practical investigations have a **common** background: the description of **viscous** liquid flow between a stationary and a moving flat **plate**. This is the reason why each of the available correlations has the same **general** form, as given in the following:

$$q_s = K \frac{d^A \, \Delta p \, \Delta d^B}{\mu \, l} \tag{4.84}$$

where:

q_s = plunger slippage, bpd,
d = plunger diameter, in,
Δp = pressure differential across plunger, psi,
Δd = plunger fit, in,
μ = liquid viscosity, cP, and
l = plunger length, in.

The constant K and the exponents A and B vary with the different authors.

As seen from the formula, plunger slippage is proportional to:

- the pressure **differential** across the plunger,
- the plunger **diameter**, and

- the diametrical **clearance** between the plunger and the barrel, the last two parameters raised to different exponents.

It follows that the natural **wear** of the barrel and plunger surfaces increases the liquid volume lost due to the progressively increasing clearance between the parts. The other important parameters are liquid **viscosity** and plunger **length**; slippage is inversely proportional to those. Therefore, pumping a more viscous fluid or using a longer plunger decreases the slippage loss past the plunger. Pumping from greater depths where well temperatures are higher, on the other hand, increases plunger slippage because of the reduced liquid viscosity.

Although plunger slippage can be estimated from dynamometer surveys, as discussed in Chapter 6, it is usually calculated from empirical formulas. An early formula proposed by the **Oilwell Division of US Steel** [62] is given below that considers slippage occurring only during the upstroke:

$$q_s = \frac{1.006\text{E}6\ d\ \Delta p\ \Delta d^3}{\mu\ l} \tag{4.85}$$

where:

q_s = plunger slippage, bpd,
d = plunger diameter, in,
Δp = pressure differential across plunger, psi,
Δd = plunger fit, in,
μ = liquid viscosity, cP, and
l = plunger length, in.

An extensive series of theoretical and experimental investigations [63–66] on pump slippage resulted in the following main **conclusions**.

- Early formulas greatly **overestimate** the amount of liquid slippage. Typical values, based on experimental data, are about **two** times greater for plunger fits less than 0.006 in and more than **three** times greater for fits larger than 0.006 in. This implies that pumps with fits **larger** than those selected on the basis of earlier predictions can be used without experiencing too-high pump leakages.
- The **eccentricity** of the plunger's lateral position in the barrel has a great effect on liquid slippage (also proved by [67]), a fact that most previous formulas disregarded. For a completely eccentric position, leakage rates 2.5 times **greater** than for concentric cases can be expected.
- Most previous correlations disregarded the effect of **dynamic** leakage in the pump.

Based on the findings of their extensive work, the authors [63–66] recommend the use of the following correlations.

Modified Robinson and Reekstin [63]
The best fit for historical data of pumping oil with pumps of 1.5 in diameter and 72 in length, plunger fits less than 0.01 in, and corrected for completely eccentric cases.

$$q_s = 7\text{E}6 \frac{d^{0.7}\ \Delta p\ \Delta d^{3.3}}{\mu\ l} \tag{4.86}$$

ARCO–HF [65]
An empirical formula for pumping water with 1.75 in pumps, valid for a wide range of plunger fits.

$$q_s = 870 \frac{d\ \Delta p\ \Delta d^{1.52}}{\mu\ l} \tag{4.87}$$

Patterson [66]
Based on extensive measurements in the Texas Tech University's test well, the authors developed the **first** empirical correlation that also includes the effects of **dynamic** leakage in the downhole pump. The correlation valid for 1.5, 1.75, and 2 in pumps operating with pumping speeds between 1 and 16.6 SPM is given here; as seen, plunger slippage including dynamic leakage proportionally increases with the pumping speed.

$$q_s = 453(1 + 0.14\ N) \frac{d\ \Delta p\ \Delta d^{1.52}}{\mu\ l} \tag{4.88}$$

In all previous formulas:

q_s = plunger slippage, bpd,
d = plunger diameter, in,
Δp = pressure differential across plunger, psi,
Δd = plunger fit, in,
μ = liquid viscosity, cP,
l = plunger length, in, and
N = pumping speed, strokes/min.

EXAMPLE 4.13: FIND THE PLUNGER SLIPPAGE RATES FROM THE AVAILABLE CORRELATIONS, IF WATER IS PUMPED WITH A 2.5 IN PUMP OF 48 IN LENGTH. PRESSURE DIFFERENTIAL ACROSS THE PLUNGER IS ASSUMED TO BE 3,000 PSI, AND THE PLUNGER FIT IS 0.003 IN. PUMPING SPEED IS 6 SPM

Solution

Table 4.2 contains calculated results where the case of an increased plunger fit of 0.01 in is also included. As seen, progressive pump wear can rapidly increase the volume of the liquid lost due to slippage.

Table 4.2 Calculated Liquid Slippage in Sucker-Rod Pumps for Example 4.13

	Pump Slippage			
Plunger Fit	**Oilwell**	**Mod. R&R**	**ARCO-HF**	**Patterson**
0.001 (in)	**(bpd)**	**(bpd)**	**(bpd)**	**(bpd)**
3	4.2	1.0	19.9	19.1
10	157.2	53.7	124.0	118.8

Other sources of liquid slippage in the pump itself are the traveling and standing **valves**, whose ball-to-seat contacts can also **wear** and allow liquids to leak. Frequent checks, conducted with surface dynamometers, are required to detect **worn** valves and to ensure proper operation. **Nolen and Gibbs** [68] present a technique to determine the **leakage** rate past plungers and pump valves, which relies on the use of the modern electronic dynamometers. Leaks in the **tubing** string or a leaky surface **check** valve can also result in production loss in the well, but the application of a regular maintenance schedule can mostly eliminate these factors.

Use of a proper plunger slippage equation, like the Patterson correlation, permits the **selection** of the required pump **clearance** for any given pumping conditions [69]. The main steps of the procedure to follow are:

- Calculate the liquid **displacement** of the pump at downhole conditions without considering plunger slippage. This usually requires the determination of the plunger's stroke length from the solution of the wave equation.
- Assuming different clearance values, find the volumes of liquid slippage and calculate their ratios to pump displacement.
- Select the pump clearance that gives a liquid slippage equal to about 5% of the pump's displacement.

4.5.3 VOLUMETRIC EFFICIENCY OF PUMPING

The volumetric efficiency of the subsurface pump is defined as the ratio of the liquid rate produced at the surface divided by the total volume of liquid displaced by the pump. This efficiency is affected by several different factors and can be estimated using different methods, depending on the kind of fluid produced by the pump: single-phase **liquid** or multiphase **mixture**.

4.5.3.1 Single-phase liquid production

In ideal conditions the downhole pump handles **liquid** only, and if it is in a **perfect** condition without a significant quantity of leakage then the volumetric efficiency is easily found from the amount of liquid measured in the stock tank and the calculated pump displacement:

$$\eta_{\text{vol}} = \frac{Q_o + Q_w}{PD} \tag{4.89}$$

where:

η_{vol} = pump volumetric efficiency, –,
Q_o, Q_w = oil and water volumetric rates, STB/d, and
PD = liquid volume displaced by the pump, bpd.

A more frequently used formula uses the measured liquid rates and reads:

$$\eta_{\text{vol}} = \frac{Q_o + Q_w}{Q_o\, B_o + Q_w\, B_w} = \frac{1 + WOR}{B_o + B_w\, WOR} \tag{4.90}$$

where:

η_{vol} = pump volumetric efficiency, –,
Q_o, Q_w = oil and water volumetric rates, STB/d, and
B_o, B_w = oil, water volume factors at pump intake pressure and temperature, bbl/STB.

As seen, under **ideal** conditions without any leakage in the pump, the pump's volumetric efficiency is a function of the **water cut** and the **volume factors**, i.e., the material properties of the produced oil and water.

In cases where fluid **leakage** in the pump is considerable, the volumetric efficiency drops because leakage decreases the fluid volume reaching the surface. If the volume leaked across the subsurface pump is known or calculated (see Section 4.5.2), then pump efficiency can be found from the following formula:

$$\eta_{\text{vol}} = \frac{Q_o + Q_w}{Q_o\, B_o + Q_w\, B_w + q_s} \tag{4.91}$$

where the only variable not defined before is:

q_s = plunger slippage, bpd.

EXAMPLE 4.14: MEASURED LIQUID RATES ARE 500 STB/D OIL AND 100 STB/D WATER, PUMP LEAKAGE WAS FOUND AS 58 BPD LIQUID. OIL AND WATER VOLUME FACTORS ARE 1.05 AND 1.00 BBL/STB, RESPECTIVELY. FIND THE PUMP VOLUMETRIC EFFICIENCY AND COMPARE IT TO A CASE WHEN PUMP LEAKAGE IS NEGLIGIBLE UNDER THE SAME CONDITIONS

Solution

Volumetric efficiency considering pump leakage is easily found from Eq. (4.91):

$$\eta_{\text{vol}} = (500 + 100)/(500 \times 1.05 + 100 \times 1 + 58) = 0.88 = 88\%.$$

If pump leakage is disregarded, then both the nominator and denominator of Eq. (4.90) must be extended to include the liquid lost due to slippage. To find the stock tank volume of the pump slippage rate, the liquid volume factor must be found first, using $WOR = 100/500 = 0.2$:

$$B_l = B_o/(1 + WOR) + B_w\, WOR/(1 + WOR) = 1.05/(1 + 0.2) + 1 \times 0.2/(1 + 0.2) = 1.04 \text{ bbl/STB}.$$
$$Q_s = q_s/B_l = 58/1.04 = 55.7 \text{ STB/d}.$$

Volumetric efficiency for the negligible leakage case is now calculated from the extended Eq. (4.90):

$$\eta_{\text{vol}} = (500 + 100 + 55.7)/(500 \times 1.05 + 100 \times 1 + 58) = 0.96 = 96\%.$$

4.5.3.2 Gassy production

Sucker-rod pumps are designed to pump a single-phase liquid only and usually cannot achieve high efficiencies when **free** gas is present at the **suction** of the pump. In addition to the effect of pump **leakage**, volumetric efficiency is basically defined by the following two main negative effects of the existing gas phase; both result in a reduced liquid volume being produced:

1. Gas occupies **space** within the barrel and reduces the space available for liquids.
2. The **valves** of the pump do not operate the way they are supposed to.

The first problem will be discussed in more detail later; the second requires a study of the operation of the traveling and standing valves when pumping a liquid containing free gas. At the start of the upstroke, the **dead space**, i.e., the space that exists between the standing and traveling valves at the bottom of the plunger's stroke, contains a gas–liquid mixture that has been **compressed** by the plunger during the downstroke. This compressible mixture starts to **expand** as the plunger begins its upstroke, and the standing valve can open only after pressure above it has decreased to the submergence pressure. Thus, the standing valve's opening is **delayed**, and the useful plunger stroke length is **decreased**, compared to the case where an incompressible liquid occupies the dead space. When the downstroke begins, in turn, the operation of the **traveling** valve is impaired by the free gas. Again, the barrel beneath the plunger contains a **compressible** mixture, which is at the pump intake pressure. As the plunger descends, the traveling valve cannot open **immediately** as it should, because it has the high hydrostatic pressure of the liquid column above it and a much lower pressure below. It opens only after the mixture has been **compressed** sufficiently to overcome the liquid column pressure, and that occurs later on the downstroke. All these effects reduce the **useful** stroke length of the plunger and decrease the pump's liquid production **capacity**.

Out of the three factors mentioned before, normally the effect of **free gas** in the barrel is the decisive one that defines the pump volumetric efficiency because pump **leakage** and the adverse effect of the **dead space** is kept at a minimum by proper field practice [70–72]. This is why in the following the pump volumetric efficiency is derived from an investigation of the conditions in the pump barrel.

In theory, three different situations can happen in the pump barrel when gassy wells are produced:

1. No gas reaches into the barrel because all free gas is **vented** into the annulus.
2. All gas coming from the formation goes through the pump.
3. Part of the formation gas is removed by a downhole gas **separator** before it gets into the pump.

These cases are illustrated for a sample well in Fig. 4.22, where efficiencies are plotted versus the pump intake pressure, *PIP*, the most important parameter governing the amount of free gas present at the pump's suction. The first case is identical to single-phase liquid production when only **dissolved** gas enters the pump; pump volumetric efficiency is found from Eq. (4.90) and is indicated in the figure by the solid straight line labeled "All Gas Vented." As shown, very high efficiencies are experienced that slightly decrease with the pump intake pressure (*PIP*) and illustrate the behavior of a sucker-rod pump operating under ideal conditions.

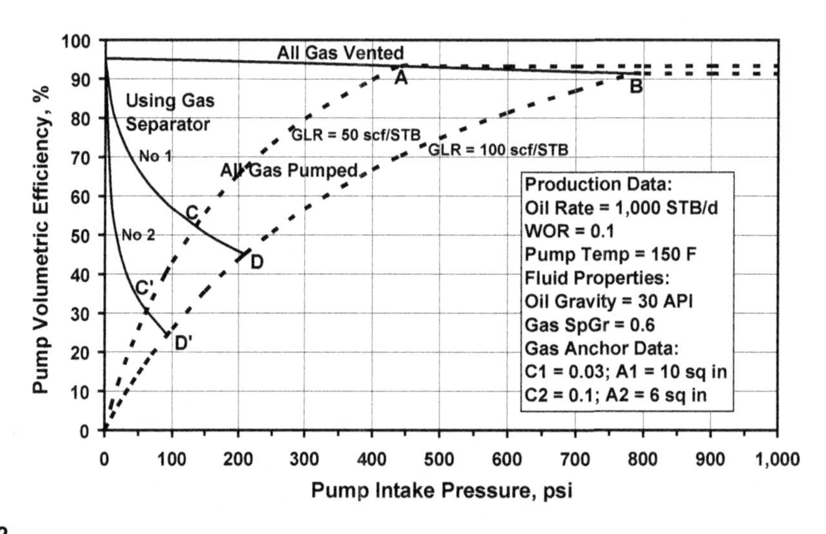

FIGURE 4.22

Pump volumetric efficiency versus pump intake pressure curves for a sample well.

If all free gas coming from the formation **enters** the pump, then pump volumetric efficiency is found from the calculated volume of the phases (liquid and gas) occupying the barrel space and can be calculated from:

$$\eta_{vol} = \frac{Q_o + Q_w}{Q_o\, B_o + Q_w\, B_w + \frac{Q_o + Q_w}{1 + WOR}[GLR(1 + WOR) - R_s]B_g/5.614} \tag{4.92}$$

where:

η_{vol} = pump volumetric efficiency, –,
Q_o, Q_w = oil and water volumetric rates, STB/d,
B_o, B_w = oil, water volume factors at pump intake pressure and temperature, bbl/STB,
WOR = production water–oil ratio, –,
GLR = production gas–liquid ratio, scf/STB,
R_s = solution gas–oil ratio at pump intake pressure and temperature, scf/STB, and
B_g = gas volume factor at pump intake pressure and temperature, cu ft/scf.

The curves labeled "All Gas Pumped" and plotted in dashed lines indicate efficiencies calculated from the previous formula for two different production gas–liquid ratios (*GLR*s). Volumetric efficiencies are very low for low *PIP* values when pump submergence below the annular liquid level is small because most of the gas phase is in **free gas** form. As *PIP* increases, e.g., by setting the pump deeper below the annular liquid level, the efficiency rapidly increases until the **bubble point** pressure is reached at point **A** or **B** for the two *GLR*s, respectively; lower efficiencies belong to greater *GLR* values. Above the bubble points all gas is in **solution** and a single-phase liquid phase is pumped; volumetric efficiency is found from Eq. (4.90).

Finally, if a gravitational gas **separator** (gas anchor) is used to remove part of the free gas from the well stream before it reaches the downhole pump, then the pump's volumetric efficiency is found from the next formula. Here the third term in the denominator represents the volume of gas that **leaves** the gas separator and enters the pump barrel; see Section 3.4.1.2.

$$\eta_{vol} = \frac{Q_o + Q_w}{Q_o\, B_o + Q_w\, B_w + C\, PIP^{0.666}\, v_{sl}^{0.5}(Q_o + Q_w)\left(\frac{B_o}{1 + WOR} + B_w\frac{WOR}{1 + WOR}\right)} \tag{4.93}$$

where the variables not defined recently:

C = separator type coefficient, –,
PIP = pump intake pressure, psia, and
v_{sl} = liquid superficial velocity in separator, ft/s.

Figure 4.22 illustrates the use of two different types of gas separators: No. 1 is a **packer-type** and No. 2 is a **poor-boy** gas anchor; see the solid curves labeled "Using Gas Separator." Both separators can remove large volumes of gas at very low *PIP*s and result in high pump volumetric efficiencies, but their performance very rapidly deteriorates as pump intake pressure increases. At some pressures (see points **C**, **D** and **C′**, **D′**) the effectiveness of the separators diminishes and **all** gas flows through the separator into the pump; the variation of pump volumetric efficiencies follows the "All Gas Pumped" cases. General conclusions on the use of gas anchors are:

- Gas anchors can increase the pump volumetric efficiency at low *PIP*s only.
- Separator effectiveness goes down to **zero** at some higher pump intake pressure.
- At higher production *GLR*s separators can be used at higher *PIP*s.
- The poor-boy separator is very inefficient; compare points **C** to **C′** and **D** to **D′**.

The **remedial** actions that can be taken to decrease the detrimental effects of free gas present at pump suction aim at two directions. At greater pump intake pressures the pump is run **deeper** so that *PIP* will be equal to or greater than the bubble point pressure. On the other hand, at low *PIPs* a gas anchor should be used to **separate** as much free gas as possible **before** it enters the pump. The reduction of the **dead space** in the pump as well as the use of **streamlined** valves are some other positive steps to consider.

Pump efficiency can also be poor when pump capacity **exceeds** well deliverability. The application of pump-off controllers to produce the well intermittently can highly increase the efficiency of the whole pumping system. Finally, **restrictions** in the pump should be mentioned as possible sources of production loss. These are especially severe when highly viscous crudes are pumped when special valves offering low resistance to flow should be used.

EXAMPLE 4.15: DECIDE WHETHER THE USE OF A PACKER-TYPE GAS SEPARATOR (C = 0.03 AND A = 10 SQ IN) WOULD IMPROVE OPERATIONS AT A PUMP INTAKE PRESSURE OF 300 PSI; OTHER WELL DATA ARE CONTAINED IN THE FOLLOWING TABLE

Oil rate = 1,000 STB/d	Water rate = 100 STB/d
Production *GLR* = 100 scf/BBL	Pump suction temperature = 150 F
Oil gravity = 30 API	Gas *SpGr* = 0.6

Solution

The decision is based on a comparison of the pump volumetric efficiencies valid for the cases of all gas pumped and the use of the gas anchor. For both cases the solution gas–oil ratio and the phases' volume factors at the pump intake pressure must be calculated first.

Oil specific gravity is found from Eq. (2.4):

$$\gamma_o = 141.5/(131.5 + 30) = 0.876.$$

The solution gas–oil ratio at suction conditions is found from Eq. (2.10):

$$y = 0.00091 \times 150 - 0.0125 \times 30 = -0.238, \text{ and}$$
$$R_s = 0.6\left(300/18/10^{-0.238}\right)^{1.205} = 34.5 \text{ scf/STB}.$$

Volume factors of water and oil are found from Eq. (2.8) and Eq. (2.11):

$$B_w = 1.0 + 1.21\text{E-}4(150 - 60) + 1\text{E-}6(150 - 60)^2 - 3.33\text{E-}6\,300 = 1.018 \text{ bbl/STB}.$$
$$F = 34.5(0.600/0.876)^{0.5} + 1.25\,150 = 216.1.$$
$$B_o = 0.972 + 1.47\text{E-}4\,216.1^{1.175} = 1.053 \text{ bbl/STB}.$$

Gas volume factor calculations necessitate the calculation of the critical parameters of the gas from Eqs (2.23) and (2.24):

$$p_{pc} = 709.6 - 58.7\,0.6 = 674 \text{ psia}.$$
$$T_{pc} = 170.5 + 307.3\,0.6 = 355 \text{ R}.$$

Pseudo-reduced parameters and the deviation factor are calculated from Eqs (2.17) and (2.18) and Eq. (2.28), respectively:

$$p_{pr} = (300 + 14.7)/674 = 0.467.$$

$$T_{pr} = (150 + 460)/355 = 1.719.$$

$$Z = 1 - 3.52\,0.467/10^{0.9813\times1.719} + 0.274 \times 0.467^2/10^{0.8157\times1.719} = 0.969.$$

Volume factor of gas is calculated from Eq. (2.21):

$$B_g = 0.0283\,0.969(150 + 460)/(300 + 14.7) = 0.053\ \text{cu ft/scf}.$$

The volumetric efficiency of the pump for the "all gas pumped" case can now be calculated from Eq. (4.92) as follows:

$$\eta_{vol1} = (1,000 + 100)/\{1,000 \times 1.053 + 100 \times 1.018 + (1,000 + 100)/(1 + 0.1)[100(1 + 0.1) - 34.5]\ 0.053/5.614\} = 0.589 = 58.9\%.$$

For the gas separator case the superficial liquid velocity in the separator body, v_{sl}, is found from Eq. (3.8):

$$v_{sl} = 9.36\text{E-}3(1,000 + 100)/10[1.053/(1 + 0.1) + 1.018 \times 0.1/(1 + 0.1)] = 1.081\ \text{ft/s}.$$

The volumetric efficiency is evaluated from Eq. (4.93):

$$\eta_{vol2} = (1,000 + 100)/\{1,000 \times 1.053 + 100 \times 1.018 + 0.03 \times 300^{0.666} \times 1.081^{0.5}.$$

$$(1,000 + 100) \times [1.053/(1 + 0.1) + 1.081 \times 0.1/(1 + 0.1)]\} = 0.397 = 39.7\%.$$

Since the pump volumetric efficiency without the gas separator, 58.9%, is greater than the value just calculated, the use of the separator is not recommended.

4.6 TORSIONAL LOAD CALCULATIONS

4.6.1 INTRODUCTION

In a sucker-rod pumping unit the prime mover's rotary movement is transmitted through the V-belts and the gearbox to the **crankshaft**, i.e., the slow-speed shaft of the speed reducer. It is here where the energy input to the system by the prime mover is **converted** into useful power to lift well fluids to the surface. The power required at the crankshaft is primarily determined by the loads acting at the polished rod, which vary greatly during the pumping cycle. The effect at the crankshaft of the polished rod loads is **torque**, which is defined as force acting on a lever arm. Thus, in order to analyze the operational conditions of surface pumping equipment, it is essential to accurately predict the torques occurring at the crankshaft.

In the pumping cycle, great forces are required to move the polished rod during the **upstroke** period, whereas no outside force is needed during the **downstroke**, since the rod string descends due to its own weight. This implies that the prime mover has to do all its useful work during only a portion of the cycle. Thus, a relatively **large**-capacity motor and speed reducer are required, which are completely unloaded on the downstroke. Such an operational pattern would mean a very **ineffective** operation. To improve the energy efficiency of the pumping system, some means of **counterbalancing** is applied to the pumping unit, which can take the form of **beam** or **rotary** counterweights or an air cylinder. The use of the proper amount of counterbalance can significantly decrease the instantaneous torque requirements during the upstroke and allows the use of smaller speed reducers and prime movers.

In addition to the torques resulting from polished rod loads and those required to turn the counterweights, **inertial** torques can also occur on the gearbox. Inertial torques are the result of the energy **stored** in and **released** from the oscillating masses of the pumping unit structure and may arise in those parts of the unit that do not move with a constant angular velocity. Even when the crankshaft turns at a constant rate, some components of the system, like the horsehead and the walking beam, have varying velocities and inertial torque in these parts cannot be neglected. Inertial torques in the rotating parts (cranks, counterweights, etc.) produced by an uneven crankshaft rotational speed can also be significant and must be properly accounted for.

4.6.2 THEORY OF GEARBOX TORQUES

In order to find the instantaneous and average energy requirements, as well as other operating parameters of pumping, instantaneous **torques** on the crankshaft have to be calculated. Generally, three kinds of torques can be distinguished on pumping unit gear reducers:

1. **Rod torque** is the result of polished rod loads and can be calculated based on the unit's kinematic parameters.
2. **Counterbalance torque** is required to move the counterweights, and is sinusoidal versus crank angle on gearboxes using rotary counterweights.
3. **Inertial torques** represent the energy stored in and released from the accelerating/decelerating parts of the pumping unit and are usually considered when the speed variation of the crankshaft, as compared to the average pumping speed, is considerable.

The **net torque** loading on the speed reducer is the **sum** of the above components. This is the torque that the power input by the prime mover must overcome. In the following, detailed methods of calculating the components of net crankshaft torque are presented.

4.6.2.1 Rod torque

The torque required to overcome the polished rod load is often called **rod torque** and is found with the use of the torque factor concept introduced in Section 3.7.5.2. Torque factors represent **imaginary lever arms** that, when multiplied with the polished rod load valid at the same crank position, give the resultant crankshaft torque. Basically, Eq. (3.64) is used in the following slightly modified form:

$$T_r(\theta) = TF(\theta)[F(\theta) - SU] \tag{4.94}$$

where:

$T_r(\theta)$ = rod torque at crank angle θ, in-lb
$TF(\theta)$ = torque factor at crank angle θ, in,
$F(\theta)$ = polished rod load at crank angle θ, lb, and
SU = structural unbalance, lb.

Structural unbalance, SU, is one of the characteristic parameters of the pumping unit and shows whether the walking beam is "**tail heavy**" or "**horsehead heavy.**" It is defined as the force required at the polished rod to keep the walking beam in a **horizontal** position with the pitmans disconnected from the cranks. Depending on its **direction**, this force is considered to be positive when acting downward and negative when acting upward; its value is supplied by the manufacturer. The pumping unit manufacturer should also provide the basic kinematic parameters for every 15 degrees of crank angle: the position of rods and the torque factor, as recommended in the API specification for pumping units, **API Spec. 11E** [73]. **Appendices B–E** of this book contain SU values along with dimensional data of sucker-rod pumping units of a major manufacturer for conventional, air-balanced, Mark II, and Reverse Mark geometries, respectively.

4.6.2.2 Counterbalance torque

4.6.2.2.1 Mechanically counterbalanced units

In the majority of cases, **rotary counterweights** attached to the crank(s) are used to counterbalance the pumping unit. The usual arrangement on a **conventional** unit is shown in Fig. 4.23. The counterweights impose a mechanical **moment** about the crankshaft, which equals the weight of the counterweights multiplied by the distance between their center of gravity and the crankshaft. Since the length of the lever arm continuously changes as the crank rotates during the pumping cycle, the torque required to turn the counterweights about the crankshaft varies with the crank **angle**. It is easy to see that counterbalance torque **opposes** rod torque in direction and varies with a sinusoidal function of the crank angle, as given here:

$$T_{CB}(\theta) = -T_{CB\ max} \sin\theta \tag{4.95}$$

where:

$T_{CB}(\theta)$ = counterbalance torque, in-lb,
$T_{CB\ max}$ = maximum moment of the counterweights, cranks, and crank pins, in-lb, and
θ = crank angle, degrees.

In the case of a **conventional**-geometry pumping unit, the **maximum** moment of the rotary counterweights, crank pins, and cranks, $T_{CB\ max}$, is measured while the cranks are **horizontal**. This is due to the fact that the counterweights' center of gravity usually falls on the center line of the cranks. Other than conventional geometries employ **phased** counterweight positions, where the counterweights' center of

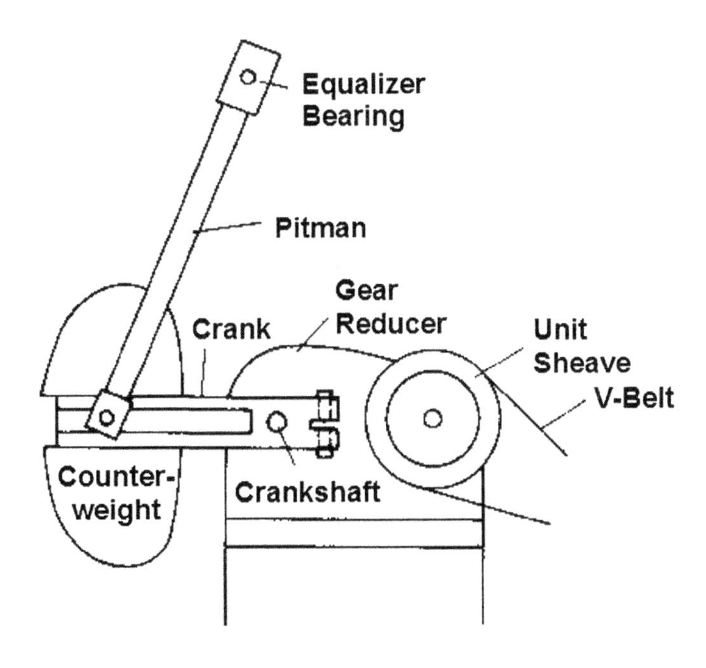

FIGURE 4.23

The arrangement of rotary counterweights on a typical conventional pumping unit.

gravity **leads** or **lags** the crank's motion. Therefore, the **phase angle** between the crank and the counterweight arm must be included in the above formula, and the following expressions can be derived:

$$T_{CB}(\theta) = -T_{CB\ max} \sin(\theta + \tau) \quad \text{for } \textbf{Mark II} \text{ units, and} \tag{4.96}$$

$$T_{CB}(\theta) = -T_{CB\ max} \sin(\theta - \tau) \quad \text{for } \textbf{Reverse Mark} \text{ (Torqmaster) units} \tag{4.97}$$

where:

$T_{CB}(\theta)$ = counterbalance torque, in-lb,
$T_{CB\ max}$ = maximum moment of the counterweights, cranks, and crank pins, in-lb,
θ = crank angle, degrees, and
τ = angle between the crank's and the counterweight arm's center line, degrees.

Maximum Counterbalance Moment

The calculation of counterbalance torque over the pumping cycle is only possible if the maximum moment, $T_{CB\ max}$, about the crankshaft caused by the cranks, crank pins, and counterweights is known. This maximum moment can be calculated either by a detailed **analysis** of the components of the counterbalance system or by **measuring** its effect on the polished rod; both methods are discussed in the following.

The typical arrangement of counterbalance components is shown in Fig. 4.24, from which it can be concluded that their maximum moment about the crankshaft is composed of:

- the moment of the **cranks** plus crank pins, and
- the moment of the **counterweights** plus auxiliary weights.

The combined moment of the cranks plus the crank pins is supplied by the manufacturer, along with the following data on counterweights and auxiliary weights:

- weight per piece, and
- lever arm M of the counterweight's center of gravity (CG) when at the **outmost** position on the crank.

The only unknown required for the calculation of the counterweights' moment about the crankshaft is the actual position of the center of gravity; this is found from the distance D of the counterweight

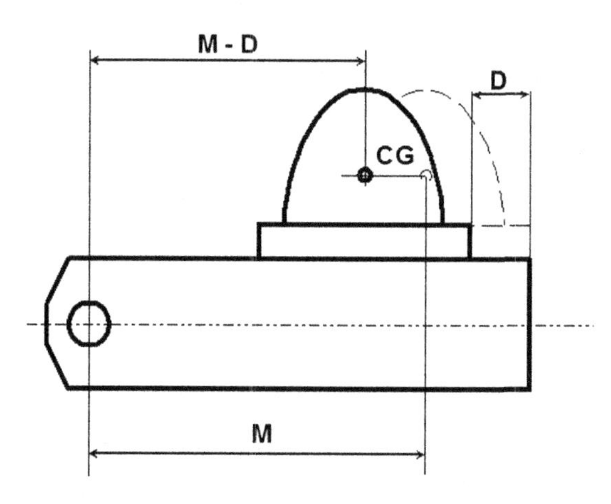

FIGURE 4.24

Typical arrangement of counterweights on the crank of a pumping unit.

measured from the **long end** of the crank. The maximum counterbalance moment can thus be derived from the next formula:

$$T_{CB\ max} = T_{crank} + (M - D)[N\ W + n\ w] \quad (4.98)$$

where:

$T_{CB\ max}$ = maximum moment of the counterweights, cranks, and crank pins, in-lb,
T_{crank} = moment of the cranks, in-lb,
M = lever arm of counterweights when at the **outmost** position on the crank, in,
D = distance of counterweight from long end of crank, in,
N = number of counterweights on crank, –,
n = number of auxiliary weights on crank, –,
W = weight of each counterweight, lb, and
w = weight of each auxiliary weight, lb.

This formula was derived for the assumptions that:

- counterweights are installed in **pairs**,
- they are on the same position on **both** sides of the crank, and
- auxiliary weights are fixed to counterweights in such a way that their centers of gravity overlap.

If these conditions are met, then the center of gravity of the combined mass of the crank plus the counterweights falls on the **center line** of the crank and the counterbalance torque is **in phase** with the rotary movement of the crank. This is the normal arrangement, but any **lead** or **lag angle** between the crank center line and the center of gravity changes the phase angle of the counterbalance torque and modifies the variation of the net torque acting on the gearbox.

For illustration purposes, Table 4.3 contains the necessary data of counterweights and auxiliary weights available from a major manufacturer for a given crank type. Figure 4.25 presents maximum counterbalance moments of the master weights (four pieces installed) of the same unit as a function of the weights' position on the crank.

Table 4.3 Mechanical Data of Master and Auxiliary Counterweights for a Sample Crank

Crank Type = 6468C				
Moment of Cranks = 134,370 in-lb				
Counterweight		Aux. Weight		
	Weight		Weight	M
Type	(lb/piece)	Type	(lb/piece)	(in)
2RO	1,708	2S	612	42.20
3CRO	1,327	3BS	572	45.26
5ARO	913	5AS	366	49.77
5CRO	662	5C	327	50.96
6RO	504	6	190	52.51
7RO	315	7	140	54.16

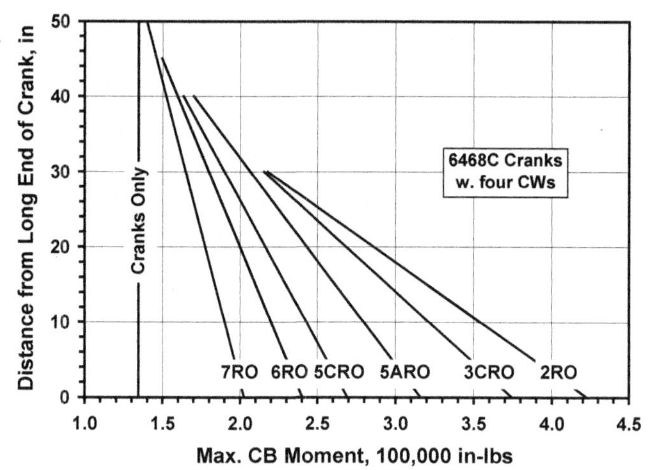

FIGURE 4.25

Maximum counterbalance moments of different master weights used on a sample crank.

EXAMPLE 4.16: FIND THE MAXIMUM COUNTERBALANCE MOMENT ON THE GEARBOX FOR A CONVENTIONAL PUMPING UNIT IF 6468C CRANKS AND FOUR 3CRO COUNTER-WEIGHTS ARE USED AT A POSITION OF $D = 12$ IN FROM THE LONG END OF THE CRANK

Solution

From Table 4.3 the moment of the cranks is 134,370 in-lb, maximum lever arm of the **3CRO** counterweight is $M = 45.26$ in, and each counterweight weighs 1,327 lb.

Maximum counterbalance moment is found from Eq. (4.98) as:

$$T_{\text{CB max}} = 134,370 + (45.26 - 12)(4 \times 1,327) = 310,914 \text{ in lb}.$$

The same value can be read from Fig. 4.25.

The other technique to calculate maximum counterbalance moment is based on the **measurement** of the unit's actual counterbalance effect, *CBE*, i.e., the **force** at the polished rod arising from the **moment** of the counterweights when the crank is horizontal. *CBE* is usually measured during dynamometer surveys using one of the procedures detailed in Section 6.3; maximum counterbalance moment is evaluated from the following formula, valid for pumping units of any geometry:

$$T_{\text{CB max}} = \frac{TF(\theta)(CBE - SU)}{\sin(\theta + \tau)} \tag{4.99}$$

where:

$T_{\text{CB max}}$ = maximum counterbalance moment about the crankshaft, in-lb,
$TF(\theta)$ = torque factor at crank angle θ, in,
θ = crank angle where *CBE* was measured (either 90° or 270°), degrees,
CBE = measured counterbalance effect, lb,
SU = structural unbalance, lb, and
τ = phase angle (offset) between the crank's and counterweight arm's center line, degrees.

4.6.2.2.2 Air-balanced units

Air-balanced units require special treatment, since they utilize an **air cylinder** that provides the necessary counterbalance. In case the force (*CBE*) imparted on the polished rod by the air pressure in the cylinder is known, the counterbalance torque is found by multiplying this force with the torque factor at the given crank angle:

$$T_{CB}(\theta) = -TF(\theta)\ CBE(\theta) \tag{4.100}$$

where:

$T_{CB}(\theta)$ = counterbalance torque, in-lb,
$TF(\theta)$ = torque factor, in, and
$CBE(\theta)$ = counterbalance effect at the polished rod, lb.

In order to find the variation of the counterbalance effect (*CBE*) versus crank angle, Fig. 4.26 is presented where a schematic drawing of an air-balanced pumping unit is shown. If the **moments** about the unit's saddle bearing of the relevant forces are equated, the following equation is reached:

$$A\ CBE = F_a\ L \tag{4.101}$$

where:

A, L = dimensional data of the pumping unit, in,
CBE = counterbalance effect at the polished rod, lb, and
F_a = air pressure force, lb.

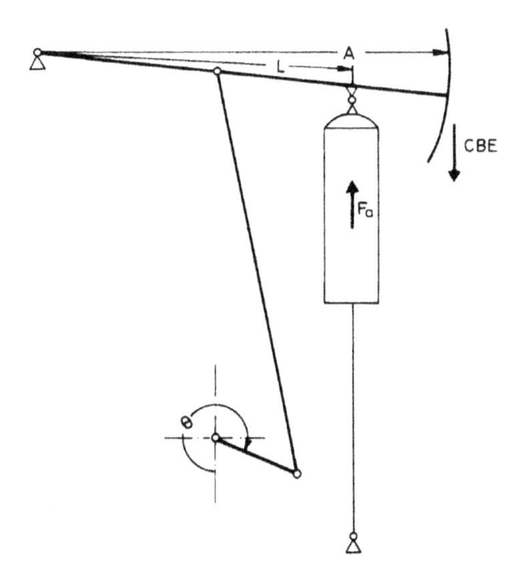

FIGURE 4.26

Schematic drawing of an air-balanced pumping unit.

Although the two forces are not always parallel to each other throughout the whole stroke, the above formula is a good **approximation** of the real conditions. Air pressure force is easily found from the cross-sectional area of the piston and the air pressure inside the cylinder:

$$F_a = A_p(p_a - p_u) \tag{4.102}$$

where:

F_a = air pressure force, lb,
A_p = cross-sectional area of air piston, sq in,
p_a = air pressure inside the cylinder, psi, and
p_u = air pressure to overcome the structural unbalance of the pumping unit, psi.

The last parameter, p_u, is analogous to the **structural unbalance**, *SU*, of the mechanically balanced pumping units and represents the air pressure necessary to keep the walking beam in a **horizontal** position. Now Eq. (4.101) is solved for *CBE*, and Eq. (4.102) is substituted in the resultant formula:

$$CBE = A_p \frac{L}{A}(p_a - p_u) \tag{4.103}$$

Designating the term $(A_p L)/A$ as M_a, we get:

$$CBE = M_a(p_a - p_u) \tag{4.104}$$

It is customary to assume (see Appendix D of **API Spec 11E** [73]) that air pressure inside the cylinder varies **linearly** with polished rod stroke length. Designating maximum and minimum air pressures by p_{max} and p_{min}, respectively, this assumption leads to the following final formula for the counterbalance effect:

$$CBE(\theta) = M_a\left[(p_{max} - p_u) - (p_{max} - p_{min})PR(\theta)\right] \tag{4.105}$$

where:

$CBE(\theta)$ = counterbalance effect at crank angle θ, lb,
M_a; p_u = constants of the given pumping unit, supplied by the manufacturer,
p_{max}; p_{min} = maximum and minimum air pressure in the cylinder, psi, and
$PR(\theta)$ = dimensionless polished rod stroke function, defined by Eq. (3.59), –.

Upon substitution of the above equation into Eq. (4.100), the counterbalance torque versus crank angle for an air-balanced pumping unit is found as:

$$T_{CB}(\theta) = -TF(\theta)M_a\left[(p_{max} - p_u) - (p_{max} - p_{min})PR(\theta)\right] \tag{4.106}$$

where the variables not defined previously are:

$T_{CB}(\theta)$ = counterbalance torque at crank angle θ, in-lb, and
$TF(\theta)$ = torque factor, in.

4.6.2.3 Inertial torques

In the preceding discussions, no considerations were made for any inertial effects that may occur in the different moving parts of the pumping unit, e.g., the crankshaft was assumed to rotate with a **constant** angular velocity. However, when the unit is not properly balanced or when high-slipelectric motors are used, then great crankshaft speed variations may occur during the pumping cycle and actual torques can be **affected** by inertia. In addition, there are some parts of the pumping unit (the walking beam, the horsehead, etc.) that make an oscillating motion even when the crankshaft turns at a constant speed; inertial torques arise in these also.

The physics background of inertial torque calculations is the relation of the **torque** acting on a body that rotates around one of its axes and the body's angular **acceleration**. The relevant formula is very similar to **Newton's second law** and can be written in consistent units as:

$$T = I\frac{d^2\alpha}{dt^2} \tag{4.107}$$

where:

T = torque,
$d^2\alpha/dt^2$ = angular acceleration, and
I = mass moment of inertia.

As seen, torque T causes angular acceleration $d^2\alpha/dt^2$, depending on the body's mass moment of inertia, I. This latter parameter is a measure of rotational **inertia** and is found from the mass **distribution** of the given body. After dividing the body into infinitesimally small dm masses whose distances from the axis of rotation are denoted by r, the mass moment of inertia is defined by the following integral:

$$I = \int_m r^2 dm \tag{4.108}$$

Inertial torques in the different parts of the pumping unit, therefore, can be calculated if the parts' mass moments of inertia and their accelerations around the same axis are known.

4.6.3 PRACTICAL TORQUE CALCULATIONS

The previous sections described the theory of gearbox torque calculations and presented the formulas to find the different kinds of torque components. When these calculation models are applied in practice, then two basic cases can be distinguished, depending on the angular **velocity** of the crankshaft: (1) those with a **constant** or nearly constant and (2) those with **varying** crankshaft velocities. Since the importance and magnitude of the different torque components for those two cases differ widely, their separate treatment is presented in the following.

4.6.3.1 Cases with constant crank speed

In the majority of pumping installations the crankshaft's angular velocity is **constant** during the pumping cycle and matches the measured pumping speed. These are the cases when the gearbox is properly counterbalanced and an electric motor with a low slip drives the pumping unit. **API Spec. 11E** [73] suggests that up to a speed variation of 15% over the average pumping speed, neglecting inertial effects does not introduce errors greater than 10% in torque calculations.

4.6.3.1.1 The API torque analysis

The **API torque analysis** model (published in the appendices of **API Spec. 11E** [73] as a recommended calculation procedure) can be applied to any class of pumping unit geometries. This calculation model uses the torque factor concept and the formulae introduced in earlier sections of this book, with the following basic **assumptions**:

- **Frictional** losses in the pumping unit structure are neglected, i.e., a torque efficiency factor of unity is used.
- **Inertial** torques are neglected.
- The **change** in structural unbalance, *SU*, with crank angle is also disregarded.

Under these conditions the **net torque** acting on the gearbox is simply found from the sum of the **rod torque** and the **counterbalance torque**. Based on the formulas derived earlier for the two torques, we receive the following expressions for mechanically balanced pumping units:

$$T_{\text{net}}(\theta) = TF(\theta)[F(\theta) - SU] - T_{\text{CB max}} \sin\theta \quad \text{for } \textbf{Conventional} \text{ geometry} \tag{4.109}$$

$$T_{\text{net}}(\theta) = TF(\theta)[F(\theta) - SU] - T_{\text{CB max}} \sin(\theta + \tau) \quad \text{for } \textbf{Mark II} \text{ geometry} \tag{4.110}$$

$$T_{\text{net}}(\theta) = TF(\theta)[F(\theta) - SU] - T_{\text{CB max}} \sin(\theta - \tau) \quad \text{for } \textbf{Reverse Mark} \text{ geometry} \tag{4.111}$$

where:

$T_{\text{net}}(\theta)$ = net torque on the gearbox at the crank angle θ, in-lb,
$TF(\theta)$ = torque factor at the crank angle θ, in,
$F(\theta)$ = polished rod load at the crank angle θ, lb,
SU = structural unbalance of the pumping unit structure, lb,
$T_{\text{CB max}}$ = maximum moment of counterweights and cranks, in-lb,
θ = crank angle, degrees, and
τ = counterweight arm offset angle, degrees.

The basic requirement for the calculation of gearbox torques is the knowledge of polished rod **loads** as a function of crank **angle**, since rod torque is found by multiplying the loads and torque factors belonging to the same crank angle; see Eq. (4.94). This condition, however, is not met if data recorded on **conventional** dynamometer cards are used to find polished rod loads because these cards record the **load** against polished rod **displacement**. Thus the load-versus-crank angle function must be derived before the torques on the speed reducer can be calculated. Also, no information is available from the conventional dynamometer survey on the variation of crank speed during the pumping cycle; this is the reason why the **API torque analysis** model neglects inertial effects when calculating gearbox torques.

For the reasons detailed, the most essential part of the API torque analysis method is the calculation of the polished rod **load**-versus-crank **angle** function. This involves execution of a hand-construction procedure on the conventional dynamometer card that consists of the following steps:

1. The portion of the horizontal axis of the card between the two extremes of the polished rod's stroke is **scaled** from 0.0 to 1.0. This scaling conforms to the definition of *PR* (position of rods) function, one of the basic kinematic parameters of the pumping unit.
2. At a given crank angle, θ, the $PR(\theta)$ value is found from manufacturer's data or calculated using a kinematic model of the unit's geometry (see Section 3.7.5.2).
3. At the $PR(\theta)$ value just determined, a **vertical** line is drawn on the dynamometer card that will cross the card at two points. The polished rod load at the given crank angle equals the ordinate of the point belonging to the actual part of the stroke (up or down).
4. After repeating Steps 2–3 for a number of crank angles, the rod load-versus-crank angle function is available at several points.

The procedure is illustrated in the following example problem.

EXAMPLE 4.17: CALCULATE THE TORSIONAL LOAD ON THE SPEED REDUCER OF A C-160D-173-86 PUMPING UNIT AT A CRANK ANGLE OF 60°. THE UNIT'S STRUCTURAL UNBALANCE IS 450 LB, THE *PR* (POSITION OF ROD) AND *TF* (TORQUE FACTOR) VALUES AT THE SAME CRANK ANGLE ARE 0.314 AND 45.2 IN, RESPECTIVELY. MAXIMUM COUNTERBALANCE MOMENT IS $T_{CB\ max}$ = 250,000 IN-LB; THE DYNAMOMETER CARD TAKEN ON THE WELL IS SHOWN IN FIG. 4.27

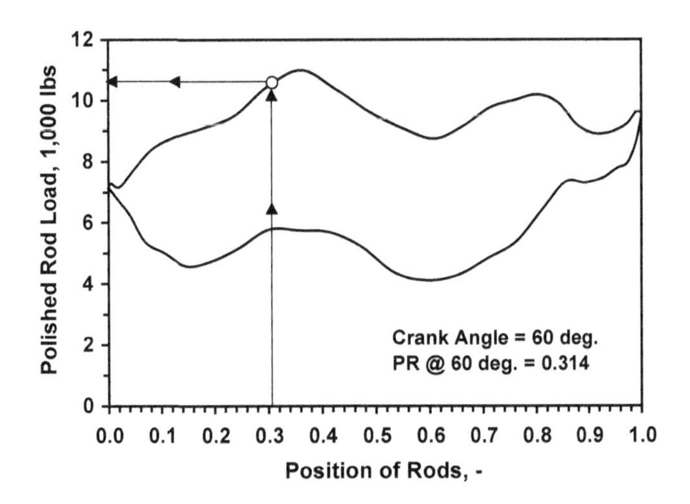

FIGURE 4.27

Graphical determination of one point of the polished rod load versus crank angle function according to the API torque analysis, for Example 4.17.

Solution

The polished rod load for the given crank angle of 60° is found from Fig. 4.27 as 10,700 lb, since the unit is on the upstroke.
Rod torque is found from Eq. (4.94):

$$T_r = 45.2(10{,}700 - 450) = 463{,}300 \text{ in lb.}$$

Counterbalance torque is calculated from Eq. (4.95), since the unit is of the conventional geometry:

$$T_{CB} = -250{,}000 \sin 60 = -216{,}506 \text{ in lb.}$$

The net torque loading on the speed reducer is the sum of the above components:

$$T_{net} = 463{,}300 - 216{,}506 = 246{,}794 \text{ in lb.}$$

The reducer is seen to be overloaded, since its torque rating is only 160,000 in-lb.

4.6.3.1.2 Computer solutions

The API torque analysis method relies on **hand** construction and **visual** read-off, which can impair the **accuracy** attainable. There are also other **drawbacks** to this method:

- The number of points that can be read off the card is **limited**, because manufacturers usually supply *PR* and *TF* values for each 15 degrees of crank angle only.
- Actual peak torque values may remain **undetected**, since these can lie between subsequent points.

The main **disadvantage** of the API procedure is that it is **tedious** and time-consuming. Consequently, computerized torque calculations are frequently used to increase accuracy and to speed up well analysis. Such calculation procedures rely heavily on an efficient handling of the dynamometer card. Either a digitizer [74] or a plotter with digitizing capabilities [75] can be used to calculate the polished rod load-versus-crank angle function.

EXAMPLE 4.18: MAKE A COMPLETE TORQUE ANALYSIS OF THE DYNAMOMETER CARD PRESENTED IN THE PREVIOUS EXAMPLE

Solution

Calculation results are presented in Exhibit 4.6, and a plot of the torques against crank angle is given in Fig. 4.28. As seen, the speed reducer is severely overloaded during the upstroke.

4.6.3.1.3 Considerations for articulating inertial torque

It is important to understand that inertial effects exist even if the prime mover speed is **constant** and the crankshaft turns with a constant angular velocity. This is caused by those structural parts of the pumping unit that move with varying **accelerations** during the pumping cycle [76] like the beam, horsehead, equalizer, etc. The gearbox torque associated with such inertial effects is called the **articulating torque**, which represents the energy stored in and released from those parts of the pumping unit structure that perform an **oscillating** motion, such as: (1) the walking beam, (2) the horsehead, (3) the equalizer, and (4) the pitmans.

In order to find the instantaneous values of articulating torque, the torque about the saddle bearing required to accelerate/decelerate the structural parts involved is computed first, based on Eq. (4.107):

$$T' = \frac{12}{32.2} I_b \frac{d^2\theta_b}{dt^2} \tag{4.112}$$

```
                 TORQUE ANALYSIS OF PUMPING UNITS

                     USING DYNAMOMETER CARDS

                    According to API Spec. 11E

    Name of Field :                  Well Identification : Ex. 4.15
    Date Card Taken :                Time Card Taken :

    Pump Setting Depth :  4608.0 ft  Dynamic Liquid Level :  4608.0 ft
    Liquid Prod. Rate  :   247.9 bpd Liquid Spec. Gravity :    0.90

                 P U M P I N G   U N I T   D A T A

                 Manufacturer : LUFKIN
API Designation :  C-160D-173-86       Rotation :          CLOCKWISE
Structural Unbalance :   450 lbs       CB Offset Angle :   0.0  deg
Meas. Stroke Length :  86.00 in        Pumping Speed :   14.86  SPM
Linkage Dimensions :  R =   37.00 in    K =  151.34 in     C =   96.05 in
                      P =  114.00 in    A =  111.00 in     I =   96.00 in

            D Y N A M O M E T E R   C A R D   D A T A

             Dynamometer Data Taken from File : demo3.CRD

Dynamometer Constant :  1000 lbs/cm    No. of Digitized Points :  75
                   Maximum CB Torque :  250000 in-lbs


            C A L C U L A T E D   P A R A M E T E R S

Min. Polished Rod Load :    3902 lbs   Hydraulic Power :           7.6 HP
Peak Polished Rod Load :   11033 lbs   Polished Rod Power :       12.4 HP
     Percent of Rating :    63.8 %     Lifting Efficiency :       60.9 %

                                            Actual           Ideal
                                       C O U N T E R B A L A N C I N G

Max. CB Torque            in-lbs           250000           309789
Min. Net Torque           in-lbs           -34985           -23117
Peak Net Torque           in-lbs           254466           200031
     Percent of Rating      %               159.0            125.0
Cyclic Load Factor          -               1.596            1.400
```

EXHIBIT 4.6

Complete torque analysis of a pumping unit based on the dynamometer card in Fig. 4.27.

where:

T' = inertial torque about the saddle bearing, in-lb,
I_b = total mass moment of inertia of the components involved, $\text{lb}_\text{m}\,\text{ft}^2$, and
$\mathrm{d}^2\theta_b/\mathrm{d}t^2$ = angular acceleration of the beam, $1/\text{s}^2$.

This torque can be converted to an equivalent **force** acting at the **polished rod** by dividing it by the distance between the saddle bearing and the polished rod, A (see Figs 3.99–3.102):

$$F = \frac{T'}{A} = \frac{12}{32.2}\frac{I_b}{A}\frac{\mathrm{d}^2\theta_b}{\mathrm{d}t^2} \tag{4.113}$$

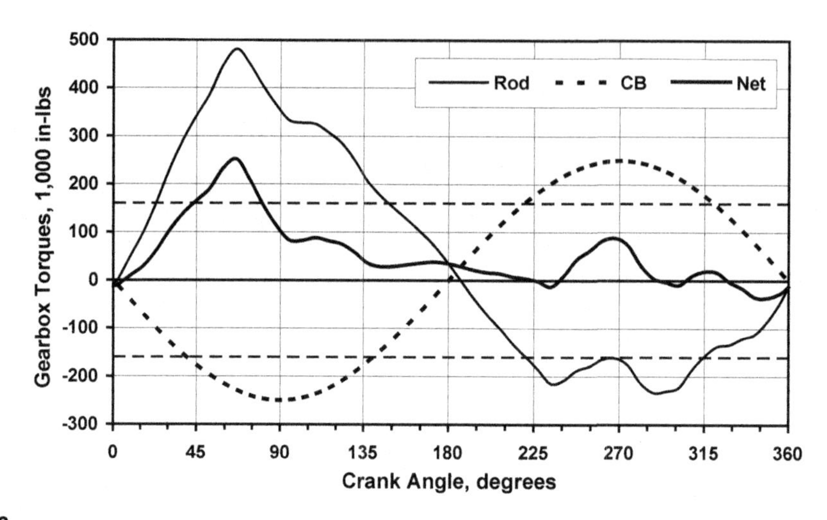

FIGURE 4.28

Variation of gearbox torque components versus crank angle for Example 4.18.

The torque required at the gearbox can now be found by multiplying this force with the pumping unit's torque factor; thus the final formula for articulating torque is found:

$$T_{ia}(\theta) = \frac{12}{32.2} TF(\theta) \frac{I_b}{A} \frac{d^2\theta_b}{dt^2} \tag{4.114}$$

where:

$T_{ia}(\theta)$ = articulating inertial torque on the gearbox at crank angle θ, in-lb,
$TF(\theta)$ = torque factor at crank angle θ, in,
A = distance between the saddle bearing and the polished rod, in,
I_b = mass moment of inertia of the beam, horsehead, equalizer, bearings, and pitmans, referred to the saddle bearing, $lb_m\ ft^2$, and
$d^2\theta_b/dt^2$ = angular acceleration of the beam, $1/s^2$.

During the polished rod's **upstroke** articulating torque **increases** the required net torque on the gear reducer for **positive** beam accelerations due to the inertia of the parts driven. For **negative** accelerations (when the beam and the connected parts are decelerating), articulating torque helps the system by **releasing** the kinetic energy stored. On the **downstroke** the role of beam acceleration is the opposite because the torque factors become negative.

The net mass moment of inertia of the structural parts that cause articulating torque and that include (1) the walking beam, (2) the horsehead, (3) the equalizer, and (4) the pitmans is usually supplied by the pumping unit manufacturer. The use of lightweight structural parts (beam, horsehead, etc.) is beneficial in reducing the torsional load on the gearbox. The walking beam's angular **acceleration** pattern, required for the calculation of the articulating torque, can be derived from the kinematic model of the pumping unit, e.g., **Svinos'** [77] method; see Eq. (3.82).

To illustrate the effects of articulating torque on the loading of gearboxes, the sample case discussed in Example 4.18 was reworked and inertial torques were included while calculating the net gearbox torque. The net mass moment of inertia of the pumping unit's parts making oscillating motion was found from manufacturer data as $I_b = 125{,}789$ lb_m ft^2. Figure 4.29 presents the variation of articulating torque with the crank angle, as well as a comparison of the net gearbox torque-versus-crank angle functions calculated with and without inertial effects. As already discussed, during the upstroke articulating torque is positive as long as the oscillating parts accelerate and store kinetic energy. Later on the upstroke, when those parts start to decelerate, the kinetic energy stored is returned to the system, resulting in negative articulating torque values; similar behavior can be observed during the downstroke.

The effect of articulating torque on the net gearbox torque is closely related to the acceleration pattern of the given pumping unit, but its magnitude is not very significant if compared to other torque components like rod torque and counterbalance torque. In the case of the example given before, as shown in Fig. 4.29, the torque load on the gearbox is only negligibly affected by including articulating inertia effects. Since this seems to be the **typical** behavior, one can conclude that **neglecting** articulating torque is fully **justified** when calculating net gearbox torques for pumping units operating at **constant** crank speeds.

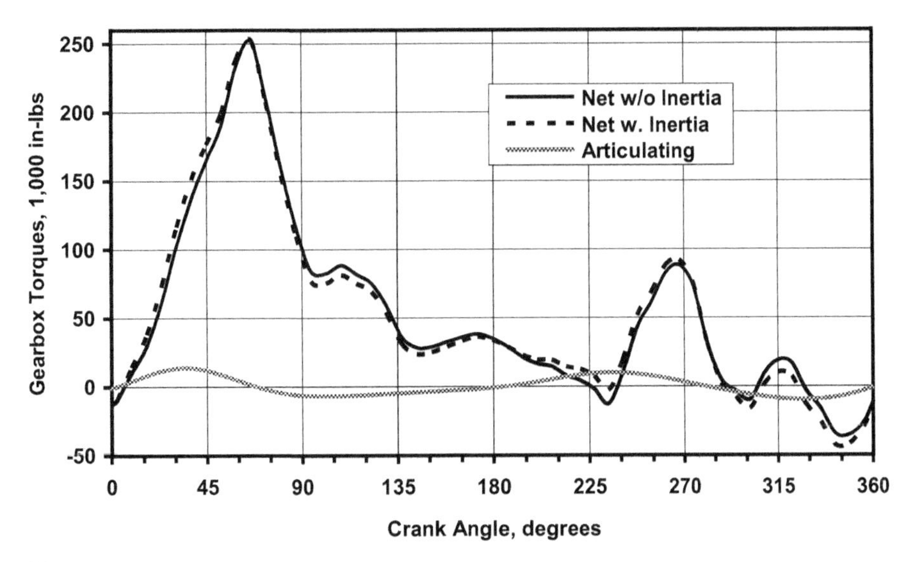

FIGURE 4.29

The effect of articulating torque on net gearbox torque for Example 4.18.

4.6.3.2 Cases with variable crank speed

When the pumping unit is driven by a multicylinder engine, a high-slip or even an ultrahigh-slip (UHS) electric motor the angular velocity of the crankshaft changes during the pumping cycle: the crank speeds up when the unit is lightly loaded and slows down as the load becomes heavier. The high accelerations/decelerations coupled with the heavy rotating masses give rise to **inertial** torques because of the flywheel effect; gearbox torque calculations must be appropriately modified to include this. In such cases, in addition to the torques normally present on the gearbox and discussed so far, **rotary inertial** torque must be also calculated, as discussed in the following [76].

4.6.3.2.1 Rotary inertial torque

Rotary inertial torque exists only when the angular speed of the crankshaft **varies** over the pumping cycle and has a greater importance than articulating torque. It can either increase or decrease the load on the gearbox: at times when crankshaft speed increases, the additional load (rotary inertial torque) on the gearbox is converted to kinetic energy stored in the rotating parts. On the other hand, if crankshaft speed decreases, then the energy stored in the cranks and counterweights is returned into the system and the torque load on the gearbox decreases. This kinetic energy **interchange** happens in the following rotating **masses** of the pumping unit:

- The cranks with crank pins,
- The counterweights and auxiliary weights, and
- The slow-speed shaft and slow-speed gear of the speed reducer.

Since all these components rotate around the crankshaft, their combined moment of inertia can be found from simple **addition** of the individual moments. Using the **net** moment of inertia of the rotating system, the rotary inertial torque on the gearbox is found from the next formula, which is based on Eq. (4.107):

$$T_{\mathrm{ir}}(\theta) = \frac{12}{32.2} I_s \frac{\mathrm{d}^2\theta}{\mathrm{d}t^2} \tag{4.115}$$

where:

$T_{\mathrm{ir}}(\theta)$ = rotary inertial torque on the crankshaft at crank angle θ, in-lb,
I_s = mass moment of inertia of the cranks, counterweights, and slow-speed gearing referred to the crankshaft, $\mathrm{lb_m\,ft^2}$, and
$\mathrm{d}^2\theta/\mathrm{d}t^2$ = angular acceleration of the crankshaft, $1/\mathrm{s}^2$.

As seen, rotary inertial torque changes with the angular **acceleration** of the crankshaft, which can be derived from the angular **velocity** of the same. Crankshaft velocity, in turn, can be directly measured with a generating **tachometer** or, as is the usual case, is inferred from electronic **dynamometer** measurements, as detailed later.

Net Mass Moment of Inertia

The individual **components** of the net mass moment of inertia, I_s, of the pumping unit's rotating parts are the following:

$$I_s = I_{\mathrm{cr}} + I_{\mathrm{CW}} + I_g \tag{4.116}$$

where:

I_s = net mass moment of inertia, $\mathrm{lb_m\,ft^2}$,
I_{cr} = mass moment of inertia of the cranks about the crankshaft, $\mathrm{lb_m\,ft^2}$,
I_{CW} = mass moment of inertia of the counterweights about the crankshaft, $\mathrm{lb_m\,ft^2}$, and
I_g = mass moment of inertia of the slow-speed shaft and gear about the crankshaft, $\mathrm{lb_m\,ft^2}$.

Since data on cranks and the slow-speed gearing (I_{cr} and I_g) are supplied by the manufacturer, only the actual moment of inertia of the counterweight system must be found by calculation.

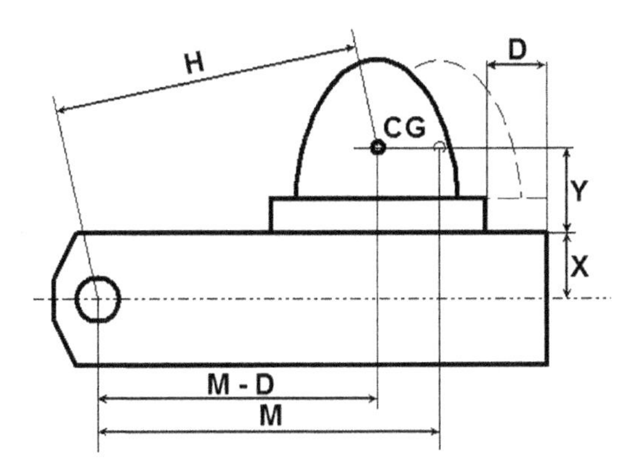

FIGURE 4.30

Principle of calculating the mass moment of inertia of counterweight components.

Figure 4.30 shows the principle of calculating the mass moment of inertia of counterweight components. If the moment of inertia of a single counterweight about its center of gravity is known, then it can be referred to the crankshaft by using the **Huygens-Steiner theorem**, as follows:

$$I_{CW} = I_{CG} + m_{CB} \left(\frac{H}{12} \right)^2 \tag{4.117}$$

where:

I_{CW} = mass moment of inertia of the counterweight about the crankshaft, $lb_m\ ft^2$,
I_{CG} = mass moment of inertia of the counterweight about its center of gravity, $lb_m\ ft^2$,
m_{CB} = mass of the counterweight, lb_m, and
H = distance between the crankshaft and the CG of the counterweight, in.

The distance H changes with the **position** of the counterweight on the crank and is easily found from:

$$H = \sqrt{(X + Y)^2 + (M - D)^2} \tag{4.118}$$

where:

H = distance between the crankshaft and the CG of the counterweight, in,
X, Y = dimensions of the crank and counterweight, in,
M = maximum lever arm of counterweights, in, and
D = distance of counterweight from long end of crank, in.

Sample data on a crank's and different counterweights' moment of inertia are given in Table 4.4 for a **160D** gearbox. The next example illustrates the calculation of net mass moment of inertia.

Table 4.4 Mass Moment of Inertia Data of Master and Auxiliary Counterweights for a Sample Crank

Crank Type = 6468C					
Mass Moment of Inertia = 22,282 lb_m ft^2					
X = 10 in					
Counterweights			Aux. Weights		
	Moment	Y		Moment	Y
Type	(lb_m ft^2)	(in)	Type	(lb_m ft^2)	(in)
2RO	2,458	14.2	2S	756	14.2
3CRO	1,384	13.3	3BS	562	13.3
5ARO	707	13.4	5A	272	13.4
5CRO	430	11.8	5C	220	11.8
6RO	229	9.9	6	83	9.9
7RO	114	8.6	7	51	8.6

EXAMPLE 4.19: FIND THE COMBINED (NET) MOMENT OF INERTIA OF THE CRANKS, COUNTERWEIGHTS, AND SLOW-SPEED GEAR ABOUT THE CRANKSHAFT IF A 160D GEARBOX IS USED WITH 6468C CRANKS AND FOUR 3CRO COUNTERWEIGHTS AT A POSITION OF D = 12 IN FROM THE LONG END OF THE CRANK

Solution

The inertia of one crank, from Table 4.4, is 22,282 lb_m ft^2, so two cranks have:

$$I_{cr} = 2 \times 22,282 = 44,564\ lb_m\ ft^2.$$

To find the inertia of the counterweights, first the distance H is found from Eq. (4.118) by using $X = 10$ in and $Y = 13.3$ in, taken from Table 4.4, and $M = 45.26$ in, from Table 4.3:

$$H = \left[(10 + 13.1)^2 + (45.26 - 12)^2\right]^{0.5} = 40.49\ \text{in.}$$

Since the mass moment of inertia of one counterweight about its center of gravity is 1,384 lb_m ft^2, as found from Table 4.4, and its mass is 1,327 lb_m, from Table 4.3, the moment about the crankshaft of four counterweights from Eq. (4.117) is:

$$I_{CW} = 4\left[1,284 + 1,327(40.49/12)^2\right] = 65,967\ lb_m\ ft^2.$$

The mass moment of inertia of the slow-speed gearing of the 160D gearbox from manufacturer data:

$$I_g = 318\ lb_m\ ft^2.$$

The net moment of inertia of the system is the sum of individual inertias, according to Eq. (4.116):

$$I_s = I_{cr} + I_{CW} + I_g = 44{,}564 + 65{,}967 + 318 = 110,849\ lb_m\ ft^2.$$

4.6.3.2.2 Calculation procedure

As already discussed, gearbox torques are normally calculated based on dynamometer measurements that provide the variations of polished rod load and position. Modern dynamometer systems register these data as a function of **time** but give no information on the crank angles valid at the measured times. This circumstance, however, prohibits a **direct** calculation of gearbox torques because all torque components depend on the crank angle, θ. **Rod** torque changes with the torque factor, which varies with the crank angle; see Eq. (4.94); **counterbalance** torque is a direct function of crank angle; see Eq. (4.95). **Inertial** torques are found from the acceleration patterns of different components of the pumping unit; these also change with the variation of the time history of crank angle. Gearbox torque components, therefore, can only be calculated if the change of crank angle, θ, with time is determined from dynamometer measurements.

Since direct calculation of crank angles from the measured polished rod positions is not possible, crank angles are inferred from the pumping unit's kinematic parameters. This can be completed several ways, but all methods are based on setting the measured polished rod positions equal to the positions determined from the kinematic analysis (see Section 3.7.5.2) of the pumping unit, i.e.:

$$s(t) = S \times PR(\theta) \tag{4.119}$$

where:

$s(t)$ = measured value of polished rod position at time t, in,
S = polished rod stroke length, in, and
$PR(\theta)$ = position of rods at crank angle θ, –.

For each measured polished rod position, $s(t)$, the corresponding crank angle, θ, is found when the above equation is satisfied; this procedure results in a series of crank angles as a function of time $\theta(t)$. This function, in turn, allows one to find the two important components of gearbox torque: rod and counterbalance torque. To calculate **rod torque** the torque factor (TF) is found for each crank angle from the kinematic analysis of the pumping unit; torque is the product of the torque factor and the measured polished rod load (see Eq. 4.94). **Counterbalance torque** varies with the sine function of the crank angle, according to Eq. (4.95). Determination of the inertial torque components, however, needs special considerations, as detailed in the following.

Articulating inertial torque, as discussed in Section 4.6.3.1.3, primarily depends on the angular acceleration pattern of the beam, i.e., $d^2\theta_b/dt^2$. This acceleration can be found from a kinematic analysis of the pumping unit, as proposed by **Svinos** [77] and detailed in Section 3.7.5.2.2. A more direct calculation model was introduced by **Gibbs** [31], which is based on the following formula, which expresses the polished rod's position with the angular position of the beam's centerline:

$$s(t) = A\ \theta_b(t) \tag{4.120}$$

where:

$s(t)$ = measured value of polished rod position at time t, in,
A = distance from the center bearing to the horsehead, in, and
$\theta_b(t)$ = angle between the center line of the walking beam and the line connecting the crankshaft to the center (saddle) bearing, radians.

Expressing the angle θ_b from this formula and then differentiating the result twice with respect to time, we receive:

$$\frac{d^2\theta_b}{dt^2} = \frac{1}{A}\frac{d^2}{dt^2}s(t) \tag{4.121}$$

where:

$d^2\theta_b/dt^2$ = angular acceleration of the beam, $1/s^2$,
$s(t)$ = polished rod position versus time, in, and
A = distance from the center bearing to the horsehead, in.

This formula permits the direct calculation of the beam's angular acceleration from the polished rod position–time data obtained from a dynamometer survey by differentiation of that function, $s(t)$. The easiest way to differentiate this function is to first fit the measured data with a truncated **Fourier** series and then find the second derivative of that. Because of the relatively smooth variation of polished rod position with time, a maximum of 10 terms in the Fourier series are recommended by **Gibbs** [31].

Rotary inertial torque varies, as shown in Eq. (4.115), with the angular **acceleration** of the crankshaft, i.e., the second derivative of crank angle with respect to time, $d^2\theta/dt^2$. Since the crank angle–time function, $\theta(t)$, is already known at several times, any numerical method of differentiation could be used to reach the required parameter. A practical solution, used by several investigators, involves fitting the crank angle-versus-time function with a truncated **Fourier** series. Differentiation of the Fourier series is quite simple, so this approach guarantees an easy solution to determine the variation of the crankshaft's angular acceleration and thereby the rotary inertial torques arising during the pumping cycle.

4.6.3.2.3 Net torque

In cases where the pumping system operates with varying crankshaft speeds, the **net** torque on the gearbox must include the **inertial** effects as well, and the formulas derived for a constant crankshaft speed (Eqs 4.109–4.111) cannot be used. The proper formula for net torque is the algebraic **sum** of all possible torque components:

$$T_{\text{net}}(\theta) = T_r(\theta) + T_{\text{CB}}(\theta) + T_{\text{ia}}(\theta) + T_{\text{ir}}(\theta) \tag{4.122}$$

where:

$T_{\text{net}}(\theta)$ = net torque on speed reducer at crank angle θ, in-lb,
$T_r(\theta)$ = rod torque at crank angle θ, in-lb,
$T_{\text{CB}}(\theta)$ = counterbalance torque at crank angle θ, in-lb,
$T_{\text{ia}}(\theta)$ = articulating inertial torque at crank angle θ, in-lb, and
$T_{\text{ir}}(\theta)$ = rotary inertial torque at crank angle θ, in-lb.

Upon substitution into this equation of the relevant formulae introduced earlier, except for the expression for counterbalance torque, a generally applicable formula is found:

$$T_{\text{net}}(\theta) = TF(\theta)\left[F(\theta) - SU + \frac{12}{32.2}\frac{1}{A}I_b\frac{d^2\theta_b}{dt^2}\right] + T_{\text{CB}}(\theta) + \frac{12}{32.2}I_s\frac{d^2\theta}{dt^2} \tag{4.123}$$

where:

$T_{\text{net}}(\theta)$ = net torque on speed reducer at crank angle θ, in-lb,
$TF(\theta)$ = torque factor at crank angle θ, in,

$F(\theta)$ = polished rod load at crank angle θ, lb
SU = structural unbalance, lb,
A = distance between the saddle bearing and the polished rod, in,
I_b = mass moment of inertia of the beam, horsehead, equalizer, bearings, and pitmans, referred to the saddle bearing, $lb_m\ ft^2$,
$d^2\theta_b/dt^2$ = angular acceleration of the beam, $1/s^2$,
$T_{CB}(\theta)$ = counterbalance torque at crank angle θ, in-lb,
I_s = mass moment of inertia of the cranks, counterweights, and slow-speed gearing referred to the crankshaft, $lb_m\ ft^2$, and
$d^2\theta/dt^2$ = angular acceleration of the crankshaft, $1/s^2$.

The first term in this equation represents **rod** torque, corrected for articulating inertial effects; the second one stands for the **counterbalance** torque; and the last term gives the rotary **inertial** torque. The formula can be applied to any mechanically counterbalanced pumping unit after substitution of the proper expression for counterbalance torque, $T_{CB}(\theta)$. For a **conventional** geometry, use Eq. (4.95); for **Mark II** and **Reverse Mark** (Torqmaster) units, use Eqs (4.96) and (4.97), respectively.

In the derivation of the rod torque equation, no allowance was made to account for frictional **losses** in the pumping unit's structural parts because the **API torque analysis** neglected all energy losses from the polished rod to the gearbox. Since friction in the wireline hanger and the structural bearings consumes energy and increases the torque required at the crankshaft, the rod torque term in the above equation must be corrected by the torque **efficiency** of the pumping unit structure, η_t. Depending on the sign of the corrected rod torque (which includes articulating inertial torque), the following final expressions are derived for the calculation of net gearbox torques:

For **positive** corrected rod torques the net torque must be increased because of the frictional energy losses:

$$T_{net}(\theta) = TF(\theta)\frac{F(\theta) - SU + \frac{12}{32.2}\frac{1}{A}I_b\frac{d^2\theta_b}{dt^2}}{\eta_t} + T_{CB}(\theta) + \frac{12}{32.2}I_s\frac{d^2\theta}{dt^2} \tag{4.124}$$

For **negative** corrected rod torques the net torque is decreased because friction consumes some of the work done by the falling rod string before it reaches the gearbox:

$$T_{net}(\theta) = TF(\theta)\left[F(\theta) - SU + \frac{12}{32.2}\frac{1}{A}I_b\frac{d^2\theta_b}{dt^2}\right]\eta_t + T_{CB}(\theta) + \frac{12}{32.2}I_s\frac{d^2\theta}{dt^2} \tag{4.125}$$

Average **torque efficiency factors** for new and worn pumping units were presented by **Gipson and Swaim** [78]; their correlation is reproduced in Fig. 4.31. Efficiency η_t is seen to **improve** as net torque approaches the torque **rating** of the speed reducer and diminishes for lower loads. For fully loaded reducers, efficiencies around 90% can be assumed.

4.6.3.2.4 Sample calculation

To illustrate the determination of gearbox torques for installations with variable crankshaft speeds, a sample calculation is presented. The well is produced with a 1.5 in pump set at 8,000 ft and using a two-taper API 76 rod string. The surface pumping unit is a conventional C-640D-365-168 unit running

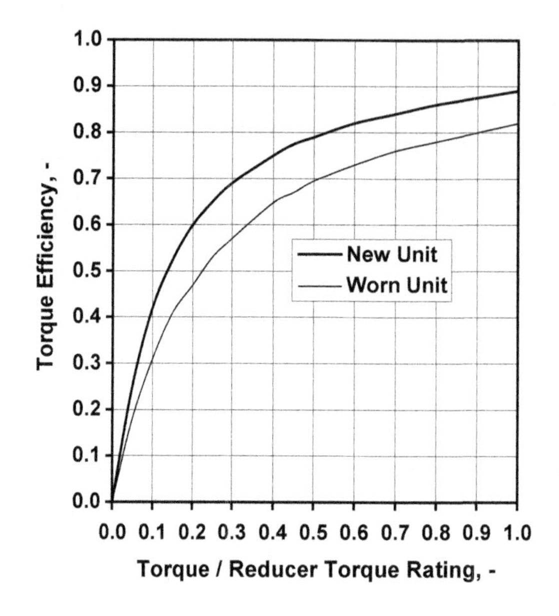

FIGURE 4.31

Average torque efficiency factor values for pumping units.

After ***Gipson and Swaim*** *[78].*

at an average pumping speed of 5.98 SPM using a polished rod stroke length of 168 in. Polished rod position, $s(t)$, and load, $F(t)$, were measured by an electronic dynamometer as a function of time; the dynamometer card constructed from those data is presented in Fig. 4.32.

Since dynamometer records do not contain information on the crank angles needed to find rod and counterbalance torques, crank angles had to be computed from the variation of the measured polished

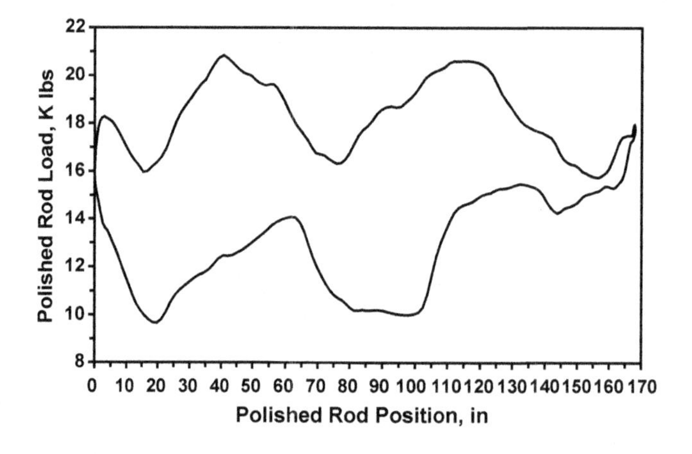

FIGURE 4.32

A sample surface dynamometer card.

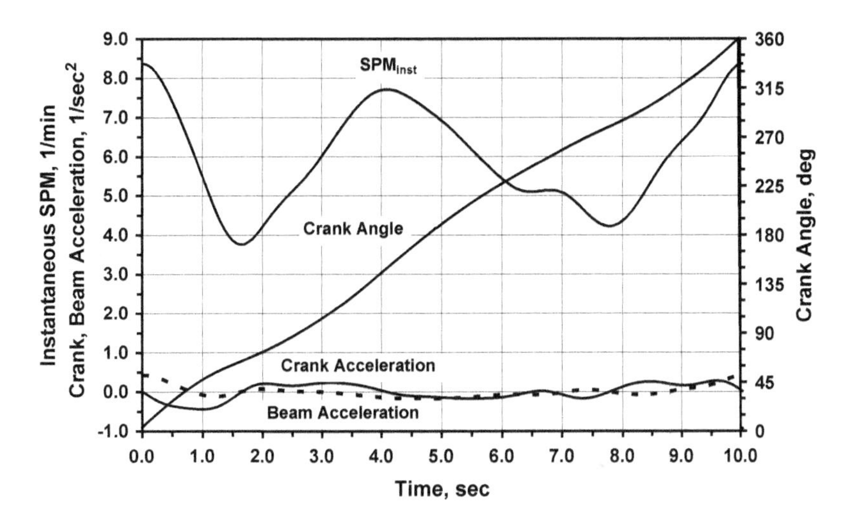

FIGURE 4.33

Kinematic parameters of the pumping unit calculated from dynamometer data represented by Fig. 4.32.

rod position, $s(t)$, as detailed previously. The calculated crank angles as a function of time, $\theta(t)$, are displayed in Fig. 4.33; the values do not fall on a straight line as they would for a constant crankshaft speed, indicating a varying crankshaft velocity. Using Fourier approximation, the first and second derivatives, $d\theta(t)/dt$ and $d^2\theta(t)/dt^2$, of $\theta(t)$ are determined. Of these, $d\theta(t)/dt$ represents the angular velocity of the crankshaft from which instantaneous pumping speeds, SPM_{inst}, are found. Compared to the average pumping speed of 5.98 SPM, instantaneous crankshaft speed fluctuates more than 50% during the pumping cycle. The angular acceleration of the crank, $d^2\theta(t)/dt^2$, is also plotted in the same figure in a solid line. Based on the derivatives and using the kinematic analysis of the pumping unit as proposed by **Svinos** (see Section 3.7.5.2.2), the angular acceleration pattern of the beam, $d^2\theta_b/dt^2$, is also found and plotted versus time in dashed line.

The kinematic parameters plotted in Fig. 4.33 form the basis of gearbox torque calculations, as discussed in the following. To illustrate the necessary calculations, let's find the gearbox torques at a time $t = 1.0$ s where measured polished rod load is 19,617 lb and the calculated crank angle is $\theta = 47.35°$.

Rod torque is relatively simply found from the polished rod load recorded by the dynamometer, $F(t)$, and the torque factor derived from the crank angle, $\theta(t)$, belonging to the same time. The unit's structural unbalance is $SU = -1{,}500$ lb and the torque factor at the given crank angle is $TF = 78.2$ in; rod torque is found from Eq. (4.94) as:

$$T_r = 78.2(19{,}617 + 1{,}500) = 1{,}651 \text{ k in lb.}$$

Articulating inertial torque is basically defined by the walking beam's acceleration, which at the given time equals $d^2\theta_b/dt^2 = -0.076$ 1/s^2. The mass moment of inertia of the beam, horsehead, equalizer, bearings, and pitmans, referred to the saddle bearing, found from manufacturer data, is

$I_b = 1{,}047{,}183$ lb_m ft^2 and the distance between the saddle bearing and the polished rod is $A = 210$ in. The torque is calculated from Eq. (4.114) as follows:

$$T_{\text{ia}} = 12/32.2 \times 78.2 \times 1{,}047{,}183/210(-0.076) = -11 \text{ k in lb.}$$

Counterbalance torque calculations necessitate knowledge of the current counterbalance conditions; the unit has 94110CA cranks and four ORO master weights positioned at 10 in from the long end of the crank. Maximum moment of the cranks is $T_{\text{crank}} = 470{,}810$ in-lb, the weight of one counterweight is $W = 3{,}397$ lb, and its maximum lever arm is $M = 77.4$ in, from manufacturer data. Maximum counterbalance moment can now be calculated from Eq. (4.98):

$$T_{\text{CB max}} = 470,810 + (77.4 - 10)4 \times 3,397 = 1,387 \text{ k in lb.}$$

Counterbalance torque at the given time is evaluated according to Eq. (4.95), as follows:

$$T_{\text{CB}} = -1,387 \sin(47.35) = -1,020 \text{ k in lb.}$$

Rotary inertial torque calculations start with the determination of the system's net mass moment of inertia; the manufacturer supplies the moment of the cranks as $I_{\text{cr}} = 247{,}244$ lb_m ft^2 with their dimension $X = 11.5$ in and the moment of the slow-speed shaft and gear as $I_g = 4{,}400$ lb_m ft^2. The moment of inertia for one counterweight about its center of gravity is $I_{\text{CG}} = 8{,}017$ lb_m ft^2, its dimension $Y = 19$ in.

Distance H is defined by Eq. (4.118) as:

$$H = \left[(11.5 + 19)^2 + (77.4 - 10)^2\right]^{0.5} = 73.98 \text{ in.}$$

Mass moment of inertia of the four ORO counterweights, according to Eq. (4.117), is:

$$I_{\text{CW}} = 4\left[8,017 + 3,397(73.98/12)^2\right] = 548,510 \text{ lb}_\text{m} \text{ ft}^2.$$

The system's net mass moment of inertia is the sum of all inertias; see Eq. (4.116):

$$I_s = 247,244 + 548,510 + 4,400 = 800.2 \text{ k lb}_\text{m} \text{ ft}^2.$$

Finally, rotary inertial torque is found from Eq. (4.115) using the crank acceleration of $\text{d}^2\theta/\text{d}t^2 = -0.439$ $1/\text{s}^2$ at the given time:

$$T_{\text{ir}} = 12/32.2 \times 800,200(-0.439) = -130.9 \text{ k in lb.}$$

Finally, the net gearbox torque is the sum of the four torque items:

$$T_{\text{net}} = 1,651 - 11 - 1,020 - 130.9 = 489 \text{ k in lb.}$$

The variation of calculated torques during the pumping cycle is presented in Fig. 4.34. As seen, articulating inertial torque (denoted Art. in the figure) is negligible if compared to other torque components. Rotating inertial torque (denoted Rot.), on the other hand, very substantially reduces the torque load on the gear reducer whenever its sign is negative. Negative rotary torque indicates that energy stored in the heavy rotating parts of the pumping unit is released and helps reducing the net torque. This behavior is caused by the great variation of crankshaft speed during the pumping cycle, as shown in Fig. 4.33.

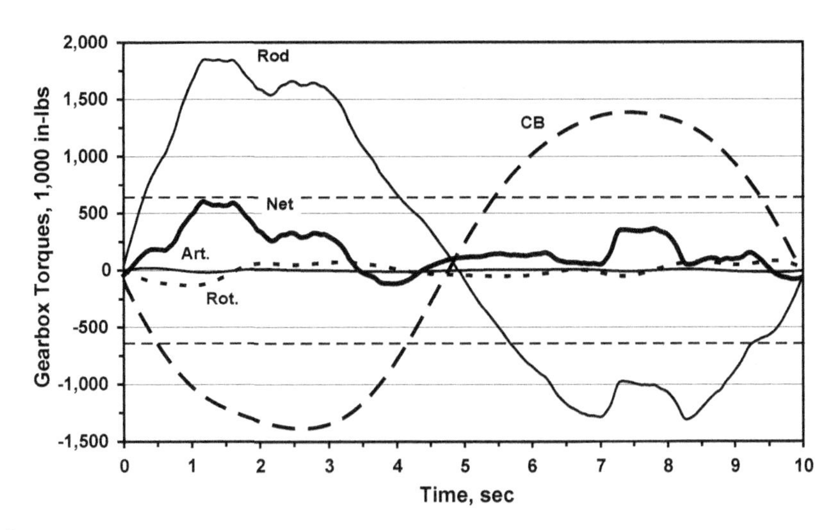

FIGURE 4.34

Variation of gearbox torque components versus time for the case given in Fig. 4.32.

4.6.3.3 Permissible load diagrams

An **alternate** way of evaluating the torsional loads on a pumping unit's speed reducer is provided by the use of the **permissible load** concept, introduced by **Gault** [79]. Permissible loads are defined as polished rod **loads** during the pumping cycle that give net torques equal to the speed reducer's torque **rating** at the actual amount of counterbalance. They can be calculated from the formulae describing net torque by solving those for the polished rod **loads** i.e., by reversing the calculation of gearbox torques from loads. After manipulating Eqs (4.109)–(4.111) valid for gearboxes operating at constant speed, we get:

$$F_p(\theta) = \frac{T_{al} + T_{CB\ max} \sin\theta}{TF(\theta)} + SU \quad \text{for } \textbf{Conventional} \text{ geometry} \tag{4.126}$$

$$F_p(\theta) = \frac{T_{al} + T_{CB\ max} \sin(\theta + \tau)}{TF(\theta)} + SU \quad \text{for } \textbf{Mark II} \text{ geometry} \tag{4.127}$$

$$F_p(\theta) = \frac{T_{al} + T_{CB\ max} \sin(\theta - \tau)}{TF(\theta)} + SU \quad \text{for } \textbf{Reverse Mark} \text{ geometry} \tag{4.128}$$

where:

$F_p(\theta)$ = permissible load at crank angle θ, lb,
T_{al} = the maximum allowed torque (rating) of the speed reducer, in-lb,
SU = structural unbalance of the pumping unit, lb,
$TF(\theta)$ = torque factor at crank angle θ, in,
$T_{CB\ max}$ = maximum moment of counterweights and cranks, in-lb,
θ = crank angle, degrees, and
τ = counterweight arm offset angle, degrees.

Permissible loads change with the actual counterbalance **moment**, the crank **angle**, unit **geometry**, and structural **unbalance**. Their values approach infinity at the start and at the end of the polished rod's stroke, since at these points the torque factors are zero. Permissible loads can be plotted directly on a **dynamometer** card and define two curves: one for the upstroke and one for the downstroke. The **envelope** between these curves represents the **permissible** polished rod loads that do not **overload** the speed reducer. If any of the permissible load curves **cuts** into the dynamometer diagram, then the gearbox is **overloaded** at those positions. Therefore, in order to prevent the speed reducer's overload, the actual dynamometer card must fit between the two curves of the permissible load diagram. Before the advent of personal computers, families of permissible load curves were plotted on transparent overlays for different pumping units and counterbalance moments; these were used to visually analyze dynamometer cards for overload conditions. Today the predominant use of computers and electronic dynamometer systems makes this procedure obsolete.

Although the definition just described for the calculation of permissible loads is universally used, several **modifications** are known, as listed in the following; for a complete treatment refer to **Gibbs'** book [31].

- Inertial effects can also be included in the calculation and the net torque formula given in Eq. (4.123) can be solved to find permissible loads. Only used if high-slip prime movers drive the pumping unit; then the vertical space between the upstroke and downstroke curves of the permissible load diagram widens as compared to cases disregarding inertia.
- The shape of the permissible load diagram changes if the unit's torque rating, T_{al}, is substituted with a negative sign in the relevant formulas [80]; for a proper analysis the downstroke curve must be calculated both ways.
- Although permissible loads indicate the torque loading of the gearbox, the unit's structural rating, i.e., the maximum polished rod load allowed on the unit, can also be included in the permissible load diagram. This is accomplished by limiting the calculated permissible load values by the unit's structural rating.

Even in the present age of diagnostic computer programs, permissible load diagrams find an important application in **matching** the pumping unit's kinematic behavior with the characteristics of the dynamometer card. Dynamometer cards can be classified as **overtravel** or **undertravel** cards depending on the general **trend** of the measured polished rod loads; overtravel cards exhibit a general **downward** slope from left to right, whereas undertravel cards slope **upward** from left to right. Permissible load diagrams, at the same time, can have envelopes between the upstroke and downstroke loads that have a definite **orientation** from left to right. As shown in Fig. 4.35 for a conventional and a Mark II unit of the same size (**160D-173-86**) and using the same maximum counterbalance moment, the behavior of the conventional unit is much more beneficial for an **undertravel** than an overtravel card. The Mark II, at the same time, would be a bad choice, since an undertravel card would easily cut in the permissible load curve, indicating an overload situation. On the other hand, an **overtravel** card would perfectly fit the Mark II unit's permissible load diagram. The selection of the proper pumping unit geometry for actual well conditions represented by a predicted dynamometer card can thus be accomplished with the help of permissible load diagrams.

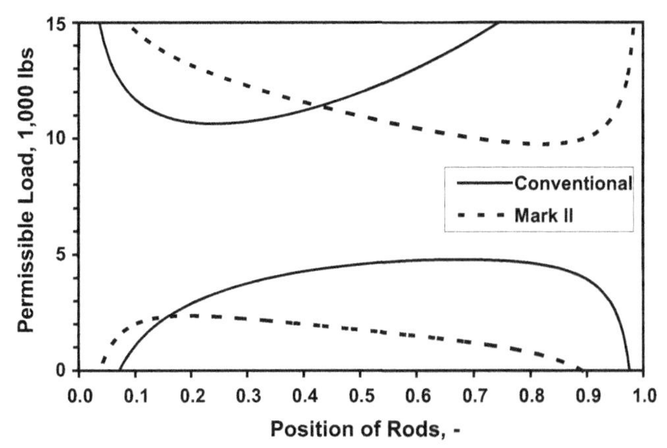

FIGURE 4.35

Comparison of permissible load diagrams for a conventional and a Mark II unit of the same size.

EXAMPLE 4.20: CALCULATE THE PERMISSIBLE POLISHED ROD LOAD AT A CRANK ANGLE OF 60° FOR THE CASE PRESENTED IN EXAMPLE 4.17

Solution

The unit is of conventional geometry and Eq. (4.126) should be applied, using data provided in Example 4.17:

$$F_p(60) = [160,000 + 250,000\ \sin(60)]/45.2 + 450 = 8,780\ \mathrm{lb}.$$

The actual polished rod load at the same crank angle was found as 10,700 lb in Example 4.17, which is much greater than the permissible load just calculated. Thus, in accordance with the conclusions of Example 4.17, the speed reducer is greatly overloaded at the given crank angle. The complete permissible load diagram for the example case, superimposed on the dynamometer card, is presented in Fig. 4.36, as indicated by the points denoted "Actual."

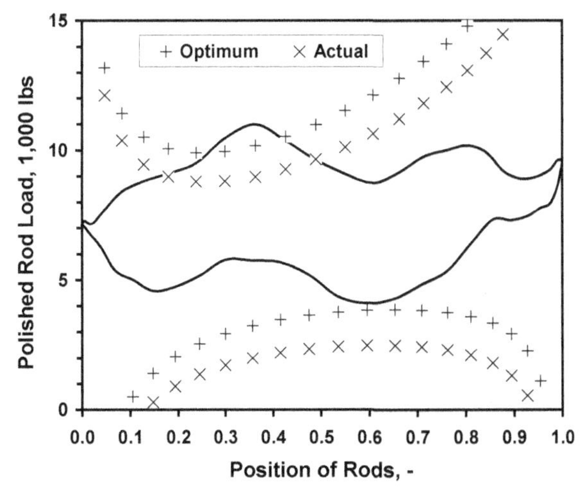

FIGURE 4.36

Permissible load values (for actual and ideal counterbalance conditions) superimposed on the dynamometer card for Example 4.20.

4.6.4 OPTIMUM COUNTERBALANCING

4.6.4.1 Introduction

As discussed in previous sections, net gearbox torque heavily **fluctuates** during the pumping cycle, making the prime mover's operating conditions far from ideal. Any kind of prime mover, be it an engine or an electric motor, operates properly if working under a constant load; the torque loading on the pumping unit gearbox, in spite of the positive effects of the counterweights, can never become steady. The unit is called "underbalanced" or "**rod-heavy**" when the peak upstroke torque is greater than the downstroke peak (for illustration see Fig. 4.28). Increasing the maximum counterbalance moment too much, on the other hand, causes an "overbalanced" or "**weight-heavy**" situation when the downstroke peak torque becomes greater.

The principal aim of counterbalancing a pumping unit is to **even out** the torsional loads on the speed reducer during the pumping cycle and to provide a relatively **smooth** loading for the prime mover. The use of either **beam** or **rotary counterweights** or an **air cylinder** produces a counterbalance torque, which **opposes** rod torque and thus reduces the fluctuations of the net torque on the crankshaft. Since the counterbalance torque is easily **changed** (by changing the **size** of counterweights and/or their **positions** on the crank or by other means), torque fluctuations normally can be reduced to an acceptable level. This means that most often an optimum or **ideal** counterbalance condition can be found; the use of the proper counterbalance has many beneficial effects, such as:

- Reducer size (torque rating) can be **decreased** significantly when compared to an unbalanced condition.
- The size of the required prime mover is also **smaller**, with an associated lower energy demand for pumping.
- The smoother operation of a properly balanced speed reducer lowers **maintenance** costs and increases equipment life.

4.6.4.2 Optimization principles

Due to its practical and economic importance, the concept of **optimum** counterbalancing has been a heavily discussed problem, and different solutions have been proposed over the years [81–83]. **Gipson and Swaim** [78] present a procedure that enables the calculation of the proper counterbalance moment according to the principle of equalizing peak torques. Using the same theory, **Keating et al.** [83] developed a field procedure that assures a minimum of work to be performed in achieving the optimum conditions; **Rowlan et al.** [84] also equalize the peak torques on the upstroke and the downstroke. Most of these methods try to find the counterbalance moment that results in:

- equal **peak** net torques on the upstroke and downstroke, or
- equal **horsepowers** for the upstroke and downstroke, or
- a minimum of the **cyclic load factor**, calculated on the net torque values.

In field practice, the first of these principles is generally used, based on the fact that the torque–amperage performance curve of electric motors is **linear**. By insertion of a clamp-on **ammeter** on one of the motor's supply wires, the **peak** currents on the upstroke and downstroke can be detected easily. Proper balancing of the unit is indicated when these peaks are about equal. This widely accepted procedure has the following drawbacks:

- It is time consuming because it involves a trial-and-error procedure requiring several stops and restarts and moving of the heavy counterweights.

- It can be **inaccurate** because during shutdowns the fluid level in the well may change and well conditions may not be stable during the balancing operation; later, as the well stabilizes, the unit may be out of balance again.
- Another source of inaccuracy is that the clamp-on ammeter measures rms current, which is always positive and does not indicate the **direction** of the current; thus measured peak currents can belong to positive or negative torques. Therefore, the ammeter cannot distinguish between the **motoring** and the **braking** operating mode of the electric motor; an unbalanced unit with a large negative peak torque may appear balanced. As suggested by **Gibbs** [31], watching the slack in the V-belts can reveal if the motor is driving or is driven and acts as a generator; in the latter case the ammeter must not be trusted.

The calculation model of **Gibbs** [31] is based on the observation that the upstroke and downstroke peak net torques occur at the same crank angles independent of the maximum counterbalance moment, $T_{CB\ max}$. For a conventional pumping unit and neglecting inertial effects, Eq. (4.109) can be used to equate the two peak net torques while the optimum counterbalance is used:

$$T_r(\theta_u) - T_{CB\ opt}\sin\theta_u = T_r(\theta_d) - T_{CB\ opt}\sin\theta_d \tag{4.129}$$

where:

T_r = rod torques at crank angles θ_u and θ_d, in-lb,
$T_{CB\ opt}$ = optimum counterbalance moment, in-lb, and
θ_u, θ_d = crank angle belonging to the up/downstroke peak net torque, respectively, radians.

Solution of this equation results in the following formula, which defines the optimum counterbalance moment, $T_{CB\ opt}$, to be used to keep the peak net torques during the upstroke and the downstroke equal:

$$T_{CB\ opt} = \frac{T_r(\theta_u) - T_r(\theta_d)}{\sin\theta_u - \sin\theta_d} \tag{4.130}$$

EXAMPLE 4.21: FIND THE OPTIMUM COUNTERBALANCE MOMENT FOR THE CASE GIVEN IN EXAMPLE 4.18

Solution

Under the existing counterbalance conditions, the crank angles where the net torque peaks occur are as follows, along with the rod torques valid at those crank angles, read from torque calculation data used to construct Fig. 4.28:

	θ	T_r
Upstroke	66.1°	480,600 in-lb
Downstroke	267.7°	−161,500 in-lb

Using Eq. (4.130) the counterbalance moment necessary to achieve optimum counterbalance conditions is found as:

$$T_{CB\ opt} = (480,600 + 161,500)/[\sin(66.1) - \sin(267.7)] = 335,530 \text{ in lb.}$$

The variation of gearbox torques at optimum conditions is presented in Fig. 4.37. As seen, the two peak net torques are equal, although slightly above the rated torque capacity of the gearbox.

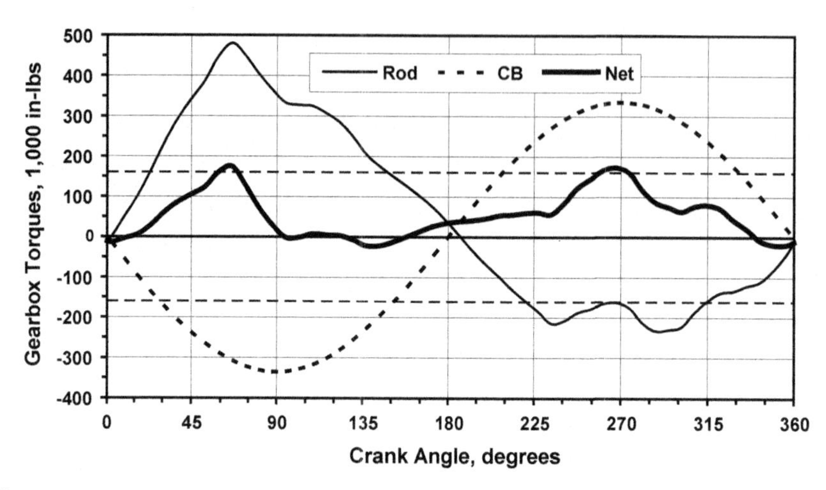

FIGURE 4.37

Calculated crankshaft torques, plotted against crank angle for the case in Example 4.21.

4.6.4.3 Minimum cyclic load factor method

The rationale of minimizing the **cyclic load factor** [75] lies in the evaluation of the energy requirements of pumping. As discussed in Section 3.10.3.3, an electric motor for pumping service must be derated by an electric cyclic load factor (*CLF*), which allows for motor overheating due to cyclic load. This *CLF* can be calculated easily if based on the net **torque** pattern, because the plots of electric current versus torque characteristics of pumping motors are **linear**. The *CLF* defined this way is the ratio of the root mean square and the average of the **net** torque over the pumping cycle:

$$CLF = \frac{\sqrt{\dfrac{\int_{\theta=0}^{2\pi} [T_{net}(\theta)]^2 d\theta}{2\pi}}}{\dfrac{\int_{\theta=0}^{2\pi} T_{net}(\theta) d\theta}{2\pi}} \quad (4.131)$$

where:

CLF = mechanical cyclic load factor, based on net torque, –,
$T_{net}(\theta)$ = net torque at crank angle θ, in-lb, and
θ = crank angle, radians.

The value of the mechanical cyclic load factor is an effective indicator of the "**smoothness**" of the gearbox's loading over the pumping cycle. For a completely flat variation of net torque (never possible on a pumping unit) a value of $CLF = 1$ is found from Eq. (4.131). Fluctuating torques on the gearbox, on the other hand, generate *CLF*s greater than unity; the greater the torque fluctuations are, the greater

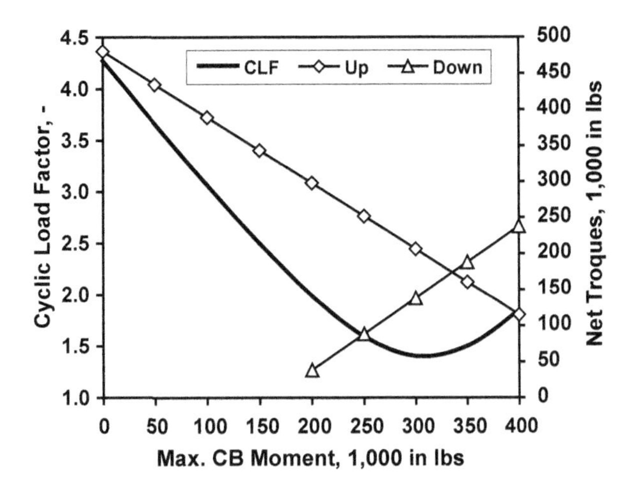

FIGURE 4.38

The variation of CLF and peak up, and downstroke net torques with the maximum counterbalance moment for Example 4.18.

the *CLF* becomes. Since the required prime mover power is directly **proportional** to the cyclic load factor, the lower the *CLF*, the lower the energy requirement of fluid lifting. But the value of *CLF* can be changed at will by **adjusting** the counterweights (changing their size or position); thus a logical solution for finding the optimum counterbalance condition is to **minimize** the cyclic load factor. This approach, therefore, assures the minimum of input power requirements and can improve the economy of pumping.

The **merits** of the procedure to minimize the cyclic load factor are shown in Fig. 4.38, which displays *CLF* values versus maximum counterbalance moment for the well used in Example 4.18. As seen, the cyclic load factor curve clearly exhibits a **minimum** as the maximum counterbalance moment, $T_{CB\ max}$, is increased. For comparison, the variations of the **peak** upstroke and downstroke net torques are also plotted in the same figure. Intersection of the two curves represents the counterbalance moment required to set the two peaks (upstroke and downstroke) equal. It can be observed that the *CLF* value valid at this point is slightly greater than the minimum *CLF* achieved by the optimization model just described.

The calculated torques for the properly balanced case of Example 4.18 are plotted in Fig. 4.39. Evaluation of additional calculated data contained in Exhibit 4.6, shows a considerable improvement of the operating parameters as the counterbalance moment was increased from the original 250,000 in-lb to the optimum of 309,790 in-lb value. Application of the optimum counterbalance moment has reduced the original overload of the speed reducer from 159% to 125%. The cyclic load factor is reduced by approximately 12%, which results in a decrease of power consumption by the same amount. Finally, the permissible loads for the optimized case (indicated by the points denoted "Optimum") are shown in Fig. 4.36, previously discussed.

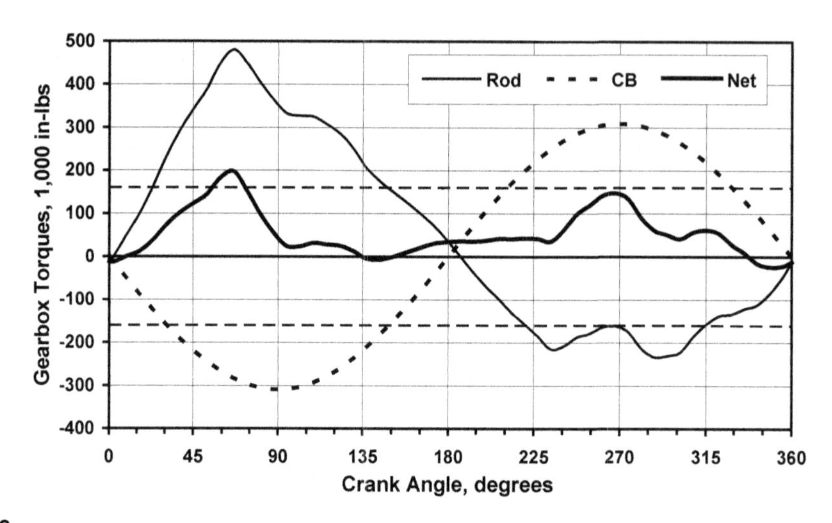

FIGURE 4.39

Calculated crankshaft torques, plotted against crank angle, using the ideal amount of counterbalance for Example 4.18.

4.6.4.4 Concluding remarks

After the optimum counterbalance moment is found, one can easily determine the necessary modifications on the current counterweight arrangement. Smaller adjustments require the **existing** counterweights to be differently **positioned** on the cranks; the necessary movements "**in**" (toward the crankshaft) or "**out**" (toward the end of the crank) are calculated and the weights are repositioned accordingly. Greater changes require the removal of existing counterweights and the use of different ones (smaller or greater) at different positions; simple calculations based on Fig. 4.24 or diagrams similar to Fig. 4.25 provide the necessary information [84,85]. All computer program packages commercially available for the analysis of sucker-rod pumping installations include this feature and help the operator maintain proper counterbalance conditions of the pumping unit.

One more topic remains to be mentioned regarding proper counterbalancing. It is customary to assume that the center of gravity of the counterweights lies on the **center line** of the crank or, in the case of **Mark II** and **Reverse Mark** units, of the counterweight arm. In some cases, however, weights of different **sizes** or at different **positions** may be installed on the two sides of the crank(s). Such arrangements can significantly affect the position of the **center** of gravity for the combined mass of the crank plus the counterweights. Under these circumstances, the **lead** or **lag angle** between the crank center line and the center of gravity (compared to the sense of the unit's rotation) must be accounted for properly in counterbalance torque calculations. To do so, the number, size, and exact position of the counterweights must be known. **Chastain** [86] showed that this phenomenon can be used to advantage in reducing the net torque on conventional-geometry pumping units.

4.6.5 THE EFFECTS OF UNIT GEOMETRY

The continual search for finding more efficient ways to drive the sucker-rod pump has brought about the evolution of pumping unit geometries. The common initiative behind these developments was the reduction of net crankshaft **torques** and their fluctuations over the pumping cycle [87]. Basically, two possibilities are available to achieve these goals:

- The first involves the reduction of the unit's **torque factors** by appropriate alterations to the geometrical arrangement of the structural parts. Such improvements can greatly reduce the magnitude of rod torques.
- The other choice, often combined with the first, modifies the **counterbalance** torque (usually by placing the counterweights on a counterweight arm, which is phased to the crank arm) so that it more closely follows the shape of the rod torque. Thus, the sum of rod and counterbalance torques produces a more even net torque over the pumping cycle.

Both of the above modifications can be clearly observed on the **Mark II** and **Reverse Mark** units when compared to the basic conventional geometry. Therefore the proper **selection** of the unit geometry to be used on a given well has a prime importance in increasing the efficiency of sucker-rod pumping. In solving this problem, the results of several published investigations [88–90] can be used to advantage, but the best approach is the utilization of the modern **predictive** analysis methods previously described, in conjunction with the solution of the damped wave equation. As already discussed, the predictive analysis allows comparing the performance of different pumping unit geometries on the same well, all other parameters (production rate, rod string data) unchanged. Reference is made here to Example 4.12, where a conventional and a Mark II unit were compared on a 6,000 ft well. After performing torque analyses on the two dynamometer diagrams calculated with the predictive model (see Figs 4.11 and 4.12), the permissible loads shown in Figs 4.40 and 4.41 were obtained, assuming perfectly balanced units. As seen in the given case, the conventional unit is overloaded both on the upstroke and on the downstroke. The Mark II unit, on the other hand, experiences a slight overload on the downstroke only and should be the right choice for the well conditions investigated.

4.6.6 INFERRED TORQUE CALCULATIONS

In previous sections the determination of gearbox torques was accomplished by using **mechanical** calculations that required a detailed description of the **kinematics** of the pumping unit as well as the proper knowledge of counterbalance conditions. One can encounter many problems when collecting the necessary information needed for such calculations, e.g., dimensional data of the pumping unit may be inaccurate or unknown, the weights and positions of counterweights may be unknown, etc. All these problems are eliminated if gearbox torques are **inferred** from measurements conducted on the electric motor driving the pumping unit. The applicability and accuracy of inferred calculations are ensured by the very fast electrical **response** time of electric motors [91] used in sucker-rod pumping service, as discussed in Section 3.10.3.4. These motors automatically and quickly adapt to the variable loading conditions during the pumping cycle by **immediately** altering their performance parameters (speed, output power, current drawn, etc.). The next sections describe the possible solutions of inferring gearbox loads from measured motor performance parameters.

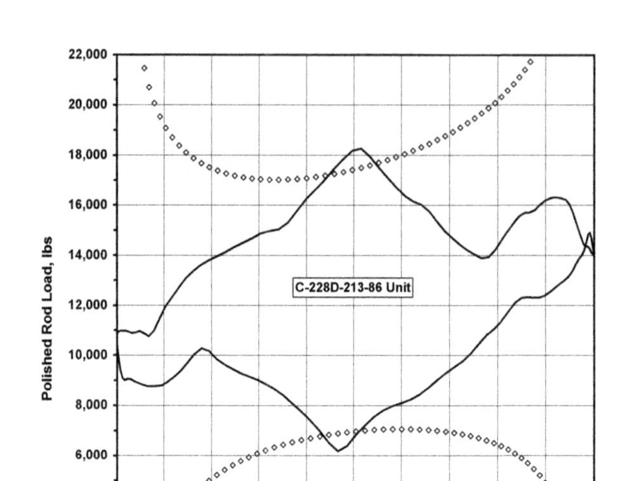

FIGURE 4.40

Permissible loads for Example 4.12, using a conventional pumping unit.

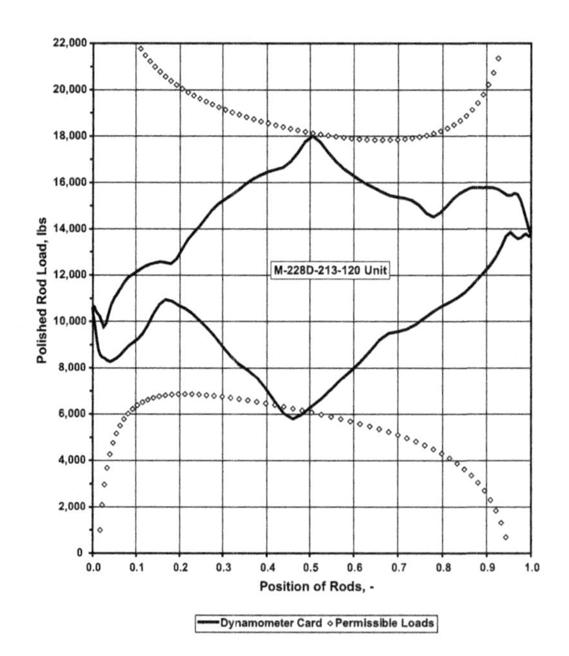

FIGURE 4.41

Permissible loads for Example 4.12, using a Mark II pumping unit.

4.6.6.1 Torque loading from power measurements

The electrical power input to the motor can be measured if **voltages** and **currents** in the three phases are recorded; normally three voltage sensors are connected to the proper phase terminals and two clamp-on current sensors are installed on two of the motor's power supply wires [51]. This setup permits the continuous recording of the motor **current** and the input electric **power** during the pumping cycle. The measured instantaneous electric power should cover the surface power requirements of pumping, which include (1) the power to drive the gearbox and (2) any losses occurring from the gearbox to the prime mover. The **useful** power of the surface system, i.e., the power required to drive the gearbox, can be calculated from the net **torque** on the gearbox and the pumping **speed**. This is because in a rotating system power equals torque times cycle frequency; thus, instantaneous **gearbox power** is found from the following expression:

$$P_{gb}(t) = T_{net}(t)\frac{N(t)\pi}{30} \tag{4.132}$$

where:

$P_{gb}(t)$ = instantaneous mechanical power on the gearbox, in-lb/s,
$T_{net}(t)$ = instantaneous net gearbox torque, in-lb, and
$N(t)$ = instantaneous pumping speed, 1/min.

The **electric power** that the motor should provide is easily found from the gearbox power, $P_{gb}(t)$, if the efficiency of power transmission in the surface system, η_{surf}, is known. This efficiency includes the power efficiencies of (1) the gearbox, (2) the V-belt drive, and (3) the electric motor. Expressing the required motor power in kW units and using Eq. (4.132), we get:

$$P_e(t) = \frac{1}{\eta_{surf}}P_{gb}(t) = \frac{1}{\eta_{surf}}T_{net}(t)\frac{N(t)\pi}{30 \times 8,851} \tag{4.133}$$

where the variables not defined before are:

$P_e(t)$ = instantaneous input electric power to the motor, kW, and
η_{surf} = power transmission efficiency, –,

Expressing net torque on the gearbox, $T_{net}(t)$, from the previous formula, we arrive at the basic expression of **inferred** torque calculations [92–94], as follows:

$$T_{net}(t) = 84{,}520\frac{P_e(t) \times \eta_{surf}}{N(t)} \tag{4.134}$$

where:

$T_{net}(t)$ = inferred net gearbox torque, in-lb,
$P_e(t)$ = instantaneous input electric power to the motor, kW,
η_{surf} = power transmission efficiency, –, and
$N(t)$ = instantaneous pumping speed, 1/min.

The use of this formula allows the calculation of instantaneous gearbox torques from the variation of motor power versus time; this is made possible by the quick response of the electric motor to changes in its loading conditions. The direct measurement of electrical **power** taken by the motor eliminates the errors introduced by improper information on pumping unit dimensions and improper or unknown counterbalance data. Inferred torques allow the analyst to investigate the loading of the gearbox and to determine optimum counterbalance conditions [95] without the burden of tedious mechanical torque calculations. It is important to note that the torques thus calculated represent the effects of **all** torque components that act on the gearbox and include **inertial** torques as well; these are normally much more complicated to find with the conventional approach.

The procedure just described is utilized in a computer program package [96] developed for the analysis of sucker-rod pumping installations by a major service company. Calculation results properly reproduce the variation of gearbox torques received from conventional mechanical torque calculations. The program uses a constant surface efficiency representing average pumping conditions of $\eta_{surf} = 0.8$ and a constant pumping speed; this speed is composed of the average pumping speed during the pumping cycle and a correction factor accounting for speed variations during the stroke. Thus Eq. (4.134) is modified as given in the following:

$$T_{net}(t) = 84{,}520 \frac{P_e(t) \times \eta_{surf}}{N_{avg}\ SV} \tag{4.135}$$

where:

$T_{net}(t)$ = inferred net gearbox torque, in-lb,
$P_e(t)$ = instantaneous input electric power to the motor, kW,
η_{surf} = surface power efficiency, –,
N_{avg} = average pumping speed, 1/min, and
SV = speed variation ($SV < 1$), –.

For illustration, Fig. 4.42 shows the variation of net torques inferred from electrical power measurements, along with those found from conventional mechanical calculations for an example case.

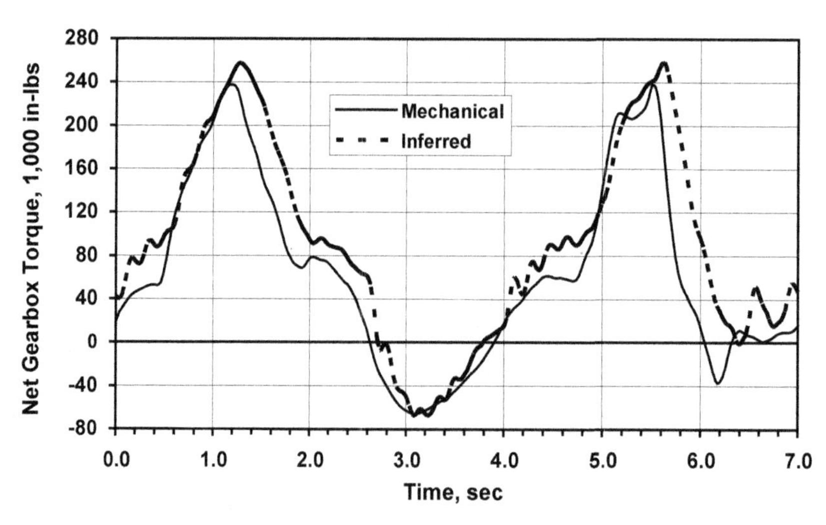

FIGURE 4.42

Comparison of calculated and inferred net gearbox torques for an example case.

4.6.6.2 Torque calculations from motor speed

The torque load on the gearbox as well as on the prime mover, as discussed in previous sections, is widely **fluctuating** during the pumping cycle. This cyclic load causes a continuous variation of the motor's **speed** because of the very rapid response of induction motors to changes in their loading conditions. As given in Section 3.10.3.4, all performance parameters of such motors are the functions of motor speed only. This circumstance enables one to **infer** operating parameters (current drawn, torque developed, etc.) from measured motor speeds without the need for electrical measurements, as first proposed by **Gibbs** [97,98]. The basic prerequisite of such inferred calculations is the proper knowledge of motor performance curves in a wide range of motor speeds, which is seldom available.

The inferred torque calculations just mentioned require an accurate measurement of instantaneous motor **speed** during the pumping cycle. This can be accomplished by using a generating **tachometer** whose output can be digitized and recorded as a function of time for further processing. A more accurate solution was proposed by **Gibbs** [97] where motor speed was digitally measured using a **Hall-effect** transducer. The transducer senses the passing of a small permanent **magnet** attached to the motor shaft or the motor sheave; its signals are processed to furnish the instantaneous speed-versus-time function. Calculation of operating parameters of the motor is straightforward: based on the measured speeds and the motor performance curves (normally available in tabulated form) and using an interpolation scheme, motor torque, current, and efficiency at the measured times are easily found.

The most important use of the procedure just described is to find the inferred gearbox **torque** from motor speed measurements. If inertial torques and energy losses in the drive train are neglected (justified in most cases, except when ultrahigh-slip motors are used), then power input at the motor equals power output at the gearbox:

$$T_m(t) \times RPM(t) = T_{\text{net}}(t) \times N(t) \tag{4.136}$$

where:

$T_m(t)$ = instantaneous motor torque, in-lb,
$RPM(t)$ = instantaneous motor speed, 1/min,
$T_{\text{net}}(t)$ = instantaneous gearbox torque, in-lb, and
$N(t)$ = instantaneous pumping speed, 1/min.

Solving the formula for the inferred gearbox torque and using Eq. 3.92 to express pumping speed from prime mover speed and the overall speed reduction of the drive train, we receive:

$$T_{\text{net}}(t) = T_m(t)\, Z\, \frac{D}{d} \tag{4.137}$$

where:

$T_{\text{net}}(t)$ = inferred gearbox torque, in-lb,
$T_m(t)$ = instantaneous motor torque, in-lb,
d = prime mover sheave pitch diameter, in,
D = gearbox sheave pitch diameter, in, and
Z = speed reduction ratio of the gearbox, –.

The advantage of this type of inferred torque calculation is that no electrical or mechanical measurements, which may be quite costly, are needed. Its limitation is the need for **accurate** information on the motor torque versus speed performance curve in a broad range of motor speeds. Since electric motor manufacturers seldom publish these data, the application of inferred torque calculations is limited.

4.6.7 PRIME MOVER TORQUE

The torsional load on the motor shaft is less than that on the gearbox because of the much higher **speed** of the motor due to the overall speed reduction ratio of the drive train. Motor torque can be found from Eq. (4.137) by including the **efficiency** of power transmission in the surface system, η_{surf}, as follows:

$$T_m(t) = \frac{T_{net}(t)}{Z_d} \frac{1}{\eta_{surf}} \tag{4.138}$$

where:

$T_m(t)$ = instantaneous motor torque, in-lb,
$T_{net}(t)$ = instantaneous net gearbox torque, in-lb,
Z_d = speed reduction ratio of the drive train, –, and
η_{surf} = power transmission efficiency, –.

Net gearbox torque, however, as defined in Eq. (4.123), does not contain the inertia of the **complete** drive train; the inertial torques related to the motor rotor, the motor and unit sheaves, and the high-speed and intermediate gearing must also be included. Manufacturers usually supply the **total** mass moment of inertia, I_t, of the drive train; the net torque formula is then corrected and the following final equation is received:

$$T_m(t) = \frac{1}{Z_d \eta_{surf}} \left[T_{net}(t) + \frac{12}{32.2}(I_t - I_s)\frac{d^2\theta}{dt^2} \right] \tag{4.139}$$

where:

$T_m(t)$ = instantaneous motor torque, in-lb,
$T_{net}(t)$ = instantaneous net gearbox torque, in-lb,
Z_d = speed reduction ratio of the drive train (see Eq. 3.91), –,
η_{surf} = power transmission efficiency, –,
I_t = mass moment of inertia of the complete drive train referred to the crankshaft, $lb_m\ ft^2$,
I_s = mass moment of inertia of the cranks, counterweights, and slow-speed gearing referred to the crankshaft, $lb_m\ ft^2$, and
$d^2\theta/dt^2$ = angular acceleration of the crankshaft, $1/s^2$.

As already demonstrated, in most cases the effect of **articulating** inertia is negligible; if **rotary** inertia is also disregarded (justified unless an ultrahigh-slip motor is used) then motor torque is found by substituting one of Eqs (4.109)–(4.111) into this formula and we receive:

$$T_m(t) = \frac{1}{Z_d \eta_{surf}} \left\{ TF(\theta)[F(\theta) - SU] - T_{CB}(\theta) \right\} \tag{4.140}$$

where:

$T_m(t)$ = instantaneous motor torque, in-lb,
Z_d = speed reduction ratio of the drive train, –,
η_{surf} = power transmission efficiency, –.
$TF(\theta)$ = torque factor at crank angle θ, in,
$F(\theta)$ = polished rod load at crank angle θ, lb,
SU = structural unbalance of the pumping unit, lb, and
$T_{CB}(\theta)$ = counterbalance torque at crank angle θ, in-lb.

4.7 POWER REQUIREMENTS OF SUCKER-ROD PUMPING

The economy of a sucker-rod pumping system or any other type of artificial lift installation can best be evaluated by considering the **lifting costs** in monetary units per volume of liquid lifted. Since most of today's rod-pumping installations are driven by **electric** motors, part of the operating costs is represented by the electric **power** bill. Because of the worldwide trend of inflating electric power **prices**, this single item has become the most decisive constituent of operating expenditures in sucker rod-pumped fields. Consequently, the never-ending pursuit of operating cost reduction can be translated to the reduction of energy **losses** both downhole and on the surface. Furthermore, the proper choice of the prime mover type and size and the right selection of the pumping parameters are decisive factors when an efficient pumping system is desired.

4.7.1 INTRODUCTION

The power consumed by the pumping unit's motor comprises, in addition to the energy required to **lift** well fluids to the surface, all the energy **losses** occurring in the well and in the surface machinery. Therefore, any efforts to reduce these losses should start with a perfect understanding of their nature and magnitude. Figure 4.43 and the following discussion present the possible sources of energy losses along the well stream's flow path, grouped into downhole and surface loss categories.

The rod-pumping system's **useful** output work is done by the downhole **pump** when it lifts a given amount of liquid from the pump setting depth to the surface. This work is usually described by the so-called **hydraulic power**, P_{hydr}, and can be calculated as the increase in **potential** energy of the liquid pumped.

At the other end of the system, the electric prime mover takes the **required** power from the surface power **supply**, that power being accurately measured. Since actual power requirements at the motor vary within the pumping cycle, an average input power value, P_e, valid for one pumping cycle is found from power **meter** readings. This power covers all requirements of the pumping system, including the **useful** power used for fluid lifting and all energy **losses** occurring in the downhole and surface systems, and it represents the total energy **input** to the system.

If the input and output powers are known from actual measurements, an **overall** efficiency for the pumping system can easily be defined. Since the system's **useful** work is represented by the hydraulic power spent on fluid lifting, and the total energy input equals the measured **electric** power, the rod pumping system's energy **efficiency** is found from:

$$\eta_{system} = \frac{P_{hydr}}{P_e} \tag{4.141}$$

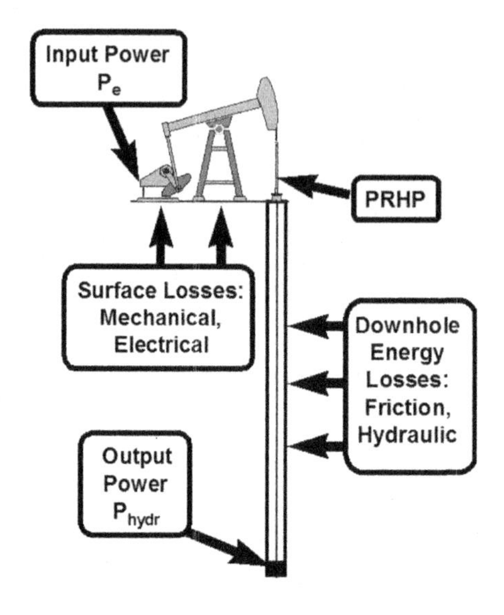

FIGURE 4.43

Schematic drawing of power flow in a sucker-rod pumping installation.

where:

η_{system} = overall energy efficiency of the pumping system, –,
P_{hydr} = hydraulic power used for fluid lifting, HP, and
P_e = electrical power input at the motor's terminals, HP.

As seen from this formula, the system efficiency can only be found if a reliable estimate for the **useful** hydraulic power developed by the pumping system is known. The most accurate way to find the hydraulic power is provided by a **measured** downhole pump **card**, but this is seldom available. As detailed later, there is some **confusion** in the industry about the calculation of the hydraulic power and available models exhibit many discrepancies, to be discussed in the following.

4.7.2 HYDRAULIC POWER CALCULATIONS

4.7.2.1 Previous models

In general, the **power** required to move fluids through a pump is found from the volumetric **rate** of the fluid pumped and the pressure **increase** developed by the pump. Using oilfield units, the following equation can be developed:

$$P = 1.7\text{E-}5\ Q\ \Delta p \tag{4.142}$$

where:

P = power requirement, HP,
Q = volumetric pumping rate, bpd, and
Δp = pressure increase through pump, psi.

When applying this formula to sucker-rod pumps, it is customary to use the liquid rate, Q, actually produced and **measured** at the surface. In this fashion, the pump's volumetric losses are automatically accounted for, since the measured rate includes the effects of the following volumetric losses along the flow path:

- Improper pump **fillage** due to gas interference or insufficient inflow from the well.
- **Leakage** losses in the barrel–plunger clearance, as well as in the pump's valves, due to mechanical wear.
- Leakage in tubing and flowline decreases the liquid amount produced by the pump.

It should be noted that all subsequent calculations assume that **incompressible** liquids are produced by the sucker-rod pump. The presence of considerable amounts of free gas in the tubing and the annulus can greatly modify the pressure conditions, and power calculations must be modified accordingly.

As pointed out by **Lea and Minissale** [99,100] the technical literature quite consistently assumes that the sucker-rod pump's useful pressure **increase**, Δp_u, equals the hydrostatic pressure calculated from the "**net lift,**" the depth of the dynamic liquid **level** measured from the surface in the well's annulus. Thus, Eq. (4.142) takes the form:

$$P_{\text{hydr}} = 7.36\text{E-6}\ Q\ SpGr\ L_{\text{dyn}} \tag{4.143}$$

where:

P_{hydr} = hydraulic power used for lifting, HP,
Q = liquid production rate, bpd,
$SpGr$ = specific gravity of the produced liquid, –, and
L_{dyn} = dynamic liquid level in the well's annulus, ft.

The power calculated from this formula includes the power required to overcome the surface **wellhead** pressure and, therefore, increases with an increase in wellhead pressure, as proved by **Lea et al.** [101]. Calculations performed on the same well producing the same liquid rate will obviously give **different** values of hydraulic power, giving different system **efficiencies** under the same conditions. The formula also prevents the comparison of the overall efficiency of different pumping wells, making it of **dubious** value. In order to properly compare different pumping conditions, a **standardized** formula for hydraulic power and overall system efficiency calculations is desirable.

4.7.2.2 The proper calculation model

The sucker-rod pump exercises its **useful** work against the Earth's **gravity** by lifting a given amount of liquid from the pump setting depth to the surface. The mechanical work against the gravitational force must overcome the hydrostatic pressure of the liquid column present in the tubing string. Since the pump's suction pressure (the pump intake pressure, PIP) helps the pump to lift the liquid, the useful pressure increase, Δp_u, developed by the pump equals:

$$\Delta p_u = 0.433\ SpGr_t\ L_{\text{pump}} - PIP \tag{4.144}$$

where:

Δp_u = useful pressure increase in pump, psi,

$SpGr_t$ = specific gravity of the produced liquid in the tubing, –,
L_{pump} = pump setting depth, ft, and
PIP = pump intake pressure, psi.

It should be noted that the pump's discharge pressure, p_d, of course, is greater than the liquid's hydrostatic pressure because the pump must overcome the wellhead pressure plus all possible hydraulic losses arising in the tubing string. The pressure and energy losses occurring in the tubing–rod string annulus, as discussed before, cannot readily be calculated and an estimation of their magnitude is possible only through the evaluation of the dynamometer card. However, since the powers used against the wellhead pressure and against the hydraulic losses are considered **wasted**, they must not be included in the calculation of the useful power.

Finally, the pump's **useful** power output is found from substituting this formula into Eq. (4.142), and the following final equation, recommended by **Lea et al.** [101], is derived:

$$P_{hydr} = 1.7\text{E-}5\, Q\left[0.433\, SpGr_t\, L_{pump} - PIP\right] \quad (4.145)$$

where:

P_{hydr} = hydraulic power used for lifting, HP,
Q = liquid production rate, bpd,
$SpGr_t$ = specific gravity of the produced liquid in the tubing, –,
L_{pump} = pump setting depth, ft, and
PIP = pump intake pressure, psi.

The pump intake pressure, *PIP*, is calculated from the following formula with the liquid specific gravity valid in the annulus, $SpGr_a$:

$$PIP = p_{wh} + \Delta p_g + 0.433\, SpGr_a \left(L_{pump} - L_{dyn}\right) \quad (4.146)$$

where:

PIP = pump intake pressure, psi,
p_{wh} = wellhead pressure, psi,
Δp_g = gas column pressure increase in annulus, psi,
$SpGr_a$ = specific gravity of the annulus liquid, –,
L_{pump} = pump setting depth, ft, and
L_{dyn} = dynamic liquid level in the well's annulus, ft.

As can be easily observed, contrary to previous models, Eq. (4.145) excludes the power **wasted** for overcoming the wellhead pressure and all hydraulic losses occurring in the well. Therefore, it represents the possible **minimum** power required to lift well fluids to the surface. Since its value is **constant** as long as the pump intake pressure is constant, it provides a **standard** way to compare the energy efficiency of the same system under different conditions or the efficiencies of **different** pumping systems. Because of its beneficial features, the general application of this equation for calculating the power efficiency of sucker-rod pumping systems is recommended.

EXAMPLE 4.22: THE PUMP IS SET AT 6,000 FT IN A 6,500-FT-DEEP WELL, THE MEASURED LIQUID RATE IS 500 BPD WITH A WATER–OIL RATIO OF *WOR* = 3. AT 200 PSI WELLHEAD PRESSURE THE DYNAMIC LIQUID LEVEL WAS MEASURED AT 4,500 FT, OIL AND WATER SPECIFIC GRAVITIES ARE 0.85 AND 1.03, RESPECTIVELY. FIND THE SYSTEM'S HYDRAULIC POWER AT THE ORIGINAL WELLHEAD PRESSURE AND AT 400 PSI, IF THE SYSTEM'S LIFTING CAPACITY IS NOT ALTERED. STATIC GAS COLUMN PRESSURE CALCULATIONS RESULTED IN GRADIENTS OF 5 PSI/1,000 FT AND 6 PSI/ 1,000 FT FOR THE TWO CASES

Solution

First find the liquid specific gravities in the annulus and tubing:

$SpGr_a = 0.85$, since the annulus contains oil above the pump, due to gravitational separation.

$$SpGr_t = SpGr_o/(1+WOR) + SpGr_w\ WOR/(1+WOR) = 0.85/(1+3) + 1.03 \times 3/(1+3) = 0.985.$$

Now the pump's intake pressure, *PIP*, at 200 psi wellhead pressure can be found from Eq. (4.146):

$$PIP = 200 + 5 \times 4,500/1,000 + 0.433 \times 0.85(6,000 - 4,500) = 774.6 \text{ psi}.$$

Because the pump operates with the same settings, the flowing bottomhole pressure, consequently the *PIP* does not change for the other wellhead pressure of 400 psi. *PIP* being fixed, the new dynamic liquid level is found from Eq. (4.146), after expressing Δp_g with the gas gradient:

$$L_{dyn2} = \frac{p_{wh} + 0.433\ SpGr_a\ L_{pump} - PIP}{0.433\ SpGr_a - grad_g} = \frac{400 + 0.433 \times 0.85 \times 6,000 - 774.5}{0.433 \times 0.85 - 0.006} = 5,065 \text{ ft}$$

The two liquid levels known, the hydraulic powers according to the conventional formula can be found from Eq. (4.143):

$$P_{hydr1} = 7.36\text{E-}6 \times 500 \times 0.985 \times 4,500 = 16.3 \text{ HP, and}$$

$$P_{hydr2} = 7.36\text{E-}6 \times 500 \times 0.985 \times 5,065 = 18.4 \text{ HP}.$$

The proper expression yields a single value for both cases, as calculated from Eq. (4.145):

$$P_{hydr3} = 1.7\text{E-}5 \times 500(0.433 \times 0.985 \times 6,000 - 774.6) = 15.2 \text{ HP}.$$

As seen, the old model estimated an increase in hydraulic power for a higher wellhead pressure and therefore, in contrast to the proposed model, cannot be used for comparison.

4.7.3 ENERGY LOSSES

4.7.3.1 Downhole losses

The sources of downhole energy losses are the **pump**, the rod **string**, and the **liquid** column in the tubing string where irreversible mechanical as well as hydraulic losses take place.

Pump Losses

- Mechanical **friction** between the sucker-rod pump's barrel and plunger is usually unknown and can only be estimated.
- **Hydraulic** losses in improperly sized valves, especially when pumping highly viscous crudes, can increase downhole losses.

Losses in the Rod String

- Mechanical **friction** takes place wherever the rod string, while reciprocating in the tubing, rubs against the tubing inside wall, significantly increasing the energy losses in highly deviated or crooked wells or in wells experiencing rod **buckling.** The magnitude and severity of these frictional forces cannot accurately be determined; only estimates based on the well's inclination data can be made.
- Mechanical **friction** in the stuffing box is usually minimal, but extreme conditions (a dried-out or too-tight stuffing box) may increase its magnitude.

Losses in the Liquid Column

- Fluid **friction** in the tubing–rod annulus adds to the irreversible losses because the pump action must overcome the resulting pressure differential between the pump setting depth and the wellhead. Since **transient** flow in an **eccentric** annulus is involved and the size of the annulus changes with depth in wells with tapered rod strings, accurate calculation of the frictional pressure drop as well as the energy losses is practically impossible.
- **Wellhead** pressure imposes an additional power loss on the downhole pump that, by nature, cannot be considered a part of the useful work performed on the well fluid.
- **Damping** forces oppose the movement of the rod string because well fluids impart a viscous force on the rods' outside surface.

4.7.3.2 Surface losses

On the surface, energy losses occur at several places from the polished rod to the prime mover's electrical connections. These can be classified according to their occurrence as **mechanical** losses in the drive train (pumping unit, gearbox and V-belt drive) and losses in the **prime mover**.

Losses in the Drive Train

- Mechanical friction in the pumping unit's structural **bearings** is usually very low, provided the unit is properly maintained.
- Mechanical friction in the **gearbox** occurs between well-lubricated gear surfaces; therefore power losses in the gearbox are usually low.
- Mechanical friction in **V-belts** and sheaves causes a minimal power loss if the right size sheaves with the proper number and tightness of V-belts are used.

Prime Mover Losses

- Mechanical losses due to friction occur in the structural **bearings** of the electric motor.
- Windage loss is consumed by the **cooling** air surrounding the motor's rotating parts.
- Electrical losses include **iron** (or core) and **copper** losses, of which the decisive one is copper loss, resulting in the heating of the motor, and is proportional to the square of the current drawn.

4.7.4 POWER EFFICIENCY OF THE PUMPING SYSTEM

In systems where energy losses of different nature in series-connected system components are involved, the system's total efficiency can be broken down into **individual** efficiencies representing the components of the system. Total or **overall** system efficiency is then calculated as the **product** of the constituting efficiency items. In our case, one would have to assign separate efficiency figures to all or many of the **individual** kinds of energy losses detailed before. In this approach, it is necessary to designate efficiencies for the effects of the rod–tubing friction, the fluid friction in tubing, etc. However, as was discussed before, **most** of the individual energy losses in the pumping system are **difficult** or even **impossible** to predict, making this solution of questionable value.

A more workable solution classifies energy losses according to their place of occurrence and utilizes two or three individual efficiencies for the description of the system's total energy efficiency [102,103]. As a natural choice, one item is assigned to describe the sum of all **subsurface** losses, with one or two additional items representing **surface** energy losses. This approach not only provides a more **reliable** solution for the determination of the rod-pumping system's energy efficiency but allows one to identify the possible ways to increase the system's total effectiveness, as will be shown later.

4.7.4.1 Lifting efficiency

The mechanical energy required to operate the polished rod at the surface is the sum of the **useful** work performed by the pump and all the downhole energy **losses** detailed previously, i.e., those occurring in the sucker-rod pump, the rod string, and the fluid column. The amount of this work is directly proportional to the power required at the polished rod, the so-called **polished rod power** (*PRHP*), a basic pumping parameter. It represents the mechanical power exerted on the polished rod and can be found in several ways. The most reliable solution involves taking a dynamometer card and performing calculations based on the area of the card. If a dynamometer card is not available, as in the case of designing a new installation, the **RP 11L** procedure discussed earlier (now published as **API TP 11L** [9]) can be used for conventional pumping units. However, the solution of the damped wave equation provides good estimates for cases using any kind of pumping unit geometry.

Based on the above considerations, the energy efficiency of the **downhole** components of the pumping system is characterized by the relative amount of energy losses in the well. This parameter is called the lifting efficiency, η_{lift}, and is the quotient of the **useful** hydraulic power and the power required at the polished rod:

$$\eta_{lift} = \frac{P_{hydr}}{PRHP} \tag{4.147}$$

where:

η_{lift} = lifting efficiency, –,
P_{hydr} = hydraulic power used for fluid lifting, HP, and
$PRHP$ = polished rod power required at the surface, HP.

The use of the lifting efficiency eliminates the need to assign **individual** efficiencies of mostly **dubious** value to each particular kind of downhole loss, since it includes the effects of them **all**. In cases where polished rod power is found from a **measured** dynamometer card, the lifting efficiency represents the true energy **effectiveness** of fluid lifting in the well. If a new installation is designed, a reliable estimate of the predicted polished rod power provided by either the **API RP 11L** procedure or by a wave analysis program can serve the same goal.

Actual values of lifting efficiency can vary in very **broad** ranges. At the lower end of possible values, consider the case of a **worn-out** pump producing a very low amount of liquid achieving a negligible hydraulic power, P_{hydr}, while still consuming a definite power at the polished rod, adding up to a lifting efficiency value of almost **nil**. On the other hand, wells with big-size pumps and low pumping speeds can require little more than the hydraulic power at the polished rod under ideal conditions. **Lea et al.** [101] give estimates of lifting efficiencies between 95% and 70%, **Kilgore et al.** [103] present measured values of 85–70% "for well designed systems."

4.7.4.2 Surface mechanical efficiency

Mechanical energy losses occurring in the drive train cover **frictional** losses arising in the **pumping unit**, in the **gearbox**, and in the **V-belt drive**. Due to their effects, the mechanical power required at the prime mover's shaft, P_{mot}, is always greater than the polished rod power, *PRHP*. It is customary to describe these losses by a single mechanical efficiency, as given below:

$$\eta_{mech} = \frac{PRHP}{P_{mot}} \tag{4.148}$$

where:

η_{mech} = mechanical efficiency of the surface drive train, –,
P_{mot} = mechanical power required at the motor shaft, HP, and
$PRHP$ = polished rod power required at the surface, HP.

For the estimation of the pumping unit's overall mechanical efficiency, **Gipson and Swaim** [78] present the correlation shown in Fig. 4.44. The curves, valid for **new** and **worn** units, are both highly

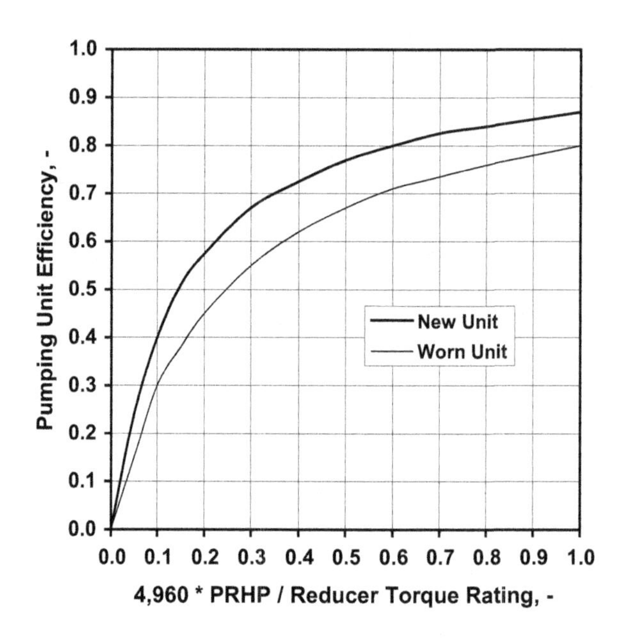

FIGURE 4.44

Mechanical efficiency factor values for pumping units.

After ***Gipson and Swaim*** *[78].*

affected by the average torque **load** on the gearbox and efficiencies improve as gearbox loading approaches the rated capacity of the unit. As found by **Bommer and Podio** [104], the figure gives unrealistically low efficiencies because it was developed for a high pumping speed of 20 SPM. According to other sources, average values of surface mechanical efficiency are high, usually over 90% in favorable conditions [101,102]. Also there is a consensus in the technical literature that efficiencies increase as gearbox **loading** increases.

4.7.4.3 Motor efficiency

To represent all losses in the motor, an **overall** efficiency factor can be used that allows the calculation of the electric power drawn from the power supply, based on the mechanical power at the motor's shaft:

$$\eta_{mot} = \frac{P_{mot}}{P_e} \tag{4.149}$$

where:

η_{mot} = overall efficiency of the electric motor, –,
P_e = required electrical power input, HP, and
P_{mot} = mechanical power required at the motor shaft, HP.

Although electric motors used in pumping service may have **full load** efficiencies close to 90% under **steady** loads, because of the motor's **cyclic** loading during pumping, actual values belong to load ranges between 30% and 80%. **Lea et al.** [101] present motor efficiencies of 78–91% for NEMA D motors of 5–60 HP sizes.

4.7.4.4 System efficiency

The rod pumping system's energy efficiency, defined in Eq. (4.141), can now be written in terms of the individual efficiencies just discussed, as follows:

$$\eta_{system} = \eta_{lift}\, \eta_{mech}\, \eta_{mot} \tag{4.150}$$

where:

η_{system} = overall efficiency of the pumping system, –,
η_{lift} = lifting efficiency, –,
η_{mech} = mechanical efficiency of the surface drive train, –, and
η_{mot} = overall efficiency of the electric motor, –.

An investigation of this formula allows one to draw important **conclusions** on the possible ways of attaining **maximum** energy efficiencies in rod pumping. To do so, the relative **importance** and the usual parameter **ranges** of the individual terms must be analyzed. Of the three parameters figuring in the equation, the possible values of the surface mechanical efficiency, η_{mech}, and the motor efficiency, η_{mot}, vary in quite **narrow** ranges. At the same time, their values can be **maximized** if the right size of gearbox and electric motor are selected. As shown before, a properly maintained pumping unit with a gearbox operated near its torque capacity ensures mechanical efficiencies greater than $\eta_{mech} = 90\%$. A properly selected electric motor can also provide relatively high η_{mot} values. Thus the

combined efficiency of the drive train and the motor can lie in the range of 70–82%, as supported by **Lea et al.** [101].

In contrast to the usual ranges of the above efficiencies, lifting efficiency, η_{lift}, can vary in very **broad** ranges depending on the **pumping mode** (plunger size, stroke length, and pumping speed) selected. For example, **Takacs** [105] reports lifting efficiencies between 94% and 38% when producing 500 bpd from 6,000 ft with different pumping modes. As supported by **Gault** [106], considerable improvements on lifting efficiencies can be realized by selecting the optimum pumping mode.

In summary, the basic requirement for achieving high overall system efficiency is to find the **maximum** possible value of the **lifting efficiency**. Since this is accomplished by the proper selection of the **pumping mode**, the choice of the right combination of pump size, polished rod stroke length, and pumping speed is of prime importance. When designing a new pumping system or improving the performance of an existing installation, this must be the primary goal of the rod-pumping specialist's efforts.

EXAMPLE 4.23: A 1 1/4 IN PUMP IS SET AT 4,329 FT AND THE DYNAMIC LIQUID LEVEL IS AT 1,449 FT. WITH A POLISHED ROD STROKE LENGTH OF 99.8 IN AND A PUMPING SPEED OF 9.4 SPM THE PUMP PRODUCES 170 BPD OF LIQUID. OIL AND WATER SPECIFIC GRAVITIES ARE 0.82 AND 1.02, RESPECTIVELY, AND THE WELL PRODUCES WITH $WOR = 0.8$. GAS PRODUCTION IS NEGLIGIBLE AND THE SURFACE PRESSURE IS 30 PSI. CALCULATE THE SYSTEM'S EFFICIENCY IF THE AVERAGE MOTOR POWER WAS MEASURED AS 6 KW, AND THE POLISHED ROD POWER WAS FOUND FROM THE DYNAMOMETER DIAGRAM AS 4.9 HP

Solution

Liquid specific gravities in the annulus and the tubing:

$$SpGr_a = 0.82, \text{ and}$$

$$SpGr_t = SpGr_o/(1 + WOR) + SpGr_w\ WOR/(1 + WOR) = 0.82/(1 + 0.8) + 1.02 \times 0.8/(1 + 0.8) = 0.91.$$

Now the pump intake pressure is found from Eq. (4.104), by assuming a static gas gradient of 0.002 psi/ft:

$$PIP = 30 + 0.002 \times 1,449 + 0.433 \times 0.82(4,329 - 1,449) = 1,055 \text{ psi}.$$

Hydraulic power is calculated from Eq. (4.145):

$$P_{hydr} = 1.7\text{E-}5 \times 170[0.433 \times 0.91 \times 4,329 - 1,055] = 1.88 \text{ HP}.$$

The system's overall power efficiency can now be found from Eq. (4.141):

$$\eta_{system} = 1.88/(6/0.746) = 0.234 = 23.4\%.$$

Also, lifting efficiency can be calculated from Eq. (4.147):

$$\eta_{lift} = 1.88/4.9 = 0.384 = 38.4\%.$$

EXAMPLE 4.24: AFTER OPTIMIZING THE PREVIOUS WELL'S OPERATION, THE SAME LIQUID PRODUCTION RATE WAS ATTAINED AT THE OPTIMUM PUMPING MODE WITH THE FOLLOWING PARAMETERS: 2 1/4 IN PUMP, A STROKE LENGTH OF 82.2 IN, AND A PUMPING SPEED OF 4 SPM. FIND THE OPTIMIZED SYSTEM'S POWER EFFICIENCY IF CALCULATED POLISHED ROD POWER IS 2 HP

Solution

The new lifting efficiency is found from Eq. (4.147) as:

$$\eta_{lift} = 1.88/2 = 0.94 = 94\%.$$

We can assume that the surface efficiency of the system remains constant and can be found from Eq. (4.150) using the data of the previous example, as follows:

$$\eta_{mech}\ \eta_{mot} = 0.234/0.384 = 0.611 = 60.9\%.$$

System efficiency for the optimum pumping mode is found from Eq. (4.150):

$$\eta_{system} = 0.94 \times 0.609 = 0.572 = 57.2\%.$$

As seen, the use of the optimum pumping mode has increased the pumping system's efficiency from 23.4% to 57.2%.

4.7.5 PRIME MOVER SELECTION

4.7.5.1 Selection of motor size

Given the value of the surface **mechanical** efficiency, the mechanical **power** required at the prime mover's shaft is easily found from the polished rod power; solving Eq. (4.148) for motor power:

$$P_{mot} = \frac{PRHP}{\eta_{mech}} \tag{4.151}$$

where:

P_{mot} = mechanical power required at the motor shaft, HP,
$PRHP$ = polished rod power, HP, and
η_{mech} = mechanical efficiency of the pumping unit, gearbox, and V-belt drive, –.

If the power demand on the motor shaft were **steady** over the pumping cycle, a motor with the above power rating would be sufficient to operate the pumping unit. The energy requirement of pumping, however, is always **cyclic** in nature because even in ideally counterbalanced cases the **fluctuations** in net gearbox torque cannot be eliminated completely. Thus, the mechanical load on the motor shaft is also fluctuating and the mechanical power P_{mot} calculated from Eq. (4.151) represents only the **average** power demand over the complete cycle. Consequently, electric motors sized for average conditions would be **overloaded** and fail prematurely.

Electric motors are designed for operation under **steady** loads and are rated based on a permissible **temperature rise**. A constant load on the motor's shaft draws a constant **current** from the electric supply. The heat generated in the motor due to this current is proportional to the thermal current, calculated as the **root mean square** (rms) value of the actual current. The motor's structural design allows for the **dissipation** of the heat generated under the rated conditions without overheating the motor above the permissible temperature. In case a variable load is applied to a motor rated to the average mechanical power required over the pumping cycle, P_{mot}, the heating effect is always **higher** than under a constant load. This is because the root mean square value of any **variable** current is inevitably **higher** than the simple average. The motor will **overheat** and eventually burn.

These facts have long been recognized [107], and electric motors intended for cyclic service are **oversized** (derated) by a **cyclic load factor** (*CLF*), introduced earlier and reproduced below:

$$CLF = \frac{\text{rms current}}{\text{avg. current}} \tag{4.152}$$

The value of the *CLF* for **unsteady** loads is always **greater** than unity, since, by definition, the rms current is greater than the average current. It is also easy to see that *CLF* is a good **indicator** of the severity of load **fluctuations** on the motor shaft. A steady load produces a value of $CLF = 1$, and the more uneven the load is, the greater the *CLF* value becomes.

The disadvantage of using the cyclic load factor as defined above is that its calculation requires the knowledge of the instantaneous motor **currents** during the pumping cycle. Since these are usually not available at the time of installation design, a **mechanical** *CLF*, based on the variation of net reducer **torques** (see Eq. 4.131), is generally used in lieu of the **electrical** *CLF*. This substitution is justified by the fact that electric motors have a **linear** relationship between torque load and the current drawn. Therefore, electrical and mechanical cyclic load factors are practically **identical** under identical conditions, as proved by **Byrd** [108]. Estimated average *CLF* values for different prime movers and pumping unit geometries, as suggested by **Lufkin Industries**, are displayed in Table 4.5.

Table 4.5 Typical CLF Values for NEMA C and D Electric Motors and Low-Speed and High-Speed Engines Used with Different Pumping Units [3]

	Typical CLF Values	
	NEMA C	NEMA D
Pumping Unit Geometry	Low-Speed Engine	High-Speed Engine
Conventional or air-balanced	1.897	1.375
Mark II	1.517	1.100

The rated (nameplate) mechanical power of the electric motor is found if the **average** power calculated from Eq. (4.151) is multiplied by the cyclic load factor:

$$P_{np} = CLF\ P_{mot} = \frac{PRHP\ CLF}{\eta_{mech}} \tag{4.153}$$

where:

P_{np} = required nameplate power of the selected motor, HP,
CLF = cyclic load factor, –,
P_{mot} = mechanical power required at the motor shaft, HP,
$PRHP$ = polished rod power, HP, and
η_{mech} = mechanical efficiency of the pumping unit, gearbox, and V-belt drive, –.

A commonly used method, credited to **R. Gault**, simply doubles the required mechanical power at the motor to find the nameplate power of the motor to be selected:

$$P_{np} = 2\,P_{mot} \tag{4.154}$$

where:

P_{np} = required nameplate power of the selected motor, HP, and
P_{mot} = mechanical power required at the motor shaft, HP.

4.7.5.2 Selection of motor type

In addition to the required nameplate **power**, the **type** of electric motor to be used as prime mover must also be decided. The choices are the low-slip **NEMA B** and **NEMA C**, the high-slip **NEMA D**, and the **ultrahigh-slip** (UHS) special motors. Low slip means a **stiff** motor that tries to maintain its speed and draws high currents when the load on the motor shaft is increased. Motors with higher slip, in turn, respond to higher loads with a decrease of their **speed** and have a lower current demand. Consequently, they operate with a considerable **variation** in motor speed over the pumping cycle, thereby enabling the utilization of **inertial** torques of the rotating parts (crank arm, counterweights) to decrease the net torque peaks. In many cases, torque reductions can be so substantial that the use of a smaller-capacity reducer is also possible. In addition, a reduction also occurs in peak polished rod loads when electric motors with high-slip properties are used [109,110].

The proper selection of the prime mover type and size is a complex task, and only general **guidelines** can be given here. Basically, both capital and operational costs should be considered in selecting the right motor to be used. Motor **investment** costs increase with an increase of motor slip and they are especially high for UHS motors because of their special rotor construction; the following average figures (relative to the cost of a NEMA D motor) are given by **Lea and Durham** [111]:

NEMA B (low-slip)	0.75
NEMA D (high-slip)	1.00
UHS (ultrahigh-slip)	2.1–2.7

A comparison of the **operational**, i.e., electrical costs of different motor types on the same well with the same liquid production was also made by the same authors. They concluded that power costs, too, **increase** with increased motor slip because UHS motors have inevitably lower efficiencies. The motors with higher slip, however, operate with lower torsional loads on the speed reducer. In practice, the operator must decide what the decisive factor on the given installation is: the reduction of electric power consumption or the reduction of torque loading on the speed reducer. Due to the ever-increasing cost of electric power, UHS motors have lost their popularity as more emphasis is placed on energy efficiency today [31].

REFERENCES

[1] Mills KN. Factors influencing well loads combined in a new formula. PE; April 1939.
[2] Szilas AP. Production and transport of oil and gas. 2nd ed. Part A. Elsevier Publishing Co.; 1985 [Chapter 4.1].

[3] Brown KE. The technology of artificial lift methods, vol. 2a. Tulsa Oklahoma: Petroleum Publishing Company; 1980 [Chapter 2].
[4] Marsh HN. High volumetric efficiency in oil well pumping and its practical results. API Production Bulletin, No. 207; 1931.
[5] Coberly CJ. Discussion of Marsh's paper. API Production Bulletin, No. 207; 1931.
[6] Gibbs SG. Assumptions of the API rod pumping design method as related to practical applications and wave equation techniques. Paper SPE 27988 presented at the University of Tulsa centennial petroleum engineering symposium held in Tulsa, OK. August 29–31, 1994.
[7] Snyder WE, Bossert AJ. Analog computer simulation of sucker rod pumping systems. Paper SPE 587 presented at the Rocky Mountain joint regional meeting of SPE, Denver, Colorado. May 27–28, 1963.
[8] Snyder WE. How to find downhole forces and displacements. OGJ August 19, 1963:96–9.
[9] API TL 11L Design calculations for sucker-rod pumping systems (conventional units). 5th ed. Washington, D.C.: American Petroleum Institute; 2008.
[10] Electric analog study of sucker-rod pumping systems. API Drilling and Production Practice; 1968. p. 232–249.
[11] Engineer R, Knight R, Davis C. Calculator program aids sucker-rod system design and optimization. OGJ August 15, 1983;11:56–63.
[12] Herlihy JD. Hand-held computer program designs rod pumping installation. PEI October 1984:26–36.
[13] Clegg JD. Rod pump design using personal computers. Proc. 33rd Annual Southwestern petroleum short course. 1986. p. 232–242.
[14] Boone DM. Program optimizes rod string design. PEI March 1990:39–40.
[15] Takacs G, Czizlavicz L, Juratovics A. Computer processing of the API RP 11L method and investigation of its application in Hungary (in Hungarian) Koolaj es Foldgaz December 1981:353–8.
[16] Griffin FD. New API design calculations for sucker-rod pumping systems. API Drilling and Production Practice; 1968. p. 220–231.
[17] Griffin FD. An update on pumping unit sizing as recommended by API RP 11L. JCPT 1976;1:45–51.
[18] Griffin FD. Eine aktuelle Uberprufung der API-Empfehlung RP 11L zur Auslegung von Tiefpumpenantrieben (in German) Erdoel-Erdgas-Zeitschrift April 1976:121–7.
[19] Pumping unit design calculations. Lufkin Industries, Inc; 1984. Form F-989-E.
[20] Clegg JD. Reducing gas interference in rod pumped wells. WO June 1979:125–9.
[21] Clegg JD. Improved sucker rod pumping design calculations. Proc. 35th Annual Southwestern petroleum short course. 1988. p. 204–221.
[22] Jennings JW, Laine RE. A method for designing fiberglass sucker-rod strings with API RP 11L. SPE PE February 1991:115–9.
[23] Jennings JW, Laine RE. Supplement to SPE 18188: a method for designing fiberglass sucker-rod strings with API RP 11L. Paper SPE 22050. 1991 [available from the SPE].
[24] Halderson MH. Artificial Brain is required to solve the sucker-rod pumping problem. API Drilling and Production Practice; 1953. p. 210–221.
[25] Norton JR. Dynamic loads in sucker rods. PE; April 1960. p. B-33–B-41.
[26] Gibbs SG. Predicting the behavior of sucker-rod pumping systems. JPT July 1963:760–78.
[27] Gibbs SG. Computer diagnosis of down-hole conditions in sucker rod pumping wells. JPT January 1966: 91–8.
[28] Gibbs SG, Nolen KB. Wellsite diagnosis of pumping problems using minicomputers. JPT November 1973: 1319–23.
[29] Hudgins TA. The computer van system: advanced well analysis technology. Paper SPE 15303 presented at the symposium on petroleum industry application of microcomputers of the SPE, Silver Creek, Colorado. June 18–20, 1986.

[30] DaCunha JJ, Gibbs SG. Modeling a finite-length sucker rod using the semi-infinite wave equation and a proof and Gibbs' conjecture. Paper SPE 108762 presented at the Annual technical conference and exhibition held in Anaheim, CA. November 11–14, 2007.
[31] Gibbs SG. Rod pumping. Modern methods of design, diagnosis and surveillance. USA: BookMasters, Inc.; 2012.
[32] Bastian M, Keating J, Jennings JW. A method to find the viscous damping coefficient and a faster diagnostic model. Proc. 37th Annual Southwestern petroleum short course. 1990. p. 255–271.
[33] Gibbs SG. Method of determining sucker rod pump performance. US Patent No. 3,343,409. 1967.
[34] Schafer DJ, Jennings JW. An investigation of analytical and numerical sucker rod pumping mathematical models. Paper SPE 16919 presented at the 62nd Annual technical conference and exhibition of the SPE, Dallas Texas. September 27–30, 1987.
[35] Everitt TA, Jennings JW. An improved finite difference calculation of downhole dynamometer cards for sucker rod pumps. Paper SPE 18189 presented at the 63rd Annual technical conference and exhibition of the SPE, Houston Texas. October 2–5, 1988.
[36] Eickmeier JR. Applications of the Delta II dynamometer technique. JCPT April–June 1966:66–74.
[37] Eickmeier JR. Diagnostic analysis of dynamometer cards. JPT January 1967:97–106.
[38] Herbert WF. Sucker-rod pumps now analyzed with digital computer. OGJ February 21, 1966:81–5.
[39] Patton LD. A computer technique for analyzing pumping well performance. JPT March 1968:243–9.
[40] Hudgins TA. Use and application of dynamometers for surface and downhole analysis. Proc. 28th Annual Southwestern petroleum short course. 1981. p. 334–341.
[41] McCoy JN, Podio AL. Integrated well performance visualization and analysis. Paper presented at the SPE workshop on microcomputer applications in artificial lift, Long Beach, California. October 16–17, 1989.
[42] Jennings JW, McCoy JN, Drake B. A portable system for acquiring and analyzing dynamometer data. Proc. 38th Annual Southwestern petroleum short course. 1991. p. 314–323.
[43] Takacs G, Papp I. Dynamometer card analysis using minicomputers (in Hungarian) Koolaj es Foldgaz March 1978:65–73.
[44] Takacs G. A new technique for data retrieval from conventional dynamometer cards. Paper SPE 26278. 1993 [available from SPE].
[45] Svinos JG. Successful application of microcomputers to analyze sucker rod pumps. Paper SPE 17789 presented at the SPE symposium on petroleum industry applications of micro-computers, San Jose, California. June 27–29, 1988.
[46] Lea JF. Boundary conditions used with dynamic models of beam pump performance. Proc. 35th Annual Southwestern petroleum short course. 1988. p. 251–263.
[47] Laine RE, Cole DG, Jennings JW. Harmonic polished rod motion. Paper SPE 19724 presented at the 64th Annual technical conference and exhibition of the SPE, San Antonio, Texas. October 8–11, 1989.
[48] Ehimeakhe V. Comparative study of downhole cards using modified Everitt–Jennings method and Gibbs method. Proc. 57th Southwestern petroleum short course, Lubbock, Texas. 2010. p. 53–70.
[49] Pons-Ehimeakhe V. Modified Everitt–Jennings algorithm with dual iteration on the damping factors. Proc. 59th Southwestern petroleum short course, Lubbock, Texas. 2012. p. 31–47.
[50] RODDIAG user manual. Theta Oilfield Services, Inc.; 2011.
[51] Well analyzer and TWM software. Operating Manual Rev. C, Echometer Company; 2007.
[52] Gibbs SG. Design and diagnosis of deviated rod-pumped wells. J Pet Technol July 1992:774–81.
[53] Gibbs SG, Dorado DM, Nolen KB, Oestreich ES, DaCunha JJ. Apparatus for analysis and control of a reciprocating pump system by determination of a pump card. US Patent No. 8,036,829. October 2011.
[54] Xu J, Nolen K, Shipp D, Cordova A, Gibbs SG. Rod pumping deviated Wells. Proc. 52nd Annual Southwestern petroleum short course. Lubbock, Texas. 2005. p. 198–209.

[55] Xu J, Nolen K, Boyer L, Gibbs SG. Diagnostic analysis of deviated rod-pumped wells. Proc. 48th Annual Southwestern petroleum short course. Lubbock, Texas. 2001. p. 133–140.
[56] RodStar-D/V manual. 2nd ed. La Habra, CA: Theta Oilfield Services, Inc.; 2011.
[57] Laine RE, Keating JF, Jennings JW. Shallow sucker rod wells and fluid inertia. Proc. 37th Annual Southwestern petroleum short course. 1990. p. 316–340.
[58] Svinos JG. Rod pumping optimization. 5th ed. Brea, California: Theta Enterprises, Inc.; January 2003.
[59] Doty DR, Schmidt Z. An improved model for sucker-rod pumping. SPE J February 1983:33–41.
[60] Csaszar AB, Laine RE, Keating JF, Jennings JW. Sucker-rod pump diagnosis with fluid inertia considerations. Paper SPE 21663 presented at the SPE production operations symposium, Oklahoma City, Oklahoma. April 7–9, 1991.
[61] Svinos JG. Use of downhole pulsation dampener to eliminate the effects of fluid inertia on a rod pump system. Paper SPE 18779 presented at the SPE California regional meeting, Bakersfield, California. April 5–7, 1989.
[62] Subsurface pumps: selection and application. Garland, Texas: Oilwell Division of US Steel; 1978.
[63] Patterson J, Williams BJ. A progress report on "Fluid slippage in down-hole rod-drawn oil well pumps." Proc. 45th Southwestern petroleum short course. 1998. p. 180–191.
[64] Patterson J, Dittman J, Curfew J, Hill J, Brauten D, Williams B. Progress report #2 on "Fluid slippage in down-hole rod-drawn oil well pumps." Proc. 46th Southwestern petroleum short course. 1999. p. 96–106.
[65] Patterson J, Curfew J, Brock M, Brauten D, Williams B. Progress report #3 on "Fluid slippage in down-hole rod-drawn oil well pumps." Proc. 47th Southwestern petroleum short course. 2000. p. 117–136.
[66] Patterson J, Chambliss K, Rowlan L, Curfew J. Progress report #4 on "Fluid slippage in down-hole rod-drawn oil well pumps." Proc. 54th Southwestern petroleum short course. 2007. p. 45–59.
[67] Chambliss RK, Cox JC, Lea JF. Plunger slippage for rod-drawn plunger pumps. J Energy Resour Technol September 2004:208–14.
[68] Nolen KB, Gibbs SG. Quantitative determination of rod-pump leakage with dynamometer techniques. SPE PE August 1990:225–30.
[69] Rowlan OL, McCoy JN, Lea JF. Use of the pump slippage equation to design pump clearances. Proc. 59th Southwestern petroleum short course. 2012. p. 155–168.
[70] Schmidt Z, Doty DR. System analysis for sucker rod pumping. Paper SPE 15426 presented at the 61st Annual technical conference and exhibition of SPE, New Orleans, Louisiana. October 5–8, 1986.
[71] Dottore EJ. How to prevent gas lock in sucker rod pumps. Paper SPE 27010 presented at the III. Latin American/Caribbean petroleum engineering conference held in Buenos Aires, Argentina. April 27–29, 1994.
[72] Robles J, Podio AL. Effects of free gas and downhole separation efficiency on the volumetric efficiency of sucker rod pumps and progressing cavity pumps. Proc. 46th Annual Southwestern petroleum short course. 1999. p. 107–122.
[73] API spec. 11E Specification for pumping units. 18th ed. Washington, DC: American Petroleum Institute; 2008.
[74] Gray CR. Dynamometer analysis using a digitizer and an IBM PC. Paper SPE 15288 presented at the symposium on petroleum engineering application of microcomputers of SPE, Silver Creek, Colorado. June 18–20, 1985.
[75] Takacs G. Torque analysis of pumping units using dynamometer cards. Proc. 36th Annual Southwestern petroleum short course. 1989. p. 366–376.
[76] Gibbs SG. Computing gearbox torque and motor loading for beam pumping units with consideration of inertia effects. JPT September 1975:1153–9.
[77] Svinos JG. Exact kinematic analysis of pumping units. Paper SPE 12201 presented at the 58th Annual technical conference and exhibition of the SPE, San Francisco, California. December 5–8, 1983.
[78] Gipson FW, Swaim HW. The beam pumping design chain. Proc. 31st Annual Southwestern petroleum short course, Lubbock Texas. 1985. p. 296–376.

[79] Gault RH. Permissible load diagrams for pumping units. Proc. 7th Annual West Texas oil lifting short course. 1960. p. 67–71.
[80] Teel L. Permissible load envelopes for beam pumping units. Proc. 38th Annual Southwestern petroleum short course. 1991. p. 375–396.
[81] Kemler EN. Counterbalancing of oil-well pumping machines. API Drilling and Production Practice; 1943. p. 87–107.
[82] Johnson DO. Counterbalancing of beam pumping units. API Drilling and Production Practice; 1951. p. 223–241.
[83] Keating JF, West JB, Jennings JW. Lifting cost reduction from dynamic balancing. Proc. 38th Annual Southwestern petroleum short course. 1991. p. 324–337.
[84] Rowlan OL, McCoy JN, Podio AL. Best method to balance torque loadings on a pumping unit gearbox. J Can Pet Technol July 2005:27–33.
[85] Ford WH, Svinos JG. Effective application of beam pumping diagnostics. Paper SPE 17444 presented at the SPE California regional meeting, Long Beach, California. March 23–25, 1988.
[86] Chastain J. Use lead/lag to reduce torque on pumping units. OGJ October 11, 1976. p. 138, 143, 144, 147.
[87] Byrd JP. Improving the torque requirements of a crank balanced pumping unit. Paper 1017-G presented at the 4th Annual joint meeting of the Rocky Mountain petroleum sections of the AIME, Denver, Colorado. March 3–4, 1958.
[88] What operators learned from competitive pumping tests. OGJ December 12, 1960:105–9.
[89] Patton LD. Comparative energy requirements of oilfield pumping units – Conventional vs Front-Mounted. JPT January 1965:26–32.
[90] Byrd JP. The effectiveness of a special class III lever system applied to sucker rod pumping. Proc. 17th Annual Southwestern petroleum short course. 1970. p. 73–87.
[91] Durham MO, Lockerd CR. Beam pump motors: the effect of cyclical loading on optimal sizing. Paper SPE 18186 presented at the 63rd Annual technical conference and exhibition of the SPE, Houston, Texas. October 2–5, 1988.
[92] McCoy JN, Podio AL, Jennings JW, Capps KS, West J. Simplified computer-aided analysis of electrical current in motors used for beam pumping systems. Paper SPE 25447 presented at the production operations symposium held in Oklahoma City. March 21–23, 1993.
[93] Podio AL, McCoy JN, Collier F. Analysis of beam pump system efficiency from real-time measurement of motor power. Paper SPE 26969 presented at the latin American/Caribbean petroleum engineering conference held in Buenos Aires. April 27–29, 1994.
[94] McCoy JN, Podio AL, Ott RE, Rowlan L, Garrett M, Woods M. Motor power/current measurement for improving rod pump efficiencies. Paper presented at the production operations symposium held in Oklahoma City. March 9–11, 1997.
[95] McCoy JN, Ott RE, Podio AL, Collier F, Becker D. Beam pump balancing based on motor power utilization. Paper SPE 29533 presented at the production operations symposium held in Oklahoma City. April 3–4, 1995.
[96] "Total well management." program package. Wichita Falls, Texas: Echometer Company; 2012.
[97] Gibbs SG. Utility of motor-speed measurements in pumping-well analysis and control. SPEPE August 1987:199–208.
[98] Gibbs SG, Miller DL. Inferring power consumption and electrical performance from motor speed in oil-well pumping units. IEEE Trans January/February 1997;33(1):187–93.
[99] Lea JF, Minissale JD. Beam pumps surpass ESP efficiency. Oil Gas J May 18, 1992:72–5.
[100] Lea JF, Minissale JD. Efficiency of artificial lift systems. Proc. 39th Annual Southwestern petroleum short course, Lubbock, Texas. 1992. p. 314–323.
[101] Lea JF, Rowlan L, McCoy J. Artificial lift power efficiency. Proc. 46th Annual Southwestern petroleum short course, Lubbock, Texas. 1999. p. 52–63.

[102] Butlin DM. A comparison of beam and submersible pumps in small cased wells. Paper SPE 21692 presented at the production operations symposium, Oklahoma City, Oklahoma. April 7–9, 1991.
[103] Kilgore JJ, Tripp HA, Hunt CL. Walking beam pumping unit system efficiency measurements. Paper SPE 22788 presented at the 66th Annual technical conference and exhibition of SPE, Dallas, Texas. October 6–9, 1991.
[104] Bommer PM, Podio AL. The beam lift handbook. 1st ed. University of Texas at Austin; 2012.
[105] Takacs G. Program optimizes sucker-rod pumping mode. Oil Gas J October 1, 1990:84–90.
[106] Gault RH. Designing a sucker-rod pumping system for maximum efficiency. SPE PE November 1987: 284–90.
[107] Howell JK, Hogwood EE. Electrified oil production. Petroleum Publishing Co; 1962.
[108] Byrd JP, Beasley WL. Predicting prime mover requirements, power costs, and electrical demand for beam pumping units. Paper 374035 presented at the 25th Annual technical meeting of the petroleum society of CIM, Calgary. May 7–10, 1974.
[109] Chastain J. How to pump more for less with extrahigh-slip motors. OGJ March 4, 1968:62–8.
[110] Hughes JW. High slip motors reduce loading on beam pumping installations. Proc. 17th Annual Southwestern petroleum short course, Lubbock, Texas. 1970. p. 109–114.
[111] Lea JF, Durham MO. Study of the cyclical performance of beam pump motors. Paper SPE 18827 presented at the SPE production operations symposium in Oklahoma City, Oklahoma. March 13–14, 1989.

CHAPTER

THE DESIGN OF THE PUMPING SYSTEM

5

CHAPTER OUTLINE

5.1 INTRODUCTION

Since all the important calculation methods available for the determination of operational conditions in sucker-rod pumping have been examined in previous chapters, now it is possible to discuss the use of these procedures in the basic **design** of a rod pumping system. System design, as defined here, involves selecting the **proper** equipment components and determining the main **parameters** of a pumping system, in order to achieve the goals set forth by the operator and to ensure an economical fluid production.

The aim of artificial lift design is to ensure the **most economical means** of liquid production within the constraints imposed by the given well and reservoir. For sucker-rod pumping, this usually means selecting the right **size** of pumping unit and gear reducer, as well as determining the **pumping mode** to

Sucker-Rod Pumping Handbook. http://dx.doi.org/10.1016/B978-0-12-417204-3.00005-4

be used (i.e., the combination of the plunger size, stroke length, pumping speed, and rod string design). The size of the pumping unit and gear reducer can be selected only if the operating conditions (loads, torques, etc.) are known, and these vary with the different pumping modes. Therefore, the basic task of a proper design lies in the optimal determination of the **pumping mode**.

For surface pumps (e.g., mud pumps), the calculation of the required plunger size, stroke length, and speed is quite straightforward. This is because pump displacement is a **direct** function of these variables, and they can be changed at will. The situation is dramatically **different** in the case of sucker-rod pumps because downhole pump stroke **length** is far from being equal to the polished rod stroke length set at the surface. This is due to the plunger's being moved by a long, **elastic rod string**. Pump displacement cannot be determined **directly** from surface parameters, and this condition is a very basic problem of a sucker-rod system design.

Section 5.2 deals with the selection of the proper **pumping mode** and covers two basic problems. If the liquid rate to be produced from the well is **prescribed**, the basic goal of the design is to lift the given rate to the surface while ensuring **optimum** operating conditions. The possible optimization criteria are presented along with a discussion of the various concepts used to rate the performance effectiveness of the pumping system. The relation between optimum pumping conditions and the direct costs of fluid lifting is investigated also. The second problem in pumping mode selection is encountered when the desired liquid rate from the well is **not limited**. In these cases, the aim of the operator is to lift the **maximum** liquid volume within the capabilities of the given pumping system.

Another basic task in rod pumping system design is to **match** the capacity of the downhole pump to the inflow rate of the well. As discussed in Section 5.3, this problem can be solved by either **continuous** or **intermittent** pumping. If the pump is operated continuously, systems analysis methods are used to find the possible fluid rates that can be achieved from the well. Possible pumping modes are determined, and a complete design of the system can be accomplished. However, the basic requirement for applying systems analysis methodology is the proper knowledge of the inflow performance of the well.

Since the expected fluid rates from the well, in most cases, are not previously known, an exact system design is not usually possible at the time of equipment installation. Usually, a pumping system with a fluid lifting capacity in excess of the **expected** well production rates is installed, and the unit is placed on **intermittent** pumping. The operation of the pumping unit is controlled by a surface device called a **pump-off controller** (POC). The various principles and performance characteristics of POC units are detailed, with a discussion of the technical and economic benefits of their use.

5.2 PUMPING MODE SELECTION

The **pumping mode** of a sucker-rod system is defined as the actual combination of pump size, polished rod stroke length, pumping speed, and rod string design. There are several standard API pump sizes and stroke lengths from which to choose; also, pumping speed can be varied within a broad range. Since, at least theoretically, all available rod string combinations can be used with these pumping modes, a fairly great number of cases may be possible. In a given case, however, most of the theoretical pumping modes turn out to be **impractical** or **uneconomical**, but the elimination of these still leaves a multitude of options to consider [1]. Given the wide choice of pumping modes, the proper design is chosen from among the remaining modes to ensure the **best** operating conditions.

Table 5.1 Pumping Modes with the Best and Worst Power Requirements of Lifting 500 bpd from 6,000 ft with a Conventional Pumping Unit

Pumping Mode	Best	Worst
API rod no.	86	86
Pump size	2 3/4 in	1 1/4 in
Stroke length	120 in	192 in
Pumping speed	7.9 SPM	13.7 SPM
PRHP	22.8 HP	54.6 HP
Lifting efficiency	97.0%	40.5%

According to ***Gault*** *[2].*

To illustrate the importance of the proper selection of pumping mode, Table 5.1 (compiled from data published by **Gault** [2]), is presented. Table 5.1 includes the best and worst cases, based on surface power requirements, of the possible pumping modes when producing 500 bpd liquid from a depth of 6,000 ft. As seen, these two extreme cases represent two widely different pumping modes, the worst one requiring more than **two** times as much polished rod horsepower (and consequently power cost) as the other. Lifting efficiencies vary from 97% for the best case to 40.5% for the worst case. Thus, by proper selection of the pumping mode, a significant reduction in power requirements and power costs can be achieved.

In the following, the selection of the proper pumping mode is discussed for two basic cases:

- when the liquid rate to be produced from the given well is **set** beforehand, and
- when the goal of the operator is to achieve the **maximum** production rate from the installation.

While the first case involves solution of an **optimization** problem, in the second case the limitations of the actual pumping unit are the governing factors.

5.2.1 OPTIMUM PUMPING MODE

5.2.1.1 Use of API Bul. 11L3

The widely used **API RP 11L** procedure [3] cannot be directly applied to optimization of the pumping mode because pump displacement is a **result** of the calculations and not an **input** variable. If one starts from the **desired** production rate and wants to find the pumping modes that achieve that rate, a tedious trial-and-error procedure has to be followed. To ease the solution of this problem, the American Petroleum Institute published **Bul. 11L3** [4], which contains several tens of thousands of **precalculated** pumping modes for conventional pumping units. This design book lists the pumping modes that would produce the given volumes for different liquid rates, pump setting depths, and rod taper combinations. Along with the details of the pumping mode, all the parameters that can be found with the **API RP 11L** procedure are given. This feature allows one to **select** the pumping mode that is considered to be the best under the conditions at hand for producing the desired liquid rate. An example of this kind of pumping system optimization is presented by **Gault** [2].

The use of **API Bul. 11L3** tables, however, introduces inherent **errors**, some of which are limitations of the original **RP 11L** calculation model. These include such assumptions as an **anchored** tubing string, **pumped-off** conditions, and 100% pump volumetric efficiency. Some problems also arise from the way rod string design was treated during the development of these tables. Namely, pumping modes were calculated utilizing the taper **percentages** given in **API RP 11L**. But these taper lengths do not reflect the effects of stroke length and pumping speed on rod design, as previously discussed in Section 3.5.3.5. Taper lengths only change with plunger size, as seen in the table of recommended taper percentages published in succeeding editions of **API RP 11L**. Although later editions of **API RP 11L** adopted the comprehensive rod design method of **Neely** [5], this situation did not change. The tables in **API Bul. 11L3**, therefore, can contain some **inherent** errors, and their use is **limited** to anchored tubing and near-pumped-off conditions.

Tables similar to those contained in **API Bul. 11L3** were published by a major pumping unit manufacturer for conventional pumping units [6]. All of the limitations previously discussed apply to these tables as well; thus they can find a limited use in pumping mode optimization.

5.2.1.2 Performance rating concepts

Before setting up any optimization procedures for sucker-rod pumping, the various **rating** methods used to **compare** and **evaluate** the performance of different pumping systems must be discussed first. Since there are several rating indices proposed in the literature, it is easy to find the appropriate indices for the case at hand. The pioneering work of **Byrd** [7] introduced several simple performance indices that describe the structural and torque loading as well as the energy consumption of different pumping modes. It was Byrd who showed that taking one of these criteria for the basis for optimization does not necessarily result in an optimum value for a different performance index [8]. Thus, for example, if a pumping mode with the minimum structural loading is selected, the torque and power requirements are not usually at a minimum. Therefore, in order to achieve an overall optimum operation, **complex** rating systems must be used, which should include several (or most) of the important performance parameters of pumping.

The next significant contribution to pumping mode optimization was made by **Estrada** [9], who used the solution of the wave equation to find the surface operational parameters of sucker-rod pumping. He developed optimization **tables** for **Mark II** pumping units and proposed the use of an **economic index** (*EI*) to rate the effectiveness of different pumping modes. The economic index includes the values of *PPRL, PT, PRHP,* and lifting efficiency into a single formula, and assigns the same importance to each of these:

$$EI = 10^{-7}\frac{PPRL\ PT\ PRHP}{\eta_{\text{lift}}} \tag{5.1}$$

where:

EI = economic index, -
$PPRL$ = peak polished rod load, lb
PT = peak net torque on the speed reducer, in lb
$PRHP$ = polished rod horsepower, HP
η_{lift} = lifting efficiency, %

After **sorting**, by ascending *EI* values, all the possible pumping modes that ensure the production of the desired rate, the mode with the **lowest** *EI* value is selected. In case this pumping mode is not feasible for any reason, the mode with the **next**-lowest economic index is selected. Application of the *EI* concept, according to **Byrd** [8], not only assures minimal **investment** costs but a minimum of **operating** costs as well.

Based on the work of Estrada and Byrd, **Lekia** [10] introduced a new parameter to evaluate the overall economy of a pumping system, the **performance index** (*PIX*), as defined below:

$$PIX = 10^{-9} \frac{PPRL\ PT\ P_{np}\ CLF}{ITE\ \eta_{lift}} \tag{5.2}$$

where:

PIX = performance index, -
$PPRL$ = peak polished rod load, lb
PT = peak net torque on the speed reducer, in lb
P_{np} = name-plate power of prime mover, HP
CLF = cyclic load factor, -
η_{lift} = lifting efficiency, %
ITE = index of torsional effectiveness, %
$ITE = 6.3 \times 10^6 \frac{PRHP}{N\ PT}$

where:

$PRHP$ = polished rod horsepower, HP
N = pumping speed, 1/min
PT = peak net torque on the speed reducer, in lb

As before, the **lowest** *PIX* value assures **maximum** economy of rod pumping.

A major contribution to the problem of pumping mode selection came from **Byrd** [1], who proposed a very comprehensive rating system. The **performance effectiveness rating system** considers most of the dominant factors of sucker-rod pumping and rates pumping modes according to their performance effectiveness (PE) values. The PE value for a given pumping mode is found from a **complex** formula not reproduced here; the larger the PE, the more effective the pumping system. Unfortunately, this method requires the use of optimization tables that are not yet available.

5.2.1.3 Maximizing the lifting efficiency

An investigation of the rating principles just presented shows that all methods assign a high **importance** to the lifting efficiency and tend to **maximize** it. Therefore, an easy-to-use and reliable optimization procedure can be based on the principle of **maximizing** the lifting efficiency [11], defined here:

$$\eta_{lift} = \frac{P_{hydr}}{PRHP} \tag{5.3}$$

where:

η_{lift} = lifting efficiency, -
P_{hydr} = hydraulic power required for fluid lifting, HP
$PRHP$ = polished rod horsepower, HP

This requirement coincides with the case of setting the polished rod horsepower (*PRHP*) to be a **minimum**. This is because lifting a given liquid volume from a given pump setting depth, i.e., for a given hydraulic power, lifting efficiency and *PRHP* are **inversely** proportional. The pumping mode determined with this principle needs the **least** amount of prime mover power, as the system's total energy requirement is a direct function of *PRHP*. Application of this optimization concept, therefore, gives the most **energy-efficient** and thus most **economical** pumping mode for the production of the required liquid rate from the given pump setting depth. As shown by **Gault** [2] also, a pumping system design utilizing this principle results in minimum operational costs and in a maximum of system efficiency.

5.2.1.3.1 Calculation procedure

This section presents a detailed description of an optimization procedure based on this rating principle. This model is valid for **conventional** pumping units only, since it utilizes the **API RP 11L** [3] calculation procedure; the original calculations, however, are modified to include the enhancements detailed in Section 4.3.6. Pumping of liquid only is assumed; and the downhole pump is assumed to completely fill during every pumping cycle.

To use the **RP 11L** technique for the calculation of sucker-rod pumping parameters, the mechanical properties of the rod string in use have to be known. Because these properties change with different **taper** percentages, the rod string must be **designed** first. But all the exact rod string design procedures require plunger size, stroke length, and pumping speed (which are about to be determined) as input variables. The selection of pumping modes, therefore, can be achieved only by an **iterative** scheme. This is not the case with **Bul. 11L3**, in which taper percentages recommended by API were used, eliminating the need for iterations.

The optimization method about to be detailed accounts for the **iterative** nature of pumping mode selection. The rod string is designed **throughout** the calculation process, thus ensuring a more exact solution of the problem. This approach reduces errors that may be present in **API Bul. 11L3** optimization tables under such conditions in which API taper lengths (used for the development of those tables) and rod percentages actually calculated **differ** considerably.

The details of the optimization method are described by the flowcharts in Figs 5.1 and 5.2. Basic input data are the following:

- the unit's API designation,
- available polished rod stroke length values,
- range of available pumping speeds,
- fluid properties,
- wellhead pressure,
- whether the tubing string is anchored or not,
- rod string data (API rod number, type of rods (coupled or continuous), service factor to be used), and
- estimated pump volumetric efficiency.

After these data, the desired production data are entered:

- desired liquid production rate, q_D,
- pump setting depth, L, and
- dynamic liquid level, L_D.

The calculations start with the selection of the smallest possible plunger **diameter**, d, and the first polished rod stroke **length**, S (see Fig.5.1(a)). Now only the pumping **speed** that will produce the desired volume has to be determined; this value is calculated in a subroutine that is detailed

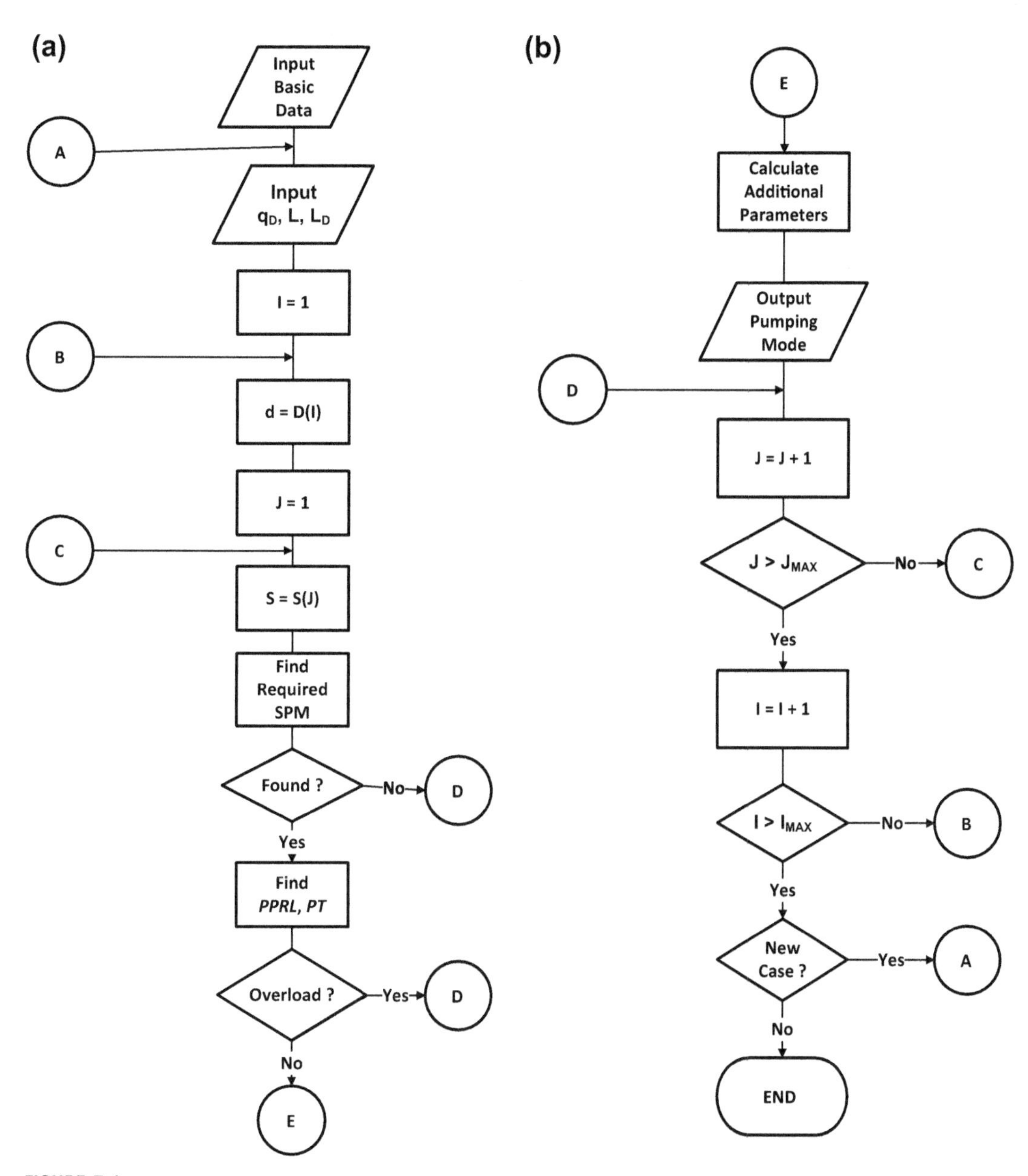

FIGURE 5.1

Flowchart for finding the pumping mode with maximum lifting efficiency.

*According to **Takacs** [11].*

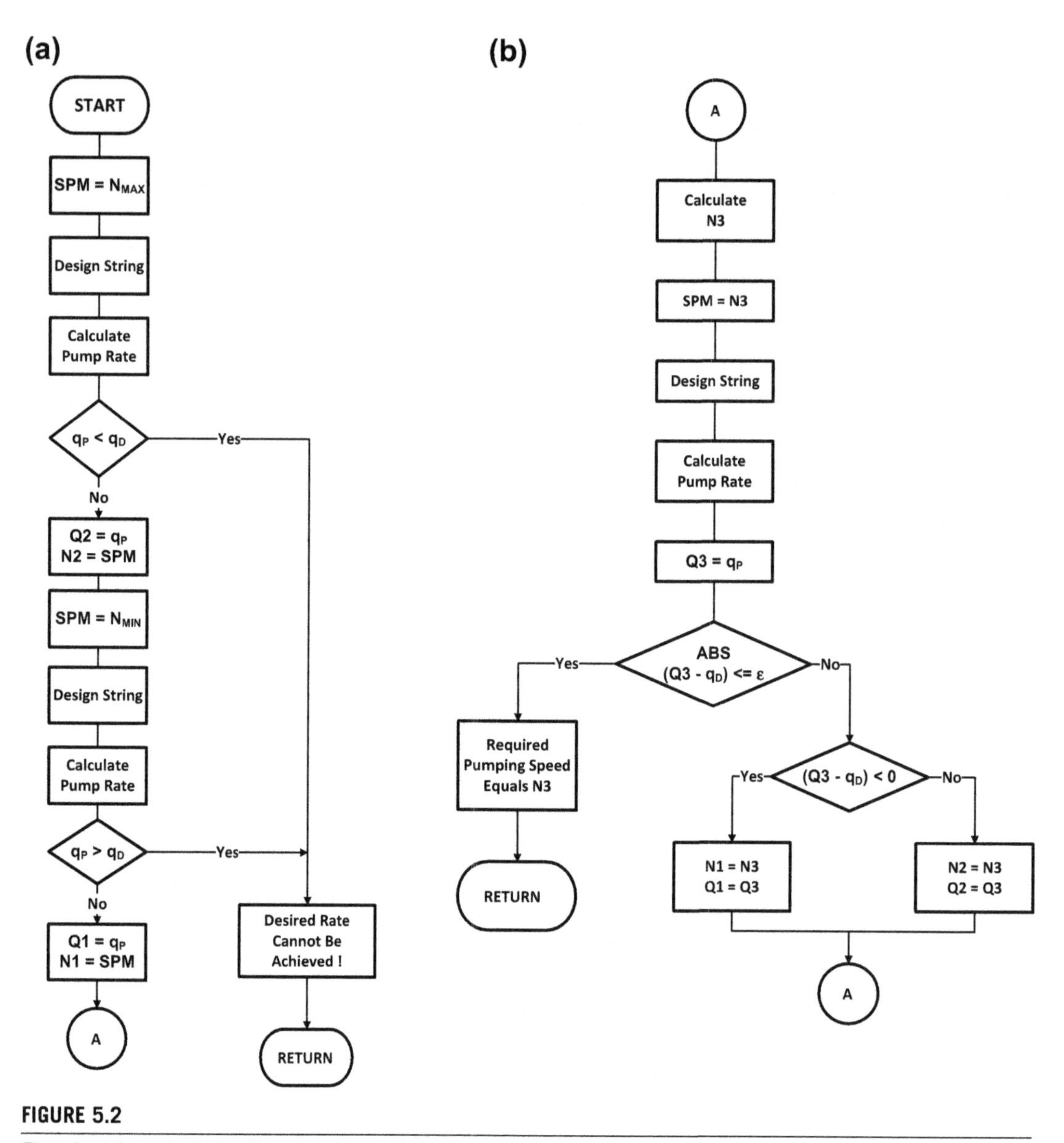

FIGURE 5.2

Flowchart for calculating the pumping speed required to produce the desired liquid rate with a given plunger size and stroke length.

According to ***Takacs*** *[11].*

later.Assume that this pumping speed is already known, which means that one pumping mode has already been found. The next step is to check the pumping unit for **overload** conditions. In case thepeak polished rod load (*PPRL*) is below the unit's structural capacity, and the resulting peaknet-torque(*PT*) does not exceed the rating of the speed reducer, one valid pumping mode has beenfound.

This calculation scheme gives one **combination** of plunger size, stroke length, and pumping speed that assures the production of the desired liquid rate. Further operational parameters are calculated, such as *PRHP*, lifting efficiency, etc. (see Fig. 5.1(b)). These parameters are then output, and another pumping mode is sought. The next value of polished-rod stroke length, *S*, is chosen, and the calculations are repeated. When all values of the pumping unit's available stroke lengths have been used, the next **plunger** size, *d*, is selected and the whole procedure repeated.

The operation of the subroutine in this optimization method is illustrated in the flowcharts in Fig. 5.2. These calculations find the pumping **speed** needed to achieve the production **rate** desired, given the plunger size and the surface stroke length. In the first part of the procedure (Fig. 5.2(a)), checks are made to ensure that the desired liquid rate, q_D, falls between the maximum and minimum pump displacements attainable with the given pumping unit. For this reason, pump displacement, q_P, is calculated for the lowest and highest values of available pumping speeds. If these pump displacements **bracket** the desired volume, an iterative calculation follows that determines the necessary pumping speed.

Figure 5.2(b) illustrates the principle of the iteration method used. The "**regula falsi**" type of solution is applied to find the required pumping **speed**, *N*3, starting from the given two points of the pump displacement–pumping speed function. With *N*3 known, the rod string can be **designed** already, since every variable that affects string design is known at this point: plunger size, stroke length, pumping speed, etc. Taper lengths having been determined by the method of **Neely** [5], the **RP 11L** procedure can be used to find pump displacement, q_P, that corresponds to the pumping speed calculated before.

If the pump displacement valid for pumping speed *N*3 equals the **desired** production rate, q_D, the required pumping speed, and accordingly the required pumping mode, is found. Otherwise, the iteration method is repeated until the necessary pumping speed is determined.

At the end of the calculations described, those pumping modes will be available that **ensure** the pumping of the required liquid rate from the given well yet do not result in **overloaded** conditions using the given pumping unit and gear reducer. From these pumping modes, the **optimal** one with the highest value of the lifting efficiency η_{lift} can be found. Application of this pumping mode results in the least amount of electric power **usage** and power **costs** as well. In case this mode is not feasible for some reason (e.g., restriction on tubing size), another one is selected that has the next-highest lifting efficiency value.

Two major design problems can be solved with the use of the above procedure:

1. Given the pumping unit and speed reducer, the **optimum** pumping mode can be selected. In this case the number of possible pumping modes is limited by the constraints imposed by the pumping unit and speed reducer, such as maximum allowed structural capacity and torque rating.
2. The **selection** of pumping unit and speed reducer sizes also can be accomplished. If, during the optimization process, the constraining parameters are replaced by their maximum possible values, a much wider range of available pumping modes will result. After the mode with the highest lifting efficiency is chosen, the proper size of the required equipment is easily found from the calculation data.

EXAMPLE 5.1: SELECT THE OPTIMUM PUMPING MODE FOR THE FOLLOWING CASE (SAME DATA AS THOSE USED BY GAULT [2]):

Pump setting depth (equal to dynamic liquid level) = 6,000 ft.

Desired liquid production = 500 bpd.

Pump volumetric efficiency = 100%.

Sp.G. of produced fluid = 1.0 (water).

Use different API tapers composed of coupled API **Grade D** sucker rods with a service factor of SF = 1.0. The tubing string is anchored.

Solution

To facilitate the selection of pumping unit and gear reducer sizes for the job, calculations were preformed with maximum limits set to the torque rating and structural capacity values. Of the results, only one sheet containing part of the calculated data for API **taper 85** is given in Exhibit 5.1.

Plunger Size	Stroke Length	Pumping Speed	Rod Size	String Length	String Loading	Polished Rod Loads Maximum	Polished Rod Loads Loading	Polished Rod Loads Minimum	Peak Net Torque		P.R. Power	Lifting Eff.
[in]	[in]	[SPM]	[in]	[ft]	[%]	[lbs]	[%]	[lbs]	[in-lbs]	[%]	[HP]	[%]
1.25	169.82	15.4*	1	1043.9	79.2	21357	70	1551	875491	96	48.5	45.7
			7/8	1175.9	79.2							
			3/4	1360.0	79.2							
			5/8	2420.2	79.2							
1.25	146.85	18.8*	1	1006.6	88.9	23402	77	1009	832763	91	60.1	36.8
			7/8	1133.9	88.9							
			3/4	1311.4	88.9							
			5/8	2548.0	88.9							
1.50	169.82	11.8	1	1276.0	74.2	21379	70	4337	809631	89	35.8	61.9
			7/8	1437.3	74.2							
			3/4	1662.3	74.2							
			5/8	1624.4	74.2							
1.50	146.85	13.4	1	1254.9	76.2	21391	70	3987	716809	79	35.6	62.2
			7/8	1413.6	76.2							
			3/4	1634.8	76.2							
			5/8	1696.7	76.2							
1.50	124.50	16.9*	1	1191.7	83.5	22187	73	3140	643890	71	43.5	50.9
			7/8	1342.4	83.5							
			3/4	1552.5	83.5							
			5/8	1913.4	83.5							
1.75	169.82	9.2	1	1610.6	76.1	23037	76	6379	806817	88	29.8	74.2
			7/8	1814.3	76.1							
			3/4	2098.3	76.1							
			5/8	476.7	76.1							
1.75	146.85	10.6	1	1574.3	77.8	23100	76	5817	728689	80	30.0	73.8
			7/8	1773.4	77.8							
			3/4	2051.0	77.8							
			5/8	601.2	77.8							
1.75	124.50	12.7	1	1524.0	80.4	23142	76	5366	618373	68	30.4	72.7
			7/8	1716.8	80.4							
			3/4	1985.5	80.4							
			5/8	773.8	80.4							
2.00	169.82	7.4	1	2000.9	80.8	25061	82	8023	844996	93	26.8	82.5
			7/8	2254.0	80.8							
			3/4	1745.1	80.8							
2.00	146.85	8.6	1	1963.5	82.2	25176	83	7334	753422	83	26.7	82.8
			7/8	2211.8	82.2							
			3/4	1824.8	82.2							
2.00	124.50	10.3	1	1914.1	84.2	25139	82	6812	628445	69	26.4	83.9
			7/8	2156.2	84.2							
			3/4	1929.7	84.2							
2.25	169.82	6.1	1	2479.3	87.4	27401	90	9421	904445	99	25.0	88.5
			7/8	2792.9	87.4							
			3/4	727.8	87.4							
2.25	146.85	7.3	1	2424.3	88.6	27462	90	8700	793932	87	25.1	88.2
			7/8	2730.9	88.6							
			3/4	844.8	88.6							

EXHIBIT 5.1

Partial calculation results for Example 5.1.

The pumping modes that did not overload the **Grade D** rods and exhibited the best and worst lifting efficiencies were selected from the several calculated modes and are given in Table 5.2. Using the best pumping mode, the energy input at the polished rod is only slightly greater than the required hydraulic power (22.1 HP), showing a very efficient operation. On the other hand, the worst mode uses almost three times as much energy as the best one. It is also interesting to see that these two extreme pumping

Table 5.2 Pumping Modes with the Best and Worst Lifting Efficiencies for Example 5.1

Pumping Mode	Best	Worst
API rod no.	86	85
Pump size	2 1/2 in	1 1/4 in
Stroke length	120 in	144 in
Pumping speed	8.3 SPM	18.4 SPM
PRHP	23.5 HP	58.3 HP
Lifting efficiency	94.1%	37.9%
Pumping unit size	C-912D-305-168	

modes both require the same size of pumping unit. As total energy costs are directly related to polished rod horsepower, big savings can be realized by choosing the right pumping mode.

Figure 5.3 shows the maximum values of calculated lifting efficiencies for different rod combinations versus pump size. Since no consideration is given here to rod overload conditions, some of the pumping modes displayed may require the use of high-strength sucker rods. It is evident that increasing the plunger size increases the attainable maximum lifting efficiencies for all rod tapers. Therefore, use of **bigger** plunger diameters with correspondingly slower pumping speeds is always **advantageous**, because these result in lower energy requirements.

Another observation, in line with practical experience, is that use of the heavier rod strings (85 or 86 instead of 75 or 76) can increase greatly the power requirements for smaller pumps. The difference is not so pronounced for larger pumps, because in those cases rod string weight becomes a smaller fraction of the total pumping load.

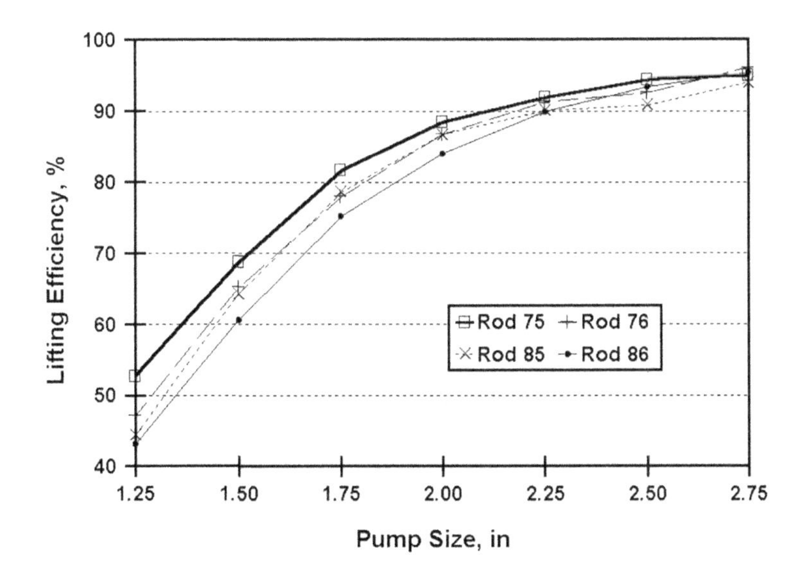

FIGURE 5.3

Maximum lifting efficiencies for different rod tapers versus pump size for Example 5.1.

Table 5.3 Pumping Modes for Lifting 1,300 blpd from 6,900 ft

Plunger Size	Stroke Length	Pumping Speed	η_{lift}	η_{system}
(in)	(in)	(SPM)	(%)	(%)
2 3/4	166	12.7	77.2	54.0
2 1/2	193	12.2	71.7	50.2
2 3/4	193	11.2	76.9	53.8
2 3/4	166	12.7	77.2	54.0
2 1/2	216	11.0	72.9	51.0
2 1/2	186	12.5	71.6	50.1
2 3/4	216	9.7	80.6	56.4
2 3/4	186	11.6	77.9	54.5

According to ***Takacs and Hadi*** *[12].*

The use of high-strength sucker rods can significantly increase the application ranges of rod pumping, as proved in a recent study [12] where the latest rod type with premium **Tenaris** connections was used. The **goal** of the optimization was to produce 1,300 blpd from a depth of 6,900 ft and the calculated pumping modes are presented in Table 5.3. As can be observed, relatively high lifting efficiencies can be obtained even for this relatively high liquid rate. System efficiencies calculated from Eq. (4.150) are also given, assuming average values of the mechanical and the motor efficiencies resulting in $\eta_{mech}\,\eta_{mot} = 0.70$. The authors reported that the sucker-rod pumping system with an efficiency of 56% was more energy efficient than an ESP installation whose efficiency reached 41% only. This example indicates that special high-strength sucker-rod materials allow the use of sucker-rod pumping installations in depth and production rate ranges where their use was possible never before.

5.2.1.3.2 Reduction in operating costs

The power costs of driving the prime mover constitute a **significant** part of the **operating** costs in rod pumping. This is mainly due to the cost of **electricity**, which, compared to earlier years, has increased extensively in the Unites States [2]. Thus, the importance of the proper selection of the pumping mode that achieves minimum energy requirements cannot be overestimated. As discussed before, the optimization procedure just detailed provides the least amount of **power** requirement at the polished rod. Since total energy usage of the pumping system is **directly** related to polished rod horsepower (*PRHP*), the optimization model automatically arrives at the most **energy-efficient** pumping system.

In order to show the merits of the optimization procedure discussed, an economic evaluation of four wells in a Hungarian oil field is presented [11]. **Actual** conditions of the wells are compared to calculated ones in Table 5.4. The rows with the well numbers contain the measured parameters; the subsequent rows display calculated pumping modes in the order of descending effectiveness. In every case, annual energy cost **savings** in percentages, related to present conditions, are given also. Evaluation of the results permits the following conclusions to be drawn.

In Well 113, the one-taper 7/8 in string is **oversized** with a low average loading and consequently has a high total weight. Decreasing the string weight and/or increasing the pump size ensures annual savings from 21% to 10.5%. The same considerations apply to Well 331, only the savings are less. In Well 126, an increase to 2 3/4 in of the pump size alone would reduce energy costs by 5.3%. In Well 429, using the present pump with a lighter string could bring about a 6.3% savings of the operating costs.

Table 5.4 The Effect of Pumping Mode Optimization on the Energy Costs of Pumping in an Example Field

	Setting	Dynamic Level	Liquid Rate	Rod No.	Pump Size	Stroke Length	Pumping Speed	*PRHP*	Lifting Efficiency	Annual Savings
Well No.	(ft)		(bpd)		(in)	(in)	(1/min)	(HP)	(%)	(%)
NI-113	1,339	984	377	77	2 3/4	78.7	7.0	5.1	54.0	
				65	3 3/4	47.2	7.0	4.0	67.9	21.1
				66	3 3/4	47.2	6.8	4.0	67.8	21.1
				65	3 1/4	63.0	6.4	4.0	67.1	21.1
				77	3 3/4	47.2	6.5	4.2	66.9	18.4
				66	3 1/4	63.0	6.3	4.2	65.8	18.4
				87	3 3/4	47.2	6.4	4.2	65.3	18.4
				87	3 1/4	63.0	6.1	4.6	60.0	10.5
NI-126	4,928	3,445	497	86	2 1/4	157.5	7.0	15.1	72.2	
				86	2 3/4	137.8	5.0	14.3	76.3	5.3
				87	2 3/4	118.1	6.0	15.0	73.3	0.9
				97	2 3/4	118.1	5.7	15.0	73.2	0.9
NI-331	3,937	2,133	535	87	2 3/4	118.1	7.0	12.1	69.7	
				75	3 1/4	118.1	5.5	11.1	75.3	7.8
				86	3 1/4	98.4	6.2	11.4	73.5	5.6
				87	3 1/4	98.4	6.0	11.5	73.2	4.4
NI-429	4,003	2,854	503	87	2 3/4	137.8	6.0	14.9	71.5	
				75	2 3/4	157.5	5.0	13.9	75.6	6.3
				86	3 1/4	118.1	5.1	14.3	73.7	3.6
				87	2 3/4	137.8	5.4	14.7	71.7	0.9

According to ***Takacs*** *[11].*

It can be observed, in general, that rod string **oversizing** results in heavier strings, which in turn requires more **power** from the prime mover. The use of **larger** pumps, on the other hand, decreases energy costs as an effect of the lower pumping speeds required. Therefore, significant operating cost savings can be realized by using the optimum pumping mode that ensures maximum lifting efficiency.

5.2.2 MAXIMIZING THE PUMPING RATE

Many times, the liquid rate to be produced from the well is not **limited**, and the operator's goal is to reach the largest possible rate. Although sucker-rod pumping is not particularly **suited** for high-rate production, modern equipment and operating practices allow fairly large volumes to be pumped. Approximate maximum rates for different types of pumping units using high-strength rod materials and maximum pump sizes are given in Fig. 5.4 [13] and may be regarded as the ultimate production capacity of present-day sucker-rod pumping installations.

It is clearly visible from these pumping capacity values that attainable rates progressively **decrease** with increasing lifting **depths**. This feature indicates that, when high volumes are lifted from deep wells, the main **limitation** of sucker-rod pumping is the available **strength** of the rod string material. In high-capacity shallow wells, this is not a major problem, since rod strength is not usually exceeded by rod stresses. In these cases, the main limitations are normally the torque **rating** of the speed reducer and the restricted **fall** of the rod string during downstroke. In the following, the major factors

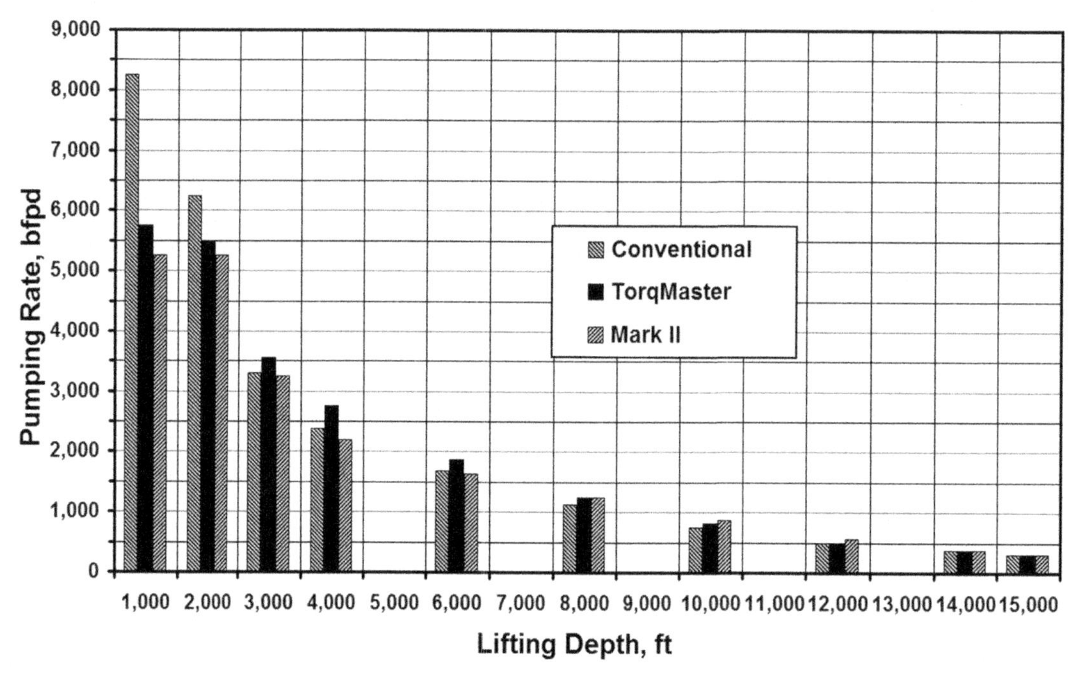

FIGURE 5.4

Calculated maximum pumping rates versus pumping depth.

According to ***Ghareeb et al.*** *[13].*

restricting the increase of pumping rates are discussed first, then considerations on attaining the maximum liquid production from an installation are given.

5.2.2.1 Limiting factors

According to **Byrd** [14], the fluid lifting capacity of a rod pumping system is limited by one or more of the following:

- the size of the bottomhole **pump**,
- the polished rod stroke **length**,
- the pumping **speed**,
- the **strength** of the rod material, and
- the structural and torsional **capacity** of the pumping unit.

Since most of the above factors are **interrelated**, a short discussion of their effects on pumping rate is necessary. Increasing pump size, stroke length, or both clearly leads to higher production rates, but the simultaneous increase in rod loads and stresses can bring about an **overload** of the rod material in use. Higher-**strength** rods, however, allow larger plungers and stroke lengths to be used and clearly improve production rates. Actual rod loads also are limited by the structural **capacity** of the pumping unit and the torque **rating** of the speed reducer. Usually, larger-capacity units can pump more liquid, but, once again, the strength of the rod material can turn out to be the limiting factor.

Pumping **speed**, which is a major governing parameter, requires a special treatment. It has long been recognized in the field that the ability to **increase** the pumping speed is **restricted** by the downward velocity of the horsehead during the downstroke. At a sufficiently high speed, the carrier bar (affixed to the horsehead by the wireline hanger) begins to move faster on the downstroke than the polished rod, which falls due to total rod weight acting on it. Then, after the start of the upstroke, the upward-moving carrier bar hits on the polished rod clamp, which still moves downwards. The high-**impact** forces and the resulting large torques can easily **overload** the pumping unit or the speed reducer, which is why this situation is always avoided. The pumping speed at which this phenomenon starts to occur is called the **critical** pumping speed.

Critical pumping speed is primarily determined by the actual **velocity** of rod fall during the downstroke. This is affected by retarding forces such as **friction** (in the stuffing box, rod-to-tubing, fluid, etc.) and **buoyancy**. An accepted critical speed for average conditions is 70% of the **free-fall** velocity of the rods in air, but unfavorable conditions (especially a high liquid viscosity) can reduce this figure considerably. Based on rod fall velocity, the critical pumping speed is calculated from the pumping unit's kinematic behavior, in particular from the polished rod velocity during the downstroke. Stroke length has a direct impact on critical speed and the longer the stroke, the lower the maximum pumping speeds that can be allowed.

Table 5.5 contains critical pumping speeds for conventional pumping units, as suggested by **Byrd** [14]. A major manufacturer proposes the following equation [15] to determine critical pumping speed:

$$N_{cr} = C\sqrt{\frac{60,000}{S}} \tag{5.4}$$

where:

N_{cr} = critical pumping speed, 1/min
S = polished rod stroke length, in

Table 5.5 Critical Pumping Speeds versus Polished Rod Stroke Lengths for Conventional Pumping Units

Stroke Length	Critical Speed
(in)	(1/min)
64	23.0
74	21.5
86	19.5
100	18.5
120	16.5
144	15.0
168	14.0

After ***Byrd*** *[14].*

The constant C in the above formula is taken as 0.7 for conventional, 0.63 for air balanced, and 0.56 for Mark II geometry pumping units.

5.2.2.2 Finding the maximum rate

In light of the previous discussion, a direct determination of the maximum pumping rate for a sucker-rod pumping system is clearly **impossible**. Because of the great number of influencing parameters and the many interactions between them, only complex calculation procedures can be used that account for all conditions involved [16]. In the following, a model is presented (based on the **RP 11L** procedure) for finding the **maximum** attainable pumping rate from a pumping system using a conventional pumping unit.

This calculation procedure properly accounts for all effects the various pumping parameters have on pumping rate and includes the **design** of the rod string as well. It is started by taking an available plunger size, a stroke length, and the minimum of available pumping speeds (Fig. 5.5(a)). At this point, the rod string is designed either by **Neely**'s method [5] for standard API rod materials or by setting the rod stresses equal at the top of each taper section, for high-strength rod materials. If the string is not **overloaded**, pump displacement (PD) and additional operating parameters are determined. Next, these parameters are checked against the **limiting** factors:

- the peak polished rod load versus the unit's **allowable** structural load, and
- the peak net torque versus the speed reducer's torque **rating**.

It is at this point that actual pumping speed is compared to its **critical** value by checking the calculated minimum polished rod load. Obviously, critical pumping speed is detected when minimum polished rod load approaches **zero** as a result of the carrier bar leaving the polished rod clamp.

If both the rod string and the pumping unit are under **normal** loads, a valid pumping mode is already found. Then the procedure is repeated with a higher pumping speed (Fig. 5.5(b)). After all feasible pumping speeds have been used, the operational parameters of the pumping mode with the **largest** liquid rate calculated so far are output. This rate represents the largest fluid volume that can be

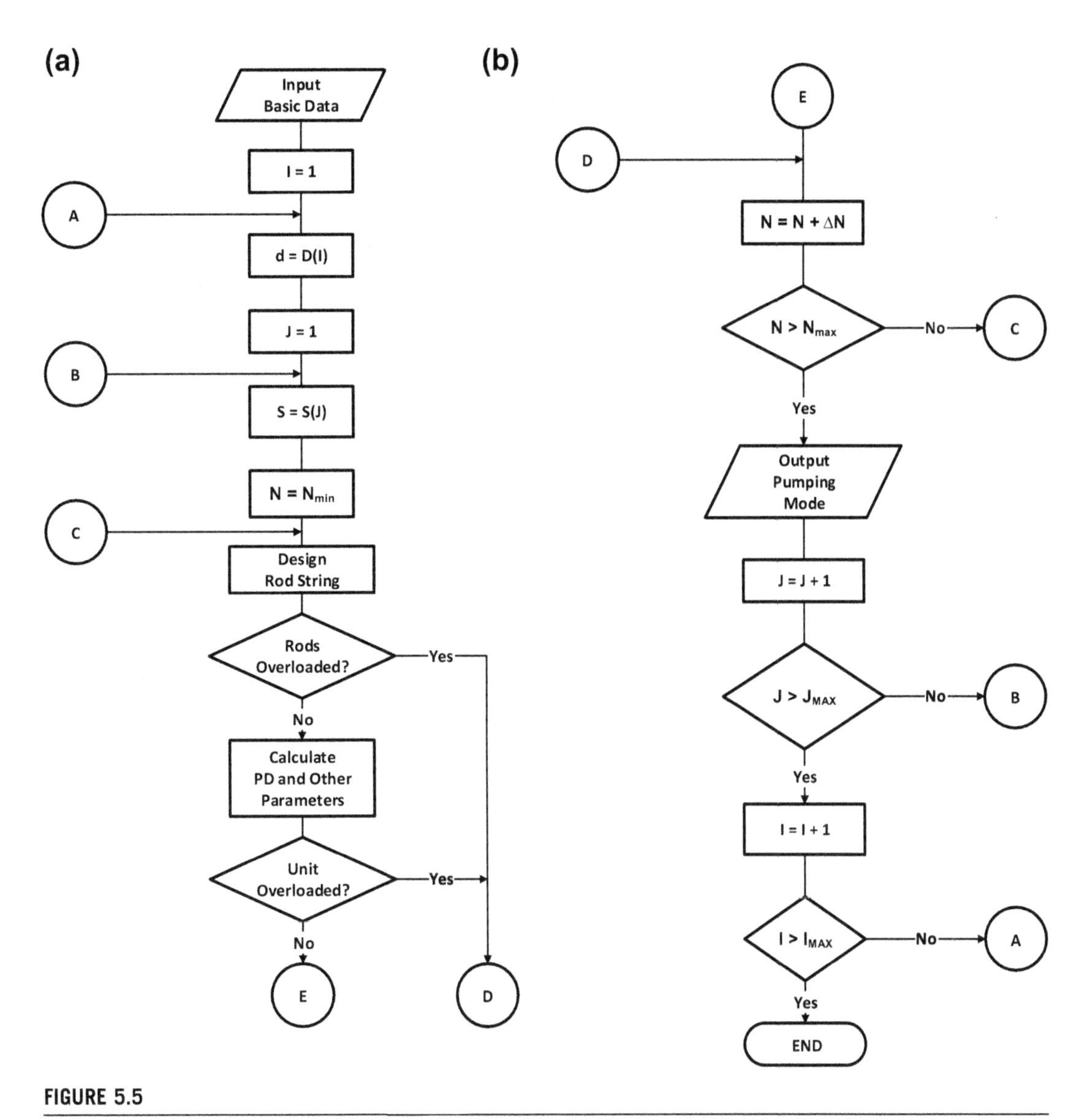

FIGURE 5.5

Flowchart of finding the maximum attainable pumping rate from a rod pumping installation, using a conventional pumping unit.

lifted with the previously set plunger size and stroke length. In the following, first the stroke length and then the plunger size is changed, and the calculations are repeated. When all the available values have been used, several pumping **rates** with their respective pumping modes are known, and it only remains to select the mode that ensures the **overall** maximum rate.

EXAMPLE 5.2: FIND THE MAXIMUM PUMPING RATE FROM A 8,000 ft WELL WHEN PUMPING WATER WITH A C-912D-365-168 UNIT. THE ROD STRING IS A THREE-TAPER 86 STRING OF GRADE D RODS. AVAILABLE PLUNGER SIZES ARE 1.75, 2, AND 2.25 in; POLISHED ROD STROKE LENGTHS ARE 100, 120, 144, AND 168 in.

Solution

Exhibit 5.2 contains the calculated pumping modes that represent the maximum liquid rates for the plunger size–stroke length combinations considered. Out of these, the overall maximum liquid volume is 446 bpd, which can be reached by using a 1.75 in plunger, a stroke length of 168 in, and a pumping speed of 9 strokes/min. Using this pumping mode, both the speed reducer and the rod string are only slightly overloaded (below 10% overload), which is acceptable in most cases.

It is interesting to note that under the same conditions, the use of a high-strength rod material with an allowable stress of 50,000 psi and a two-taper 76 string would allow a maximum liquid production of 981 bpd. This clearly illustrates that the limiting factor in maximizing liquid production from deep sucker-rod-pumped wells is the strength of the rod material.

Plunger Size	Stroke Length	Pumping Speed	Rod String Size	Rod String Length	Rod String Loading	Max.P.R. Load	Peak Net Torque	P.R. Power	Lifting Eff.	Pumping Rate
in	in	SPM	in	ft	%	lbs	in-lbs	HP	%	bpd
1.75	100.00	14.0	1	2448.1	107.0	30181	508826	37.5	68.6	435.6
			7/8	2552.0						
			3/4	3000.0						
1.75	120.00	12.0	1	2440.6	107.6	30091	595423	35.2	68.7	409.7
			7/8	2533.6						
			3/4	3025.8						
1.75	144.00	10.0	1	2440.6	106.0	29691	709627	31.4	73.0	388.5
			7/8	2533.6						
			3/4	3025.8						
1.75	168.00	9.0	1	2425.1	107.1	29913	919026	33.6	78.3	445.9
			7/8	2512.6						
			3/4	3062.2						
2.00	100.00	11.0	1	2899.9	107.0	31114	468752	24.1	80.1	327.5
			7/8	2945.6						
			3/4	2154.6						
2.00	120.00	10.0	1	2862.5	109.0	31459	555356	26.8	79.8	362.3
			7/8	2895.3						
			3/4	2242.2						
2.00	144.00	8.0	1	2884.1	108.3	31462	669561	25.0	87.0	369.0
			7/8	2916.0						
			3/4	2200.0						
2.00	168.00	7.0	1	2871.0	108.6	31490	870345	26.8	90.9	413.5
			7/8	2905.6						
			3/4	2223.4						
2.25	120.00	5.0	1	3534.4	108.6	33594	469358	10.7	98.9	179.0
			7/8	3729.9						
			3/4	735.7						

EXHIBIT 5.2

Calculated pumping modes for Example 5.2.

5.2.2.3 Casinghead pressure reduction

In sucker-rod-pumped wells with low liquid production rates, typical in **mature** fields, the pressure maintained at the **casinghead** negatively effects the well's producing **bottomhole** pressure and hence its liquid rate. Even relatively **low** pressures at the casinghead due to the flowline pressure, coupled with the static pressure of the tall gas **column** present in the well's annulus, may add up to flowing bottomhole pressures that **restrict** the drawdown across the formation. With declining reservoir pressures the surface back pressure needed to pass fluids into the surface gathering system becomes a greater percentage of the flowing bottomhole pressure and well production will **decrease**. In such cases, one very effective way to **increase** liquid production rates is the continuous **removal** of produced gas from the annulus, thereby maintaining a pressure as low as **possible** (even down to vacuum) at the casinghead.

In the past, casinghead gas was simply **vented** so as to keep a low **back pressure** on the formation and to maximize liquid production. Recently, environmental regulations and the increase of natural gas prices **prohibit** this solution and casinghead pressure reduction units are getting increasingly popular in mature oil fields, which suck gas from the casinghead and discharge it to the flowline. Such units must handle the relatively low volumes of **wet** natural gas and have suction pressures close to **vacuum**; they discharge compressed gas with a properly selected pressure into the **flowline** or a separate **gas** line. Several different kinds of compressors can be used (rotary vane, rotary screw, reciprocating, etc.); their selection is based on the produced gas volume, the required suction/discharge pressures, and the compressor's ability to handle wet streams. Typical applications of casinghead compression involve a **test** period to evaluate the economics of the procedure using a **portable**, gas-driven unit; if feasible, a final electric **motor-driven** unit is installed on the well or at a central position to service several wells [17,18].

The need for a central compressor system and the necessary piping is eliminated if gas compressors mounted on the pumping unit are used. They benefit from a simple construction like the **beam-mounted gas compressor** (BMGC), a relatively simple device that is fitted on the pumping unit and utilizes the normal pumping **action** to compress casinghead gas. Several different types using single-acting or double-acting gas compressors were developed and used in the oil field long before [19–21].

In the following a BMGC is described whose heart is a **double-acting** positive-displacement **compressor**, with main parts such as a cylinder, a piston, and four check valves [22]. The compressor's cylinder is attached to the **Samson post** or the pumping unit's skid, and the piston rod is fixed to the walking **beam**; both connections use **clamps** for easy mounting and removal (see Fig. 5.6). Both sides of the cylinder are connected, via flexible hoses, to the **casinghead** to suck gas from the well and to the **flowline** to discharge gas after compression. The unit can be used on pumping units of any known geometry.

As the walking beam moves up and down, one side of the piston **draws** wellhead gas into the cylinder, while the other side compresses and **discharges** gas into the flowline downstream of the pumping tee. This double-acting operation is provided by the sequential and **automatic** opening and closing of the four **check valves** on the BMGC unit. Figure 5.7 shows the operation of the system during the **downstroke** of the pumping unit with the current status of the check valves. Here the gas that was sucked in during the upstroke below the moving piston is being compressed and **transferred** to the flowline while gas is sucked in on top of the piston. During the upstroke of the beam, the status of the check valves is reversed automatically and gas above the piston is discharged into the flowline. The direction of gas flow is **constant** during the complete pumping cycle; this feature ensures that casinghead gas is **continuously** compressed and transported into the flowline.

FIGURE 5.6

General arrangement of a beam mounted gas compressor (BMGC) on a pumping unit.

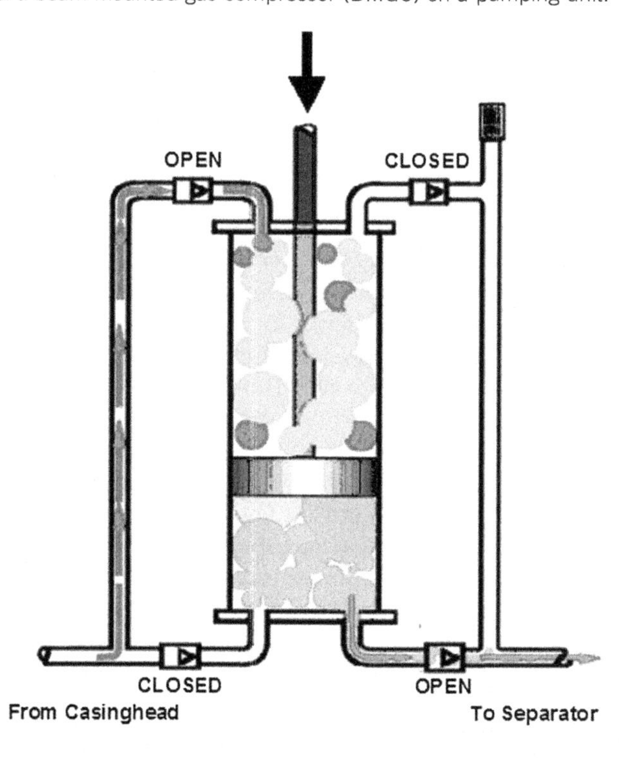

FIGURE 5.7

Schematic operation of a double-acting BMGC on the downstroke of the pumping unit.

The general **benefits** of casinghead gas compression are:

- Liquid and gas flow rates usually **increase** due to the reduction of flowing bottomhole pressures; the **profitability** of wells improves.
- The operating conditions of the subsurface **pump** are improved; gas locking may be eliminated.

The use of BMGC units provides the following **additional** benefits [23–25] over other types of wellhead gas compression:

- Since the unit is driven by the pumping unit, it is **reliable** and energy **efficient**.
- **Capital/rental** costs are less than those of motor-driven compression units.
- Installation is simple; unit is **portable** and can be easily moved to a new well.
- It provides an **emission-free** solution.
- It can compress **wet** casinghead gas without scrubbing and can be used also in an H_2S environment, due to proper selection of materials.

5.3 MATCHING PUMPING RATE TO WELL INFLOW

Up to this point, it was tacitly assumed that the fluid volume lifted to the surface by the sucker-rod pump was **available** at the pump intake in the well. Putting it more technically, **well inflow performance** was not considered in the design of the pumping system. This approach cannot withstand practical application, since, naturally, the pumping system and all components thereof should be chosen in accordance with well **deliverability**. Installing a low-capacity system in a well capable of high production rates or running a high-capacity pump in a low producer will inevitably result in inefficient overall operation, no matter how well the system components were selected. Thus, in order to achieve an economical rod pumping operation, the design of the lifting equipment and the selection of the operational parameters must primarily be based on the well's deliverability.

The basic task of rod pumping design is to achieve a pumping rate **matching** the inflow rate to the given well by selecting the right equipment components and by adjusting them properly. A prerequisite of this is the exact knowledge of the well's inflow performance relationship (**IPR curve**), which gives attainable production rates against flowing bottomhole pressure. Improper or unreliable information on IPR is a very common cause of poor design and inefficient operation. Therefore, every effort has to be made to ensure that enough information is available on well deliverability at the time of installation design.

There exist basically three options for ensuring that pumping rate **matches** well inflow:

1. Select a pumping mode that delivers a liquid volume equal to the anticipated well production rate by operating a pumping unit **continuously** at a fixed pumping speed in its normal range of speeds.
2. Provided the lifting capacity of the pumping system is higher than the well inflow rate, the system is operated by controlling its daily pumping **time**. This is the case of **intermittent** pumping, when during the production period the pumping unit runs at a fixed pumping speed.
3. With the pumping system having a greater capacity than well inflow, the pumping unit is run **continuously** under a **controlled** variable pumping speed.

Continuous pumping is the obvious choice for wells with **sufficient** previous information on their liquid production. Wells with well-known and stable IPR curves are produced this way, after the choice is made on the proper pumping mode to meet well inflow. In many cases, however, well inflow is **overestimated**, and a pumping system with a higher-than-required capacity is installed on the well; then the problem of reducing the system's lifting capacity arises. At such times, very often, **intermittent** operation is chosen, mainly due to the many uncertainties in well inflow performance information. The best use of **intermittent** pumping is on wells with **unknown** or rapidly **changing** inflow properties. An efficient **intermittent operation**, however, requires an optimum selection of the daily total **pumping** time; otherwise, well production may be lost or severe operational problems may occur. The application of some means of pumping **speed control** can greatly alleviate these hindrances while enabling the well to produce **continuously**.

In the following, continuous, intermittent, and speed-controlled pumping are discussed separately, and system design with well inflow properties properly accounted for will be emphasized.

5.3.1 CONTINUOUS PUMPING

5.3.1.1 Systems analysis

The simultaneous operation of a productive **formation** and an oil well drilled into it is easily described by **systems analysis** [26,27] methods. In the following, this methodology is applied to sucker-rod pumping for developing a pumping system design that involves proper considerations for the inflow performance of the well.

A schematic drawing of a pumping well, showing the different components of the **production system** along with the various **nodes** separating them, is shown in Fig. 5.8 Systems analysis

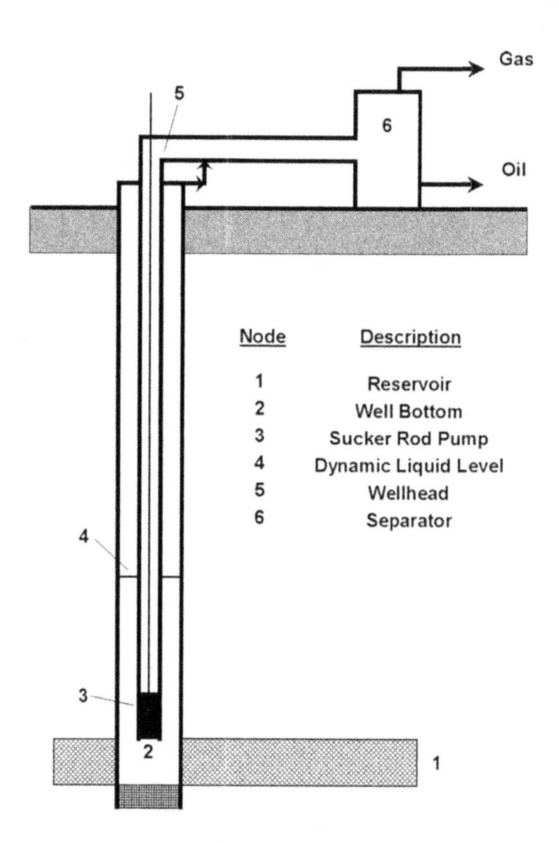

FIGURE 5.8

The production system of a rod pumping well.

principles have shown that the production equipment and the formation are connected in **series** at the well bottom. This implies that the same liquid rate must flow through both components of the system. Therefore, in order to find the production rate under some specific conditions, a common solution of the performance of the two **components** must be sought. Since liquid inflow from the formation is characterized by the familiar **IPR equation** (see **Chapter 2**), only the performance of the pumping system has to be described. This involves the determination of a pumping rate–bottomhole pressure relationship, which can be superimposed on the IPR curve to find the total system's production rate.

As seen from Fig. 5.8, two **paths** are available to describe the pumping system's performance: one in the **tubing** and the other in the **casing annulus**. In the first case, methods developed for flowing or gas-lifted wells can be applied, but this approach necessitates the calculation of two-phase pressure drops in the tubing/rod string **annulus** with increased calculation requirements and reduced **accuracy**. The work of **Schmidt and Doty** [28] followed this line of thought and developed performance curves for the pumping system producing gassy fluids. Other investigators [29–31] developed calculation models to determine the cooperation of a producing formation and the rod pumping installation based on the **API RP 11L** [3] procedure. The procedure developed earlier by the author [32] is based on modeling the phenomena occurring in the well's **annulus**. It is well known that the depth of the dynamic liquid level is a direct **indication** of the well's flowing bottomhole pressure. This fact is utilized in the calculation procedure, and the concept of **system performance curves** for describing the operation of the pumping system is introduced.

5.3.1.1.1 System performance curves

To describe the performance of a rod-pumping system, the pumping rates **attainable** at different conditions have to be found. The conditions having the most **significant** impact on the production rate are the pumping **mode** used (i.e., the combination of pump size, polished rod stroke length, pumping speed, and rod string design), and the pump setting **depth**. In light of these conditions, the performance of the pumping system can be described effectively by plotting pumping **rates** versus pump setting **depth** for different pumping modes. These plots are defined here as **system performance curves** and can be constructed by utilizing any calculation procedure that properly approximates the operational parameters of pumping. The curves thus created represent the rates attainable with different pumping modes from different depths, and they show the performance of the pumping system alone.

In the following explanation, system performance curves are developed with the basic **assumptions** as listed herein. Although these restrictions can affect considerably the applicability of the analysis method, there are several conditions in which these requirements are met (e.g., wells produced from a strong water-drive reservoir):

- **Conventional** pumping units are considered, allowing for **RP 11L** procedures to be used.
- Pumping of a single-phase **liquid** is assumed, and the effects of any free gas in the annulus are disregarded.
- **Pumped-off** conditions are presumed, with the dynamic liquid level being at pump setting depth.

The procedure to construct system performance curves is illustrated on the flowchart shown in Fig. 5.9 After the input of the necessary basic data, the parameters of a pumping mode are input: plunger size, *d*; polished rod stroke length, *S*; and pumping speed, *N*. A value for pump setting depth, *L*, is assumed and the **RP 11L** calculations are used to find the operational parameters of pumping at the given conditions. The most important of these is pumping **rate**, PD, which, if plotted in the function of

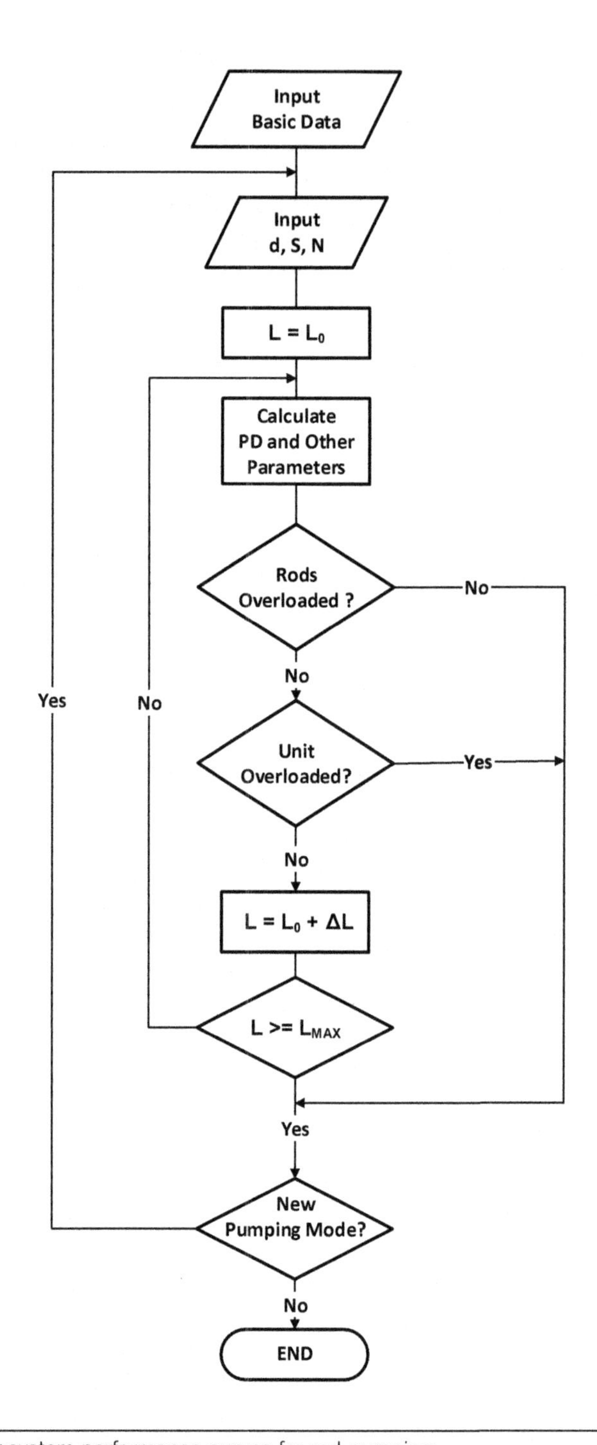

FIGURE 5.9

Flowchart of calculating system performance curves for rod pumping.

setting **depth**, L, gives one point on the **system performance curve**. Next, the calculated operating parameters are checked for **overload** conditions by evaluating the loading on the **rods** and on the **pumping unit**. In case the fatigue limit of the rods is exceeded, or either the unit structure or the speed reducer is **overloaded**, the **maximum** attainable pump setting depth has been reached and a new pumping mode is processed. In all other cases the pump setting depth, L, is increased, and calculations are repeated for each new depth. After a prescribed maximum depth, L_{max}, is reached, another pumping mode is selected, and the whole procedure is repeated. The described method yields several **corresponding** pumping rate–setting depth points for every assumed pumping mode and the points then define the **system performance curves**.

Figures 5.10 and 5.11 show example system performance curves for a pump diameter of 1.25 in and for API tapers of 76 and 86, respectively. (An **anchored** tubing string, 100% pump volumetric efficiency, and pumping of **water** also were assumed). Each sheet contains calculated pumping rates plotted against pump setting depth for selected polished rod stroke lengths and pumping speeds. Pump size, API taper number, and rod material are held constant on every sheet. The different curves, therefore, represent different pumping modes.

Every performance curve starts (at a pump setting depth of **zero**) from the pumping rate that could be calculated with a plunger stroke length **equal** to polished rod stroke length. As pump setting depth

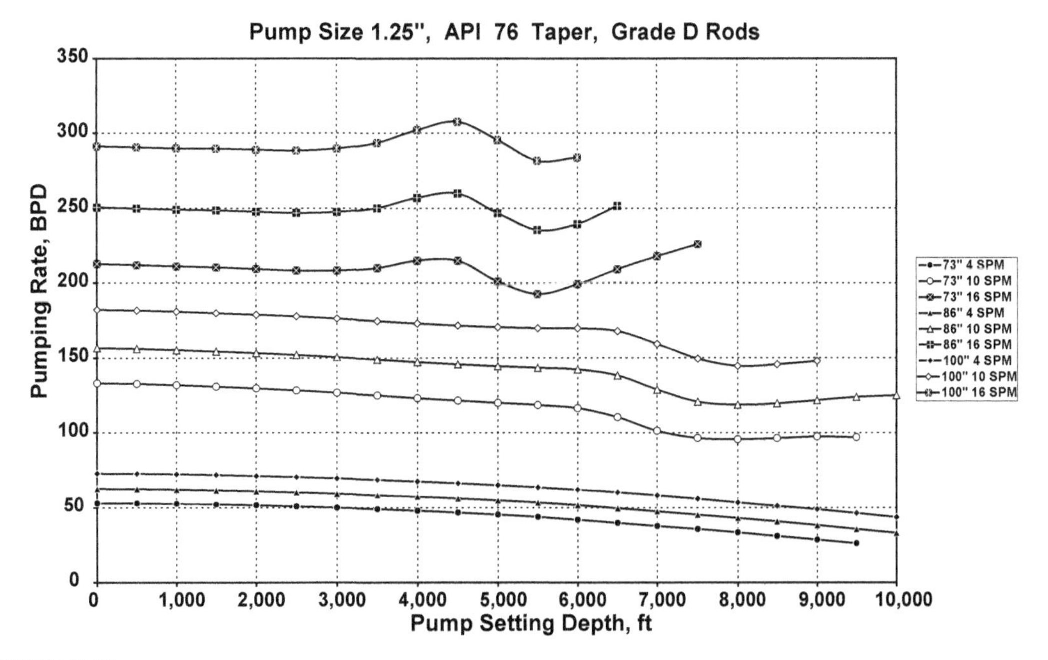

FIGURE 5.10

System performance curve sheet for a pumping system with a 1.25 in plunger, API 76 rod string, and Grade D rods.

After ***Takacs*** *[32].*

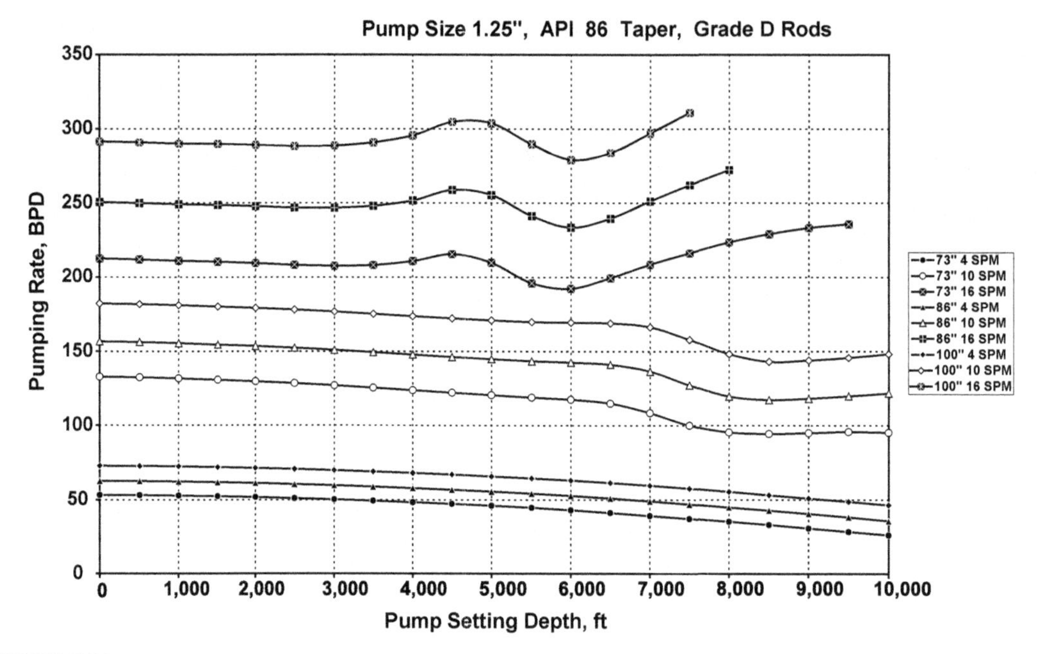

FIGURE 5.11

System performance curve sheet for a pumping system with a 1.25 in plunger, API 86 rod string, and Grade D rods.

*After **Takacs** [32].*

increases, a point is reached where maximum rod **stress**, due to the combined effects of fluid load, rod string weight, and dynamic forces, exceeds the **allowable** stress for the rod material. This depth gives the **maximum** pump setting depth that can be reached with the given rod material (Grade D in the examples). Comparison of Figs 5.10 and 5.11 shows that the use of a stronger string (86 taper instead of 76 taper) allows pumping from greater depths. But this holds only if rod strength is considered to be the sole limiting factor.

In order to properly describe the system's performance, the operational characteristics of the **surface** equipment must also be taken into account. Therefore, the structural **capacity** of the pumping unit, as well as the allowable torque **rating** of the speed reducer, must not be exceeded by the actual peak polished rod load and the peak torque, respectively. These effects are considered in Figs 5.12 and 5.13, valid for a **C-228D-213-100** pumping unit. Compared to the previous figures, the individual curves end at smaller depths than rod strength alone would allow. This is caused by the fact that the mechanical **limitations** of the pumping unit (maximum structural load and maximum allowed gearbox torque) are reached before the rod string is **overloaded**. As can be seen, the use of the heavier string with the API 86 taper overloads the pumping unit at much shallower depths than the API 76 string. Thus, in contrast to the conclusions drawn from Figs 5.10 and 5.11, the lighter rod string can be operated at greater depths than the heavier one. Figure 5.14 shows similar curves for a 2 in plunger and an API 76 taper string.

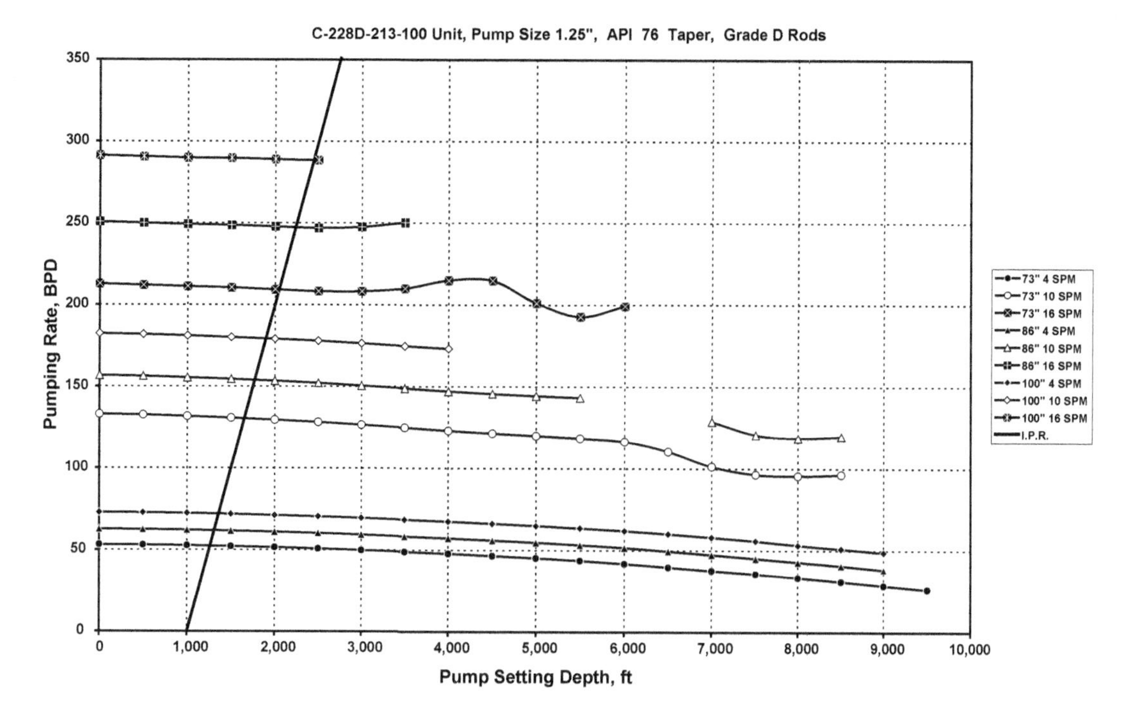

FIGURE 5.12

System performance curve sheet for a pumping system with a C 228D-213-100 unit, a 1.25 in plunger, API 76 rod string, and Grade D rods.

*After **Takacs** [32].*

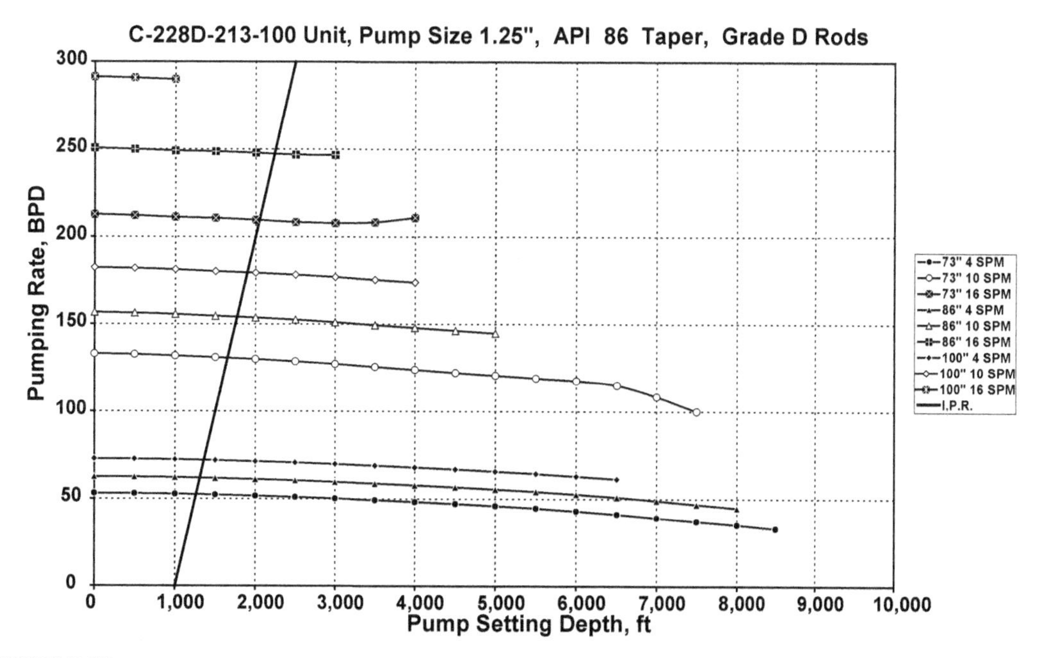

FIGURE 5.13

System performance curve sheet for a pumping system with a C 228D-213-100 unit, a 1.25 in plunger, API 86 rod string, and Grade D rods.

*After **Takacs** [32].*

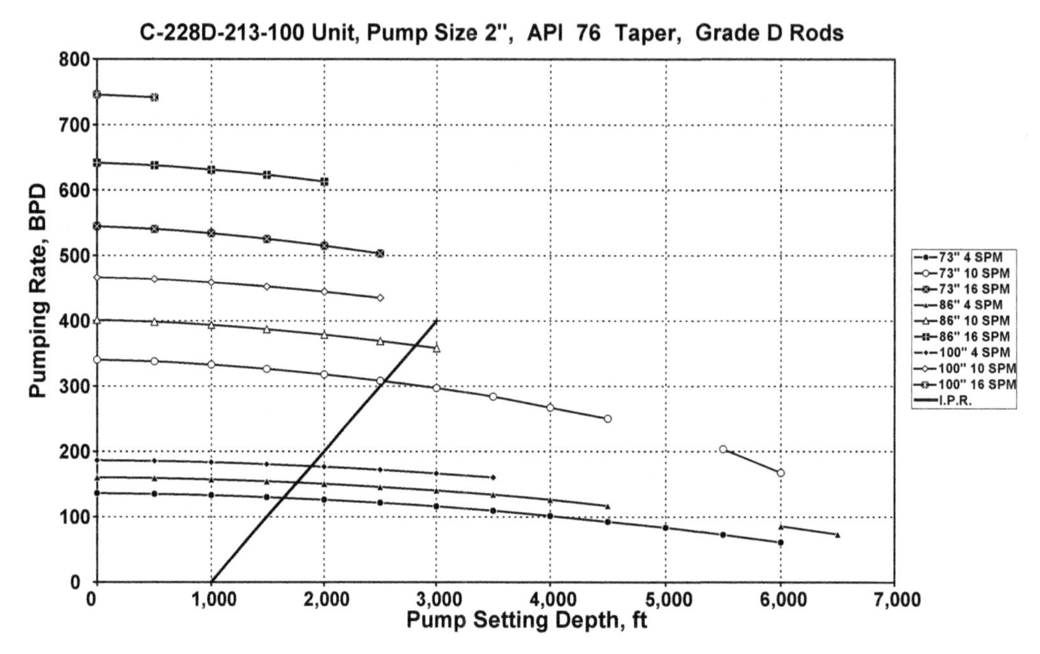

FIGURE 5.14

System performance curve sheet for a pumping system with a C 228D-213-100 unit, a 2 in plunger, API 76 rod string, and Grade D rods.

*After **Takacs** [32].*

5.3.1.1.2 The use of performance curves

After the performance of the lifting **equipment** has been determined, the operation of the other system (the productive **formation**) must be described. It can be shown that IPR curves plot as **straight** lines in the coordinate system production rate versus dynamic liquid level, used for displaying **system performance curves**. Systems analysis, therefore, can be accomplished by **superimposing** the IPR of the well on the system performance curve sheet valid for the given conditions. Intersections of these curves give pumping **rates** attainable under different conditions and allow the determination of the various possible common **operating** points of the formation and the rod pumping system.

To illustrate the procedure, an example well with a productivity index of PI = 0.46 bpd/psi and the following data were used to construct the straight-line IPR curves in Figs 5.12–5.14:

- Well depth = 3,000 ft
- Static liquid level = 1,000 ft
- Dynamic liquid level at 400 bpd liquid production = 3,000 ft.

The **intersections** of the IPR line with the system performance curves indicate, for different pumping modes, the rates **attainable**, along with the corresponding pump setting depths required to achieve those rates. Provided that **system performance curve** sheets for different pumping units are available, one has only to select the right sheet and to **plot** the actual IPR on it, in order to analyze the performance of different pumping systems. Therefore, this method can be utilized for **designing** a sucker-rod pumping system with due regard to well **deliverability** and to ensure an efficient pumping operation. The use of this procedure, however, is limited by the basic assumptions indicated earlier, and for cases in which sufficient information is available on well inflow performance.

EXAMPLE 5.3: FIND THE PUMPING RATES FOR THE AVAILABLE PUMPING MODES INDICATED IN FIGS 5.12 AND 5.14. DETERMINE THE MAXIMUM RATE THAT CAN BE ACHIEVED FROM THE WELL, IF AVAILABLE PLUNGER SIZES ARE 1.25 AND 2 in, AND AN API 76 ROD STRING IS USED

Solution

Table 5.6 contains pumping rates read off from Figs 5.12 and 5.14, at the intersection of the IPR and system performance curves. The maximum rate is found as 370 bpd with the following pumping mode: 2 in plunger, 86 in stroke length, and 10 strokes/min pumping speed. The dynamic fluid level is at 2,780 ft in this case, showing near pumped-off conditions to prevail.

Table 5.6 Pumping Rates Determined for the Pumping Modes and Well Data Given in Example 5.3

Stroke Length	Pumping Speed	Pumping Rate, bpd for Plunger Size	
(in)	(1/min)	(1.25 in)	(2 in)
73	4	52	130
	10	130	310
	16	210	–
86	4	65	150
	10	155	370
	16	245	–
100	4	75	180
	10	180	–
	16	290	–

5.3.2 INTERMITTENT PUMPING

In cases where **insufficient** or no **inflow** performance data are available, an optimum selection of the pumping mode is **impossible** prior to equipment installation. The usual solution is to install a pumping system of ample fluid lifting **capacity** on the well and to operate it **intermittently**. By reducing total daily pumping **time**, lifting capacity of the pumping system is thus reduced to meet well inflow rate. During shutdowns, well fluids move up the casing annulus, while during production periods the fluid level drops to the pump intake. Proper intermittent pumping practices do not allow the fluid level to drop **below** the pump setting depth because severe operational **problems** can arise. But, at the same time, the desire to produce as much fluid as **possible** requires a dynamic level **near** the setting depth. A proper design of intermittent pumping, therefore, ensures a **minimum** average fluid level during the pumping cycle while **preventing** the dynamic level from dropping below the pump.

5.3.2.1 Early methods

In early pumping practice, **manual operation** of intermittent wells was used with a pumper in charge of a group of wells. The motor was started and stopped **manually**, which involved several visits to the well and lengthy trials to arrive at proper operation. The next step came with the introduction of **time clocks**, which allowed for **automatic** cyclic operation and required manual settings of the start and stop times only. Although more reliable than simple manual operation, the use of time clocks still needed a lot of **adjustment** and was not very flexible. Later, the **percentage timer** was introduced, which worked in automatic cycles of 15–30 min with pumping time being a preset **percentage** of cycle time. Timers are inexpensive and simple to operate, as compared to pump-off control devices; their proper setting, however, is possible by trial-and-error only [33]. Following the changes in well inflow performance is both a time-consuming and an **unreliable** procedure. Since all timers have inherent drawbacks, the evolution of better automatic controls was inevitable.

5.3.2.2 Pump-off control

An ideal intermittent control for a rod pumping system provides an **automatic** operation and ensures the lowest possible flowing bottomhole **pressure** to achieve maximum fluid production from the well. Therefore, the basic **goal** of any control principle is to **lower** the fluid level to the pump intake by the end of the operational period. With the fluid level at the pump intake, and pump capacity in excess of well inflow, the pump barrel does not **completely** fill up with fluids during the upstroke. Then, on the downstroke, the plunger will hit the fluid level in the barrel, producing the phenomenon known as **fluid pound**. The well is said to be **pumped off**. Since pump-off is a direct **indication** of the fluid level in the well, most of the control devices include a means to **detect** this condition. Hence, the generic name of intermittent pump controllers: **pump-off controllers**.

5.3.2.2.1 Early types of pump-off controller units

Pump-off controllers were introduced in rod pumping technology in the early 1970s. Early units were capable only of starting and stopping the pumping unit and worked on several different principles, as discussed in the following.

Fluid Level Controllers. If a direct detection of the fluid **level** in the well is made, it can be used to control the pumping unit's operation. Early methods used permanently installed **downhole** pressure transducers to sense fluid levels and to start or stop the pumping unit [34]. Other units utilized a **sonic** device on the surface, which automatically **monitored** the depth of fluid level and started the prime mover when a set fluid column height has accumulated over the pump [35]. The unit is stopped when the level drops to pump intake depth. These types of POCs have not gained popularity because of their **complexity** and high maintenance requirements.

Flow/No-Flow Detectors. Installed on the flowline, the detector senses the **reduction** in pumping rate when pump-off is reached and **shuts** down the unit for a specified time. Today, these detectors are obsolete because they are difficult to adjust and can be confused by well heading.

Vibration Sensing. Mechanical **vibrations**, always present in a pumping unit, **increase** in intensity when a pumped-off condition occurs. This increase is detected by a device that controls the starting and stopping of the pumping unit. Due to its **unreliability**, this system is seldom used.

Motor Current Detection

When pump-off occurs, the rod string descends with the full fluid **load** acting on it, since the traveling valve does not open on the start of the downstroke. This means that less **energy** is required from the prime mover to **lift** the counterweights, because rod load provides most of the power. It follows that the **current** drawn by the motor significantly **decreases** during the downstroke when compared to a normal pumping cycle. This phenomenon is utilized in the control principles that use electrical transducers to sense motor current—a low-cost solution without much material requirements. Possible variations of this POC principle are depicted in Fig. 5.15, modified after **Westerman** [36].

One of the possible solutions monitors the **difference** in peak currents, which should be equal (provided the unit is perfectly balanced) for normal pumping. After this difference reaches a preset value, the unit is shut down for a predetermined period of time. On the next controller, a **point** can be set on the downstroke with a specified current value. In case of pump-off, actual current falls **below** this value and the controller shuts down the unit. **Average** current over the pumping cycle also can be

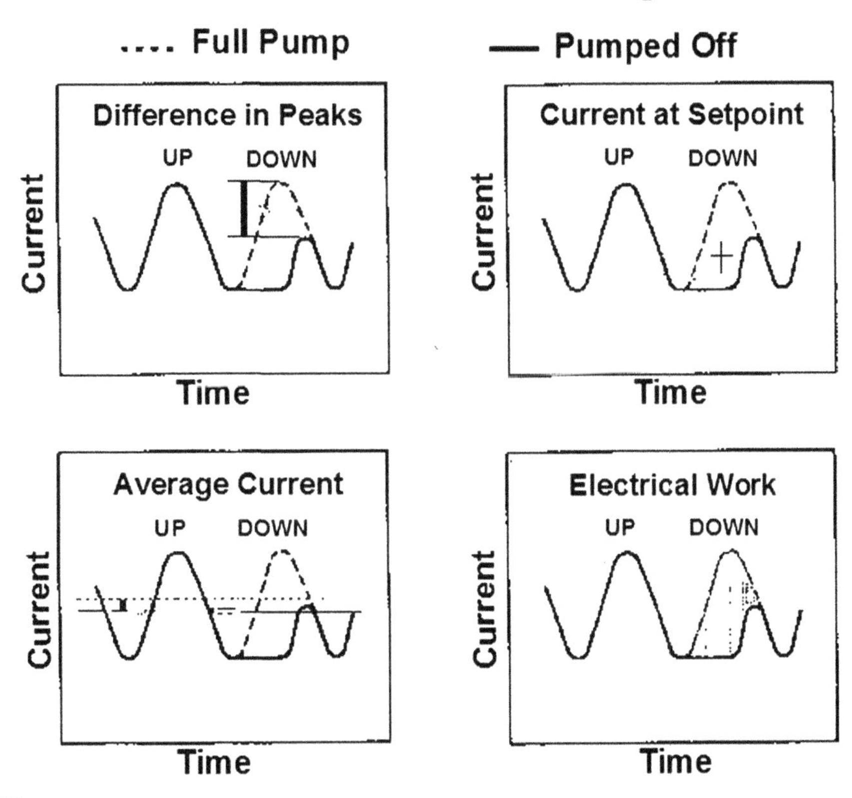

FIGURE 5.15

Operational principles of pump-off control devices utilizing the detection of motor current.

*Modified after **Westerman** [36].*

monitored, and a pumped-off condition can be identified easily from the drop in actual average current. Finally, electrical **work** during the cycle is **reduced** in pumped-off cases. Integrating instantaneous currents throughout one pumping period, the controller is set to stop the pumping unit when a prescribed reduction in electrical work is sensed.

It can be concluded, in general, that motor current detection has **inherent** errors that made the use of this principle obsolete in modern POC units [37]. One of the main drawbacks is that pump-off conditions and a high liquid level cannot be **distinguished** by the unit and this may cause improper operation. An increasing liquid level in the well results in less energy requirement to lift well fluids and the POC erroneously shuts down the pumping unit. Another problem is that the controller is **sensitive** to the mechanical counterbalancing of the pumping unit, because the current transducer cannot differentiate between the motoring and generating current. Since a well with a parted rod string may take a greater average current than under normal conditions, the POC will not shut down the unit when it would be necessary to do so. In addition, these systems do not provide **further** information on the operation of the pumping system and are, therefore, seldom used in modern pumping control [38,39].

5.3.2.2.2 Modern pump-off controller units

Polished Rod Load and Position Monitoring

One of the most **popular** classes of POC devices utilizes the monitoring of polished rod **loads** and **positions** because polished rod loads are good **indicators** of the operation of the downhole pump. Pumping malfunctions, including pump-off, are easily **detected** by an evaluation of surface loads. An added, but very important, advantage is that polished rod load data, along with rod position information, allow a complete **analysis** of the pumping system's operation through the solution of the wave equation. Due to the many **advantageous** features, many present-day pump-off controllers work on the principle of polished rod load monitoring.

These POC units require two principal parameters to be measured over the pumping cycle: **rod load** and **rod position**. Load can be sensed either directly at the polished rod with a **load cell**, or indirectly at the beam. Polished rod load cells are more **accurate** but are subject to damage by workover personnel. Beam-mounted transducers offer less accuracy but are more **robust** and have been used since the first controllers of this class appeared [40]. Rod **position** information can be gained from a continuous **potentiometer**, which provides a direct measurement but **limited** life. In some controllers, polished rod position is **approximated** by a simple sine function of pumping time [41], instead of a direct measurement. Only a position **switch** without any moving parts is required to identify one point of the position–time function. These switches detect the passing of the crank and send a signal when the crank reaches a given position. Because of the approximate nature of the position data their output cannot be reliably used for solving the wave equation; advantages are the easy installation, reliable service, and no moving parts [42]. The latest development is the use of an **accelerometer**, which gives a signal that is integrated twice to get polished rod position. The load and position data from these transducers is utilized to construct and to store dynamometer cards during operation for later or real-time analysis. For a complete treatment of load and position measuring devices used in POC units, consult **Gibbs'** book [37].

Figure 5.16, modified after **Westerman** [36], illustrates the basic pump-off detection systems that monitor polished rod **load** and **position**; the figure contains two dynamometer cards for each case: one representing a full pump and the other valid at pump-off conditions. If a **set point** is properly selected

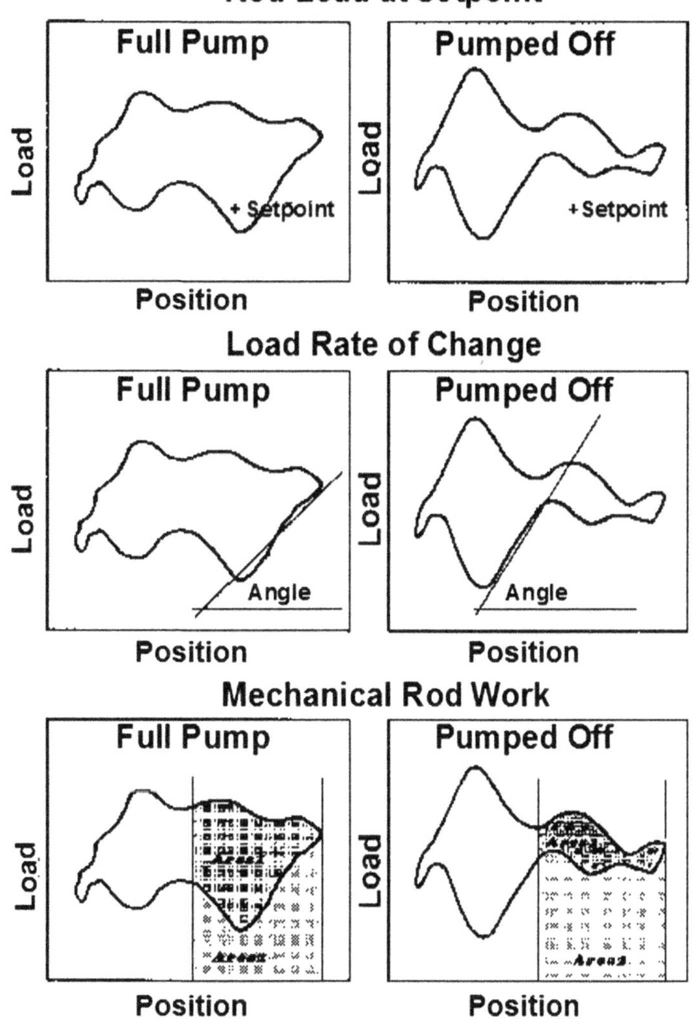

FIGURE 5.16

Operational principles of pump-off control devices utilizing the monitoring of rod loads.

Modified after ***Westerman*** *[36].*

on the dynamometer card, it will fall **below** the actual downstroke loads when pump-off occurs. This is because the transfer of fluid load from the rod string to the tubing is delayed by the late opening of the traveling valve. The POC unit should shut down the power supply to the pumping unit for a preset time so that liquid can accumulate in the well. One very successful type of POC measures the rate of

change, with respect to time, of the polished rod load during the downstroke. At pump-off, rod load experiences a more rapid decrease at the downstroke; if the controller detects a higher rate than previously set, then it shuts down the pumping unit.

Another principle of pump-off control involves calculation of the mechanical **work** done at the polished rod. Since work is proportional to the **area** enclosed by the card, and pump-off involves a definite reduction in energy input to the system, only a calculation of the card's **area** or a part thereof must be accomplished to detect a pumped-off condition [43]. This approach is not successful when the full card area is monitored because power used during the pumping cycle depends on the reservoir pressure as well, and an increasing liquid level in the well decreases power usage, causing the controller to malfunction. A better solution sets two vertical lines on the card and calculates the area inside or below the card between these two rod position lines. If the inside area of the card is used, the decrease of **Area 1** in Fig. 5.16 is monitored, whereas the increase of **Area 2** is monitored if the area below the card is used. This is based on the fact that the area under the dynamometer card on the downstroke (**Area 2** in the figure) is very sensitive to pump-off and it increases rapidly as pump-off progresses.

Pump Card Monitoring

All the previously discussed control algorithms can be applied with the utilization of pump cards [44] calculated from surface data. Polished rod load and position data being continuously fed to the POC by the transducers installed on the pumping unit, a microprocessor in the POC is programmed to solve the damped wave equation and to create pump cards. Since the pump card's shape changes from stroke to stroke, the development of a pump-off condition is easily detected by any of the methods presented in Fig. 5.16 and the POC can decide accordingly when to shut down the pumping unit. The advantages of using pump cards instead of surface dynamometer data can be summarized as follows:

- Since the pump card represents the conditions at the pump without the effects of downhole friction along the rod string, which are contained in surface dynamometer cards, all errors related to this effect are eliminated.
- Malfunctions difficult to find from surface cards are easily detected (e.g., tubing leaks, slipping tubing anchor) if the displacement of the plunger is calculated from the pump card.
- By calculating the pump intake pressure (PIP) from the pump card, the POC unit can distinguish between gas interference and fluid pound because gas interference is associated with high PIPs, whereas fluid pound occurs at low PIPs. This feature allows the POC to continue pumping a well during gas interference, even with a high liquid level that caused earlier POCs to shut down the pumping unit by mistake.

Motor Speed Monitoring

As discussed in Section 3.10.3, electric motors used in sucker-rod pumping service react **immediately** to any changes in load and change their rotational speeds according to their torque–speed performance curve. This feature makes it possible that motor **power** can be calculated accurately, provided the following parameters are known: (1) instantaneous rotational speed of the motor, and (2) the steady-state torque–speed performance curve of the given motor. Since motor input power decreases as the well produces less liquid in a pumped-off condition, POC devices operating on this principle were developed [45,46]. Such types of POCs utilize a **Hall-effect** transducer to sense the instantaneous rotational **speed** of the prime mover's shaft or the V-belt sheave. Speed data over the pumping cycle, combined with motor performance data versus motor speed, allow the calculation of the output **power** at the motor's shaft. This information is then used to detect a **decrease** in motor power, a primary **sign** of pump-off conditions. The main advantages of POC units developed along this principle are their **simplicity**, their reliability, and the elimination of cabling to moving parts. Their basic shortcoming, however, is the need for reliable motor performance curves that are seldom available from motor manufacturers.

Monitoring of motor speed is also used in POC devices that utilize load cells to measure polished rod loads and require the measurement of polished rod position during the pumping cycle. In general, position transducers can be **analog** (potentiometers, inclinometers, angle transducers) or **digital** devices (proximity switches, reed switches), but all have their practical limitations. A reliable and time-proven solution involves the monitoring of motor speed similarly to that just discussed [47]. Two **Hall-effect** transducers are used: one tracks the passage of a small permanent magnet attached to the motor shaft or the motor sheave; the other signals the passing of a magnet fixed to the pumping unit's crank. The signal from the first transducer represents instantaneous motor **speed** while the second signals the **bottom** of the polished rod stroke. Those two parameters, combined with the kinematic parameters of the pumping unit geometry, allow the calculation of polished rod position versus time. The advantages of this arrangement are numerous because they: (1) are easy to install, (2) are more reliable than analog devices, and (3) supply data sufficiently accurate for wave equation calculations.

5.3.2.2.3 Benefits of pump-off controllers

Pump-off controllers may be used in different configurations in the oil field. A **stand-alone** POC provides local control at the wellhead, with an inherent low cost and the necessity of periodic on-site adjustments. POCs connected to a central **computer** transmit data through communications channels and receive control from the central machine. Field adjustments are mainly eliminated, and detection of pumping malfunctions is highly efficient (but at a much higher cost). Finally, POCs can be hooked to a field-wide supervisory control and data acquisition (**SCADA**) system. In such systems, the local unit accomplishes data processing and basic local control functions. The central supervisory computer provides further data processing and **analysis** as well as operations **control**.

The availability of inexpensive, high-capacity microprocessors in the early 1980s has brought about basic changes in the functions of a modern POC unit. The ample memory storage capability of these processors allows the storage of large amounts of data for later analysis or transmission. Usually, **several** dynamometer diagrams are stored before the pumping unit is shut down and when operating malfunctions occur. This feature ensures that operational problems are easily **detected** and analyzed. The present-day POC, therefore, in addition to providing the basic functions of intermittent pumping control (starting and stopping the pumping unit), becomes a well **management** tool as a result of its data acquisition and data processing capabilities.

The basic economic benefits of the application of all types of POC units are similar and stem from several different features:

- Energy consumption is **reduced** due to shorter pumping times. In time cycle control, it is almost inevitable that wells are pumped longer than necessary, for fear of production losses. The POC units eliminate **overpumping** by shutting down the unit when pump-off occurs. An average energy conservation of 20–30% can generally be achieved.
- Well production rates moderately **increase** because lower average fluid levels can be reached. Lower fluid levels mean lower-flowing bottomhole pressures and, consequently, higher fluid inflow into the well. Usual figures of production increase are 1–4%.

- Maintenance costs of downhole and surface equipment can be **reduced** by 25–30%. This is mainly attributed to the elimination of fluid-pound conditions, which are responsible for a great number of pulling jobs. Additionally, earlier detection of pumping malfunctions by the POC units further decreases the need for workover operations.
- Last but not least, the use of POC devices significantly increases the amount and reliability of available **information** on system performance. The efficiency of field personnel is facilitated, and the overall economic **parameters** of rod pumping can be improved significantly.

5.3.3 PUMPING AT CONTROLLED SPEEDS

As discussed in the previous section, intermittent pumping reduces the pumping system's capacity by limiting its daily **run time**. When the unit operates it runs at a constant pumping **speed** and then it is shut down for an extended period. This kind of intermittent operation has several **drawbacks**, as follows:

- The daily average flowing **bottomhole pressure** is relatively high because well fluids rise in the tubing during the off-time, restricting the production rate that can be achieved.
- Every time the unit is started up, high instantaneous **power** is needed because of the high **inertia** of the surface and downhole equipment.
- Sand or solids produced along with well fluids can **settle** during the **downtime** on the sucker-rod pump, causing several operational problems.
- Static and dynamic pumping loads are very similar to those during continuous pumping operations, limiting the **fatigue** life of the equipment.

The solutions to overcoming the just-described disadvantages of intermittent pumping utilize **continuous** operations with a **controlled** pumping speed. Pumping speed can be set at a **constant** low level unattainable from a normal pumping unit, or it can be **adjusted** automatically by utilizing a variable speed drive (VSD) unit.

5.3.3.1 Low-speed continuous pumping

In cases when the pumping system's lifting capacity using the available **lowest** pumping speed is still **greater** than the well's inflow rate, running the pumping unit at an even lower **speed** would be desired. Pumping unit gearboxes, however, usually have a recommended **minimum** pumping speed below which they must not be operated because of loss of lubrication to the gears. Lower-than-normal pumping speeds can be reached if a **jackshaft** is fitted between the prime mover and the gearbox; to maintain proper lubrication of the gearbox extra oil **wipers** must be installed in the gearbox. As shown in Fig. 5.17, the two **sheaves** on the ends of the jackshaft are of different **diameters** and provide a substantial **reduction** of pumping speed. For the example case, the sheaves of 8 and 24 in diameters can reduce the unit's speed to one-third of the original.

Experience with more than 500 wells [48] proved the following **advantages** of low-speed continuous pumping:

- Investment and installation costs are **moderate**; they are much less than the cost of a single tubing pulling operation.
- Rod stress **fluctuations** are much reduced; the **fatigue** life of rods is increased with greatly reduced rod and pump repair costs.

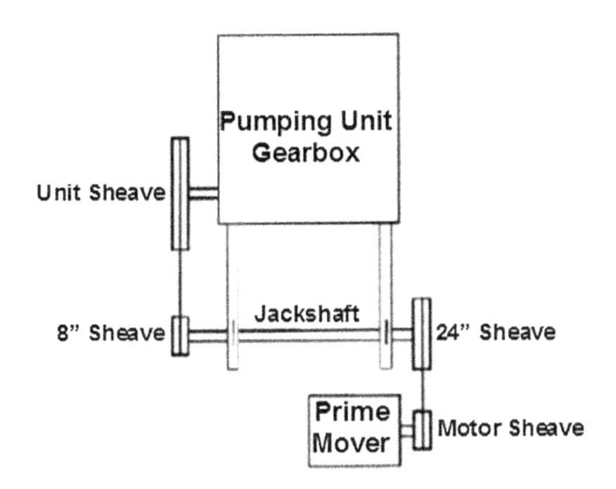

FIGURE 5.17

Schematic arrangement of a jackshaft to reduce pumping speed.

- As the result of continuous operation, **power** consumption is reduced while the well's production rate is **maximized**; sand problems are minimized at the same time.

5.3.3.2 Use of variable speed drive units

The displacement of the sucker-rod pump also can be changed by varying the **speed** of the prime mover driving the pumping unit. Since pumping units are driven normally by induction-type AC motors and the speed of those motors mainly depends on the frequency of the electricity applied to the motor's terminals (see Section 3.10.3), pumping speed can be changed at will by changing the **frequency** of the power supply. As pump displacement varies proportionally to speed, a proper control of the driving frequency ensures that pumping rate always **matches** the inflow rate from the formation. Electric frequency is usually adjusted by VSDs that are widely used in many industries as well as in the oil patch, e.g., to control the performance of surface pumps or electric submersible pumps.

5.3.3.2.1 Basics of variable speed drive operation

A VSD unit receives **alternating** voltage from the power **supply** at a constant (usually 60 Hz) frequency, internally rectifies this to a direct voltage, and then **synthesizes** an alternating voltage of any **frequency** at its output; output current fluctuates according to the load on the unit. Connecting the device between the power system and the pumping unit's prime mover, motor speed, and, consequently, pumping **speed** can be changed at will within a fairly broad range. VSD units have the following three basic components (see Fig. 5.18):

1. the **rectifier** section converts the 60 Hz AC voltage into a DC voltage,
2. the **DC control** section provides a smooth DC waveform to the next section, and
3. the **inverter** section creates an AC voltage at a selected frequency, simulating a sine wave.

The first VSD units manufactured in the late 1970s used **thyristors** to create the AC waveform and produced a very crude approximation of a sine wave. The typical voltage and current output is shown in Fig. 5.19, from which the name of the unit came: "**six-step**" VSD. Motor current, as seen, is more

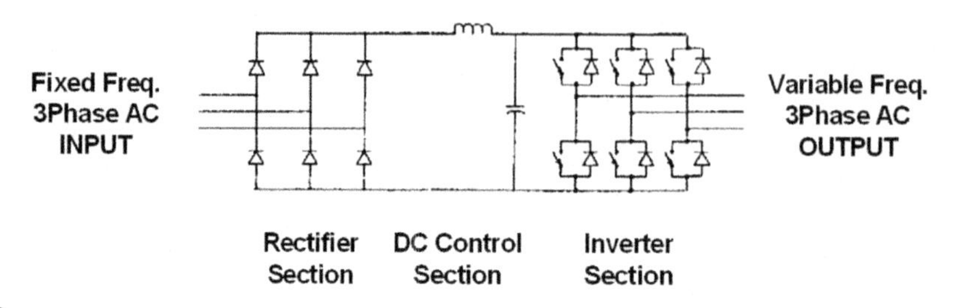

FIGURE 5.18

Main components of a VSD unit.

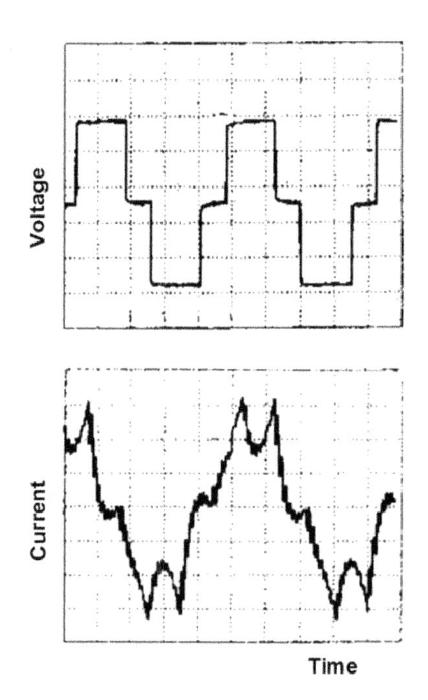

FIGURE 5.19

Waveforms of a "six-step" VSD unit.

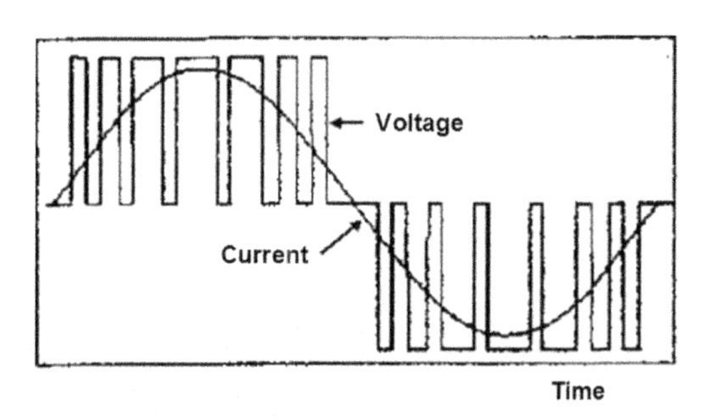

FIGURE 5.20

Waveforms of a VSD unit using pulsed-width modulation (PWM).

sinusoidal than the square voltage wave because of the inductive characteristics of the AC motor. Six-step VSDs were bulky and unreliable and were almost completely replaced by the early 1990s with units using "pulsed-width modulation," or **PWM**, in their inverters to create a more sinusoidal current. These devices used, instead, thyristors, extremely fast semiconductor switches like insulated gate bipolar transistors to create the output frequency. As shown in Fig. 5.20, the output voltage of these PWM-based VSDs takes the form of a series of voltage pulses, each of the same magnitude but having a different, controlled width. The output frequency is controlled by adjusting the number of pulses per cycle (called the carrier frequency) and the output voltage is controlled by the width of the pulses. The current in the motor very effectively approximates a sine wave because the motor is primarily an inductive device.

Motor speed control with early VSD units was accomplished using the "volts per Hz" principle. This is based on the fact that both the magnetic flux density inside the motor and the torque developed by the motor change if the motor is supplied with variable frequency at a constant voltage. In order to keep the torque **constant**, the ratio of the voltage and the frequency has to be kept constant because motor torque equals voltage per frequency squared; the constant **volts/frequency** ratio makes the motor a **constant-torque**, **variable-speed** device. This is the reason why such VSD units' output frequencies and voltages satisfy the following equation:

$$\frac{U_2}{f_2} = \frac{U_1}{f_1} \tag{5.5}$$

where:

f_1, f_2 = AC frequencies, Hz, and
U_1, U_2 = output voltages at f_1 and f_2 Hz, V.

Up-to-date VSDs utilize the "**sensorless flux vector**" principle [49,50], which is designed to control the AC motor's output torque. This eliminates the major limitation of the "volts per Hz" control mechanism that it does not control the torque of the motor. Torque control in flux vector control is achieved by adjusting the current supplied to the motor; the unit's microprocessor continuously calculates the total required motor current (the sum of the magnetizing and the torque-producing currents) so that the motor develops a torque matching the load exerted on it during every pumping cycle. These VSDs allow a more precise control of motor parameters than previous units and can sustain speeds **regardless** of the torque load. Since the processors used for this purpose sense (or emulate) the motor's torque output, torque peaks are detected and gearboxes can be protected against overload. One of the advantages of this control method is that **NEMA B** motors can be used, which, if compared to the universally used **NEMA D** motors (1) are about 10% more efficient, (2) are less expensive, and (3) have power factors close to unity. In summary, VSDs using this new technology allow a much enhanced control of rod pumping operations.

5.3.3.2.2 Use of variable speed drives in pump-off control

Application of VSD units in rod pumping control reached a level today that can be called "well management" where POC and VSD units work together (sometimes in a single unit) to optimize the operation of rod pumping installations. Optimization is meant here to obtain the maximum possible liquid rate from the well while meeting all mechanical and economic restrictions of the pumping system. The usual arrangement of POC and VSD devices is illustrated in Fig. 5.21. Operational data on the pumping cycle are collected first; this can involve any of the previously detailed methods, like surface load, position, or motor speed data measurements. Based on these data, fed to the POC, the POC signals to the VSD to adjust the pumping speed; the VSD, receiving power at a constant frequency, outputs a variable frequency to the motor, thereby changing the speed of the motor for the next pumping cycle. The change in motor speed alters the pumping speed of the unit and, consequently, the production **capacity** of the system to **match** the inflow liquid rate from the formation.

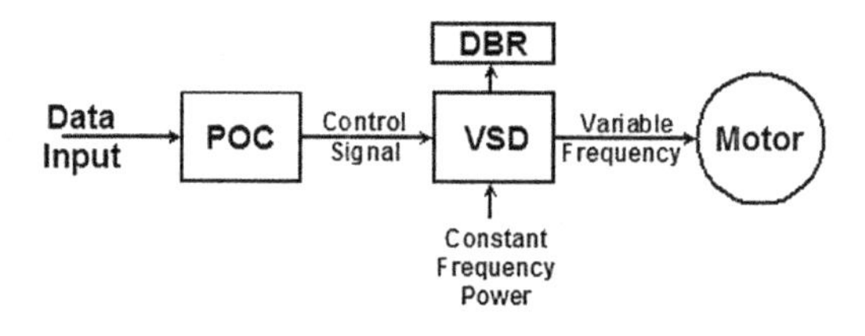

FIGURE 5.21

The usual arrangement of POC and VSD devices for pumping speed control.

There is a basic problem with this arrangement during those periods of the pumping cycle when the torque on the motor is **negative**. As discussed in Section 4.6.4, negative torques cannot be eliminated completely, even when using the optimum counterbalance on the gearbox. Because of the negative torques the motor becomes an electric **generator** for portions of the pumping cycle, and creates regenerative power; this power, however, cannot be returned through the VSD to the power supply. This is the reason why in most types of VSDs **regenerative** power must be **dissipated** as heat through the dynamic braking resistors (DBRs), increasing the wasted energy. A special VSD unit offered by a major manufacturer overcomes this problem and allows regenerative power to be returned to the power supply [51].

The first application of a VSD unit to control a sucker-rod pumping system is reported by **Guffey et al.** [52]. The system continuously **monitors** the downstroke portion of the dynamometer card because the area under the card on the downstroke increases rapidly while pump-off progresses, as shown in Fig. 5.16. At pump-off, pump displacement **exceeds** well inflow and the microprocessor-based controller signals to the VSD unit to **decrease** the electric **frequency** and to slow down the prime mover. On the other hand, if the fluid level **rises** above the pump, pumping speed is increased to cope with the higher inflow rate to the well. Thus, input frequency to the motor is continuously **controlled** so as to stabilize the **fluid level** in the well near to pump setting depth. This operating mechanism ensures a continuous **matching** of the system's pumping capacity to changing well inflow conditions.

A major manufacturer's well manager unit [53] adjusts the pumping speed at every stroke in order to change the pumping system's **capacity** to match the inflow from the well. The device senses pump-off conditions based on the actual liquid fillage of the downhole pump from the surface dynamometer card or the calculated downhole pump card. In the latter case, as shown in Fig. 5.22, the current **fillage** of the pump is detected at the lower right corner of the pump card and is defined as the ratio of the plunger's net stroke to its gross stroke. For a full pump (the dashed curve) liquid fillage is 100% and it decreases as pump-off conditions are approached; see the solid curve. As long as pump fillage in any pumping cycle falls within a deadband around a **set point**, the VSD unit keeps the pumping speed unchanged. Every time that actual pump fillage falls to the left of the deadband, **pump-off** conditions are approached, and the VSD unit slows down the unit. Because the **capacity** of the pumping system is thus decreased, liquid level in the well's annulus rises and pump fillage increases. On the other hand, fillage values to the right of the deadband indicate increased fluid inflow from the formation and the pumping unit is sped up to increase its capacity to match well inflow and to reduce the liquid level in the annulus. This **automatic** operation ensures that the capacity of the pumping system always matches the inflow from the well and that the **maximum** possible liquid rate is produced from the well.

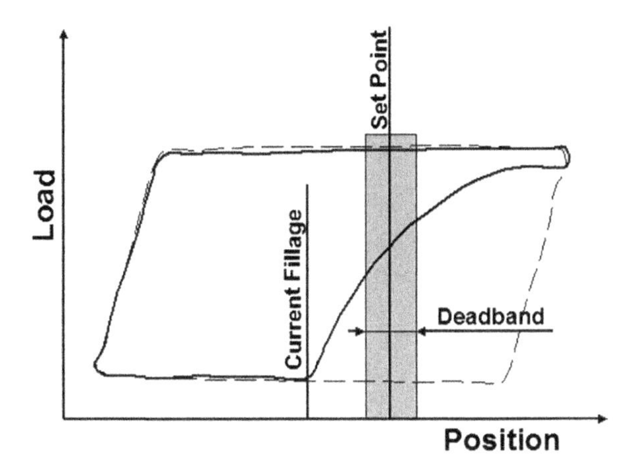

FIGURE 5.22

Operational principle of a well manager unit.

5.3.3.2.3 Latest developments

All well management systems discussed so far change the pumping speed from stroke to stroke so that the pumping unit operates at different, but **constant** speeds during every subsequent cycle. **Beck et al.** [54] was the first to devise a control unit that makes multiple adjustments to the prime mover's speed **within** every single cycle. This kind of operation improves system efficiency and increases pump displacement; use of this controller increased well rates by an average of 21% in a 100-well field [55].

The use of well controllers with incorporated VSD units in sucker-rod pumping was raised to a new level by a Canadian invention described by **Palka and Czyz** [56,57]. They utilized the extremely **quick** response time of electric induction motors to frequency changes and created a controller that varies the pumping speed **throughout** the pumping cycle. With the use of certain transducers and specially developed software, their controller permits the **optimization** of the total pumping system according to several different criteria. The most important optimization goal is to **maximize** the well's liquid **production** rate while **minimizing** sucker-rod **loading** and **energy** consumption by finding the proper **variation** of the instantaneous pumping speed during the pumping cycle.

In order to solve the multivariable nonlinear optimization problem, the authors relied on the exact description of:

- the kinematic **behavior** of the pumping unit to calculate the movement of the polished rod,
- the solution of the damped **wave equation** to find the plunger's stroke length and the stresses along the rod string, and
- the **dynamics** of the surface system to calculate static and inertial torques on the motor's shaft.

All problems listed above are solved by the utilization of state-of-the-art theories and calculation models and the pumping controller's logic was developed accordingly. The optimization process may include several **constraints**, such as the following:

- rod stresses in any rod section at any time must not exceed the **limits** found from the modified Goodman diagram,
- maximum gearbox torque should be less than gearbox **rating**, and
- energy consumption can be limited.

The remote cloud-based software does all necessary calculations and the optimum motor speed versus time function is presented in the form of a **Fourier series**. Calculated instantaneous motor speeds for the whole pumping cycle are then converted to the appropriate electric **frequencies** and these are sent to the VSD unit to change the instantaneous motor speed at the appropriate time or rod position throughout the cycle. To achieve the optimum speed variation, rapidly changing **periods** of speeding up and slowing down follow each other during the up- and downstroke of the polished rod. These **rapid** speed changes make it possible to **maximize** the liquid production of the pump while meeting all constraints set beforehand because the controller accounts for all static and dynamic effects occurring downhole and on the surface.

To illustrate the capabilities of this new control method, a sample case is presented [58] where the objective was to maximize the liquid production from an existing well while maintaining the same pumping equipment loading. The 1.75 in rod pump was set at 8,546 ft and a composite rod string was used; the polished rod stroke length was 144 in. Originally the pumping unit was operated at a constant speed of 5.12 SPM and the well produced about 191 bpd of liquid. After the well manager was installed, the well's liquid production rate had increased to 357 bpd while the loading of the rod string and the gearbox remained constant. The controller varied the pumping speed within every pumping cycle with an average value of 7.89 SPM.

The substantial increase of the pumping rate is the consequence of the longer plunger stroke length made possible by the continuous adjustment of the prime mover's speed during the pumping cycle. As shown in Fig. 5.23, measured surface dynamometer cards and calculated pump cards for the constant- and variable-speed cases differ considerably. Plunger stroke length has increased from 108 in, valid for the constant speed, to 128 in when using the variable-speed controller. The variation of instantaneous motor speed is given in Fig. 5.24; it is changed very rapidly by the controller between 750 and 1,920 rpm with an average speed of 1,220 rpm. The pumping speed varies accordingly, from as low as 4.9 SPM to higher than 12 SPM. This extremely rapid adjustment of

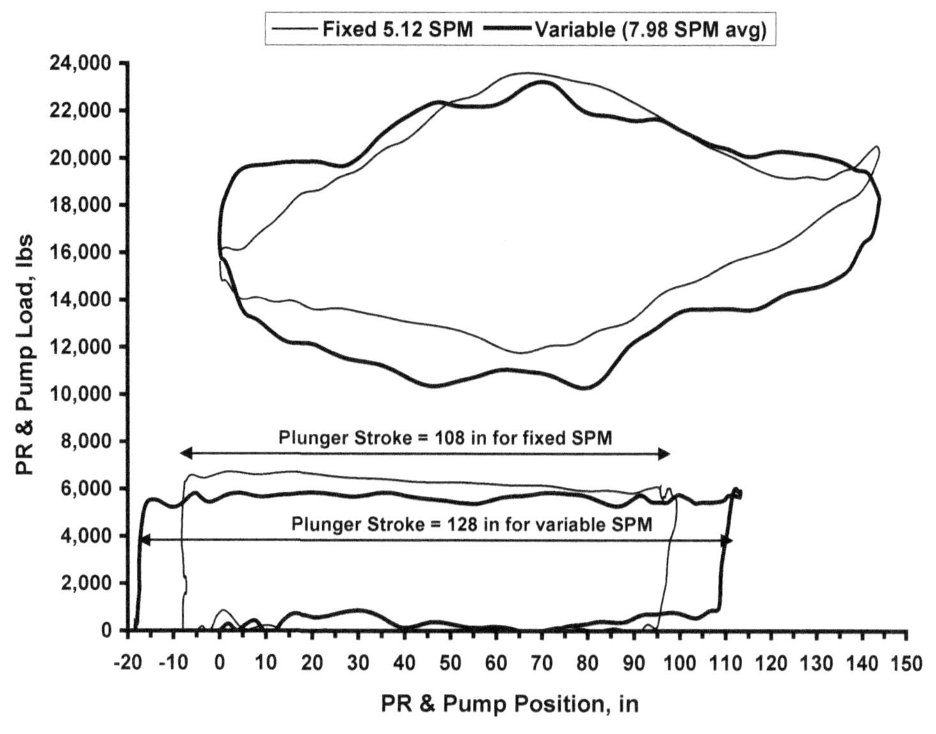

FIGURE 5.23

Comparison of surface and pump cards for constant- and variable-speed operation.

According to Ref. [58].

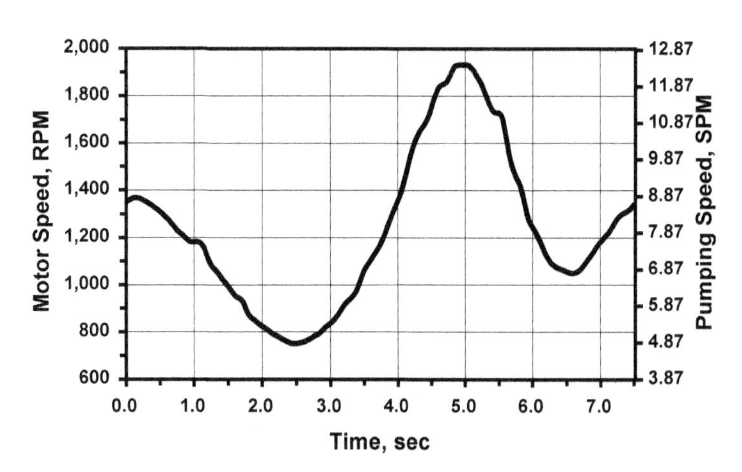

FIGURE 5.24

Variation of motor and pumping speeds during the pumping cycle.

According to Ref. [58].

the pumping speed, coupled with the accurate description of the rod string's behavior, allows the controller to maximize the plunger's stroke length and, consequently, the liquid rate produced by the downhole pump.

5.3.3.2.4 Benefits of using variable speed drives

The general benefits of using VSD units, especially the modern "**sensorless flux vector**" types, in pump-off control can be summed up as follows [59]:

- The use of **NEMA B** motors with higher efficiencies than those of **NEMA D** motors is possible with an associated lower required motor size.
- They allow a "**soft start**" of the pumping unit by slowly increasing the pumping speed from zero at system start-up, eliminating the high instantaneous current demands and the shock loads on the pumping unit and the rod string, compared to conditions when the motor is started across the line.
- The "**sensorless flux vector**" VSD can limit the **peak torque** on the gearbox to the unit's allowed rated torque; no gearbox overload can occur.
- The need for unit/motor **sheave** changes to adjust pumping speed is eliminated, with the associated reduction of production costs.
- Damage caused by sand and/or solids settling on the downhole pump during shutdowns (stuck and/or damaged pumps) is eliminated because of the continuous operation.
- The accuracy of pump-off control is enhanced as compared to a stand-alone POC unit.

REFERENCES

[1] Byrd JP. Mathematical model enhances pumping unit design. OGJ January 29, 1990:87–93.

[2] Gault RH. Designing a sucker-rod pumping system for maximum efficiency. SPE PE November 1987: 284–90.

[3] API TL 11L. Design calculations for sucker-rod pumping systems (conventional units). 5th ed. Washington, D.C.: American Petroleum Institute; 2008.

[4] API Bul. 11L3. Sucker rod pumping system design book. 1st ed. Dallas, Texas: American Petroleum Institute; 1970.

[5] Neely AB. Sucker rod string design. PE March 1976:58–66.

[6] Lufkin conventional pumping system design manual. Lufkin Industries, Inc; 1983.

[7] Byrd JP. A technique for measuring and evaluating the performance of a beam pumping system. JCPT January–March 1973:33–7.

[8] Byrd JP. Pumping deep wells with a beam and sucker rod system. Paper SPE 6436 presented at the deep drilling and production symposium of the SPE in Amarillo, Texas. April 17–19, 1977.

[9] Estrada ME. Design and optimization tables for the mark II oil field pumping unit (M.S. thesis). The University of Tulsa; 1975.

[10] Lekia S, Keteh D. An improved technique for evaluating performance characteristics and economy of the conventional and mark II beam and pumping systems (M.S. thesis). The University of Tulsa; 1982.

[11] Takacs G. Program optimizes sucker-rod pumping mode. OGJ October 1, 1990:84–90.

[12] Takacs G, Hadi B. Latest technological advances in rod pumping allow achieving efficiencies higher than with ESP systems. J Can Pet Technol April 2011:53–8.

[13] Ghareeb MM, Shedid SA, Ibrahim M. Simulation investigations for enhanced performance of beam pumping system for deep, high-volume wells. Paper SPE 108284 presented at the international oil conference and exhibition held in Vera Cruz, Mexico. June 27–30, 2007.

[14] Byrd JP. High volume pumping with sucker rods. JPT December 1968:1355–60.
[15] Pumping unit catalog 1990–91. Lufkin Texas: Lufkin Industries, Inc; 1990.
[16] Nolen KB. High volume pumping analysis. PE August 1970:60–70.
[17] Richards L, Sidebottom J. Systems reduce casinghead pressure. The American Oil and Gas Reporter; May 2005.
[18] www.hy-bon.com.
[19] Al-Khatib AM. Improving oil and gas production with the beam-mounted gas compressor. J Pet Technol February 1984:276–80.
[20] Neuvar KR. Effects of a beam-mounted gas compressor on a conventional pumping unit. Proc. 35th Southwestern petroleum short course, Lubbock, Texas. 1988. p. 278–287.
[21] McCoy C. Increasing production on rod pumping wells. Proc. 37th Southwestern petroleum short course, Lubbock, Texas. 1990. p. 350–358.
[22] McCoy CD. Pump Jack operated compressor. US Patent 4,530,646. 1985.
[23] Huff C, Huff J, Day T, Sipes K, McGiunis M, McCoy CH. Walking beam-operated compressor offers solution for wellhead compression. World Oil March 2001:99–101.
[24] McCoy C. Beam gas compressor: the greenest of compression systems. Proc. 57th Southwestern petroleum short course, Lubbock Texas. 2010. p. 123–126.
[25] www.beamgascompressor.com.
[26] Brown KE. The technology of artificial lift methods, vol. 4. Tulsa, Oklahoma: PennWell Publ. Co; 1984.
[27] Brown KE, Lea JF. Nodal systems analysis of oil and gas wells. JPT October 1985:1751–65.
[28] Schmidt Z, Doty DR. System analysis for sucker rod pumping. SPE PE May 1989:125–30.
[29] Pope CD. Optimizing high-volume sucker rod lift. Paper SPE 25421 presented at the production operations symposium held in Oklahoma city. March 21–23, 1993.
[30] Guirados CD, Ercolino JM, Sandoval JL. Nodal B: a unique program for optimum production of sucker rod pumping oil wells. Paper SPE 30183 presented at the petroleum computer conference held in Houston. June 11–14, 1995.
[31] Han D, Wiggins ML, Menzie DE. An approach to the optimum design of sucker rod pumping systems. Paper SPE 29535 presented at the production operations symposium held in Oklahoma city. April 2–4, 1995.
[32] Takacs G. Application of systems analysis to sucker rod pumped wells. Proc. 38th annual Southwestern petroleum short course. Lubbock, Texas. 1991. p. 366–374.
[33] McCoy JN, Becker D, Podio AL. Timer-controlled beam pumps reduce operating expense. OGJ September 13, 1999:86–97.
[34] Szilas AP. Production and transport of oil and gas. 2nd ed. Elsevier Publishing Co.; 1985 [Part A, Chapter 4.1].
[35] Godbey JK. New controller optimizes pumping-well efficiency. OGJ September 24, 1979:161–5.
[36] Westerman GW. Pump-off controllers match pump capacity to production. OGJ November 21, 1977: 131–5.
[37] Gibbs SG. Rod pumping. Modern methods of design, diagnosis and surveillance. USA: BookMasters, Inc.; 2012.
[38] Lea JF. New pump-off controls improve performance. PEI December 1986:41–4.
[39] Westerman GW. Pump-off control, state of the art. Proc. 36th annual Southwestern petroleum short course. Lubbock, Texas. 1989. p. 384–391.
[40] Stoltz JR. The application of well load monitors to beam pumping. Paper SPE 4537 presented at the 48th fall meeting of the SPE of AIME, at Las Vegas, Nevada. September 30–October 3, 1973.
[41] Patterson MM. A pump-off detector system. JPT October 1979:1249–53.
[42] Rod pump control transducers. Lufkin Automation brochure; 2011.

[43] Gibbs SG. Method for determining the pump-off of a well. US Patent 3,951,209. April 1976.

[44] Gibbs SG. Monitoring and pump-off control with downhole pump cards. US Patent 5,252,031. October 1993.

[45] Gibbs SG. Method for monitoring an oil well pumping unit. US Patent 4,490,094. December 1984.

[46] Gibbs SG. Utility of motor-speed measurements in pumping-well analysis and control. SPE PE August 1987: 199–208.

[47] Dugan L, Howard L. Beyond pump-off control with downhole card well management. Proc. 49th Southwestern petroleum short course, Lubbock, Texas. 2002. p. 55–61.

[48] Bommer PM, Shrauner D. Benefits of slow-speed pumping. J Pet Technol October 2006:76–9.

[49] Lovelance J. Beam pumping with variable speed drives – past, present and future. Proc. 60th annual Southwestern petroleum short course, Lubbock, Texas. 2013. p. 83–89.

[50] Poythress M. True intelligence at the wellsite. Proc. 50th annual Southwestern petroleum short course, Lubbock, Texas. 2003. p. 177–188.

[51] REGEN. Rod pumping and progressing cavity pump applications. Lufkin TX: Lufkin Automation brochure, LUFKIN Industries; 2011.

[52] Guffey CG, Rogers JD, Hester LR. Field testing of variable-speed beam-pump computer control. SPE PE May 1991:155–60.

[53] SAM well manager. Variable speed drive rod pump control. Houston TX: User Manual, Lufkin Automation; 2009.

[54] Beck TL, Peterson RG, Garlow ME, Smigura T. Rod pump control system including parameter estimator. US Patent 7,168,924. January 2007.

[55] Peterson RG, Smigura T, Brunings CA, Quijada S, Wilfredo A, Gomez AJ. Production increase at PDVSA using an improved SRP control. Paper SPE 103157 presented at the annual technical conference and exhibition held in San Antonio. September 24–27, 2006.

[56] Palka K, Czyz JA. Optimizing downhole fluid production of sucker-rod pumps with variable motor speed. SPE Prod. Oper. May 2009:346–52.

[57] Palka K, Czyz JA. Method and system for optimizing downhole fluid production. Canadian Patent 2,526,345. March 2011.

[58] Palka K. Personal communication. Calgary, Canada: PumpWell Solutions Ltd.; March 2013.

[59] Anderson S, Franklin J, Hawthorn D. Application of variable frequency drives to the rod pump system. Proc. 51st annual Southwestern petroleum short course, Lubbock, Texas. 2004. p. 11–14.

CHAPTER

THE ANALYSIS OF SUCKER-ROD PUMPING INSTALLATIONS

6

CHAPTER OUTLINE

Sucker-Rod Pumping Handbook. http://dx.doi.org/10.1016/B978-0-12-417204-3.00006-6

6.1 INTRODUCTION

This chapter deals with the methods and procedures that are available to **analyze** the operation of sucker-rod pumping wells. The first broad topic discussed covers the **inflow** performance testing of wells placed on rod pumping. As discussed in this chapter, special **well testing** procedures are required to conduct a production test because the methods developed for flowing or gas-lifted wells are not applicable. Among these, the **acoustic** determination of annular liquid levels is described, since such measurements form the basis of most of the testing procedures. The results of a properly conducted well test provide valuable information on actual well **potential** and help the pumping analyst to decide on possible changes to the pumping system.

The most common procedures to measure and analyze the operating conditions of rod-pumping installations use polished rod **dynamometers**; this chapter discusses the **hardware** used and the **procedures** followed when making dynamometer surveys. The use of hydraulic and electronic dynamometers to take surface dynamometer cards as well as to test the downhole pump's valves is detailed. Interpretation of surface and downhole dynamometer cards is also presented, with an emphasis on the evaluation of **pump cards** calculated from surface data using the solution of the damped wave equation.

6.2 WELL TESTING

The purpose of well testing is the determination of the well's **inflow performance** behavior. This involves the measurement of production rates for **several** bottomhole pressures and the construction of the well's inflow performance relationship (**IPR**) curve. Since the performance of any producing well constantly changes with the **depletion** of the reservoir and with other factors, wells must be tested **regularly** for changes in their inflow performance. In case of flowing and gas-lifted oil wells, an inflow performance test is conducted by running a pressure element (pressure bomb) to the bottom inside the tubing string and thereby recording flowing and static bottomhole pressures. This procedure cannot be applied to rod-pumped wells because the rod string present in the tubing prevents the running of wireline tools in the well. Thus the well must be **killed** and the rods pulled out before a downhole pressure survey can be conducted involving high workover costs, which are usually prohibitive. Although theoretically possible, pressure measurements in the well annulus are also ruled out as a result of many practical implications, such as restricted passage area in the annulus, a high probability of wireline breaks, etc. Therefore, the inherent need for running inflow performance tests also in rod-pumped wells necessitated the development of techniques and procedures different from those applied in wells placed on any other type of lift.

6.2.1 PRODUCTION TESTING

6.2.1.1 Conventional production tests

As discussed before, knowledge of the well's **liquid rate** is absolutely necessary to estimate its inflow performance during well testing. A production test, a standard part of oil field operations, supplies the liquid and gas production **rates** obtained from the well through periodic **measurements**. Liquid rates are conventionally measured in **storage tanks** by gauging; modern field management utilizes **lease automatic custody transfer** (LACT) units for this purpose. The liquid rates obtained by a production test are used for several reasons: to fulfill regulatory obligations, to facilitate reserve estimations, to help reservoir management, etc. Since wells are tested only **periodically**, production rates between tests are assumed to be unchanged; this is the basis of production allocation between several wells.

In order to establish an accurate IPR curve the production rate used in well testing must come from the **latest** production test, but for best results **simultaneous** measurements of the production rate and the flowing bottomhole pressure are recommended.

6.2.1.2 Inferred production calculations

Conventional well tests have an inherent **inaccuracy** due to the errors in liquid rates because these rates are measured only **periodically**. The concept of **inferred** production rates [1–3] eliminates this problem and uses the downhole pump as a **metering** device to calculate the pump's liquid and gas production rate for **every** pumping cycle. Cycle volumes accumulated for 24 h indicate the well's daily production rate, which is a very accurate estimation of the actual rate obtained from the well. This novel technique requires the continuous calculation of **downhole** pump cards from surface dynamometer data, which some of the advanced pump-off controller (**POC**) devices are capable to accomplish. The use of inferred production, if applied in sucker-rod pumped fields, may **reduce** or completely **eliminate** traditional production testing procedures and the need for related facilities.

The **background** of inferred production rate calculations is very simple: if the **effective** plunger stroke length used for liquid production is known, then the liquid volume passing through the pump during a pumping cycle is equal to the product of the plunger's cross-sectional area and the effective stroke length. In any consistent units this is written as:

$$\Delta V = A_p S_{\mathrm{eff}} \tag{6.1}$$

where:

ΔV = liquid volume per pumping cycle,
A_p = plunger area, and
S_{eff} = effective stroke length.

The only unknown in the formula is the **effective** plunger stroke; this must be **estimated** from the pump card, which, in turn, is calculated from the measured surface dynamometer card using the damped wave equation. In case a negligible gas rate flows through the pump, the **effective** plunger stroke is found by reducing the gross plunger stroke length by the following items, read from the pump card (see Fig. 6.1):

- Tubing stretch, e_t, calculated from **Hooke's law** using the fluid load. This term is zero if the tubing is anchored.
- Stroke loss, S_{leak}, due to fluid **leakage** through the pump. This parameter is determined from the actual leakage rate in the pump, the cross-sectional area of the plunger, and the pumping speed.

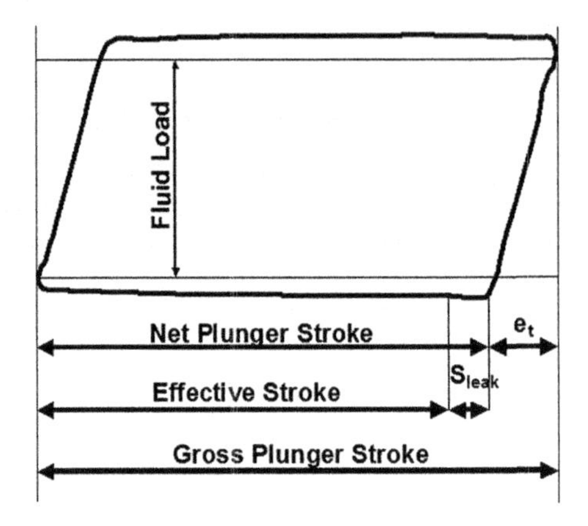

FIGURE 6.1

Schematic pump card in a well with negligible gas production.

Plunger leakage is usually calculated from the evaluation of the loads and velocities of the plunger or found from the data of valve tests.

If there is a considerable amount of **free gas** passing through the pump, the effective stroke needed for inferred production calculations is based on the **net** stroke. This, in turn, is determined from the shape of the pump card, as shown in Fig. 6.2. The **effective** stroke length used for liquid production is less than the net stroke; the difference is the **sum** of the following terms:

- Stroke loss, S_{leak}, due to fluid **leakage** through the pump; calculated as before.
- Stroke loss, S_{comp}, to compensate for the **compression** of free gas that enters the pump while the standing valve is open. Determination of the stroke loss due to gas compression requires complex calculations.

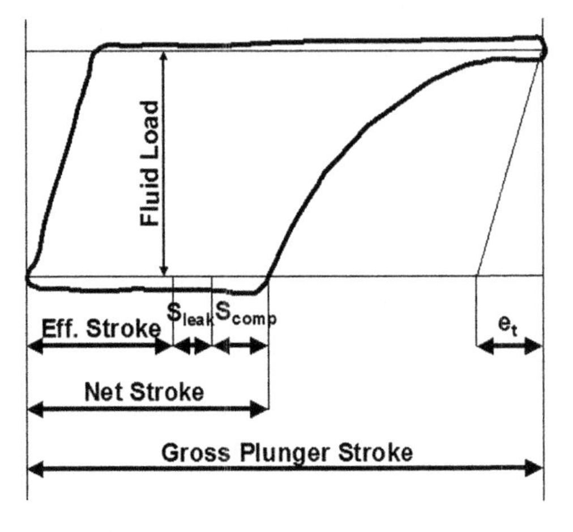

FIGURE 6.2

Schematic pump card in a well with considerable free gas production.

In any case (with or without considerable gas production) the **in situ** liquid volume passing through the pump is simply found from the **effective** plunger stroke length and the plunger cross-sectional **area**; see Eq. (6.1). Since inflow performance calculations require production rates to be expressed in standard units, the in situ volumes must be corrected by the appropriate **volume factors** to get stock-tank barrels.

As detailed previously, utilization of inferred production rates involves quite **complex** calculations; these have to be preprogrammed in the memory of the surface control device installed on the wellhead. Up-to-date POC devices usually include the necessary instrumentation and provide continuous information on the well's production rates, thus facilitating well testing with a high accuracy; they can even replace conventional testing procedures. The **accuracy** of inferred production (IP) calculations was compared to stock-tank volume measurements for three wells and the difference was found to be less than 2% [2]. A different source [4] reported similar accuracies for a case of 15 wells; inferred production rate calculations can thus provide a reliable alternative to conventional well testing.

6.2.2 DETERMINATION OF ANNULAR LIQUID LEVELS

As mentioned previously, a **packer** is not commonly run in a rod-pumped well; thus, well fluids may freely enter the casing–tubing annulus. It was also shown that the **height** above the formation of the fluid column is a direct indicator of the well's actual **bottomhole** pressure. This fact is utilized in most of the well testing procedures developed for pumping wells, which rely on the measurement of annular liquid **levels**. After the annular liquid level is known, static and flowing **bottomhole pressures** can be found by calculation. In the following, the determination of annular liquid levels is discussed first, and then the available calculation models are detailed that permit the calculation of bottomhole pressures.

6.2.2.1 Acoustic surveys

The most common method of finding the liquid level in a pumping well's annulus is by conducting an acoustic well survey, also called **well sounding**. This survey operates on the principles of propagation and reflection of **pressure waves** in gases. Using a wave source, a pressure **pulse** (usually an acoustic impulse) is produced at the surface in the casing–tubing annulus, which travels in the form of pressure **waves** along the length of the annular gas column. These pressure waves are **reflected** (echoed) from every depth where a change of cross-sectional area occurs, caused by tubing collars, casing liners, well fluids, etc. The reflected waves are picked up and converted to electrical signals by a **microphone**, also placed at the surface, and **recorded** on paper or by electronic means. An evaluation of the reflected signals allows the determination of the depth to the liquid level in the well.

The acoustical well sounder consists of two basic components: the wellhead assembly and the recording and processing unit. The wellhead assembly is easily connected to the casing annulus by means of a threaded nipple. It contains a mechanism that **creates** the sound wave and a **microphone** that picks up the signals. Conventional well sounders utilize blank **cartridges** with black powder, which are fired either manually or by remote control. Modern well sounder units employ so-called **gas guns**, which provide the required pressure impulse by suddenly **discharging** a small amount of high-pressure gas (CO_2 or N_2) into the annulus. The recording unit processes the electric signals created by the microphone, by filtering and amplification. The processed signals are then **recorded** on a chart recorder in the function of time. The depth of the liquid level in the annulus is found by the proper **interpretation** of the acoustic chart.

An example acoustic chart is given in Fig. 6.3, which shows analog recordings of reflected sound signals on a **strip chart**. As seen, every tubing **collar** is identified by a local **peak** of the signal, whereas the liquid level is clearly marked by a much larger reflection. Determination of the liquid level depth is possible by either counting the number of **collar signals** and comparing these data to well records or finding the reflection **time** from the liquid level [5,6]. Time readings are made possible with the use of 1-s timing signals that are also recorded on the chart. If reflection time and the acoustic velocity are known, then liquid level **depth** is found from the formula given below:

$$L = \frac{\Delta t \, v_S}{2} \tag{6.2}$$

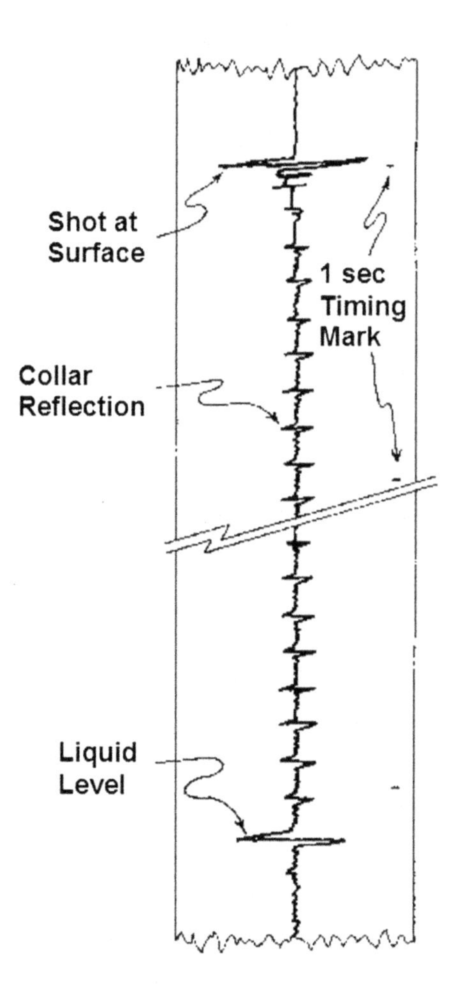

FIGURE 6.3

An example strip-chart recording from an acoustic survey.

where:

L = depth to liquid level from surface, ft,
Δt = time between generation and reflection, s, and
v_s = acoustic velocity in the gas, ft/s.

The accuracy of acoustic liquid level surveys is highly dependent on an accurate knowledge of the **acoustic velocity** valid under the actual conditions. The use of some form of the **equation of state** for the gas can supply these values [7]. In wells with **uniform** gas composition in the casing annulus, reliable acoustic velocities are calculated from this approach. The majority of wells that **vent** annulus gas to the flowline contain uniform gas columns.

The casing annulus of wells not venting gas or shut in for longer periods usually contains a gas column whose composition **varies** with depth. In such cases, great discrepancies in local acoustic velocities may occur and special precautions are necessary.

The advantages of acoustic well surveys over the **direct** determination of bottomhole pressures are the much smaller **costs** involved, the elimination of well **killing** and workover operations, and the reduced time requirement. Therefore, acoustic surveys are widely used in rod-pumping analysis. However, in case the annular fluid has a high tendency to **foam**, no firm signals on the liquid level can be attained. The latest developments in acoustic survey techniques include the automatic liquid level **monitor** [8], which automatically runs acoustic surveys and can also conduct pressure buildup and drawdown tests on pumping wells. Modern acoustic units employ microcomputers and advanced digital data acquisition techniques and ensure high accuracy and reliability of liquid level determinations [9–12].

6.2.2.2 Calculation of liquid levels

Instead of directly **measuring** liquid levels, the procedure proposed by **Alexander** [13,14] allows the indirect determination of the actual liquid level in pumping wells with a sufficient gas production. His method is based on a **mass balance** equation written on the gas volume contained in the casing–tubing annulus (see Fig. 6.4). In steady-state normal operations, the gas rate **entering** the annulus from the formation, q_1, equals the rate of gas **vented** into the flowline, q_2. Thus a constant gas volume of V is maintained in the annulus above the liquid level. If either of the gas volumetric **rates** will change, a corresponding change in the annular **volume** will occur. This phenomenon is easily described, if a mass **balance** is written on the annular gas, by equating the mass rates flowing **in** and **out** of the system with the change in the system's mass:

$$q_{m1} - q_{m2} = \frac{\mathrm{d}}{\mathrm{d}t} m \tag{6.3}$$

where:

q_{m1} = mass rate of annular gas influx, lb/min,
q_{m2} = mass rate of gas leaving the annulus, lb/min, and
m = mass of gas contained in the annulus, lb.
t = time, min

Mass flow rates are expressed with the **volumetric** gas flow rates, using the **engineering equation of state**:

$$q_m = q \frac{p_{sc} M}{R T_{sc}} \tag{6.4}$$

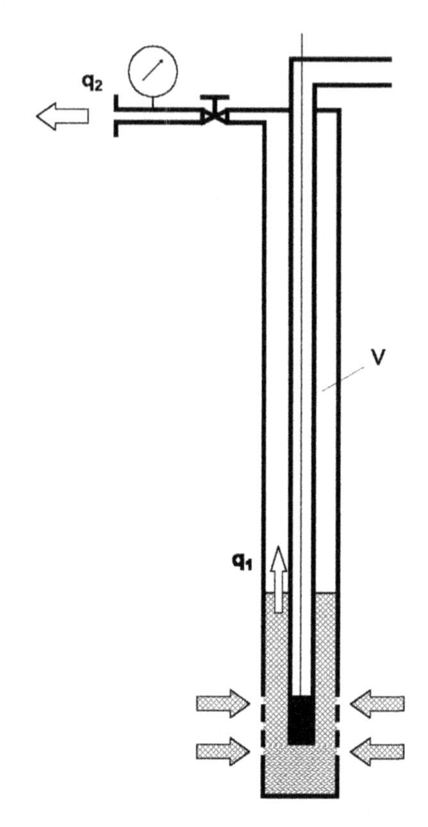

FIGURE 6.4

Illustration for writing up the mass balance equation for the annular gas volume of a pumping well.

where:

q_m = mass flow rate, lb/min,
q = volumetric flow rate at standard conditions, cu ft/min,
p_{sc} = standard pressure, 14.65 psia,
M = molecular weight of the gas, lb/mole,
T_{sc} = standard temperature, 520° Rankine, and
R = universal gas constant.

The system's mass is found from the **engineering equation of state** as:

$$m = \frac{pVM}{RTZ} \tag{6.5}$$

where the parameters not defined above are:

m = mass of gas, lb,
p = pressure, psia,
V = annular gas volume, cu ft,
T = absolute temperature, degrees Rankine, and
Z = gas deviation factor, –.

In order to find the final form of the basic equation (Eq. 6.3), the simplifying **assumption** is made that gas temperature, T, and deviation factor, Z, are **independent** of time. Then, after substitution of Eq. (6.4) into the basic equation and after differentiation of Eq. (6.5), we get:

$$q_1 - q_2 = \frac{50.8}{TZ}\left(V\frac{dp}{dt} + p\frac{dV}{dt}\right) \tag{6.6}$$

where:

q_1, q_2 = gas volumetric rates, Mscf/d,
p = pressure, psia,
V = annular gas volume, cu ft,
T = absolute temperature, degrees Rankine, and
Z = gas deviation factor, –.

Alexander uses this mass balance equation to find the annular liquid **level** for two basic physical models. The constant **volume** model assumes that casing gas volume does not considerably change during short shut-in periods when the surface casing valve is closed. The constant bottomhole **pressure** model, on the other hand, accounts for changes in annular gas volume but assumes a constant bottomhole pressure to persist during short-time fluid level buildup periods. Both models are used with the following experimental procedure illustrated in Fig. 6.4.

First, the casing **wing** valve at the surface is **closed**, which entails a gradual increase in casinghead pressure due to the continuous inflow of gas, q_1, from the formation into the annulus. The **pressure buildup** is recorded with a high-precision device, a dead-weight tester or an electronic transducer. Usually, a 5 psia pressure **increase** or a 10-min testing **period** is employed to ensure a negligibly small change in annular liquid level. Then, the casing valve is **opened** and gas is bled from the annulus at a measured rate. A critical flow **prover** provides an ideal tool for this purpose. During the venting period, casinghead pressure continuously changes because gas is removed from the system. Again, changes in casinghead pressure are **measured** and recorded with high accuracy. The test just described can be accomplished with the pumping unit either in **operation** or **shut down**, and the formulae detailed below are utilized to find the required parameters.

Using the constant casing **volume** approach, the change in casing volume is assumed to be negligible, i.e., dV/dt is zero in Eq. (6.6). Now, if the casing valve is closed, gas volumetric rate q_2 equals zero because no gas is vented from the annulus, and the basic equation is solved for the annular gas inflow rate, q_1:

$$q_1 = \frac{50.8}{TZ} V \left(\frac{dp}{dt}\right)_1 \tag{6.7}$$

where:

q_1 = annular gas flow rate, cu ft/d, and
$(dp/dt)_1$ = pressure buildup rate during casing valve shut-off, psia/min.

The next step of the experimental procedure involves the measurement of the gas **rate**, q_2, through the critical flow prover, after the casing valve has been opened. The basic equation for these conditions is written as:

$$q_1 - q_2 = \frac{50.8}{TZ} V \left(\frac{dp}{dt}\right)_2 \tag{6.8}$$

where: $(dp/dt)_2$ = pressure change rate during gas bleed-down from the casing, psia/min.

Equations (6.7) and (6.8) contain two **unknowns** and can simultaneously be solved for those parameters:

$$V = \frac{q_2 TZ}{50.8\left[(\mathrm{d}p/\mathrm{d}t)_1 - (\mathrm{d}p/\mathrm{d}t)_2\right]} \tag{6.9}$$

$$q_1 = q_2 \frac{(\mathrm{d}p/\mathrm{d}t)_1}{(\mathrm{d}p/\mathrm{d}t)_1 - (\mathrm{d}p/\mathrm{d}t)_2} \tag{6.10}$$

where:

V = volume of gas contained in the annulus, cu ft,
q_2 = gas vent rate from annulus, Mscf/d,
T = average absolute temperature of annulus gas, degrees Rankine,
Z = average gas deviation factor in annulus, –,
q_1 = annular gas flow rate, Mscf/d,
$(\mathrm{d}p/\mathrm{d}t)_1$ = time rate of change of casinghead pressure during casing shut-in, psia/min, and
$(\mathrm{d}p/\mathrm{d}t)_2$ = time rate of change of casinghead pressure during casing vent period, psia/min.

Since the annular gas volume, V, is calculated with **average** pressure and temperature in the above formulae, a **correction** is necessary if surface pressure data are used. Based on the **engineering equation of state**, the corrected volume is found from the value just calculated:

$$V_{\mathrm{corrected}} = V \frac{p_{\mathrm{surf}}}{p_{\mathrm{avg}}} \tag{6.11}$$

The pressure ratio in the above formula is derived from the equation used to calculate the pressure distribution in a static gas column:

$$\frac{p_{\mathrm{surf}}}{p_{\mathrm{avg}}} = \frac{L\left[0.001877 \frac{SpGr}{T_{\mathrm{avg}} Z_{\mathrm{avg}}}\right]}{\exp\left[0.001877 L \frac{SpGr}{T_{\mathrm{avg}} Z_{\mathrm{avg}}}\right] - 1} \tag{6.12}$$

where:

p_{surf}, p_{avg} = surface and average gas pressure, respectively, psi,
L = uncorrected depth to liquid level, ft,
$SpGr$ = gas specific gravity, –,
T_{avg} = average absolute temperature of gas column, degrees Rankine, and
Z_{avg} = gas deviation factor at average conditions, –.

It can be shown [13,14] that the constant bottomhole pressure model results in calculated gas volumes **smaller** than those received from the constant volume model. The relationship of the two gas volumes is given below:

$$V' = V - 5.61p \frac{C}{grad} \tag{6.13}$$

where:

V' = annular gas volume from the constant bottomhole pressure model, cu ft,
V = annular gas volume from the constant volume model, cu ft,
p = surface pressure, psia,

$grad$ = hydrostatic gradient in the liquid column, psi/ft, and
C = capacity of the annulus, bbl/ft.

The depth of the liquid **level** from the surface equals the length of the gas **column**, which, in turn, is found from the calculated annular gas volume and the capacity of the annulus:

$$L = \frac{V}{5.61C} \quad (6.14)$$

where:

L = depth of the liquid level, ft,
V = annular gas volume, cu ft, and
C = annular capacity, bbl/ft.

In conclusion, **Alexander**'s method discussed above allows the calculation of liquid level depths in pumping wells by following a simple testing procedure. A similar calculation model was developed by **Hasan and Kabir** [15].

EXAMPLE 6.1: FIND THE DYNAMIC LIQUID LEVEL IN THE ANNULUS WITH ALEXANDER'S PROCEDURE, FOR THE FOLLOWING CONDITIONS:

Depth of perforations = 5,000 ft
Pump setting depth = 5,000 ft
Annular capacity = 0.019 bbl/ft
Average annulus temperature = 70 F
Average deviation factor in annulus = 0.89
Gas specific gravity = 0.78
Oil gradient = 0.3 psi/ft
Surface casing pressure = 40 psi
$(dp/dt)_1 = 0.05$ psi/min
$(dp/dt)_2 = -0.1$ psi/min
Casing gas vent rate = 4.0 Mscf/d

Solution

First, the constant pressure model is used to calculate annular gas volume, using Eq. (6.9):

$$V = [4(70+460)0.89]/[50.8(0.05+0.1)] = 1,886.8/7.62 = 248 \text{ cu ft.}$$

The liquid level depth is found from Eq. (6.14):

$$L = 248/(5.61 \times 0.019) = 2,327 \text{ ft.}$$

The pressure correction factor is calculated from Eq. (6.12):

$$p_{surf}/p_{avg} = 2,327[0.001877 \times 0.78/(70+460)/0.89]/\{\exp[0.001877 \times 2,327 \times 0.78/(70+460)/0.89] - 1\}$$
$$= 0.00722/0.00725 = 0.996.$$

Corrected gas volume is by Eq. (6.11):

$$V_{corrected} = 248 \times 0.996 = 247 \text{ cu ft.}$$

The corrected fluid level is thus:

$$L = 247/5.61/0.019 = 2,317 \text{ ft.}$$

Now the constant bottomhole pressure model is used to calculate the gas volume, and Eq. (6.13) gives:

$$V' = 247 - 5.61 \times 40 \times 0.019/0.30 = 247 - 14 = 233 \text{ cu ft.}$$

Use Eq. (6.14) to find the liquid level corresponding to this gas volume:

$$L = 233/5.61/0.019 = 2,186 \text{ ft.}$$

In conclusion, the actual liquid level is expected to lie between 2,186 and 2,317 ft.

6.2.3 BOTTOMHOLE PRESSURE CALCULATIONS

6.2.3.1 Early methods

This section briefly discusses the bottomhole pressure determination methods developed in **early** production practice. The accuracies to be expected from these procedures can be substantially **lower** than those achieved with the more modern approaches detailed later.

6.2.3.1.1 Walker's method

Walker's well testing method [16] (see also **Nind** [17]), is based on the following theoretical considerations. Under steady-state conditions, i.e., pumping a **constant** liquid rate, the pumping well's flowing bottomhole pressure is **independent** of the actual depth of the liquid level in the casing annulus. This is easy to see because a constant liquid removal from a well means a stable pressure at the bottom. Since the annulus is directly connected to the pump intake (no packer is present), the pressure at the bottom of the annular gaseous fluid column should be equal to the well's bottomhole pressure. However, annulus pressure at that depth is the sum of three components: surface pressure and the pressures of the gas and fluid columns that exist above the formation. Therefore, the **same** bottomhole pressure can exist with a **shorter** fluid column if surface gas pressure plus gas column weight is greater, and vice versa.

Figure 6.5 illustrates the basics of **Walker's method**. At the beginning, the well is pumped **normally** and the surface casinghead pressure, p_{c1}, is noted. At the same time, the stabilized fluid column **height** above the formation, h_1, is measured via an acoustic **survey**. Then, by the use of a pressure regulator on the surface, casinghead pressure is **increased** to a higher value, p_{c2}. After a sufficiently long pumping period to reach **stabilized** conditions, the depressed liquid **level** is measured, and the new fluid column height, h_2, is calculated. According to the above considerations, the **pressure** at formation depth is the **same** for the two cases and can be written as follows:

$$p_{wf} = p_{c1} + p_{g1} + h_1 grad \quad (6.15)$$

$$p_{wf} = p_{c2} + p_{g2} + h_2 grad \quad (6.16)$$

where:

p_{wf} = flowing bottomhole pressure, psi,
p_{c1}, p_{c2} = casinghead pressures in the two cases, psi,
p_{g1}, p_{g2} = gas column pressures in the two cases, psi,
h_1, h_2 = fluid column heights above the formation in the two cases, ft, and
$grad$ = hydrostatic gradient of the gaseous fluid column in the annulus, psi/ft.

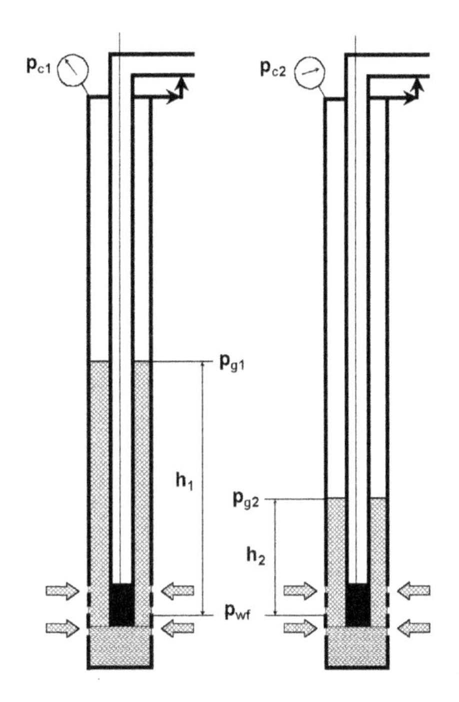

FIGURE 6.5

Walker's method for determining the flowing bottomhole pressure in a pumping well.

It can be shown that the implied **assumption** of a constant fluid gradient, *grad*, in the two cases is **justified**, and the above two equations can simultaneously be solved for the two unknowns: the fluid gradient and bottomhole pressure. Fluid gradient is found from the formula:

$$grad = \frac{(p_{c2} + p_{g2}) - (p_{c1} + p_{g1})}{h_1 - h_2} \tag{6.17}$$

Flowing bottomhole pressure, p_{wf}, is then calculated from either of the two equations given above (Eq. (6.15) or (6.16)).

Since **Walker's method** is only valid for **stabilized conditions**, great care must be taken to reach steady-state fluid **levels** in both tests. Since this process can take considerable time, sometimes even several days, the use of this bottomhole pressure determination method is limited.

EXAMPLE 6.2: CALCULATE THE FLOWING BOTTOMHOLE PRESSURE WITH WALKER'S METHOD IN A WELL WITH THE FOLLOWING DATA:

Depth of perforations = 4,000 ft
Pump setting depth = 4,000 ft
Gas specific gravity = 0.8

Measured acoustic liquid levels and casinghead pressures are:

	First Case	Second Case
Acoustic level, ft	2,000	3,600
Casing pressure, psi	200	490

Solution

The fluid column heights above the formation for the two cases (see Fig. 6.5) are:

$$h_1 = 4,000 - 2,000 = 2,000 \text{ ft.}$$

$$h_2 = 4,000 - 3,600 = 400 \text{ ft.}$$

The pressure at the bottom of the gas column is calculated from gas gradient curves, resulting in pressure gradients of 6 psi/1,000 ft and 16 psi/1,000 ft, for casinghead pressures of 200 psi and 490 psi, respectively. Gas column pressures are thus:

$$p_{g1} = 2,000 \times 6/1,000 = 12 \text{ psi.}$$

$$p_{g2} = 3,600 \times 16/1,000 = 58 \text{ psi.}$$

The average fluid gradient in the annulus is found from Eq. (6.17):

$$grad = [(490 + 58) - (200 + 12)]/[2,000 - 400] = 336/1,600 = 0.21 \text{ psi/ft.}$$

Flowing bottomhole pressure can now be calculated by Eq. (6.15):

$$p_{wf} = 200 + 12 + 2,000 \times 0.21 = 632 \text{ psi.}$$

A check is made in the second case, using Eq. (6.16):

$$p_{wf} = 490 + 58 + 400 \times 0.21 = 632 \text{ psi.}$$

As expected, the two flowing bottomhole pressures are equal.

6.2.3.1.2 Other methods

Agnew [18,19] proposed an **approximate** calculation model based on dynamometer measurements. He utilized the data of the well-known standing and traveling valve **tests** to arrive at an approximate bottomhole pressure. The accuracy of the procedure is highly dependent on the proper execution of these tests and also can be affected by downhole problems such as excessive friction, which considerably change measured rod loads.

One old practical method is the so-called **gas blow-around method**, which can be applied in wells with **considerable** gas production. Its principle is that if the fluid level is **depressed** to the pump setting depth, then pump intake pressure is found from the hydrostatic pressure of the gas **column**. Since the **length** of the gas column is precisely known (it is at pump level), its pressure can be calculated very accurately. The inaccuracies of bottomhole pressure calculations, introduced by errors in fluid **level** and fluid **gradient** values, are thus eliminated. The testing procedure requires the use of a pressure regulator and a dynamometer at the surface. The casing pressure is **increased** with the regulator until gas is **blown** around from the annulus into the pump. This occurs when the dynamometer card shows typical gas interference symptoms. The main drawback of this procedure is the usually long time required for **stabilized** conditions to occur [20].

6.2.3.2 Static conditions

If a pumping well is **shut in** for a sufficient period of time, i.e., the pumping unit is stopped and the casing valve is closed, then stabilized conditions in the well will occur. Gas, if produced from the reservoir, **accumulates** in the upper portion of the casing annulus, and liquid **fills** up the lower part of it. The well develops its static bottomhole pressure (SBHP) at the formation, which pressure balances the combined hydrostatic pressures of the liquid and gas **columns**. Therefore, static bottomhole pressure is easily found by adding to the casinghead pressure, measured at the surface, the pressures exerted by these columns, as given below:

$$p_{ws} = p_c + p_g + p_l \tag{6.18}$$

where:

p_{ws} = static bottomhole pressure (SBHP), psi,
p_c = casinghead pressure, psi,
p_g = hydrostatic pressure of the annular gas column, psi, and
p_l = hydrostatic pressure of the annular liquid column, psi.

Surface casing pressure, p_c, is measured with a high-**precision** pressure gauge or a dead-weight tester. Accurate measurements are **required** since, in most cases, this component accounts for a large part of the final bottomhole pressure. The hydrostatic pressure in the gas **column**, p_g, can only be determined if the **depth** to the static liquid level in the well is known, usually found from an **acoustic** survey. A proper evaluation of this pressure must account for gas composition and temperature distribution in the annulus and involves an iterative calculation procedure. For approximate calculations the universal gas gradient chart developed in Fig. 2.9 can be used; the chart is reproduced in **Appendix A**.

The last component of the bottomhole pressure, the liquid column's **hydrostatic** pressure, p_l, is mainly governed by the hydrostatic gradient of the annular liquid. Determination of this gradient poses no problems if **pure** oil is produced, but water-cut oil production needs special considerations. As shown by **McCoy et al.** [21], the **composition** of the annular liquid column depends on the relation of the dynamic liquid level to pump setting depth, valid just **before** the well was shut down. A general rule is that in steady-state pumping conditions, due to gravitational **separation** of the oil from the water, all liquid above the pump is pure **oil**. After the well is shut down, a water-cut liquid with approximately the same water–oil ratio (WOR), as measured in a test separator, flows below the oil column.

Figures 6.6 and 6.7 show the pumping and shut-in conditions in the annulus when the pump is set at the depth of the **formation**. If, during pumping, the stabilized liquid level is at the pump **intake**, then after shut-in, all liquid above the formation will be with the same WOR as found in a previous production test (Fig. 6.6). Although this liquid column will, with time, separate into **separate** water and oil sections, its hydrostatic **pressure** does not change. On the other hand, if the dynamic liquid level is above the pump before shut-in, the liquid column contains only **oil**. As seen in Fig. 6.7, this oil column remains on the top of the water-cut liquid after the well is shut down. Figures 6.8 and 6.9 illustrate the same pumping and static conditions for wells that have the pump run above the depth of the perforations.

After the WOR **distribution** in the liquid column has been established, the hydrostatic pressure is calculated with the help of static **gradients** for the water and the oil. For approximate calculations, average gradients taken at an average column temperature and pressure may be used. More accurate bottomhole pressure predictions require **step-wise** calculations when the effects of actual pressure and temperature on oil and water densities can properly be accounted for [21].

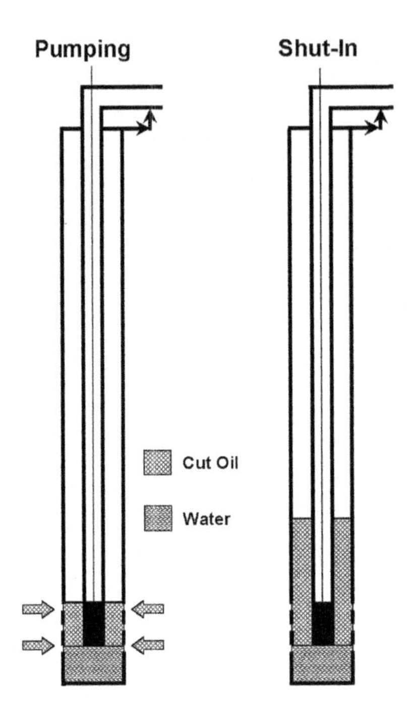

FIGURE 6.6

Pumping and shut-in conditions in the annulus when the pump is set at the formation and the dynamic liquid level is at the pump.

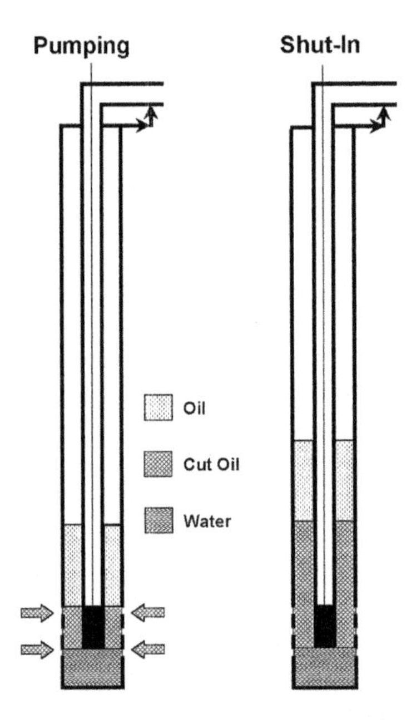

FIGURE 6.7

Pumping and shut-in conditions in the annulus when the pump is set at the formation and the dynamic liquid level is above the pump.

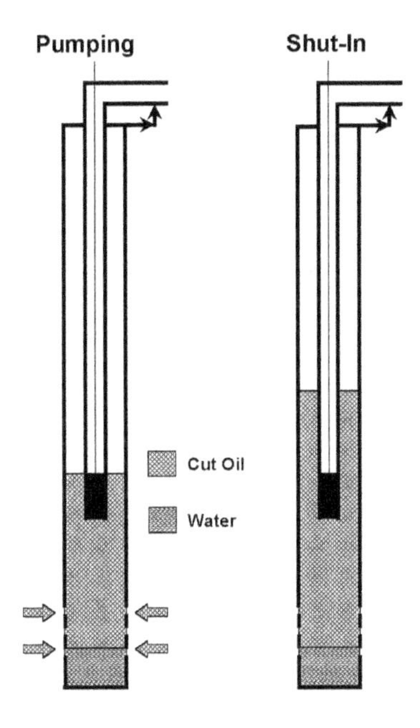

FIGURE 6.8

Pumping and shut-in conditions in the annulus when the pump is set above the formation and the dynamic liquid level is at the pump.

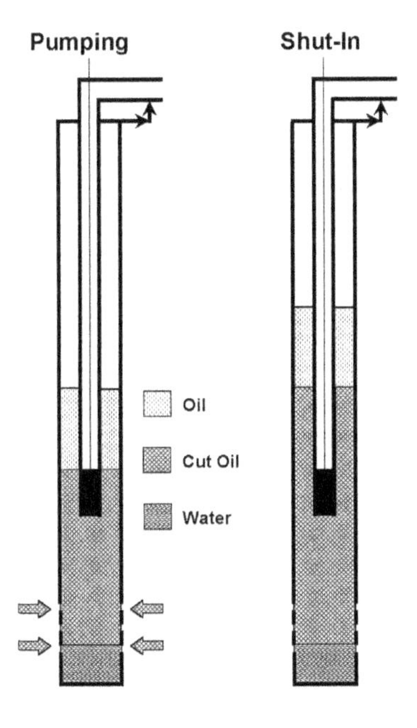

FIGURE 6.9

Pumping and shut-in conditions in the annulus when the pump is set above the formation and the dynamic liquid level is above the pump.

EXAMPLE 6.3: THE FOLLOWING CASE WILL BE USED IN SUBSEQUENT EXAMPLE PROBLEMS

Well depth is 6,000 ft with the pump set at the perforations. During pumping, the liquid level is at 5,500 ft, casinghead pressure is 100 psi. After shut-in, these parameters were measured as 4,500 ft and 450 psi, respectively. The well was previously tested and produced 300 bpd liquid with a WOR of 2. For simplicity, assume an average temperature of 80 F in the well, an oil gradient of 0.32 psi/ft, and a water gradient of 0.44 psi/ft. Specific gravity of the annulus gas is 0.75.

Calculate the static bottomhole pressure.

Solution

The gas column pressure, p_g, is calculated with a gas gradient of 14 psi/1,000 ft, read from Appendix A, corresponding to a surface pressure of 450 psi. Gas pressure is thus:

$$p_g = 4,500 \times 14/1,000 = 63 \text{ psi.}$$

To find the liquid column pressure, the composition of the static column must be determined. The 500 ft oil column present in the pumping case remains on the top of the static column. Liquid fill up from the formation equals the difference of pumping and static levels, i.e., 5,500 − 4,500 = 1,000 ft. This additional column in the annulus contains the same amount of water as found in the previous production test. The static gradient of this column is found from the gradients of oil, water and the produced WOR:

$$1/3 \times 0.32 + 2/3 \times 0.44 = 0.4 \text{ psi/ft.}$$

Liquid column pressure is composed of the pressures due to 500 ft oil plus 1,000 ft water-cut oil columns:

$$p_l = 500 \times 0.32 + 1,000 \times 0.4 = 160 + 400 = 560 \text{ psi.}$$

Static bottomhole pressure is found from Eq. (6.18) as:

$$p_{ws} = 450 + 63 + 560 = 1,073 \text{ psi.}$$

6.2.3.3 Flowing conditions

In wells with negligible gas production, the calculation of flowing bottomhole pressures is quite simple and is accomplished with the method just described for static conditions. Usually, surface **casinghead pressure** is the main contributor to bottomhole pressure, but **liquid column pressure** can also be considerable. Again, it must be remembered that, under stabilized pumping conditions, all liquid above the pump is **oil** due to gravitational separation.

If formation **gas** is also produced, the calculation of bottomhole pressures is not so straightforward because, in these cases, the casing annulus contains a **gaseous liquid column**. It is customary to assume that most of the gas produced from the formation enters the **annulus**, and this gas continuously **bubbles** through the annular liquid column and leaves the well at the surface. The main effect of the produced gas is the **reduction** of the hydrostatic gradient of the annular liquid. In light of this, the accurate calculation of flowing bottomhole pressure (FBHP) heavily depends on the proper description of this phenomenon. In the following, the available correlations for liquid gradient correction are discussed first.

6.2.3.3.1 Annular pressure gradient

In a pumping well's annulus, a special case of multiphase **flow** takes place, since formation gas continuously bubbles through a **stagnant**, static liquid column. Depending on the gas volumetric rate, the flow patterns observed are **bubble flow** at low gas volumes and **slug flow** at higher gas rates. In both cases, the **density** of the mixture is found from the basic formula:

$$\rho_m = \rho_l H_l + \rho_g H_g \tag{6.19}$$

where:

ρ_m = two-phase mixture density, lb/cu ft,
ρ_l, ρ_g = liquid and gas densities, lb/cu ft,
H_l = liquid holdup, –, and
H_g = gas void fraction, –.

The contribution of gas density to **mixture** density is negligible and, therefore, can be deleted from the equation. Since hydrostatic pressure gradient is directly proportional to liquid density, the **gradient** of the gassy liquid column can be expressed with the gradient of the gas-free liquid as:

$$grad_m = grad_l\, H_l \quad (6.20)$$

where:

$grad_m$ = hydrostatic gradient of the gassy annular liquid, psi/ft,
$grad_l$ = hydrostatic gradient of the gas-free annular liquid, psi/ft, and
H_l = liquid holdup, –.

The liquid holdup, H_l, can be considered as a **correction** of the liquid gradient; hence it is also called **gradient correction factor**. Due to its practical importance in flowing bottomhole pressure calculations, several correlations are available to determine its value. The earliest correlation, the **Gilbert S curve** [22], is shown in Fig. 6.10. Gradient correction factor (H_l) is plotted against the following group of parameters:

$$Q/A/p^{0.4}$$

where:

Q = annular gas flow rate, Mscf/d,
A = annular cross-sectional area, sq in, and
p = actual column pressure, psi.

Godbey and Dimon [8] conducted a thorough investigation of the different possible flow patterns and gave two equations:
for **bubble** flow, $v_{sg} \leq 2$ ft/s:

$$H_l = 1 - \frac{v_{sg}}{1,2\, v_{sg} + 0.183} \quad (6.21)$$

for **slug** flow, $v_{sg} > 2$ ft/s:

$$H_l = 1 - \frac{v_{sg}}{v_{sg} + 0.305} \quad (6.22)$$

where:

H_l = liquid holdup (gradient correction factor), –, and
v_{sg} = superficial velocity of the gas phase, ft/s.

With both methods, either the gradient correction factor can be calculated for the **average** conditions of the gassy liquid column or a step-wise procedure can be used to find different correction factors for different liquid levels.

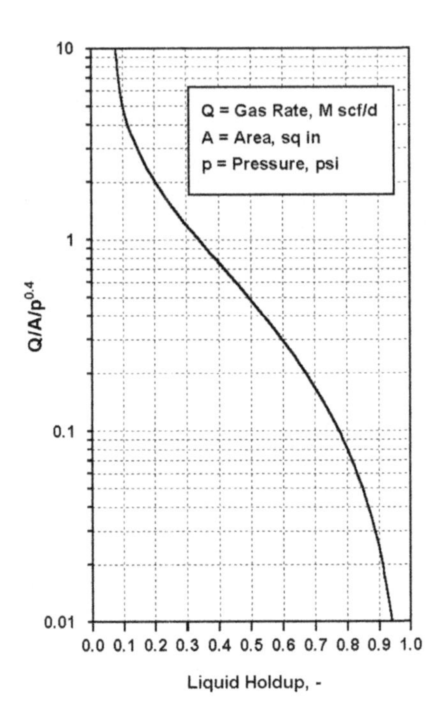

FIGURE 6.10

Gilbert's gradient correction factor correlation, after **Gipson and Swaim** [22].

EXAMPLE 6.4: CALCULATE THE GRADIENT CORRECTION FACTOR FOR THE WELL DATA GIVEN IN EXAMPLE 6.3, WITH BOTH THE GILBERT AND THE GODBEY-DIMON CORRELATIONS. ADDITIONAL DATA ARE: ANNULAR CROSS-SECTIONAL AREA OF 15 SQ IN, MEASURED ANNULAR GAS FLOW RATE OF 40 MSCF/D

Solution

No step-wise calculations are used for the **Gilbert** correlation; only a single correction factor is sought for the entire liquid column. For this purpose, a first guess of the average pressure is the pressure at the mid-height of the gas-free liquid column. Gas column pressure is assumed as 30 psi, and average pressure is the sum of surface, gas, and liquid column pressures:

$$p = 100 + 30 + 500/2 \times 0.32 = 130 + 80 = 210 \text{ psi.}$$

The correlating group of Gilbert is found as:

$$Q/A/p^{0.4} = 40/15/210^{0.4} = 0.31.$$

From Fig. 6.10, the correction factor is read off as:

$$H_l = 0.58.$$

For the second iteration, the average pressure is adjusted by including the effect of annular gas in the liquid gradient, which is $0.32 \times H_l = 0.32 \times 0.58 = 0.185$ psi/ft:

$$p = 100 + 30 + 500/2 \times 0.185 = 130 + 46 = 176 \text{ psi.}$$

The new value of the correlating group is:

$$Q/A/p^{0.4} = 40/15/176^{0.4} = 0.34.$$

The new correction factor from Fig. 6.10 is:
$H_l = 0.57$, which is close enough to the previous value and is the result of the calculations.

For the **Godbey-Dimon** correlation, the average liquid column pressure calculated previously is used. Additionally, gas deviation factor is assumed to equal $Z = 0.8$, for speeding up the calculation process. Gas superficial velocity, by definition, equals the actual gas volumetric rate divided by the cross-sectional area. The volume factor of the gas is found from:

$$B_g = 0.0283 \times Z \times T/p = 0.0283 \times 0.8(80 + 460)/176 = 0.07.$$

Superficial velocity of gas is:

$$v_{sg} = Q \times B_g/A = 40,000/86,400 \times 0.07/(15/144) = 0.31 \text{ ft/s}.$$

Since $v_{sg} < 2$ ft/s, bubble flow pattern exists and Eq. (6.21) is to be used:

$$H_l = 1 - 0.3/(1.2 \times 0.3 + 0.6) = 0.69.$$

Of the several **other** published correlations [23–25], the technique proposed by **McCoy et al.** [26] and used by the **Echometer** company should be discussed here. The gradient correction factor of these authors was based on field **measurements** and is reproduced in Fig. 6.11. Its use eliminates the often difficult determination of annular gas **flow** rates. For this purpose, the casing valve is closed in for a short period of time, while the pump is operating and the casing pressure **buildup** rate is recorded. For best results, a minimum pressure increase of 10 psi or a 10-min test period should be used. From these data, the following parameter group is calculated:

$$L' \, dp/dt$$

where:

dp = casing pressure buildup, psi,
dt = pressure buildup time, min, and
L' = corrected dynamic liquid level, ft.

The gradient **correction** factor (H_l) is then found from Fig. 6.11, where the definition of the corrected liquid level is:

$$L' = L + (1 - H_l)h_l \tag{6.23}$$

where:

L = measured depth of the dynamic liquid level, ft,
H_l = gradient correction factor, –, and
h_l = height of the gaseous liquid column, ft.

As seen above, calculation of the correction factor requires an **iterative** process, because H_l figures in the corrected dynamic liquid level that, in turn, is used to correlate H_l in Fig. 6.11. Usually, only a few iterations are necessary to converge to the proper value of the gradient correction factor, if a starting value of $H_l = 1$ is used.

FIGURE 6.11

The gradient correction factor correlation of echometer, after **McCoy et al.** [26].

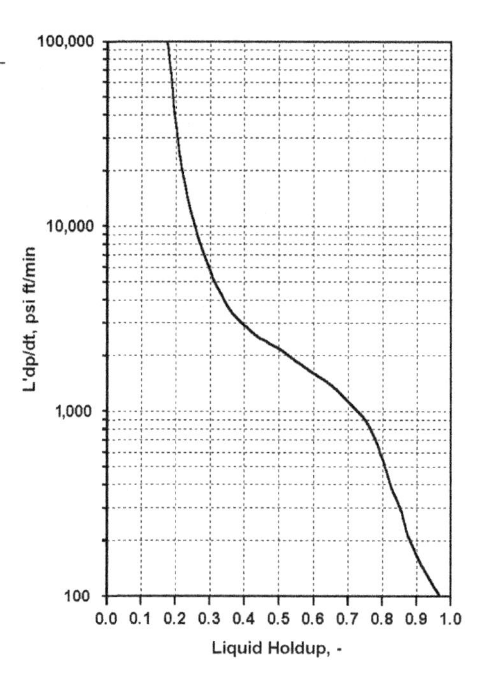

EXAMPLE 6.5: CALCULATE THE GRADIENT CORRECTION FACTOR WITH THE ECHOMETER TECHNIQUE FOR THE WELL DATA GIVEN IN EXAMPLE 6.3, IF THE MEASURED PRESSURE BUILDUP AT THE CASING HEAD IS 7 PSI IN 10 MIN

Solution

Assume a starting correction factor of $H_l = 1$. The corrected liquid level is by Eq. (6.23):

$$L' = 5,500 + (1 - 1)500 = 5,500 \text{ ft.}$$

The correlating group is found as:

$$L' \, dp/dt = 5,500 \times 7/10 = 3,850$$

Correction factor is read off from Fig. 6.11 as:

$$H_l = 0.36.$$

Use the calculated H_l to find a new corrected liquid level:

$$L' = 5,500 + (1 - 0.36)500 = 5,500 + 320 = 5,820 \text{ ft.}$$

The correlating group is thus:

$$L' \, dp/dt = 5,820 \times 7/10 = 4,074$$

Reading off from Fig. 6.11:

$$H_l = 0.35.$$

One more iteration is needed with a new corrected liquid level:

$$L' = 5,500 + (1 - 0.35)500 = 5,500 + 325 = 5,825 \text{ ft.}$$

The new correlating group is:

$$L' \, dp/dt = 5,825 \times 7/10 = 4,078$$

The correction factor that belongs to this value is $H_l = 0.35$, which is the final converged value of the gradient correction factor. The correction factors calculated in the previous example are much higher than the value found with the **Echometer** model, also observed by **McCoy et al.** [26].

6.2.3.3.2 Flowing bottomhole pressure

After the pressure gradient of the gaseous liquid column present in a pumping well's annulus has been determined, the flowing bottomhole pressure is calculated analogously to the static pressure:

$$p_{wf} = p_c + p_g + p_l \tag{6.24}$$

where:

p_{wf} = FBHP, psi,
p_c = casinghead pressure during normal pumping, psi,
p_g = hydrostatic pressure of the annular gas column, psi, and
p_l = hydrostatic pressure of the annular gaseous liquid column, psi.

As before, surface casing pressure, p_c, is **measured** at the surface, and gas column pressure, p_g, is found from the measured liquid level **depth**, L, by calculating the increase of pressure with depth in the static gas column. The last term of the equation is evaluated with the use of the gradient correction factor:

$$p_l = h_l \, grad_l \, H_l \tag{6.25}$$

where:

h_l = the height above the formation of the gaseous liquid column in the annulus, ft,
$grad_l$ = hydrostatic pressure gradient of the gas-free annulus liquid, psi/ft, and
H_l = gradient correction factor, –.

EXAMPLE 6.6: FIND THE BOTTOMHOLE PRESSURES FOR THE DIFFERENT CORRELATIONS USED IN THE PREVIOUS EXAMPLES

Solution

The hydrostatic pressure of the gas column is 30 psi, as calculated in Example 6.4. Liquid column pressure is found from Eq. (6.25), and using the different values of H_l we get:

$$p_l = 500 \times 0.32 \times 0.57 = 91 \text{ psi for the Gilbert,}$$

$$p_l = 500 \times 0.32 \times 0.69 = 110 \text{ psi for the Godbey} - \text{Dimon, and}$$

$$p_l = 500 \times 0.32 \times 0.35 = 56 \text{ psi for the Echometer correlations.}$$

Flowing bottomhole pressures are calculated by using Eq. (6.24):

$$p_{wf} = 100 + 30 + 91 = 221 \text{ psi for the Gilbert,}$$

$$p_{wf} = 100 + 30 + 110 = 240 \text{ psi for the Godbey} - \text{Dimon, and}$$

$$p_{wf} = 100 + 30 + 56 = 186 \text{ psi for the Echometer correlations.}$$

6.2.4 DETERMINATION OF WELL INFLOW PERFORMANCE

Knowledge of the static and flowing (producing) **bottomhole** pressures allows the calculation of the well's **inflow performance curve**. Two basic models can be used: a **constant** productivity index (PI) or **Vogel's** IPR curve. The productivity index, by definition, is found from:

$$PI = \frac{q}{(p_{ws} - p_{wf})} \tag{6.26}$$

where:

PI = productivity index, bpd/psi,
q = measured liquid production rate, bpd,
p_{ws} = static bottomhole pressure, psi, and
p_{wf} = flowing bottomhole pressure, psi.

The use of **Vogel's** IPR concept [27] results in the ratio of the actual to maximum possible liquid flow rate:

$$\frac{q}{q_{max}} = 1 - 0.2\frac{p_{wf}}{p_{ws}} - 0.8\left(\frac{p_{wf}}{p_{ws}}\right)^2 \qquad (6.27)$$

where the parameter not defined above is:

q_{max} = theoretical maximum production rate, bpd.

After q_{max} is determined from the above formula, the well's **IPR curve** can be constructed, and that was the purpose of well testing. Well test data can be used in the evaluation of the capabilities of the given well, as well as in investigating any changes in rod-pumping system design.

EXAMPLE 6.7: CALCULATE THE PARAMETERS OF THE INFLOW PERFORMANCE RELATIONSHIPS FOR THE CASES IN THE PREVIOUS EXAMPLE

Solution

Liquid production rate was measured as 300 bpd and p_{ws} was found as 1,073 psi in Example 6.3. Using the different p_{wf} values, the following *PI* values are found from Eq. (6.26):

$$PI = 300/(1,073 - 221) = 0.35 \text{ bpd/psi for the Gilbert,}$$

$$PI = 300/(1,073 - 230) = 0.36 \text{ bpd/psi for the Godbey – Dimon, and}$$

$$PI = 300/(1,073 - 186) = 0.34 \text{ bpd/psi for the Echometer correlations.}$$

If the Vogel IPR formula is used (Eq. 6.27) then the ratio of the actual to maximum flow rates is:

$$q/q_{max} = 1 - 0.2(221/1,073) - 0.8(221/1,073)^2 = 1 - 0.041 - 0.034 = 0.925 \text{ for the Gilbert,}$$

$$q/q_{max} = 1 - 0.2(230/1,073) - 0.8(230/1,073)^2 = 1 - 0.043 - 0.037 = 0.920 \text{ for the Godbey – Dimon,}$$

$$q/q_{max} = 1 - 0.2(186/1,073) - 0.8(186/1,073)^2 = 1 - 0.035 - 0.024 = 0.941 \text{ for the Echometer correlations.}$$

A comparison of the well's **absolute open flow potential**, as calculated with the use of the different correlations, follows:

Correlation	Open Flow Potential	
	From PI equation	From Vogel IPR
Gilbert	376 bpd	324 bpd
Godbey–Dimon	386 bpd	326 bpd
Echometer	365 bpd	319 bpd

EXAMPLE 6.8: PERFORM A WELL TEST ANALYSIS ON THE WELL PRESENTED IN EXAMPLE 6.3, USING THE ECHOMETER PROCEDURE

Solution

Exhibit 6.1 contains input data and calculation results for this case.

```
                 WELL TESTING OF A PUMPING WELL

               USING ECHOMETER'S TESTING PROCEDURE

        D A T A   of   W E L L   Ex. 6.8   tested on Nov.17, 1991

Perforations at :  6000.0 ft         Casing ID :            5.000 in
Pump Set at :      6000.0 ft         Tubing OD :            2.375 in

Wellhead Temp. :     80.0 F          Bottomhole Temp. :     100.0 F

                 F L U I D   P R O P E R T I E S

Oil Sp.Gr. :           0.8           Gas Sp.Gr. :           0.750
Water Sp.Gr. :         1.0              H2S Content :          0%
                                        CO2 Content :          0%

              L A S T   P R O D U C T I O N   T E S T

Liquid Rate :      300.0 bpd         Water Cut :              66%

                M E A S U R E D   T E S T   D A T A

                                 Pumping            Static

Measured Liquid Levels :         5500.0 ft         4500.0 ft
Casing Pressures :                100.0 psi         450.0 psi
dp (Pressure Buildup) :            7.00 psi
dt (Buildup Time) :               10.00 min


                 C A L C U L A T E D   R E S U L T S

             Calculated Annular Gas Rate :   37.5 M scf/d

                                  Pumping           Static

Oil Column Height, ft              500.0             500.0
    Gradient, psi/ft              0.3418            0.3419
    Holdup                         0.363
Liquid Column Height, ft             0.0            1000.0
    Gradient, psi/ft              0.0000            0.3992
    Holdup                         0.363

Gas Column Pressure, psi           117.2             531.6
Oil Column Pressure, psi            62.0             171.0
Liquid Column Pressure, psi          0.0             398.9
Bottomhole Pressure, psi           179.2            1101.4

           Calculated Productivity Index :   0.325 bpd/psi

      Data of Vogel IPR : q/qmax =  94.6%   qmax =   317.0 bpd
```

EXHIBIT 6.1

Detailed well test results for the well given in Example 6.8.

6.3 DYNAMOMETER SURVEYS

The most valuable tool for **analyzing** the performance of the pumping system is the **dynamometer**, which records the loads occurring in the rod string. These loads can be measured either at the surface with a polished rod dynamometer or at pump depth with a special downhole measuring device. In both cases, rod loads are recorded versus rod displacement or pumping time, during one or more complete pumping cycles. Since the variation of rod loads is a result of all the **forces** acting along the rod string and reflects the operation of the pump as well as the surface pumping unit, an evaluation of these loads reveals valuable information on downhole and surface conditions. Accordingly, the performance of the downhole and surface pumping equipment is usually analyzed by running a **dynamometer survey** (also called **well weighing**) on the given well.

The first "weight indicators" or **dynamometers** were used in the early 1920s. Since then, both the hardware and the evaluation methods have considerably improved. Thus the early and mostly **qualitative** interpretations, which relied heavily on the analyst's **skill** and previous experience, have evolved into the sophisticated, highly reliable, and **exact** analysis methods of today. Interested readers are advised to consult the many fine books devoted entirely to the subject [28–30].

The **proper** use of dynamometer techniques and the correct **interpretation** of measurement data are of utmost importance for the production engineer when he or she tries to increase the **profitability** of sucker-rod pumping. It is the **evaluation** of dynamometer surveys that forms the basis for accomplishing the following basic tasks:

- Reduction of lifting **costs**,
- Detection and prevention of equipment **failures**,
- Improvement of the **selection** and **application** of pumping equipment, and
- Increase of well **production**.

In the following, first the **hardware** of dynamometry is discussed and the basic types of available dynamometers along with their operational principles are detailed. The use of dynamometers follows, where the different procedures are described that allow the determination of various operational parameters of pumping. Finally, a basic treatment of the **interpretation** of dynamometer diagrams is presented. Torque loading of the gearbox, as well as the power requirements of pumping based on measured dynamometer cards, were discussed in Chapter 4.

6.3.1 BASIC DYNAMOMETER TYPES

6.3.1.1 Polished rod dynamometers

Polished rod dynamometers, as the name implies, are instruments recording polished rod **loads** during the pumping cycle. The conventional types are the **mechanical** and the **hydraulic** dynamometer, which both produce a continuous plot of polished rod **load** versus polished rod **displacement**, the so-called **dynamometer diagram** or card. Modern dynamometers are **electronic** devices that record the loads and displacements at the polished rod as a function of **time** and enable the analyst to accurately analyze the downhole conditions of sucker-rod pumps.

6.3.1.1.1 Conventional dynamometers

The **mechanical dynamometer** employs a steel **ring** as its load measuring device, which, being placed between the carrier bar and the polished rod clamp, carries the full polished rod load. The ring's

deflection is directly proportional to the load applied, which is **recorded** (after mechanical magnification) on **paper** attached to a rotating drum. Since the rotation of the drum is controlled by the polished rod's vertical movement, the resultant record is a **trace** of polished rod loads against displacement. The mechanical dynamometer is a rugged device, and the **Johnson-Fagg** version was extensively popular in the oil field. The major **disadvantage** of its use is the need to stop the pumping unit before the dynamometer is installed on the polished rod.

The **hydraulic dynamometer** (a well-known version is the **Leutert** dynamometer [31]) can be installed without the need to **stop** the pumping unit and, therefore, has a definite advantage over the mechanical one. Its operational principle is shown in the schematic drawing in Fig. 6.12. Before the first application on a well, a special **spacer** is installed on the polished rod between the carrier bar and the polished rod clamp. The dynamometer with its two load-sensing hydraulic **pistons** can easily be installed, even while the unit is pumping, between the shoulder of the spacer and the carrier bar. After the dynamometer is in place, hydraulic **pressure** is applied to the pistons by activating the hand pump connected to the system. The pistons lift the spacer off the carrier bar and the polished rod load hereafter is fully **supported** by the hydraulic pistons only. Thus, changes in polished rod **loads** entail changes in the hydraulic **pressure**, which are recorded by a stylus that magnifies the displacement of a spring-retarded piston. The record is made on **paper** attached to a drum **rotated** by a pull cord, one end of which is affixed to a stationary point. The rotational angle of the drum, therefore, is directly proportional to the polished rod's instantaneous **position**, and the record obtained is a plot of polished rod **load** versus polished rod **displacement**.

To accommodate different well conditions, different-**size** drums can be used for different stroke lengths. In addition, the retarding **spring** can be changed to alter the dynamometer's **load** range. These

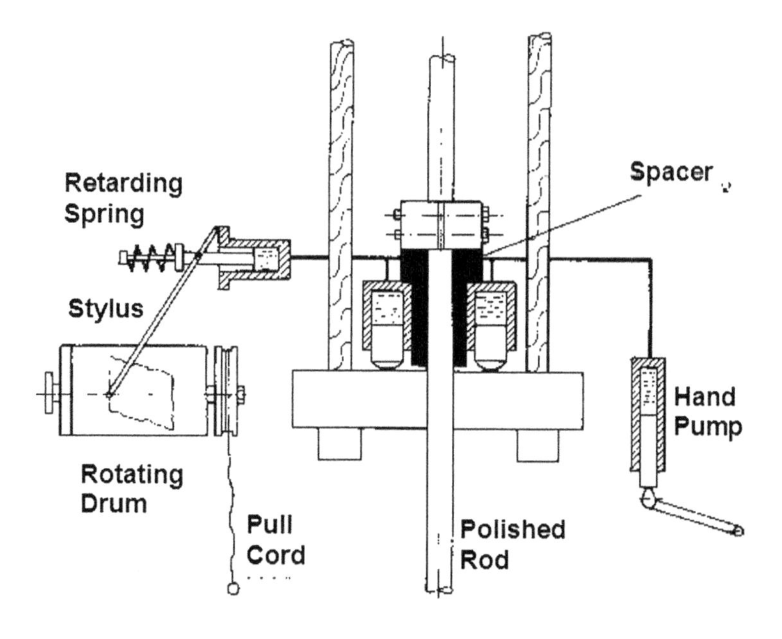

FIGURE 6.12

Schematic construction of a hydraulic dynamometer.

adjustments are easily carried out in the field, an added advantage of the use of such dynamometers. Thus, the recorded dynamometer diagram (or **card** for short) can be produced in a size that is easily interpreted.

The main **limitations** of the hydraulic (Leutert) dynamometer are the following [32]:

- Accuracy may deteriorate with the age of the instrument because the measuring spring weakens.
- They have a high amount of hysteresis caused by the drag (a) on the small piston that works against the retarding spring and (b) arising in the registration unit. Drift of measured loads can be expected.
- Load and position data are not available as a function of time, a basic requirement for the solution of the wave equation.

6.3.1.1.2 Electronic dynamometers

The **electronic** dynamometer's basic feature is that electronic **transducers**, rather than mechanical or hydraulic devices, are used for measuring well loads and rod displacements. As shown in Fig. 6.13, the main parts of such a dynamometer unit are the **load** transducer (load cell), the **position** transducer, and the **electronics**, which provides interfacing, signal recording, and processing.

The signals of both transducers, in forms of electric potential changes, are connected to data acquisition circuitry, which produces smoothed electric signals for recording and further processing. Polished rod load and position can thus be recorded on the optional portable **recorder** as a function of time. As seen earlier, this type of recording is a basic requirement when the solution of the damped **wave equation** is desired. Thus, a dynamometer of this type not only allows **surface** cards to be obtained but supplies the basic data for the construction of **downhole** cards as well.

The first electronic dynamometer was the **Delta II**, developed by **Shell Oil Co.** and widely used for obtaining data for downhole card calculations [33,34]. Later units included **microcomputers**

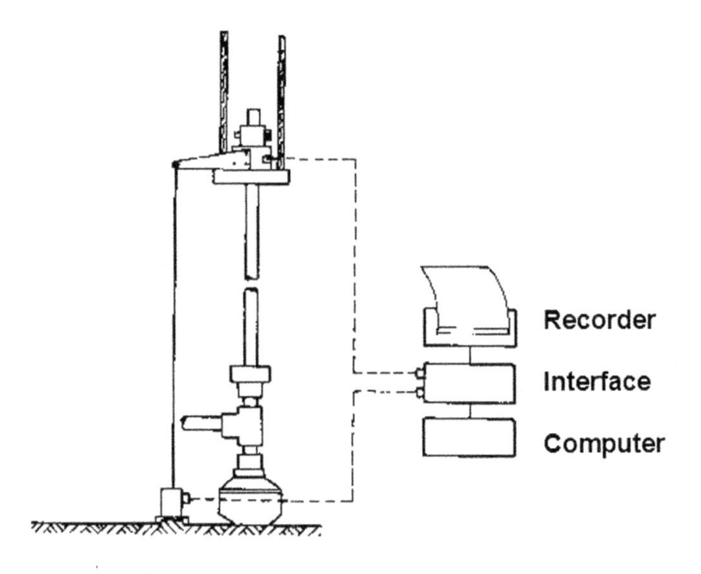

FIGURE 6.13

The main components of an electronic dynamometer.

and permitted the online analysis of measurements as well as easy data storage and retrieval operations. Today several companies offer **portable** electronic dynamometer systems controlled by **notebook** computers that perform real-time data acquisition functions and can also be used to calculate downhole cards [35–37].

Position Transducers

The earliest position transducers were spring-loaded **potentiometers** rotated by a string attached to the carrier bar; later on, **inclinometers** attached to the walking beam became popular. Both of these devices produce an **analog** signal directly proportional to polished rod displacement that is recorded as a function of time. Their common limitations are the insufficient resolution at the top and bottom of the polished rod stroke length, susceptibility to wear, and maintenance and calibration difficulties.

Accelerometers used in dynamometer systems eliminate the resolution and other problems; this is why they are becoming the preferred method to determine the variation of polished rod position. They are extremely small units permanently attached to the load sensing device. Such transducers accurately measure the polished rod's **instantaneous** acceleration, which, after being integrated twice with respect to time by the proper instrumentation, provides an accurate determination of the polished rod position versus time function [38]. The resolution of accelerometers is a function of the data sampling frequency and can thus be almost **continuous**; abrupt changes in polished rod velocity are easily recognizable.

Load Transducers

There are several versions of load transducers (aka load cells) used in electronic dynamometers and for pump-off control [39]. The **horseshoe**-type load cell, if properly calibrated, indicates **actual** loads and is the most **accurate**, with an accuracy of 0.5% or better of its rated load. During measurement it is placed between the carrier bar and the polished rod clamp. It contains several (usually 12) **strain gauges** distributed cylindrically around the polished rod. The circular distribution of the strain gauges provides an averaging effect; side loading resulting from a tilted carrier bar does not affect measurement accuracy. Horseshoe transducers are quite expensive and are not used in pump-off control applications.

The **disadvantages** of using horseshoe load cells are that (1) the pumping unit must be **stopped** while placing the dynamometer on the polished rod, and (2) while mounting the cell on top of the carrier bar the rod string must be **raised** by about 3 inches, the height of the dynamometer. Since at the same time the downhole plunger (in respect to the pump barrel) is raised by an identical vertical distance, the **spacing** of the downhole pump will also change. This may bring about changes in pumping conditions and the dynamometer cards obtained will not represent normal pumping operations. Use of a properly dimensioned **spacer** on the polished rod can solve this problem.

Polished rod transducers (PRTs) measure the changes in **diameter** of the polished rod due to pumping loads using solid-state strain gauges [40,41]. As shown in Fig. 6.14, the transducer is lightly **clamped** to the polished rod below the carrier bar by tightening its screw; it contains the strain gauges measuring the loads as well as the **accelerometer** to determine polished rod positions. These transducers are very easy to mount on the polished rod and can provide a much **safer** application than horseshoe types.

The operational principle of PRTs is that **axial** loads on the polished rod create **radial** strain that is recorded by the unit and is converted to electric voltage signals. Output signal is **linear** as a function of load and the transducer has very little **hysteresis**; temperature effects are eliminated by a compensation circuit. The loads calculated from the output of the device are, however, **relative** in nature

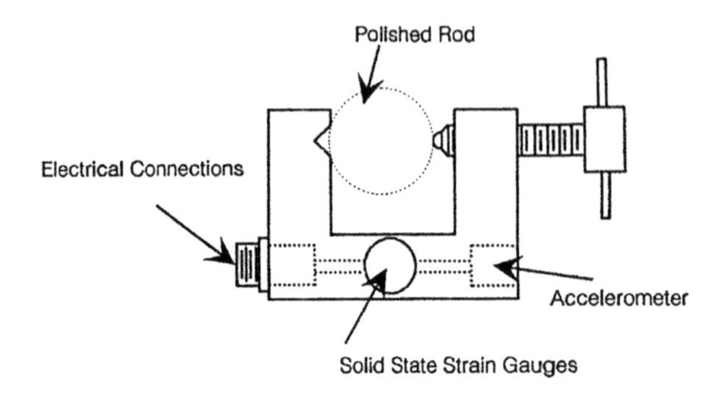

FIGURE 6.14

The general arrangement of a polished rod transducer.

because the polished rod is already under tension when the load cell is installed. This requires the application of an **offset** to the raw data, which must be obtained during calibration.

Calibration is done using a **reference** load like the buoyant rod weight, or it is automatically performed by using the proper software. The software utilizes the solution of the wave equation and calculates the downhole dynamometer card; loads are then calibrated so as to have the **bottom** of the downhole card lie on the **zero** load line. After applying the same correction to the surface load data, the estimated surface dynamometer card is drawn.

The biggest **advantages** of PRTs are their **easy** application and the fact that the **spacing** of the downhole pump is not affected during the dynamometer survey because the polished rod need not be raised. An important benefit is that the sensitivity of the transducer does not significantly change throughout its life. The accuracy of PRTs as compared to the accurate readings of horseshoe load cells is quite high: maximum deviations are below 2%, a small price for the increased **ease** of use and the much higher **safety** of operation [42].

Permanently mounted load cells come in two versions: attached to the polished rod or to the walking beam. Polished rod load cells in cylindrical shape (donut shape) are placed between the carrier bar and the polished rod clamp; they utilize strain gauges, too. Measurement with such devices can be in error because they or their cabling are easily damaged and can drift out of calibration. Beam-mounted transducers are **welded** or **clamped** to the top of the walking beam near the center bearing and sense the flexing of the beam due to pumping loads. Their use eliminates the possibility of mechanical damage of signal cables, but measurement accuracy changes with ambient temperature because tension in the beam varies with the temperature of the beam. Permanently mounted load cells are most popular in conjunction with **pump-off** controllers (POCs) or as end devices in **supervisory control and data acquisition** systems.

6.3.1.2 Downhole dynagraphs

Dynamometer cards taken at the surface can seldom be used directly to detect the operating **conditions** of the downhole pump, because they also reflect all **forces** (static and dynamic) that occur from the pump up to the wellhead. If, however, a dynamometer is placed just **above** the pump, the recorded card is a true **indicator** of the pump's operation. This is what **Gilbert's dynagraph** (a mechanical dynamometer) [43] accomplished in the 1930s. Rod loads immediately above the pump, recorded as a function of pump position, give **dynagraph cards**, a name used to distinguish them from surface cards. Although the application of **Gilbert's** dynagraph allowed a direct investigation of pumping problems, the practical **implications** associated with the necessity of running the instrument in the well had far outweighed its advantages. This seems to be true even for a modern, microprocessor-based version of the original instrument [44].

In order to provide the industry with high-quality measurements of sucker-rod pump dynamics, **Sandia** National Laboratories in 1997 conducted a series of measurements with **downhole** memory tools in sucker-rod-pumped wells [45]. The downhole sensors recorded pressure, temperature, load, and acceleration at several points along the rod string, including one immediately above the pump. A total of six wells were investigated under widely different operating conditions; their depths ranged from 2,600 to 9,300 ft. Sandia developed a downhole dynamometer **database** that includes all surface and downhole measurements and allows the construction of dynamometer cards at the different depths where sensors were installed [46]. Measured loads are "true loads" and pump cards obtained from the survey provide an efficient tool to compare the merits of different diagnostic and predictive computer program packages.

6.3.2 THE USE OF CONVENTIONAL DYNAMOMETERS

6.3.2.1 Recording a dynamometer card

Although this section is not intended as an operational guide for dynamometer measurements, some basic considerations for **conducting** a proper dynamometer survey will be discussed. By following the basic guidelines given below, dynamometer cards **representing** the current conditions can be taken, which, after being properly interpreted, provide a firm basis for an **analysis** of the pumping system [47]. It must be stressed, however, that the analysis of pumping wells does not consist of dynamometer measurements alone but includes other testing and measurement procedures as well. Therefore, to arrive at a final evaluation of the performance of the pumping system, all information gained from several analysis methods must be properly taken into account.

The basic rules to be followed **before running** a dynamometer survey when using a conventional dynamometer are:

1. Prior to dynamometer measurements, a fluid **level** survey must be run to ascertain the depth of the working fluid level present in the well.
2. Data of a current production **test** on the given well must be available, which give the latest oil, water, and gas production rates.
3. All relevant **data** on the given well and the pumping equipment must be collected and assembled. These include data on the equipment run in the well with types, sizes, depths, etc. Specific details of the tubing and rod strings, anchors, gas separators, etc. are a prime source of information. Equally important are surface equipment data: type, size of pumping unit, gear reducer, etc. Finally, the parameters of the current pumping mode (pump size, stroke length, pumping speed) must be recorded.

When **making** the dynamometer survey, the main points to remember are:

1. The **zero** load line should be recorded on the dynamometer card. This is accomplished by pulling the cord attached to the recording drum while no polished rod is applied.

2. The polished rod stuffing box should be **examined** for over-tightness and adjusted to keep friction on the polished rod at a minimum.
3. Before a representative card is recorded (the one taken under stabilized conditions), **several** cards on the same chart paper should be recorded.
4. After a correctly taken card is made, the subsurface pump's **valves** should be tested, and the actual **counterbalance effect** (*CBE*) must be recorded on the card. Both of these procedures are detailed in subsequent sections.
5. All relevant data (surface and subsurface) must be **recorded** on the card, along with the well's proper identification, to be used in later evaluations. A list of these parameters is **printed** on most types of dynamometer chart papers, which should be filled out by the analyst.

6.3.2.2 Checking the condition of pump valves

The proper operation of a sucker-rod pump is heavily affected by the **condition** of the standing and traveling **valves**. Both of these valves act as simple **check valves**, their operation depending on a proper **seal** between their seats and balls. However, due to mechanical damage, fluid erosion, corrosion, or other operational problems, valves can easily **lose** the perfect seal required for proper pump operation. Even the slightest **leak** can reduce the liquid rate lifted by the sucker-rod pump because of the high hydrostatic pressure valid at pump depth. The lifting capacity of the pump can thus considerably decrease with an associated deterioration of overall pumping efficiency. Therefore, it is of the utmost importance that the condition of the pump valves be **frequently** checked, which is fortunately quite easily accomplished with the use of a dynamometer.

6.3.2.2.1 The standing valve test

The standing valve test (**SV test**) is used to check the standing valve for **leaks** and is conducted with the dynamometer in place on the polished rod. At the start of the testing procedure, the pumping unit is **stopped** well into the **downstroke** (at about three-quarters of the way down) by switching off the motor and applying the brake. The unit must be stopped **gently** to eliminate any dynamic load effects. The actual polished rod load is immediately **recorded** on the dynamometer chart by pulling the cord of the dynamometer. As shown in Fig. 6.15, the standing valve is in a **closed** position, whereas the traveling valve is open. Since liquid load is completely carried by the **standing** valve, the polished rod load recorded at the commencement of the test represents the **buoyant weight of the rod string** only. In case the standing valve is in a **perfect** condition and holds well, the polished rod load remains **steady**. Thus, a repeated recording of the load by pulling the dynamometer cord again results in a line that falls on the first measurement. After **repeatedly** pulling on the cord at regular time intervals, no changes in polished rod load occur.

If the standing valve is **defective**, then it will leak fluids from the tubing due to the high pressure differential across its seat. As fluids leak from the space between the standing and traveling valves, the pressure immediately below the traveling valve will decrease, causing the traveling valve to **close** slowly. As time progresses, the traveling valve finally assumes the fluid load, which was originally carried by the standing valve. This **transfer of fluid load** will entail an **increase** in polished rod load, which is recorded at regular intervals (usually every second) on the dynamometer chart. As seen on the example card in Fig. 6.15, the original polished rod load (long horizontal line) **increases** as time progresses (indicated by the shorter lines), the rate of load increase being directly proportional to the **severity** of the standing valve leak.

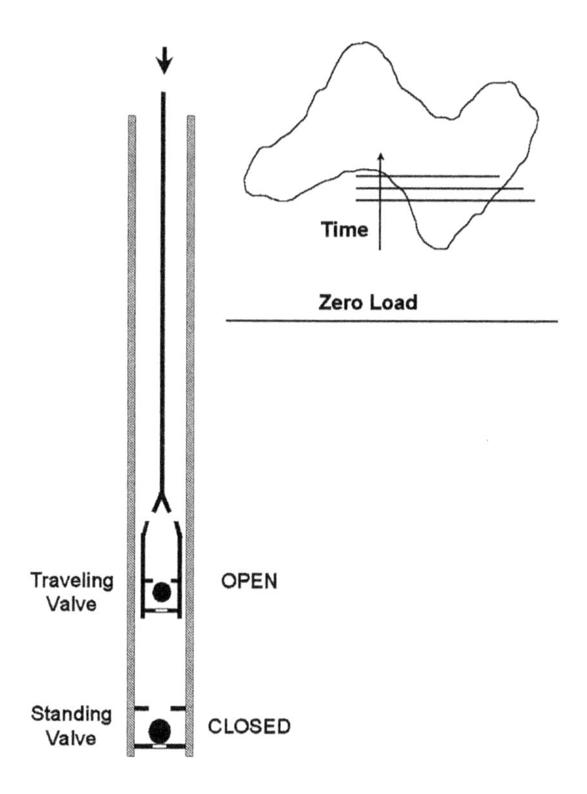

FIGURE 6.15

The principle of the standing valve test.

In case of leaking standing valves the valve test should be **repeated** several times to check for consistency. If the same results are obtained, then the **seat** of the standing valve is cut, whereas a damaged valve **ball** is indicated if the valve seems to hold on **some** trials. Usually, a leaking standing valve shows a load increase in about **20 s**. In no case should the measured load **decrease**; this indicates an **invalid** test, usually caused by stopping the unit before the traveling valve could open.

6.3.2.2.2 The traveling valve test

The traveling valve test (**TV test**) checks, with the same arrangement as used in the standing valve test, whether the traveling valve **and/or** the barrel–plunger fit is **leaking**. The pumping unit is **gently** stopped on the **upstroke** of the polished rod, near the top of the stroke. As the polished rod comes to a stop, the load on the dynamometer is immediately **recorded**. At this moment, the status of the pump valves is as shown in Fig. 6.16: the traveling valve is **closed** and the standing valve is **open**. The initial polished rod load, as recorded, represents the sum of the rod string weight in well fluids and the fluid load acting on the plunger. The standing valve is considered to be **open** and carries no load. If a perfect **seal** between the traveling valve seat and ball, and additionally, a perfect **fit** of the pump barrel and plunger, are assumed, then no **change** in polished rod load with time should

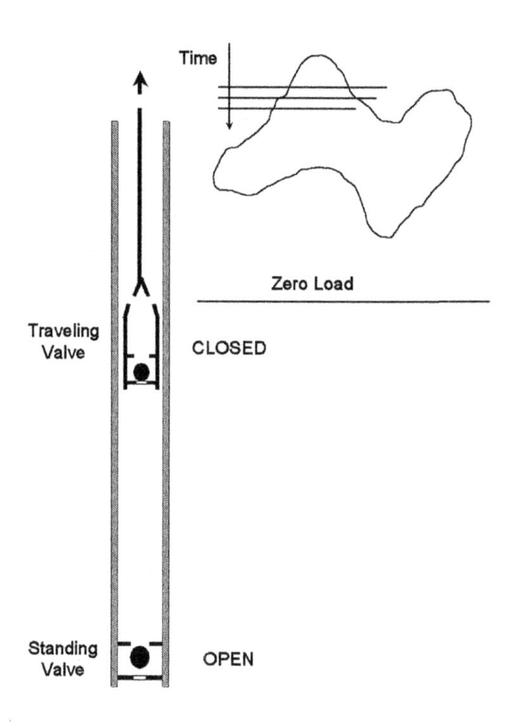

FIGURE 6.16

The principle of the traveling valve test.

occur. This condition is checked, as in the **SV test**, by pulling the cord of the dynamometer at several times in order to record the change in polished rod loads.

Any leakage in the traveling valve or between a worn pump barrel and plunger permits fluids to pass from **above** the traveling valve. This fluid leakage will slowly increase the pressure in the space between the pump's two valves, and the standing valve is **slowly** forced to close. As soon as it closes, the fluid load is no longer carried by the plunger and the rod string, but it is completely **transferred** to the standing valve and the tubing. The process of load **transfer** can be observed at the surface where polished rod loads are recorded on the dynamometer card at regular intervals. The original polished rod load, representing rod string weight plus fluid load, as recorded on the card (the long horizontal line on the example card in Fig. 6.16), is always **greater** than the loads measured **later** on. The rate of load decrease is, again, directly proportional to the leakage rates.

It should be clear from the above discussion that a **TV test** cannot differentiate between the effects of a leaking **traveling** valve and the leakage in the **pump** due to a worn barrel or plunger. As a rule, a measured loss in polished rod loads in about **5 s** is a positive indication of traveling valve leaks and/or an increased **slippage** past the plunger. As is the case with **SV tests**, the traveling valve test should also be **repeated** at different points in the upstroke. Differences in load loss rates observed during subsequent tests can indicate uneven wear or the position of a split in the pump barrel.

6.3.2.3 Measuring the counterbalance effect

With the use of a dynamometer it is possible to measure the **counterbalance effect**, i.e., the **force** at the polished rod arising from the **moment** of the counterweights. The proper knowledge of this parameter is a prime requirement for calculating the variation of **torques** on the speed reducer's slow-speed shaft. Two methods can be applied, depending on the actual degree of counterbalancing, which can show either an **over**- or **underbalanced** situation. If the pumping unit is overbalanced, or **weight heavy**, then, after stopping the motor and releasing the brake, the counterweights **lift** the polished rod. On the other hand, in a severely underbalanced or **rod-heavy** condition, the **polished rod** will **lift** the counterweights after the unit is stopped.

The counterbalance effect is measured with the dynamometer in place, carrying the full polished rod load. First the prime mover is shut down, and then the brake is applied slowly to **stop** the pumping unit with the **cranks horizontal** (this happens at crank angles of either 90° or 270°). At this position, the torque exerted on the crankshaft by the counterweights is at a **maximum**. This torque produces a vertical **force** on the polished rod, which is the *CBE* to be measured. In a **weight-heavy** situation, as shown in Fig. 6.17, the polished rod is **secured** by suitable means (a polished rod clamp and a chain) to the stuffing box, in order to prevent its vertical movement. Now the brake is completely **released**, and thus the mechanical moment of the counterweights is allowed to exert its effect on the polished rod. Since the polished rod is held stationary by the chain, the **force** that can be recorded with the dynamometer will give the actual *CBE* value. In **rod-heavy** cases, movement of the polished rod is prevented by installing a polished rod **clamp** directly over the stuffing box (Fig. 6.17). Again, after releasing the brake and pulling the cord of the dynamometer, the load recorded on the

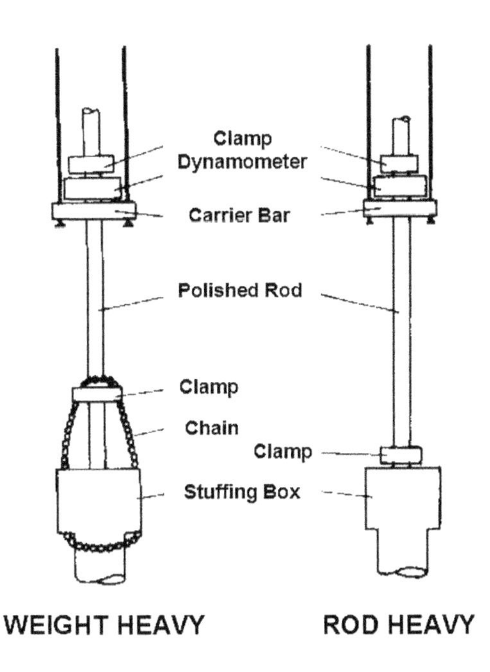

FIGURE 6.17

Two arrangements for measuring the counterbalance effect.

dynamometer card will correspond to the counterbalance effect. For greater accuracy measurements at both crank angles (90° and 270°) are made and the average of the two results is accepted.

From the *CBE* value measured above, the **maximum counterbalance torque** valid at the slow-speed shaft of the speed reducer is easily derived for any pumping unit geometry:

$$T_{CB\ max} = \frac{TF(90)(CBE - SU)}{\sin(90 + \tau)} \quad \text{if cranks stopped at } 90° \tag{6.28}$$

$$T_{CB\ max} = \frac{TF(270)(CBE - SU)}{\sin(270 + \tau)} \quad \text{if cranks stopped at } 270° \tag{6.29}$$

where:

$T_{CB\ max}$ = maximum counterbalance moment about the crankshaft, in lb,
$TF(\theta)$ = torque factor where *CBE* was measured (either 90° or 270°), in,
CBE = measured counterbalance effect, lb,
SU = structural unbalance, lb, and
τ = phase angle between the crank's and counterweight arm's centerline, degrees.

It is well known that the counterweights of pumping units of other than conventional **geometry** are affixed to counterweight **arms** placed with an **offset** angle to the cranks. Hence, counterbalance torque is at a maximum when the counterweight **arms**, rather than the cranks, are in a **horizontal** position. In spite of this fact, the American Petroleum Institute (API) recommends [48] that *CBE* be determined with the **cranks** held **horizontal** during the measurement. Equations (6.28) and (6.29) account for this effect for nonconventional (**Torqmaster** and **Mark II**) pumping unit geometries.

EXAMPLE 6.9: FIND THE MAXIMUM COUNTERBALANCE MOMENT FOR AN M-228-213-86 UNIT, IF THE MEASURED VALUE OF THE *CBE* EQUALS 13,000 LB, AND THE UNIT'S STRUCTURAL UNBALANCE IS −2,040 LB. THE *CBE* WAS MEASURED AT A CRANK ANGLE OF 90° AND THE UNIT'S OFFSET ANGLE EQUALS 24.5°

Solution

The torque factor at a crank angle of 90° is found from Fig. 3.55 as 38 in.

The counterbalance moment is found from Eq. (6.28) as:

$$T_{CB\ max} = 38(13,000 + 2,040)/\sin(90 + 24.5) = 628,070 \text{ in lb.}$$

6.3.3 THE USE OF ELECTRONIC DYNAMOMETERS

6.3.3.1 Introduction

As detailed earlier, the modern electronic dynamometers record the polished rod load and displacement as a function of **time**. The **analog** signals of load and position transducers are converted to **digital** form by the proper electronic circuitry, then the digital signals are **sampled** at a fixed frequency. The usual sampling frequency is 50 ms, i.e., 20 samples per second; this ensures an almost continuous **resolution** and a very accurate description of the pumping system's operating conditions.

Load and position signals as received from the transducers are shown in Fig. 6.18 for an example case during several pumping cycles. Also plotted is the surface dynamometer card that can be constructed using these data for a given pumping cycle.

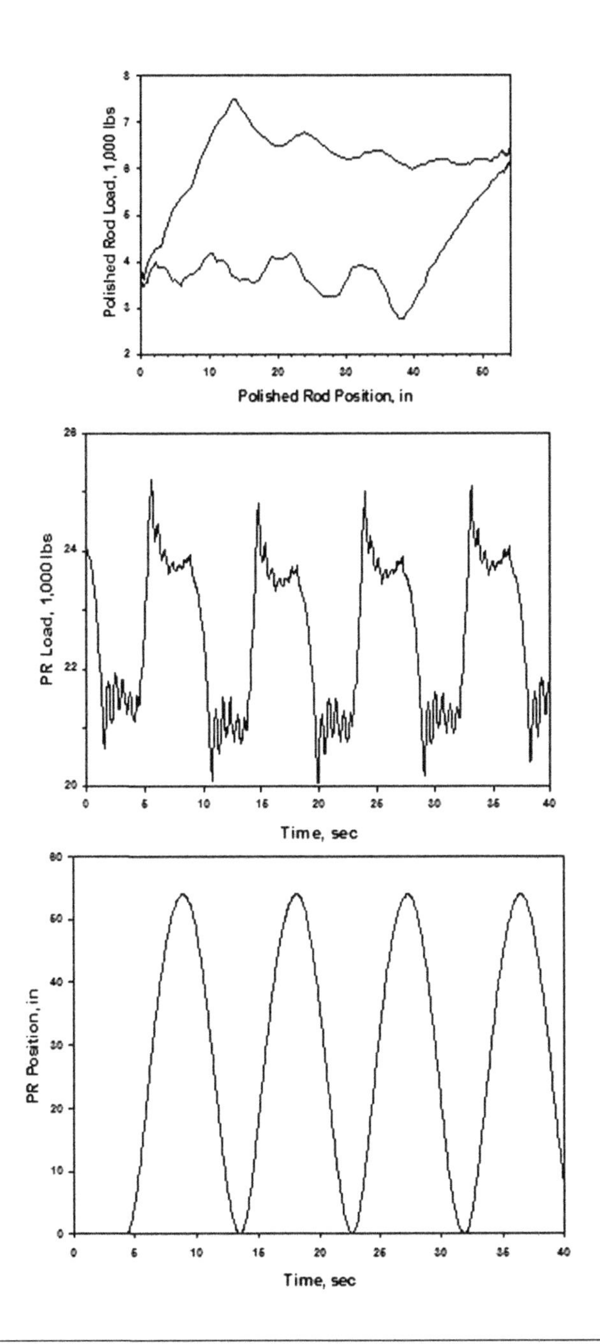

FIGURE 6.18

Typical load and position signals received from an electronic dynamometer.

The technique of running a dynamometer survey with the use of electronic dynamometers is very **similar** to that described in conjunction with the use of **conventional** dynamometers. Most of the **considerations** detailed earlier are still **valid** and the background theories of similar measurements are the same. This is the reason why the following sections should be read when a clear understanding of the principles and techniques detailed for conventional dynamometers has been achieved.

6.3.3.2 Recording a representative card

The main purpose of a dynamometer survey is to obtain a **representative** dynamometer card that properly characterizes the **stabilized** operation of the pumping system. Such cards represent the steady-state operation of the sucker-rod pump as well as other subsurface and surface equipment. It is quite easy to judge whether the pumping system's operation is stable when an electronic dynamometer is used by comparing the cards recorded during **consecutive** cycles. If the traces of several cards taken after each other fall closely on each other, then steady-state operation is reached and a representative card for further investigations can be acquired. Such a situation is illustrated in Fig. 6.19, where several cards closely overlap each other.

6.3.3.3 Checking the condition of pump valves

The most important components of the sucker-rod pump are the standing and traveling **valves**, and their actual mechanical condition greatly affects the operation of the pump. If they maintain a perfect **seal** between their seats and balls, then pump operation is basically **sound**; but the slightest **leak** can reduce the pump's liquid rate and the overall pumping efficiency. The mechanical condition of the valves of a sucker-rod pump is confirmed by running valve tests, i.e., the **standing valve test** and the **traveling valve test**.

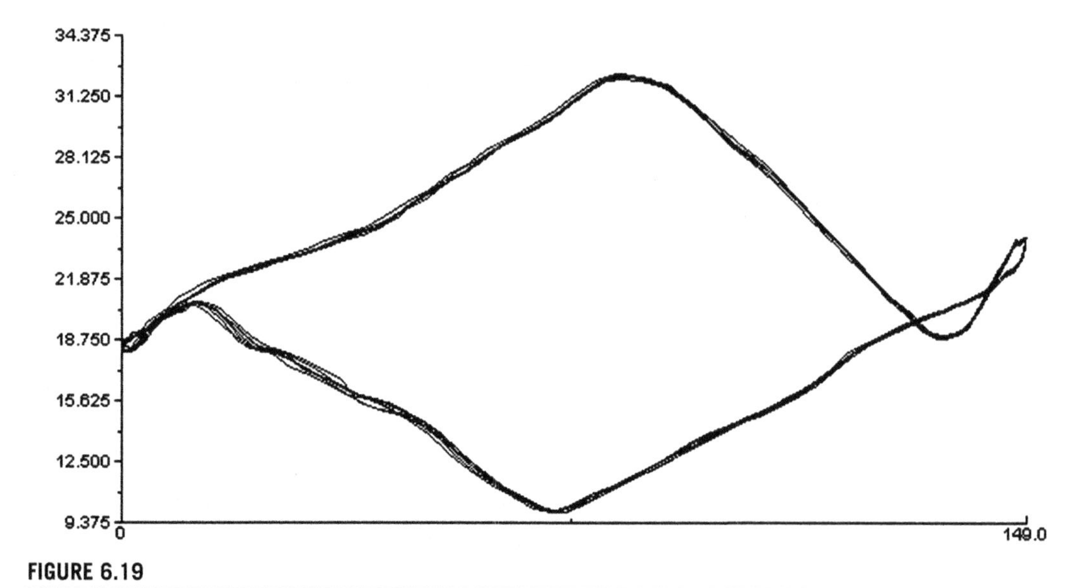

FIGURE 6.19

A sample steady-state, representative dynamometer card.

Standing Valve Test

This test is used to check the **condition** of the standing valve and it detects any **leaks** between the seat and the ball of the standing valve. The **SV test** is conducted with the electronic dynamometer installed on the polished rod while it continuously **records** the polished rod loads; test results are found from the **variation** of the recorded loads over a time period. It should be mentioned, however, that test results are valid only for negligible leakage in the pump, i.e., if the traveling valve and the plunger–barrel fit are in perfect condition.

At the start of the testing procedure the pumping unit is **stopped** on the **downstroke** when the polished rod is in the lower half of the downstroke by switching off the motor and applying the brake. In cases when there is an indication of a low pump fillage, the unit must be stopped very close to the bottom of the downstroke. When bringing the movement of the polished rod to a standstill, the application of the brake must be done **gently** and slowly to eliminate vibrations and dynamic load effects. As seen in Fig. 6.20, at the start of the **SV test** the traveling valve is in the **open** position, whereas the standing valve is in the **closed** position. The load recorded by the dynamometer at this time equals the rod string's **buoyant weight**, since the fluid load is completely carried by the standing valve. For a **perfectly** sealing standing valve this load does not change with time, as indicated on the dynamometer recording; measured loads will lie on a horizontal belonging to the **buoyant rod string weight**, seen at the position labeled **SV** in the figure. The measured rod string weight may be used to check the reliability of rod string data on file as well as to indicate the **accuracy** of the dynamometer's load cell by comparing it to the **calculated** buoyant weight.

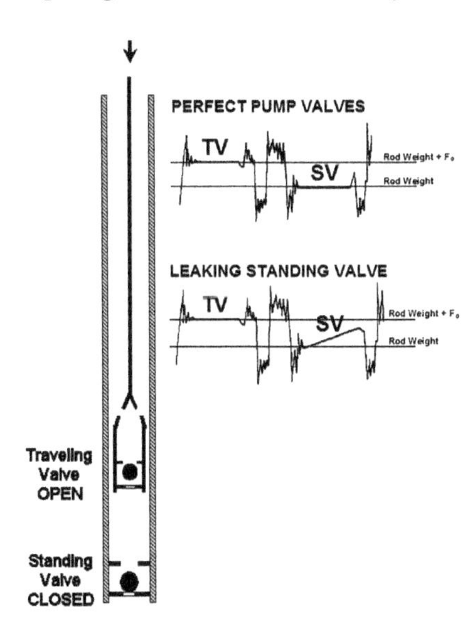

FIGURE 6.20

Typical standing valve test results from an electronic dynamometer.

A **leaking** standing valve, on the other hand, releases fluids from the tubing due to the high hydrostatic pressure differential across its seat, causing the pressure below the traveling valve to decrease. The traveling valve, being a simple check valve, starts to **close** because of the increasing pressure differential from above; thus, the fluid load originally carried by the standing valve is gradually **transferred** to the traveling valve. As a result, the load recorded continuously by the dynamometer starts to **increase** from its original level (the buoyant rod string weight), as seen in Fig. 6.20 under the label **SV**. Any change in load indicates a leaking standing valve; the time rate of load increase is directly proportional to the **severity** of the standing valve leak. The SV test should be **repeated** several times to check for consistency.

Traveling Valve Test

This procedure detects **leaks** in the traveling valve **and/or** in the barrel–plunger fit due to pump wear; the **TV test** cannot differentiate between the effects of a leaking **traveling** valve and the leakage in the **pump** due to a worn barrel or plunger. The necessary arrangement is the same as used in the SV test and the test can be conducted before or after an SV test. At the beginning the pumping unit is stopped **smoothly** and slowly by switching off the motor and applying the brake; it must be stopped on the **upstroke** and near the top of the stroke. At this position, the downhole pump's valves are in the following state: the traveling valve is **closed** and the standing valve is **open**, as shown in Fig. 6.21. At this moment the dynamometer records a polished rod load equivalent to the sum of the **buoyant rod string weight** and the **fluid load** acting on the plunger, F_O, because the traveling valve carries all loads. In case the traveling valve operates **perfectly** and there is no leakage of fluids between the pump barrel and plunger, then the polished rod load must not change with time, at least not in a short period. This can be clearly detected on the electronic dynamometer's recording as a horizontal on the rod weight plus fluid load line under the label **TV**.

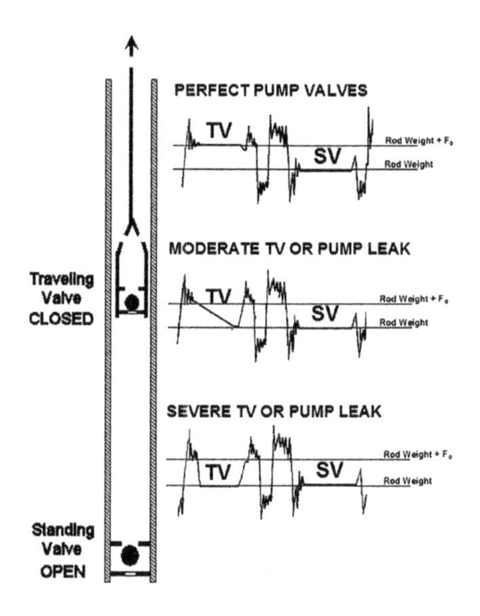

FIGURE 6.21

Typical traveling valve test results from an electronic dynamometer.

If there is a considerable **leakage** in the plunger–barrel fit of the sucker-rod pump and/or in the traveling valve itself, the fluid leaked increases the pressure above the standing valve. Because of this the standing valve begins to **close** and the **transfer** of the fluid load from the traveling to the standing valve takes place. The polished rod load measured at the start of the TV test (sum of the buoyant rod string weight plus fluid load) will finally **drop** to the buoyant rod string weight. Since the decrease in polished rod load is continuously recorded by the dynamometer, observation of this change over time gives valuable information on the mechanical condition of the downhole pump. The rate of decrease in the measured load is, of course, directly proportional to the leakage rate. In case of a **moderate** leak, as shown in Fig. 6.21, it takes some time until polished rod load decreases to the level of the buoyant rod weight, as indicated by the recording under the label **TV**. A **severely** worn-out pump or a badly leaking traveling valve, on the other hand, is indicated by an **immediate** drop of load down to the level of the buoyant rod weight.

One has to consider that the effectiveness of leak detection by the **TV test** depends on the actual fillage of the downhole pump. With a fully charged barrel indicated by a **full** pump card, even a **slight** leak in the traveling valve and/or the plunger–barrel fit leads to a considerable drop in measured load, making the **TV test** a very **effective** leak detection method. On the other hand, in a pump with **low** fluid fillage more fluid must leak into the barrel before the fluid load is transferred to the standing valve, so leak detection is less effective. As a rule of thumb, a drop in polished rod load in about **5 s** is a positive indication of traveling valve **leaks** and/or an increased **slippage** past the plunger.

The **accuracy** of a **TV test** is related to stopping the pumping unit **smoothly** because the rod string may bounce back, which leads to erroneous readings. Also, the **TV test** gives reliable results only if the standing valve is holding perfectly during the test; otherwise, calculations would represent the valve that leaks at the greater rate. Similarly to the standing valve test, the traveling valve test should also be **repeated** by stopping the pumping unit at different points during the upstroke of the polished rod. Differences in load loss rates observed during subsequent tests can indicate uneven **wear** of the plunger or the barrel as well as the position of a **split** in the pump barrel.

6.3.3.3.1 Estimation of leakage rates

The change of the measured surface loads with time, as recorded during the TV and SV tests, allows the **estimation** of leakage rates in the pump. The following procedure, first proposed by **Nolen and Gibbs** [49], can be used to **infer** fluid leakage rates across the plunger–barrel fit and either the traveling valve or the standing valve; the calculation of the leakage in the pump will be detailed in the following [3].

As already discussed, during the **TV test** the fluid load is completely **transferred** from the rod string to the tubing if the plunger or the traveling valve leaks. As the load is removed from the plunger, the rod string loses its stretch caused by the fluid load and the plunger moves upward by a distance equal to that stretch. The volume of fluid that leaks past the pump needed to equalize the pressures above and below the plunger is the product of the net stretch and the pump's cross-sectional area. In case of an unanchored tubing string, the stretch of the tubing must also be taken into account and the leakage volume is derived as:

$$V = \frac{d^2 \pi}{4} (e_r + e_t) \tag{6.30}$$

where:

V = leaked volume, cu in,
d = diameter of the downhole pump, in, and
e_r, e_t = elongations due to fluid load of the rod string and the tubing, respectively, in.

Leakage occurs only while the traveling valve is **closed** and when there is a pressure differential across it, i.e., at the start of the **TV test**. At that moment the load loss rate is at a **maximum**, $(\Delta F/\Delta t)_{max}$, which can be determined from the plot of the measured load versus time curve obtained during the **TV test**. Considering that, at least for a full pump, the traveling valve carries the fluid load during the **upstroke** only, and converting the rate into bpd units, we get the final formula for pump leakage rate:

$$q_s = 3.5d^2 \left(\frac{\Delta F}{\Delta t}\right)_{max} \left(\sum_{i=1}^{Taper} L_i E_{ri} + L_t E_t\right) \tag{6.31}$$

where:

q_s = pump leakage (slippage) rate, bpd,
d = diameter of the downhole pump, in,
$(\Delta F/\Delta t)_{max}$ = maximum load loss rate, lb/s,
E_{ri}, E_t = elastic constant of the ith rod taper and the tubing, respectively, in/(lb ft),
L_i, L_t = length of the ith rod taper and the tubing, respectively, ft, and
Taper = number of rod tapers in string, –.

As pointed out by many investigators, the above procedure yields only an approximate value for the leakage in the pump.

6.3.3.4 Counterbalance effect measurement

The **counterbalance effect** is the **force** arising at the polished rod from the **moment** of the counterweights installed on the pumping unit; knowledge of this value is necessary for the calculation of the gearbox torque required to turn the counterweights. It can be easily measured with a dynamometer installed at the polished rod at the moment when the pumping system is at static equilibrium. At that moment the rod torque balances the counterbalance torque because inertial effects have vanished, and the counterbalance effect is found from the balance of the two torques.

The method frequently used with electronic dynamometers is based on the fact that polished rod loads will **inevitably** drop due to fluid **leakage** in the downhole pump after the pumping unit is stopped on the upstroke. Immediately after stopping the unit, the polished rod load equals the sum of the buoyant rod string weight and the fluid load ($W_{rf} + F_o$), but later on, due to fluid leakage in the pump, it drops down to the buoyant rod weight, W_{rf}. If the actual *CBE* lies somewhere **between** these two values, then at one moment it will **balance** the changing load on the polished rod; this time is found when no movement of the polished rod is detected. The main steps of the procedure are as follows:

1. The pumping unit is operated normally and the dynamometer is started to record polished rod position and load; a stopwatch is simultaneously started to measure elapsed time from this moment on.
2. The prime mover is shut down and then the brake is gently applied to **stop** the unit somewhere on the upstroke.
3. Now enough time is allowed for pump leakage to **reduce** the polished rod load and the brake is shortly **released** to check the movement of the polished rod.
4. If movement of the polished rod is detected, the brake is again applied to stop any movement. Since only **short** movements are allowed, these are much easier to detect on the brake **drum** than on the polished rod.

5. Periodic and momentary **release** of the brake is continued until no polished rod **movement** is detected with a completely released brake.
6. With the polished rod at rest, system equilibrium is reached and the load measured at the polished rod equals the *CBE*. The time when equilibrium occurs is noted on the stopwatch.
7. Using the polished rod position and load recordings of the dynamometer, the load *CBE* and position *s* valid at the time marked are determined.

The recorded polished rod position, *s*, allows the determination of the position of rods (*PR*) in the upstroke from:

$$PR = \frac{s}{S} \tag{6.32}$$

where:

PR = "position of rods" function, –,
s = recorded polished rod position at equilibrium, in, and
S = polished rod stroke length, in.

Using the kinematic parameters of the pumping unit (see Section 3.7.5), the crank angle, θ_e, and the torque factor, $TF(\theta_e)$, belonging to the *PR* value just calculated are determined. Using these parameters and the measured *CBE*, the **maximum counterbalance torque** at the slow-speed shaft of the speed reducer is easily derived similarly to Eq. (6.28):

$$T_{CB\ max} = \frac{TF(\theta_e)(CBE - SU)}{\sin(\theta_e + \tau)} \tag{6.33}$$

where:

$T_{CB\ max}$ = maximum counterbalance moment about the crankshaft, lb,
θ_e = crank angle where equilibrium occurs, degrees,
$TF(\theta_e)$ = torque factor at crank angle θ_e, in,
CBE = measured counterbalance effect, lb,
SU = structural unbalance, lb, and
τ = phase angle between the crank's and counterweight arm's centerline, degrees.

This procedure cannot be applied if the counterbalance effect is (1) greater than the sum of the buoyant rod weight and the fluid load ($W_{rf} + F_o$), or (2) less than the buoyant rod string weight, W_{rf}. These situations occur when the unit is very much **weight** or **rod** heavy. Another limitation is posed by fluid **slippage** in the downhole pump; installations with excessively high pump slippage cannot be analyzed because of the rapid drop in the loads measured by the dynamometer.

6.4 INTERPRETATION OF DYNAMOMETER CARDS

The correct **interpretation** of surface or downhole dynamometer cards reveals a wealth of **information** on the operation of the sucker-rod pumping system. The most **important uses** of dynamometer cards are the following:

- Determination of **loads** acting on the pumping unit structure and in the rod string.
- Based on dynamometer card data, the torsional **loading** on the pumping unit's speed reducer can be calculated.

- From the area of the card, the **power** required to drive the pumping unit is found.
- By checking the actual counterbalance effect, the degree of the unit's counterbalancing can be determined.
- The condition of the sucker-rod pump and its valves can be determined.
- Many downhole **problems** can be detected by studying the **shape** of the dynamometer card, making the analysis of dynamometer cards a powerful **troubleshooting** tool.

6.4.1 CONVENTIONAL DYNAMOMETER CARDS

In order to understand the basic characteristic shapes of dynamometer cards taken at the surface with conventional dynamometers, cards for **simplified** conditions are discussed first. Assume a rigid, inelastic rod string; a sufficiently **low** pumping speed to eliminate dynamic forces; and pumping of an **incompressible** liquid; neglect all energy losses along the string; and assume an anchored tubing string. In this case the dynamometer card, i.e., the variation of polished rod load versus polished rod position, is represented by the parallelogram 1–2–3–4 shown in Fig. 6.22. At point 1, the upstroke begins and the traveling valve immediately closes. Polished rod load, equal to the buoyant weight of the string at point 1, suddenly **increases** to the load indicated by point 2, as the fluid load is **transferred** from the standing valve to the traveling valve. The plunger and the polished rod move **together** until point 3 is reached, while a constant load is maintained. In point 3, the end of the **upstroke** is reached, and the downstroke begins with the immediate **opening** of the traveling valve. Rod load suddenly **drops** to point 4, since fluid load is no longer carried by the traveling valve. The rod string, with the open traveling valve at its lower end, **falls** in well fluids from point 4 to 1, while polished rod load equals the buoyant weight of the rod string. At point 1, a new cycle begins.

Now, in a more realistic case, an **elastic** rod system is considered with all the other assumptions unchanged, and the shape of the dynamometer card changes to the **rhomboid** 1–2′–3–4′ in Fig. 6.22. It is due to rod **stretch** that, from point 1, rod load only **gradually** reaches its maximum value at point 2′, while the pump ascends with a **closed** traveling valve. Similarly, at the end of the

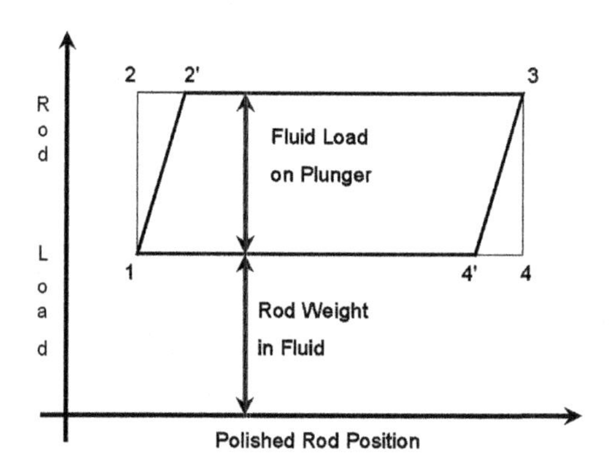

FIGURE 6.22

Theoretical surface dynamometer card shapes for low pumping speeds.

upstroke, the **transfer** of fluid load from the traveling valve to the standing valve is also **gradual** from point 3 to point 4′, since the rod string contracts to its original length. This theoretical dynamometer card shape is seldom encountered and can only be found on shallow wells when slow pumping speeds are used [50].

It follows from the above description of the pumping cycle that, under the **simplified** conditions, maximum polished rod load equals the load measured during a traveling valve test, whereas minimum load equals that of the standing valve test. Another important conclusion, in line with the considerations detailed in Section 6.3.2.2.2, is that plunger **travel** is less than the stroke **length** of the polished rod, the difference being taken up by rod string **elongation**. Hence, plunger stroke length 2–3, which is equal to polished rod stroke length for a rigid string, decreases to 2′–3 if the elasticity of the rod string is accounted for.

In a real well, the previous simplifying assumptions are **seldom** met because of the following:

- **Dynamic** rod loads occur due to the acceleration pattern of the rod string's movement;
- Stress **waves** are induced in the rod string by the polished rod's movement and by the operation of the downhole pump. These waves are **transmitted** and **reflected** in the rod string and can considerably affect the polished rod loads measured;
- The frequency of induced stress waves can **interfere** with the resonant (fundamental) frequency of the string, causing considerable changes in rod loads;
- The action of the pump valves is heavily affected by the **compressibility** of the fluids lifted; and finally;
- Downhole **problems** can exist that alter rod loads.

The **combined** effect of the above conditions changes the **shape** of the dynamometer card very significantly, as illustrated in Fig. 6.23. As shown, maximum and minimum loads differ from the values valid for the slow-speed elastic rod model, and the general shape of the card is also distorted.

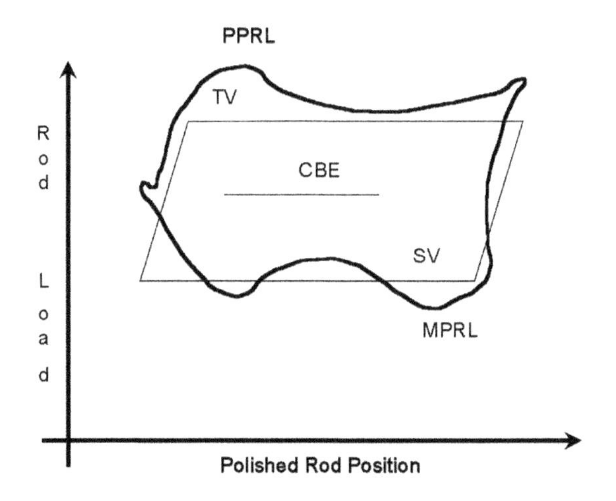

FIGURE 6.23

Comparison of an actual dynamometer card with a theoretical one valid at low pumping speeds.

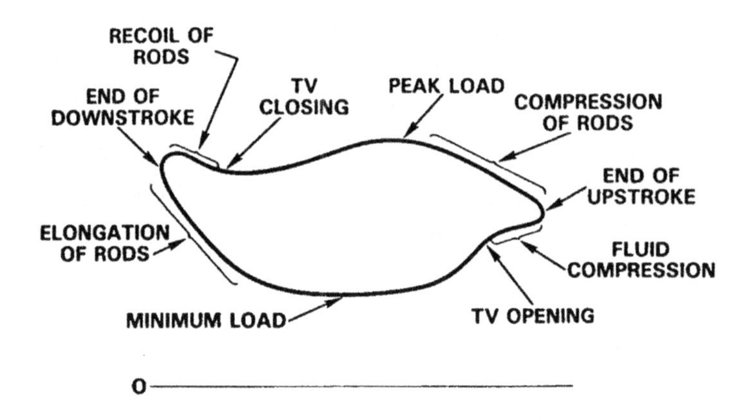

FIGURE 6.24

Characteristic features of a general dynamometer card, after **Gipson and Swaim** [51].

Explanation of a **general** dynamometer card shape is given in Fig. 6.24, after **Gipson and Swaim** [51]. At the end of the polished rod's downstroke, the plunger still travels down due to the time delay in the string's stress transmission; hence the traveling valve (TV) closes only after the start of the polished rod's upstroke. After the traveling valve closes, the rods stretch and polished rod load increases until the peak load is reached. Toward the end of the upstroke, dynamic effects tend to compress the rod string, and polished rod loads decrease. The operation of the traveling valve is again delayed and it only opens after the polished rod starts its downstroke. The rods now start to contract and polished rod loads decrease to a minimum. Close to the end of the downstroke, dynamic effects dominate, causing the polished rod loads to increase again.

6.4.1.1 Basic loads

From a properly recorded dynamometer card, six basic **loads** can be determined, as pointed out by **Gipson and Swaim** [51]. These are shown on the example card in Fig. 6.23 and include:

1. The **zero load** or baseline from which all loads are measured.
2. The *SV load*, found from a standing valve test. In an ideal case, with the standing valve not leaking, the *SV load* equals the **buoyant weight** of the rod string.
3. The *TV load*, as measured during a traveling valve test. In cases where the plunger and the valves of the sucker-rod pump are in perfect condition, this load is the **sum** of the **buoyant rod weight** and the **fluid load** on the plunger.
4. The **peak polished rod load** (PPRL), which is the maximum load during the pumping cycle and reflects the *TV load* plus the maximum of the dynamic loads occurring during the upstroke.
5. The **minimum polished rod load** (MPRL), which represents the *SV load* minus the maximum downstroke dynamic load and is found on the dynamometer card as the minimum load during the cycle.
6. The *CBE*, which represents the force at the polished rod derived from the maximum counterweight moment.

The magnitude of the above loads is found from measuring the respective **ordinate** values above the zero load line on the conventional dynamometer **card**, and taking into account the spring constant of the dynamometer used.

6.4.1.2 Polished rod horsepower

Since the product of force and distance is mechanical **work**, the area enclosed by the dynamometer card represents the **work** done on the polished rod during a pumping cycle. From this, the **power** input at the polished rod is easily expressed with the pumping speed [52]:

$$PRHP = \frac{A_c\, K\, S\, N}{396{,}000\, L_c} \tag{6.34}$$

where:

$PRHP$ = polished rod horsepower, HP,
A_c = area of the dynamometer card, sq in,
K = dynamometer constant, lb/in,
S = polished rod stroke length, in,
N = pumping speed, strokes/min, and
L_c = length of the dynamometer card, in.

The polished rod **power** (*PRHP*) thus calculated is the **total** power required at the polished rod, which covers the **useful** power of fluid lifting and the sum of all energy **losses** occurring downhole. It is a very important parameter of sucker-rod pumping and clearly indicates the system's downhole **efficiency**, as already discussed in Section 4.7.

6.4.1.3 Troubleshooting

The detection of pumping **malfunctions** from the **visual** interpretation of surface dynamometer card shapes is a task for highly specialized analysts. Due to the many **interactions** of influencing parameters and the great number of possible pumping problems, an **infinite** number of dynamometer card **shapes** can exist, making the analysis of surface dynamometer cards more an **art** than an exact science. A proper troubleshooting analysis of the pumping system, therefore, heavily relies on the analyst's **expertise** and **skill**. A complete treatment of dynamometer card analysis, if it is possible at all, is beyond the scope of this section and only some basics are discussed in the following.

The shape of a **healthy** dynamometer card is mainly governed by the pumping **speed** and the **fluid load** on the plunger, though it is also influenced by additional parameters like the composition of the rod string, prime mover slip, etc. This fact was recognized by **Sucker Rod Pumping Research Inc.** when they developed an **analog** computer to **simulate** the operation of the pumping system. Their results, later adopted by the API in the **RP 11L**, made it possible to construct dynamometer cards for a multitude of **simulated** cases. These cards are published in **API Bul 11L2** [53], where dynamometer cards for good pumping conditions (100% fluid fillage of the pump, no gas interference, pump in perfect mechanical condition) are given. The cards are classified according to the values of two **dimensionless** parameters—N/N_o' and $F_o/S/k_r$—because installations having identical such parameters have similar dynamometer cards shapes.

The use of **API Bul 11L2** in the analysis of dynamometer cards involves the calculation of the above dimensionless parameters, which are also used in the **RP 11L** procedure. Then, the analog card

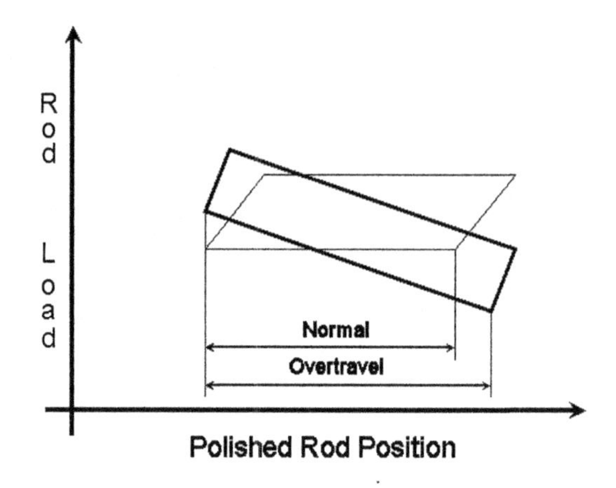

FIGURE 6.25

The general shape of an overtravel card.

closest to these conditions is selected from the collection of cards found at the respective page of **API Bul 11L2**. If the measured card closely resembles the analog card just found, no apparent pumping problems are present; otherwise, further expert analysis is required.

According to their general shape, dynamometer cards are usually categorized as normal, overtravel, or undertravel cards. A typical **overtravel card** exhibits a general **downward** slope from left to right, like the theoretical card shown in Fig. 6.25. When compared to an ideal card shape, it is apparent that plunger stroke is **increased** for the same polished rod stroke length. This can be attributed to increased **dynamic** effects at higher pumping speeds when the increased acceleration of the heavy rod string results in greater stretch along the string. **Overtravel** has two effects on the operation of the pumping system: pump displacement increases (this is the main factor when fiberglass rod strings are used), but at the same time the frequency of rod and other failures also increases. Overtravel of the downhole pump normally occurs under the following conditions:

- higher-than-normal pumping speeds,
- when using a fiberglass rod string,
- if the rod string has parted downhole,
- if the pump unseats during the pumping cycle, and
- for gas-locked and worn-out pumps.

In **undertravel** situations (see Fig. 6.26), the card slopes **upward** from left to right and plunger stroke length is **decreased** when compared to polished rod stroke length. The reduced travel of the plunger is usually due to greater than normal downhole friction. Among the possible causes of undertravel are:

- **heavy** rod loads from the use of large plungers,
- excessive downhole friction because of sand or paraffin accumulation,
- a stuck pump,
- a too-tight stuffing box.

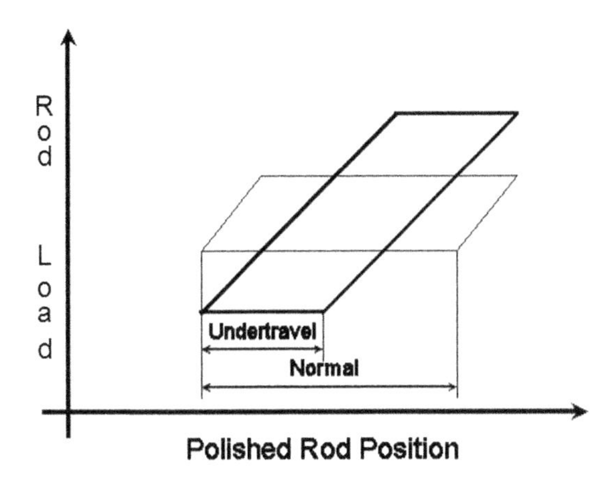

FIGURE 6.26

The general shape of an undertravel card.

6.4.2 CARDS FROM ELECTRONIC DYNAMOMETERS

If an **electronic** dynamometer is used, then the dynamometer survey provides not only the **surface** card but also the calculated **downhole** card derived from surface dynamometer measurements by solving the **damped wave equation** that describes the behavior of the rod string. The use of downhole dynamometer cards offers a more **direct** detection of pumping **malfunctions** than the use of surface cards. This is mainly due to the fact that downhole cards, calculated at the depth of the sucker-rod pump, reflect the **operating** conditions of the pump alone. All other factors, e.g., rod string behavior and surface conditions, do not affect the loads and displacements occurring at this point. Pump cards usually have more **consistent** and more **characteristic** shapes than the surface cards they are derived from. Therefore, **pump cards** offer very efficient and reliable **indications** on the operational conditions of the sucker-rod pump and are widely used in troubleshooting the rod-pumping system [54–56].

Since pump cards are available immediately after the acquisition of surface dynamometer data, the use of electronic dynamometers offers a much higher level of **accuracy** and **reliability** than those provided by conventional dynamometers. This is the reason why today's sucker-rod pumping analysis relies heavily on the use of electronic dynamometers and the solution of the damped wave equation. The following sections discuss the basic concepts of dynamometer card analysis and the different ways of finding pumping parameters from dynamometer surveys.

6.4.2.1 Ideal pump cards

Before the interpretation of troublesome pump cards is attempted one must understand the features of cards obtained under **ideal** pumping conditions. The characteristics of ideal conditions for a sucker-rod pump can be summed up as follows:

- The pump's barrel and plunger are in **perfect** condition.
- The standing and the traveling valves are **not leaking**.

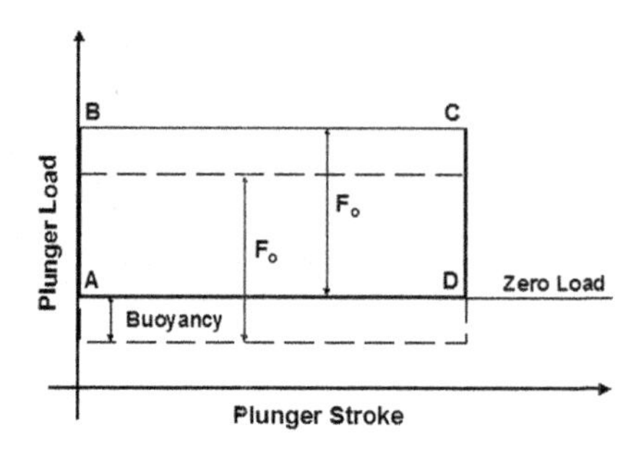

FIGURE 6.27

Ideal pump card for an installation with an anchored tubing string.

- All friction forces along the rod string are due to **viscous** damping.
- Only single-phase **liquid** enters the pump barrel.
- The barrel fills up **completely** during the upstroke.

Under such conditions pump cards calculated from the solution of the wave equation exhibit very characteristic and simple shapes, from which the most important pumping parameters are easily found. An ideal pump card for an installation with an **anchored** tubing string is given in Fig. 6.27. The loads plotted can be either **true** or **effective** rod loads, depending on the analyst's preference. Most investigators use effective loads and the card (the parallelogram A–B–C–D) sits on the zero load line; if true loads are considered then the card (indicated inside dashed line) is shifted downward by the **buoyancy** force acting on the bottom rod taper.

Investigation of the plunger's **movement** (position, velocity) during the pumping cycle can provide useful information on the behavior of the downhole pump. Figure 6.28 presents the variation of the polished rod's and the plunger's position for a complete pumping cycle for an ideal case with an **anchored** tubing string. The polished rod position is a very regular function of the elapsed time, as found from the **kinematic** analysis of the pumping unit's movement (see Section 3.2.5). The plunger with the closed traveling valve, on the other hand, does not start to move at the beginning of the upstroke (point **A**), while the load on the rods increases and the rod string gradually stretches. As soon as the load on the rods assumes the fluid load on the plunger (at point **B**) the plunger starts to move upward; the standing valve opens at the same time. The elapsed time for point **B** is found where the polished rod's position equals the stretch of the rod string due to fluid load:

$$e_r = \frac{F_o}{k_r} \tag{6.35}$$

where:

e_r = stretch of the rod string due to fluid load, in,
F_o = fluid load on plunger, lb, and
k_r = spring constant of the rod string, lb/in.

The upstroke portion of the plunger's movement lies between points **B** and **C** where the plunger's position follows that of the polished rod. Downstroke starts at point **C**; the position of the plunger at this point defines the plunger's **net** stroke length. This stroke length is less than that of

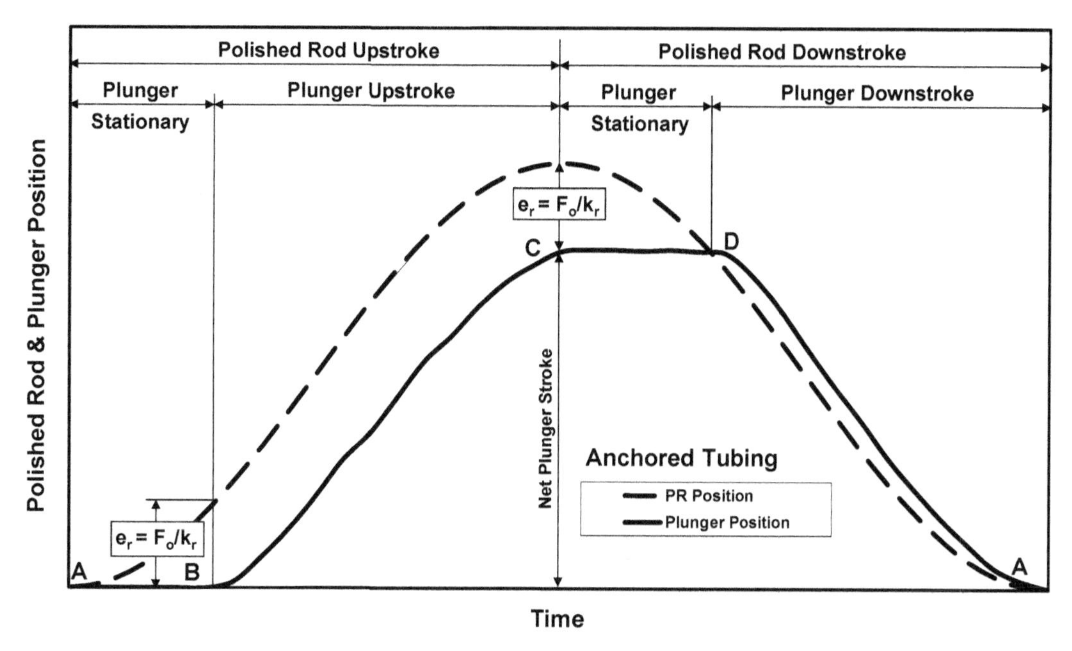

FIGURE 6.28

Comparison of the polished rod's and the plunger's movement for an anchored tubing string.

the polished rod; the **difference** is the stretch of the rod string, e_r, defined previously. At the start of the downstroke the plunger stays **stationary** again while the rod string unstretches and the fluid load from the plunger is slowly transferred to the standing valve between points **C** and **D**. At point **D** the standing valve **closes** and the rod string is completely unloaded; the plunger falls until the end of the stroke is reached at point **A** again.

It can be concluded from the previous description of the plunger's movement that under ideal conditions, i.e., (1) nonleaking standing and traveling valves, (2) anchored tubing string, and (3) perfect fit between the plunger and the barrel, the plunger must not move between points **A–B** and **C–D**. Any change in plunger position or velocity in those ranges may indicate that some of the conditions are not met: valves may leak; tubing anchor may slip, etc.

The shape of the ideal pump card changes when the tubing string is **not anchored**, as shown in Fig. 6.29. From the start of the upstroke at point **A** the tubing string unstretches because the fluid load is transferred from the standing to the traveling valve. This is why the plunger moves upward by a distance equivalent to the stretch of the tubing before it carries the fluid load F_o at point **B**. Tubing stretch is found from the following formula:

$$e_t = \frac{F_o}{k_t} \tag{6.36}$$

where:

e_t = stretch of the tubing string due to fluid load, in,
F_o = fluid load on plunger, lb, and
k_t = spring constant of the tubing string, lb/in.

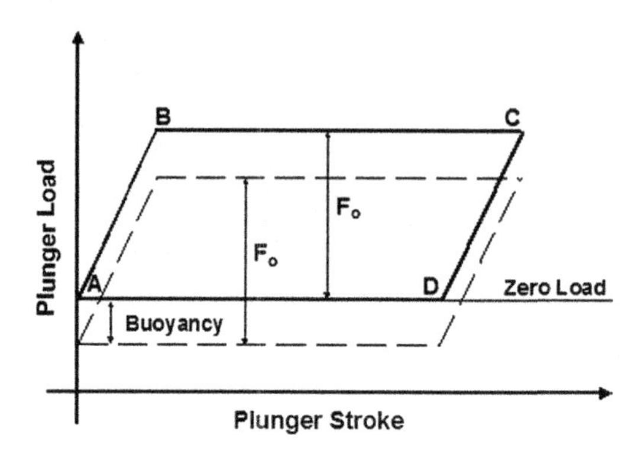

FIGURE 6.29

Ideal pump card for an installation with an unanchored tubing string.

At the end of the upstroke at point **C** the fluid load is transferred to the standing valve when the tubing stretches the amount given in Eq. (6.36); the plunger must move downward by the same distance as shown between points **C** and **D**.

The comparison of polished rod and plunger positions is presented in Fig. 6.30. At the start of the upstroke both the tubing and the plunger are moving between points **A** and **B** but in **opposite** directions. The tubing string gradually shortens because the fluid load on it is transferred to the

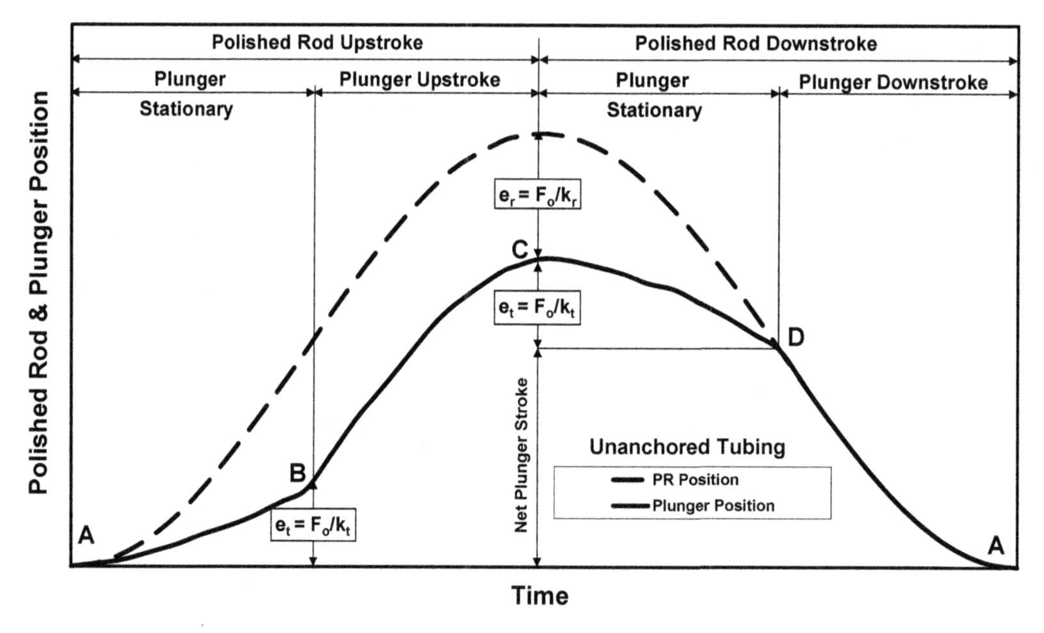

FIGURE 6.30

Comparison of the polished rod's and the plunger's movement for an unanchored tubing string.

plunger; the rod string, at the same time, is elongated. Due to the interaction of these movements the plunger stays **stationary** with respect to the pump barrel. When the plunger starts its upstroke at point **B**, it has covered a distance equal to the stretch of the tubing string, e_t. The plunger's movement attains its maximum at point **C**; plunger position here is less than the polished rod stroke length by the stretch of the rod string, e_r, found from Eq. (6.35).

Between points **C** and **D** the plunger does not move in relation to the barrel because its downward movement is eliminated by the tubing string's stretch, e_t, as the fluid load is transferred from the rod string to the tubing string. The **net** plunger stroke length is reached at point **D** and it is less than the maximum movement of the plunger (point **C**) by the stretch of the tubing string, e_t, found from Eq. (6.36). From point **D** on, the fluid load on the plunger is zero because the fluid load was transferred to the standing valve; the rod string is fully unstretched and moves the plunger during the rest of the downstroke in sync with the polished rod.

The investigation of plunger position versus time, as shown previously for ideal conditions, assists the analyst to evaluate and troubleshoot downhole problems. In addition to position data, evaluation of plunger **velocities** during the pumping cycle can facilitate the detection of pump problems. These methods of analysis are usually available in commercial wave equation program packages [57].

6.4.2.2 Basic analysis

When analyzing surface and downhole cards obtained from an electronic dynamometer survey, plotting of several **reference lines** on the cards greatly facilitates the analysis of the pumping system's operating conditions [58,59]. In the following we introduce these basic references and discuss how their use can improve the explanation of the pumping system's behavior.

6.4.2.2.1 Surface cards

The definitions of the basic reference lines to be used on the surface card are given in the following.

Buoyant weight of the rod string, W_{rf}: This represents the rod string's calculated weight in the produced liquid.

$$\begin{aligned} W_{rf} &= W_r(1 - 0.128\,SpGr) \\ &= (1 - 0.128\,SpGr)\sum_{i=1}^{Taper} L_i w_{ri} \end{aligned} \tag{6.37}$$

where:

W_{rf} = buoyant rod string weight, lb,
W_r = rod string weight in air, lb,
$SpGr$ = specific gravity of the fluid pumped, –,
L_i = length of the ith rod taper, ft,
w_{ri} = weight of the ith rod taper, lb/ft, and
$Taper$ = number of rod tapers in string, –.

Maximum upstroke load, $W_{rf} + F_{o\,max}$: This load represents the estimated maximum polished rod load when **pump-off** conditions are assumed. The theoretical **maximum fluid load** $F_{o\,max}$ is calculated by using the dynamic liquid level at the pump setting depth, i.e., by assuming a pump intake pressure of zero when the following formula is valid:

$$F_{o\ max} = A_p\left(WHP + 0.433\ L_{pump}\ SpGr\right) \tag{6.38}$$

where:

$F_{o\ max}$ = fluid load on plunger during pump-off, lb,
A_p = cross-sectional area of the plunger, sq in,
WHP = wellhead (tubing head) pressure, psi, and
L_{pump} = pump setting depth, ft.

Spring constants calculated for unanchored and anchored tubing strings: These parameters indicate the probable **slope** of the surface dynamometer card at the start of the upstroke.

$$k_{r+t} = \frac{1}{\sum_{i=1}^{Taper} L_i E_{ri} + L_t E_t} \quad \text{for unanchored tubing} \tag{6.39}$$

$$k_t = \frac{1}{\sum_{i=1}^{Taper} L_i E_{ri}} \quad \text{for anchored tubing} \tag{6.40}$$

where:

k_{r+t}, k_t = spring constant of the rod + tubing and the tubing string, respectively, lb/in,
E_{ri}, E_t = elastic constant of the ith rod taper and the tubing, respectively, in/(lb ft),
L_i = length of the ith rod taper, ft,
L_t = length of the tubing string, ft, and
$Taper$ = number of rod tapers in string, –.
TV load: The measured load during the **TV test**.
SV load: The measured load during the **SV test**.

The use of the reference loads just discussed is illustrated in Fig. 6.31 where an example surface card is presented. The spring constant of the system gives an approximation of the slope of the card at

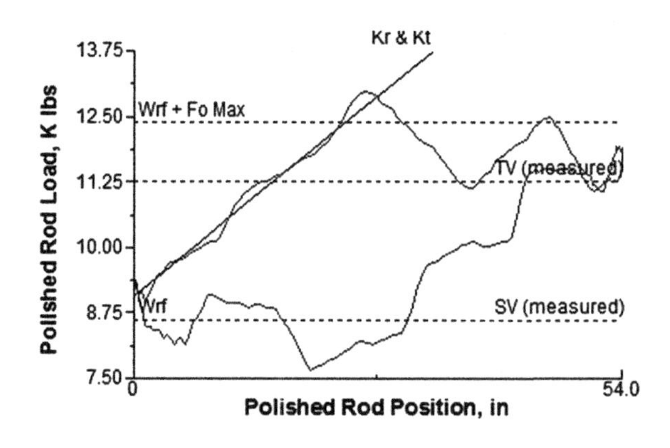

FIGURE 6.31

Reference load lines on a surface dynamometer card.

the start of the upstroke. In the given case a composite spring constant, k_{r+t}, representing the **concurrent** stretch of the tubing and the rod string, is depicted because the well's tubing string is not anchored. The results of the pump's valve tests can be used to estimate the actual fluid load on the plunger by subtracting the *SV load* from the *TV load*. These two loads can be used to detect any **unaccounted** friction loads in the pumping system; such loads cannot be described by the conventional wave equation, which assumes viscous damping only. The indications for high unaccounted friction are that *SV load* is less than the buoyant weight of the rod string, i.e., *SV load* $< W_{rf}$, while *TV load* is greater than $W_{rf} + F_{o\ max}$.

6.4.2.2.2 Pump cards

Downhole pump cards are calculated at the pump setting depth using the solution of the damped **wave equation**. Most software packages plot the pump card using **effective loads** in order to facilitate the detection of **buckling** conditions in the rod string. If effective loads are used, as discussed in Section 3.5.3.2.1, the pump does not carry any load during the downstroke and the pump card sits on the zero load line. The reference lines on calculated downhole cards are defined as follows.

Zero load line: If energy losses along the rod string are **entirely** due to viscous damping the downstroke loads should lie on the zero load line.

Maximum fluid load, $F_{o\ max}$: As before, this theoretical load on the plunger is found by assuming **pumped-off** conditions and using Eq. (6.38).

Spring constant of the tubing string, k_t (calculated for cases of **unanchored** tubing): A line with the corresponding slope represents the **stretch** of the unanchored tubing string

$$k_t = \frac{1}{E_t L_t} \tag{6.41}$$

where:

k_t = spring constant of the tubing string, lb/in,
E_t = elastic constant of the tubing, in/(lb ft), and
L_t = length of the tubing string, ft.

Fluid load from fluid level, $F_{o\ FL}$: This load is calculated from the differential pressure acting on the plunger on the **upstroke** and includes the effect of the pump intake pressure (*PIP*), which can be found from the measured dynamic liquid level in the annulus. *PIP* calculations involve running of an acoustic survey and the determination of annual liquid gradients as discussed in Section 6.2.3.3.2. Based on those parameters the fluid load on the plunger is calculated as follows:

$$F_{o\ FL} = A_p \left(WHP + 0.433\ L_{pump}\ SpGr_t - PIP \right) \tag{6.42}$$

where:

$F_{o\ FL}$ = fluid load on the plunger, as calculated from the fluid level, lb,
A_p = cross-sectional area of the plunger, sq in,
WHP = wellhead (tubing head) pressure, psi,
$SpGr$ = specific gravity of the liquid in tubing, –,
L_{pump} = pump setting depth, ft, and
PIP = pump intake pressure, psi.

Fluid load from valve tests, F_o VT: An approximation of the plunger's fluid load is obtained from the loads measured during a **valve test** procedure, as given here:

$$F_o\ \mathrm{VT} = TV\ Load - SV\ Load \quad (6.43)$$

where:

F_o VT = fluid load on the plunger, as calculated from valve tests, lb.
TV Load, SV Load = loads measured during the TV and SV tests, respectively.

Estimated loads, F_{UP}, F_{DWN}: These loads and the corresponding reference lines are **estimated** by the analyst and their correct placement requires some experience. As discussed in Section 4.4, if the correct damping factor is used in wells with **negligible** unaccounted friction, the top and bottom sides of the pump card must be near **horizontal**. Since there is almost always some amount of unaccounted friction in the downhole environment, the cards always exhibit **uneven** top and bottom sides and the proper placement of the F_{UP} and F_{DWN} loads is intended to **separate** those frictional loads from the actual loads on the plunger. It is usually assumed that friction is about **equal** on the up- and the downstroke; therefore about the same loads are removed from the top and the bottom sides of the pump card by the two reference lines.

Gross plunger stroke: The gross plunger stroke length is found as the difference of the two **extreme** positions of the plunger's travel during the pumping cycle.

Net plunger stroke: This is the part of the plunger stroke length during which the plunger moves through fluid in the barrel on the downstroke.

The utilization of the reference lines just described is illustrated in Fig. 6.32, which depicts the downhole card calculated from the surface dynamometer card previously discussed along with the proper reference lines. The card is typical for an **unanchored** tubing string and the stretch of the tubing is properly described by the slope equal to the spring constant k_t of the tubing string. The loads on the downstroke being near zero clearly indicate that there are insignificant **unaccounted** friction loads present in the well.

The actual **fluid load** on the plunger can be estimated by different ways and the following results are obtained. An acoustic survey of the dynamic liquid level in the annulus allowed the

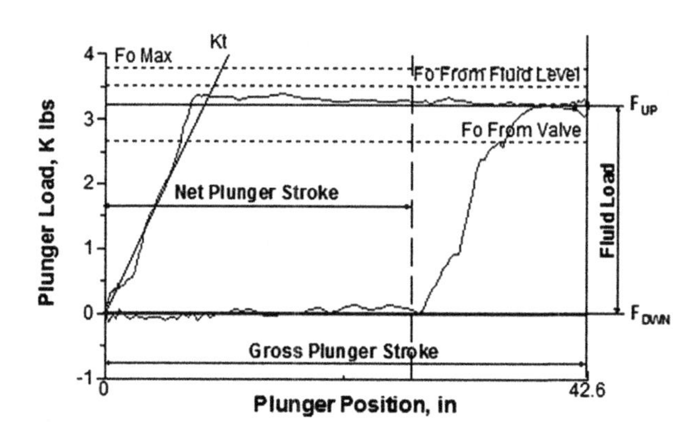

FIGURE 6.32

Reference load lines on a downhole pump card.

calculation of $F_{o\ FL} = 3{,}508$ lb; the valve tests resulted in $F_{o\ VT} = 2{,}656$ lb; whereas the placement of the F_{UP} and F_{DWN} lines gave a load of 3,285 lb, which is the accepted value. A comparison of the estimated fluid load to the theoretical maximum fluid load $F_{o\ max}$ shows that the pump supplies most of the energy needed to produce the fluids i.e., the fluid level in the annulus is near the pump and the well's producing bottomhole pressure is low. Finally, the plunger's gross stroke length is read from the card as 42.6 in; net plunger stroke is estimated as 33.2 in; the difference is due to the fact that the barrel does not completely fill during the upstroke.

6.4.2.3 Detection of common pump problems

This section discusses the ways common pump problems may be detected from calculated pump cards. Such problems include incomplete pump fillage as well as a variety of mechanical troubles affecting the pump and/or its valves.

6.4.2.3.1 Incomplete pump fillage

Under **ideal** conditions the barrel of the downhole pump **completely** fills up with **incompressible** liquid during the upstroke of the pumping cycle and the full stroke length of the plunger is utilized to produce well fluids. Pump displacement, therefore, is found from the **gross** plunger stroke length, the pump size, and the pumping speed; in this case pump **fillage**, the ratio of the net and the gross plunger stroke lengths, is 100%. Liquid fillage rates below 100% are normally experienced in the field and this condition is basically caused by not meeting one or both of the basic **assumptions** for a perfect pump operation: the barrel does not fill up **completely**, and/or the barrel contains a **compressible** medium. Pump **displacement** in such cases is much less than for a complete pump fillage and is found from the net or **effective** rather than from the gross plunger stroke length.

The possible reasons for **incomplete** or less than 100% pump fillage are as follows [60]:

- **Gas interference** means that the barrel contains, in addition to liquid, formation **gas** when the plunger reaches its top position at the end of the upstroke. At the start of the downstroke the traveling valve cannot open immediately and it takes part of the downstroke until the gas is **compressed** and the valve opens; pump fillage becomes less than perfect.
- **Fluid pound** happens when there is insufficient **inflow** from the well as compared to the displacement of the downhole pump and the barrel does not fill completely during the upstroke. The upper part of the pump contains a low-pressure gas, the compression of which greatly delays the opening of the traveling valve. Because the traveling valve opens only when the plunger is very close to the liquid level in the barrel, the **impact** of the plunger on the liquid may induce high compression loads in the rod string.
- A **choked or blocked pump** has some restriction at the pump intake that restrains the fluid inflow rate to the barrel during the upstroke, resulting in poor fillage of the pump.

Identification of the possible **causes** of incomplete pump fillage is facilitated by the analysis of the pump card's **shape** and by the results of a **fluid level** survey. As will be shown later, a proper detection of the root cause requires the application of both of these procedures because different problems may create cards of very similar shapes.

Gas Interference

At the start of any discussion on gas interference it must be made clear that the **presence** of gas or a compressible fluid in the barrel affects mainly the **downstroke** portion of the plunger's movement.

Under ideal conditions the traveling valve opens as soon as the downstroke starts because the barrel contains **incompressible** liquid that forces the valve to open. However, a compressible medium in the barrel has to be **compressed** first to a pressure greater than the pressure above the traveling valve so that the valve can open. This kind of operation is explained by the fact that both valves in the pump are simple **check** valves that open and close according to the **balance** of pressures above and below the valve. The result is that part of the plunger's downstroke is used for compressing the fluid in the barrel and does not contribute to the pump's displacement. The loss of displacement is proportional to the portion of the plunger stroke used for compression, i.e., the pump **fillage** defined before.

When modeling the downstroke conditions in the pump one usually assumes a **free** gas phase occupying the top of the barrel space. The behavior of this gas follows the **engineering gas law** while **isothermal** conditions are assumed to prevail because the average temperature in the pump is constant. Although the gas is compressed to quite high pressures during the downstroke, no part of it is **dissolved** in the liquid because of the short time available; the volume of gas at standard conditions, therefore, does not change and is equal to the volume entering the barrel at the pump's suction pressure.

Applying the **engineering gas law** for two conditions—(1) at pump suction where the gas volume occupying the barrel is V_1 and the pressure equals the pump intake pressure, and (2) at any intermediate position of the plunger between the top and the bottom of the stroke where the pressure in the barrel is p_b and the volume available to the gas has decreased to V_2—one can express the elevated barrel pressure as given in the following expression. Note that solving the formula requires **iterations** because the deviation factor Z_2 is a function of the pressure p_b that is about to be determined.

$$p_b = PIP \frac{V_1}{V_2} \frac{Z_2}{Z_1} \tag{6.44}$$

where:

p_b = pressure in the pump's barrel, psi,
PIP = pump intake pressure, psi,
V_1 = gas volume at suction conditions, cu ft,
V_2 = gas volume at pressure $\boldsymbol{p_b}$, cu ft,
Z_1 = gas deviation factor at suction conditions, –, and
Z_2 = gas deviation factor at pressure p_b, –.

Knowledge of the compressed gas pressure, p_b, permits the calculation of the variation of plunger **load** versus plunger **stroke** length for the downstroke, i.e., the construction of a part of the pump card. Fluid load on the plunger at any position is found from the **differential** pressure acting on it, as shown here:

$$F_o = A_p \left(p_d - p_b\right) \tag{6.45}$$

where:

F_o = fluid load on plunger, lb,
A_p = cross-sectional area of the plunger, sq in,
p_d = pump discharge pressure, psi, and
p_b = compressed gas phase pressure in the barrel, psi.

Pump **discharge** pressure is **constant** during the pumping cycle and equals the pressure at the bottom of the tubing string just above the traveling valve. It is found from the wellhead (tubing head) pressure and the hydrostatic pressure of the liquid in the tubing string:

$$p_d = WHP + 0.433\, L_{\text{pump}}\, SpGr_t \tag{6.46}$$

where:

p_d = pump discharge pressure, psi,
WHP = wellhead (tubing head) pressure, psi,
$SpGr_t$ = specific gravity of the liquid in tubing, –, and
L_{pump} = pump setting depth, ft.

Utilizing the equations just discussed, one can completely describe the conditions of pumping a compressible fluid during the plunger's downstroke. Since the pump intake, *PIP*, and the pump discharge, p_d, pressures are constant, the variation of the pressure in the barrel below the traveling valve, p_b, must be properly estimated to simulate the loads on the plunger.

For illustration purposes a sample case is introduced with the following basic data:

Plunger size = 2 in	Plunger stroke length = 144 in
Pump discharge pressure = 1,600 psi	Pumping temperature = 150 F
Gas specific gravity = 0.6	Pump fillage = 50%

Calculation results are contained in Fig. 6.33, where plunger **loads** during the gas compression are displayed for various pump intake pressure values as a function of plunger **position**. Note that plunger

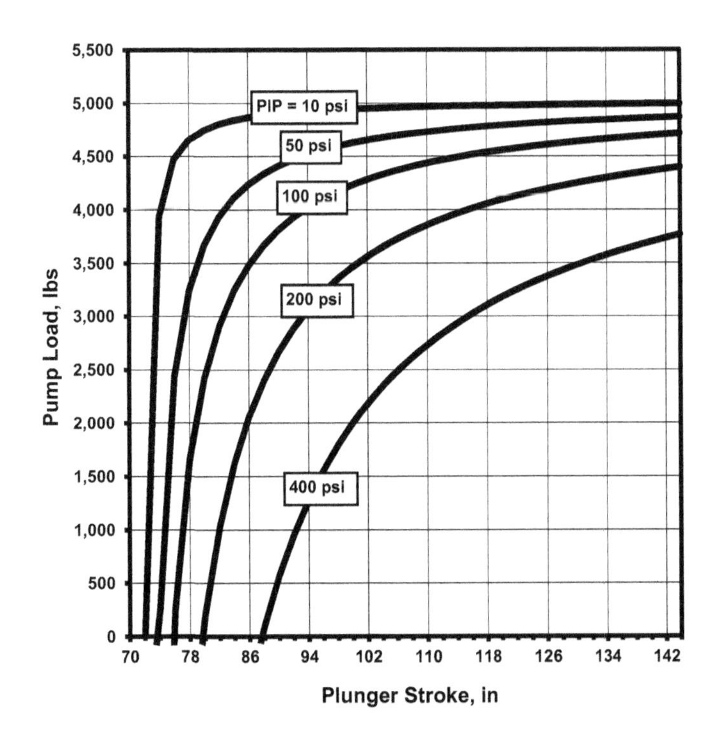

FIGURE 6.33

Calculated pump loads during gas compression for different pump intake pressures.

position is directly proportional to the plunger's swept **volume** because of the constant diameter of the barrel. Plunger loads are found from Eq. (6.45) using the barrel pressures, p_b, calculated from Eq. (6.44) describing the gas compression process.

Curves for each *PIP* value have characteristic **hyperbolic** shapes typical of gas **compression** processes. The traveling valve opens when the plunger has compressed the gas to a pressure **equal** to the discharge pressure of the pump and the load on the plunger goes down to zero (see Eq. 6.45). As can be seen, higher pump intake pressures require smaller portions of the plunger stroke to compress the gas in the barrel, while for extremely low intake pressures the plunger has to reach down very close to the original liquid level in the barrel.

The results of this example allow the construction of **synthetic** pump cards for various cases. This is accomplished by taking one of the curves valid for a given *PIP* in Fig. 6.33 and constructing the remaining parts of the pump card. The synthetic pump cards prepared for an **anchored** tubing string are given in Fig. 6.34 for various pump intake pressures. The figure also contains the reference lines discussed previously: these are the maximum possible fluid load, $F_{o\ max}$, and the fluid load found from the fluid level, $F_{o\ FL}$, both calculated from the basic formula Eq. (6.45):

$$F_{o\ max} = (2^2\pi)/4(1,600 - 0) = 5,027\ \text{lb; since}\ PIP = 0.$$

$$F_{o\ FL} = (2^2\pi)/4(1,600 - 400) = 3,770\ \text{lb; since}\ PIP = 400\ \text{psi}.$$

Let's first describe the sucker-rod pumping cycle for the case with a pump intake pressure of 400 psi. When the upstroke starts at point **A** both valves are closed; now the plunger moves a very short distance and the fluid load is **immediately** transferred from the standing to the traveling valve when the standing valve opens at point **B**. The plunger carries the fluid load $F_{o\ FL} = 3{,}770$ lb calculated from Eq. (6.45) by substituting $p_b = PIP$. The immediate opening of the standing valve from point **A** to point **B** happens only if the pump has a small **dead space**; then a very small movement of the plunger can sufficiently decrease the pressure in the barrel to the level of the pump intake pressure. Otherwise, especially if the pump is spaced too high, the opening of the standing valve is **delayed** and point **B** moves horizontally to the right.

The barrel is filling up with a mixture of gas and liquid during the plunger's travel from point **B** to point **C**. At point **C** (the top of the plunger's stroke) the barrel is full with a multiphase mixture whose pressure is at the pump intake pressure, $p_b = PIP$. From point **C** the **downstroke** starts and the plunger descends with a still-closed traveling valve continuously increasing the pressure inside the barrel. The gas contained at the top of the barrel is **compressed** from point **C** to point **D** and the pressure below the traveling valve progressively increases. Close to point **C** but farther on the downstroke, the increased pressure forces the standing valve to close. The load on the plunger decreases simultaneously with the increased barrel pressure because load is found from the differential pressure across the plunger (see Eq. 6.45). As soon as gas pressure reaches the pump discharge pressure of 1,600 psi at point **D** (at a plunger stroke length of 87.5 in) the traveling valve opens. The plunger with the open traveling valve now moves down in the compressed gas until it reaches the liquid at a stroke length of 72 in, which is 50% (the original fillage value) of the total stroke length of 144 in. From this position on, the plunger moves in liquid and reaches the bottom of its stroke at point **A**. The **effective** plunger stroke that defines the pump **displacement** is the length of the downstroke while the plunger moves in liquid, which is 72 in in this case. As seen from the relatively high *PIP* value as well as from the great

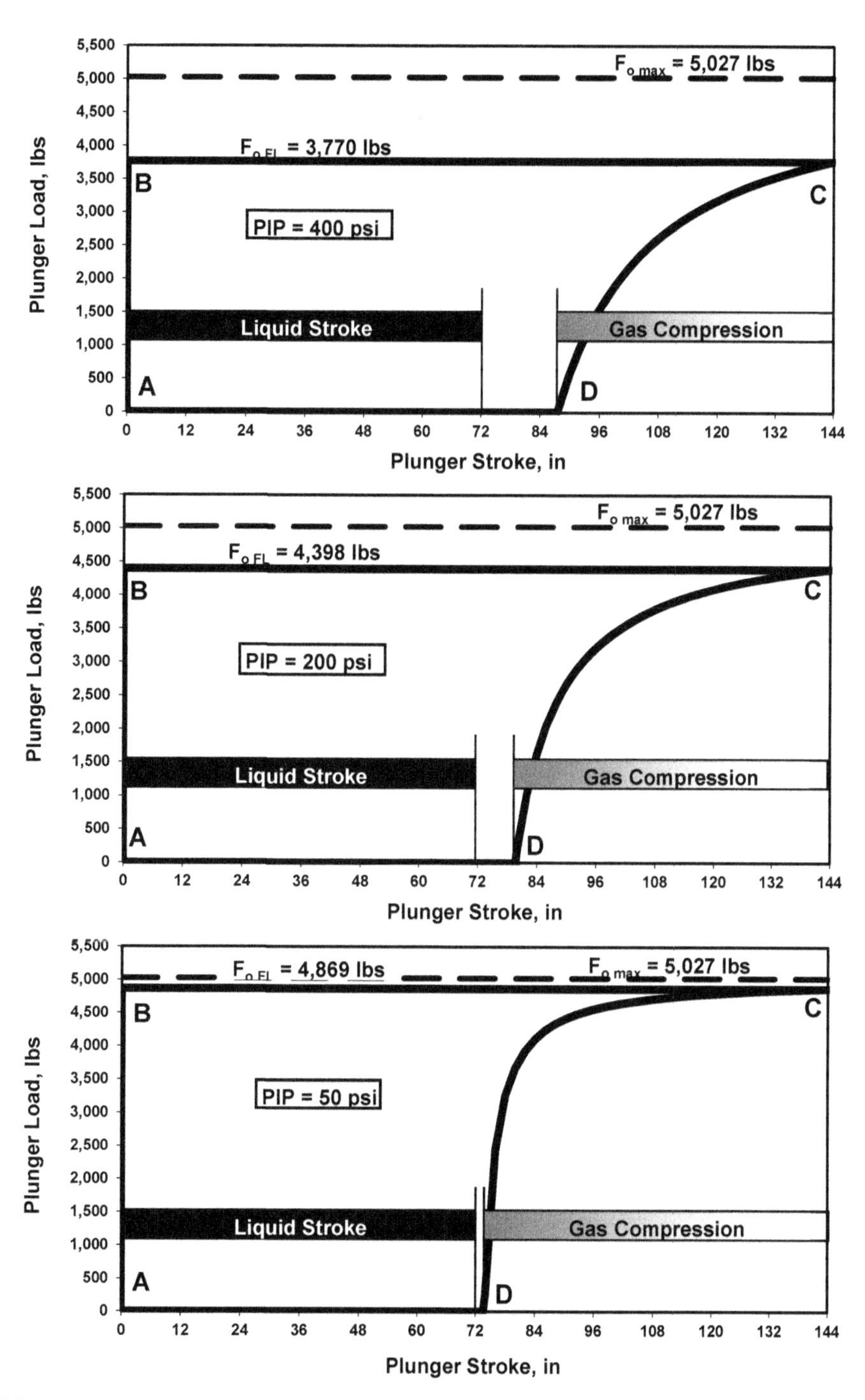

FIGURE 6.34

Synthetic pump cards for an anchored tubing string and different pump intake pressures.

difference between the $F_{o\ max}$ and the $F_{o\ FL}$ reference lines, the annular liquid level is well above the pump setting depth.

Comparison of the pump cards valid for the different cases given in Fig. 6.34 shows that fluid load during the upstroke, $F_{o\ FL}$, increases and approaches the possible maximum load $F_{o\ max}$ as the pump intake pressure decreases. This indicates that the liquid level in the annulus drops toward the pump. It is also interesting to note that the actual liquid **fillage** of the pump is not at point **D** where the traveling valve opens but always at a lower position. As clearly seen from the figures, the difference is greater at higher *PIP* values and is negligible for extremely low *PIP*s. A general **conclusion** for pump card diagnosis is that at higher *PIP*s the plunger's **effective** stroke length is less than the value estimated from the plunger position at the opening of the traveling valve.

The most important problem with gas interference is the **reduction** of effective plunger stroke length and the associated **loss** of liquid production. Additional troubles are related to the pump's reduced **speed** at the downstroke when the fluid is compressed in the barrel. As the pump slows down, rods above it that were falling at a high speed now go into **compression** and start to **buckle**; the tubing and the rods are damaged. Couplings on the rods may loosen, leading to pin failures; all these difficulties are eliminated if a properly sized gas separator is used to maintain a high liquid fillage of the pump.

Fluid Pound

Fluid pound condition is experienced when the lifting capacity of the pump **exceeds** the liquid inflow rate to the well and the pump barrel is not completely filled with liquid by the time the plunger reaches its upstroke. The fluid level in the annulus is at the pump intake, as seen in Fig. 6.35, and the barrel space above the liquid is occupied by low-pressure **gas** sucked in from the annulus. This situation is very **similar** to the gas interference problems described previously, with the only difference that very **low** pump intake pressures are involved; gas interference and fluid pound, therefore, can be **distinguished** on the basis of the actual pump intake pressure.

The description of the pumping cycle, because of the great **similarity** to gas interference problems, follows that detailed previously. Figure 6.36 presents a **synthetic** pump card prepared for the same case as before but using a very low pump intake pressure of $PIP = 10$ psi. As seen, the plunger load calculated from the fluid level, $F_{o\ FL} = 4{,}995$ lb, is almost identical to the maximum possible fluid load of $F_{o\ max} = 5{,}027$ lb, valid at zero pump intake pressure. This indicates that the liquid level in the annulus is at the pump intake; the well is being **pumped off**. The transfer of the fluid load from the traveling to the standing valve is very late in the downstroke and it takes half of the 144-in stroke length of the plunger to compress the low-pressure gas occupying the barrel. Pump **fillage**, as read at point **D**, is identical to the fraction of liquid in the barrel at the end of the upstroke, 50%.

The opening of the traveling valve at point **D** happens just **before** the plunger reaches the liquid level in the barrel; for the example case calculations similar to those presented before give a distance of about 0.5 in between the plunger and the liquid before impact. This means that, contrary to previous theories, the plunger hits the liquid with an **open** traveling valve [61] and never with a closed one; impact forces, therefore, are much smaller than those a closed traveling valve would cause. Most of the time these impact forces do not even show up on the pump card, so they cannot cause rod compression and buckling problems at the bottom of the rod string, contrary to previous beliefs.

Fluid pounding, however, causes compressive forces and may induce buckling of the rod string **above** the plunger; this can be explained from an analysis of plunger **velocity** during the gas

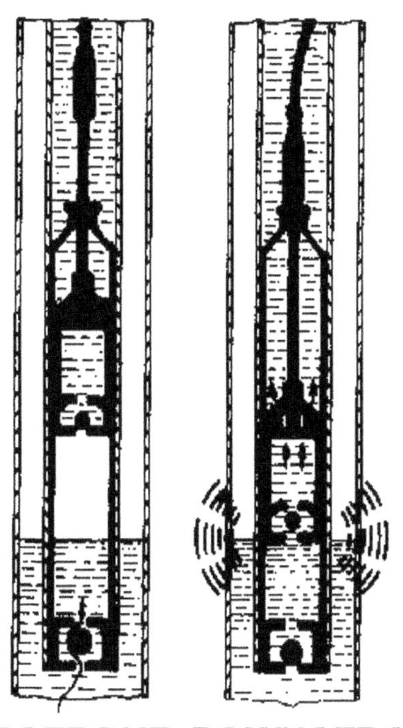

FIGURE 6.35

Illustration of pump action for fluid pound conditions.

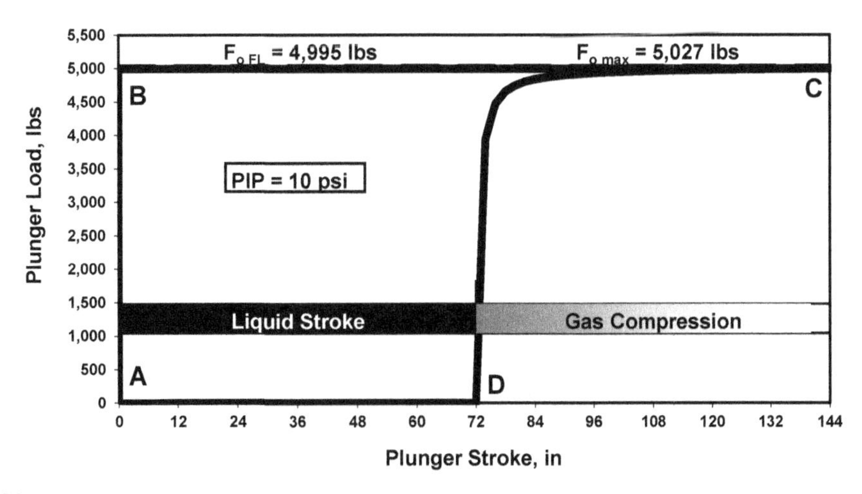

FIGURE 6.36

Synthetic pump card for pumped-off conditions.

compression period. At the beginning of this period at point **C** plunger velocity approximates the velocity of the polished rod, but it rapidly **decreases** as the plunger nears the liquid level in the barrel and compresses the gas above the liquid. The plunger may even **stop** moving for a short time when reaching the liquid level in the barrel as the traveling valve opens at around point **D**. Thus, while the rest of the rod string falls with a high velocity the bottom part of it is almost stopped; the great velocity difference causes compressive forces in the elastic rod material. The compressive forces thus generated may cause **buckling** in the rod string above the downhole pump. It is important to note that these compressive forces cannot usually be detected on the pump card; only the variation of **effective loads** along the rod string gives a proper indication [62].

As seen in Fig. 6.36, a fluid pound is often clearly indicated by the **shape** of the pump card. Its characteristic feature is the very **steep drop** in plunger loads well after the start of the downstroke, which occurs deeper and deeper in the downstroke as the fluid pound gets more and more severe during consecutive strokes. Another characteristic feature is the **pumped-off** condition: the liquid level in the annulus being at the pump intake, the fluid load on the plunger is close to the maximum possible value, $F_{o\ max}$, indicating a **low** pump intake pressure.

The dynamic loads occurring during fluid pounding can have several **detrimental** effects on the downhole equipment: the rod string can experience helical **buckling** leading to rod breaks, rod-to-tubing **wear** is increased, shock loads contribute to coupling **failure** due to unscrewing, and pump parts can be damaged, as well as the tubing, if unanchored. Dynamic loads are transmitted to the surface across the rod string and can reach the polished rod; hence the name **fluid pound**.

Choked Pump

In some cases, although a sufficiently high liquid column exists above the pump intake the pump operates with a very **low** liquid fillage. This kind of operation is caused by a **restricted** inflow into the pump barrel that prevents the barrel from filling up completely by the end of the plunger's upstroke. The flow of well fluids may be restricted (**choked**) by **solid** materials such as paraffin, scale, trash, and sand **plugging** the pump intake and/or the standing valve, and the pump is starved of liquid. Pumping of a highly **viscous** crude causes similar problems because the high frictional pressure losses across the standing valve and the pump intake may greatly reduce the liquid rate into the pump. Plunger displacement, therefore, may **exceed** the inflow rate, and as a result the barrel is only **partially** filled during the upstroke. The liquid volume filling up the barrel comes from two sources: from the annulus, and from the tubing through slippage in the barrel–plunger fit and/or from traveling valve leaks. Thus, in a completely blocked pump when no liquid reaches the wellhead, the only liquid in the barrel is due to slipping and leaking from the tubing.

As a result of choked inflow into the pump, the **pressure** below the plunger is extremely **low**. In case of a total **blockage** it can go down to the vapor pressure of the crude valid at ambient temperature; otherwise, it is normally close to **atmospheric** pressure. Due to this low pressure below the plunger, the pump load during the upstroke (see Eq. 6.45) is very high; it usually reaches $F_{o\ max}$, the maximum fluid load belonging to a **pumped-off** condition. Thus, although there is a sufficient amount of liquid in the annulus the pressure below the plunger is very low.

As the downstroke starts, the plunger falls with the **closed** traveling valve in the barrel containing an extremely low pressure gas. Since there is not much gas to be compressed, the traveling valve cannot open before **hitting** the liquid level; the impact produces great **compression** forces at the bottom of the sucker-rod string. The resulting vibrations force the traveling valve to open and close

rapidly several times before the valve finally opens and reaches the bottom of the stroke in the liquid. So, in contrast to the usual conditions during **fluid pounding**, the plunger hits the liquid level with a **closed** traveling valve, causing rod **buckling** and **damage** to the pump as well as the tubing string.

The general shape of the pump card is displayed in Fig. 6.37, representing a **synthetic** card for choked pump conditions with an **anchored** tubing string. The fluid load calculated from the fluid level, $F_{o\ FL}$, represents the conditions for the relatively high *PIP* value corresponding to the liquid column existing in the annulus. Since the actual pressure below the plunger, due to the restricted inflow to the barrel, is much lower than the *PIP*, plunger loads during the upstroke are greater and close to $F_{o\ max}$. This is exactly what is observed in **fluid pound** conditions when there is not enough inflow from the well; but now there is a sufficient liquid column in the annulus. The contradiction is explained by the fact that the choked pump creates a great pressure drop across the pump intake.

As can be seen by comparing Fig. 6.37 to Fig. 6.36, pump cards for a **choked pump** are very similar to cards found in **pumped-off** conditions. Proper diagnosis of the case is only possible if, simultaneously to the dynamometer measurements, a **fluid level** survey is also made to detect the dynamic liquid level in the well's annulus; a **high** liquid level is a positive indication for restrictions at the pump intake. Another sign for a choked pump could be the relatively high **negative** load caused by the plunger hitting the liquid level in the pump barrel during the downstroke. When hitting the liquid, the plunger's **velocity** decreases dramatically and the plunger may even **stop** for a short period. Then, just like in fluid pound conditions, the rod string experiences high **compressive** forces, leading to helical **buckling** and severe operational problems.

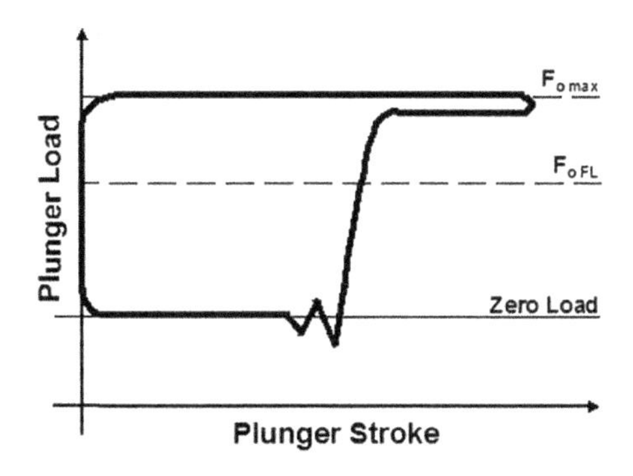

FIGURE 6.37

Synthetic pump card for a choked (blocked) pump.

6.4.2.3.2 Unaccounted friction

Let's start with the definition of **unaccounted** friction: any kind of friction along the rod string that is not accounted for during the solution of the damped wave equation. As discussed in Chapter 4, the conventional wave equation includes energy losses of a **viscous** nature only and cannot calculate any losses except those due to **viscous damping** that take place between the rod string and the well fluid. However, mechanical (a.k.a. Coulomb) **friction** occurs in the polished rod as it moves inside the stuffing box, between the rods and the tubing, etc. The effect of Coulomb friction is especially important in **deviated** wells where rod–tubing friction may become dominant; treatment of such cases requires the **modification** of the wave equation.

Since the **diagnostic** analysis of surface dynamometer cards is normally accomplished by using the conventional wave equation, cases with **negligible** Coulomb friction along the rod string give **reliable** results and the calculated pump cards are fully representative. The effect of increased mechanical friction in the well is clearly indicated in the calculated pump card: pump loads increase because the wave equation cannot remove all energy losses (viscous plus friction) from the power introduced at the polished rod. A **synthetic** pump card (for a perfect pump and an anchored tubing string) is shown in Fig. 6.38, where the typical symptoms of unaccounted friction are clearly indicated: downstroke loads are negative, upstroke loads are higher than normal, and the load range is excessive. Normal pump loads (**effective** loads) should fall between zero and the fluid load $F_{o\ FL}$ calculated from the pump intake pressure, which, in turn, is found from the dynamic liquid level. Cards with loads outside this range clearly indicate that the conventional wave equation is not able to **remove** all downhole losses from the pumping system; calculated pump loads are increased by the **unaccounted** friction loads.

Possible **sources** of unaccounted friction may include the following:

- Well **deviation** is the prime source of mechanical friction along the rod string; highly deviated wells cannot be diagnosed with the conventional wave equation.
- An **overtightened** stuffing box generates extra load on the polished rod; this can be detected on the surface dynamometer diagram if vertical changes in load are observed at the start and the end of the polished rod's stroke.
- **Deposition** of paraffin, scale, etc. on the rod or tubing.
- Friction in the **pump** due to a bent or sticking barrel.

Another effective way to detect unaccounted friction is the analysis of **valve tests**. In these cases the measured loads for the traveling and the standing valve tests are greater and respectively lower than under normal conditions due to the friction forces working opposite the direction of movement. If the plot of the traveling valve load versus time is investigated, then a definite **drop** in load can be observed some time after the unit is **stopped**. At this point the frictional load felt by the polished rod during the upstroke is released because the rod string's movement ceases.

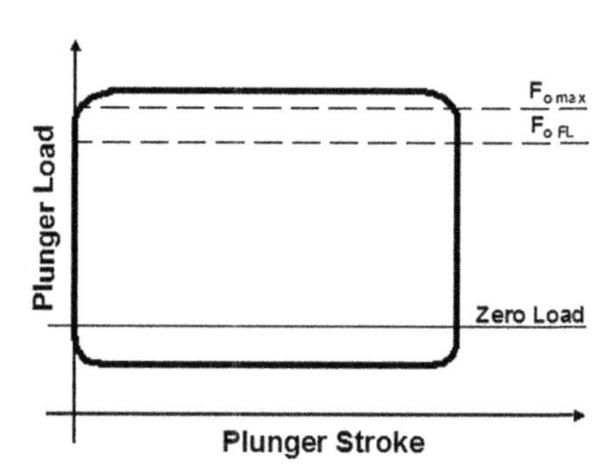

FIGURE 6.38

Synthetic pump card showing unaccounted friction.

6.4.2.3.3 Typical synthetic pump cards

Different downhole problems **distort** the shape of the ideal pump card in different but usually **typical** ways. Some possible **synthetic** pump card shapes are given in Table 6.1, which display pump cards for **anchored** and **unanchored** tubing strings for the equipment malfunctions and downhole pump problems listed. These shapes allow the **identification** of the actual pumping difficulty if caused by a

Table 6.1 Possible Synthetic Pump Card Shapes for Anchored and Unanchored Tubing Strings

Anchored Tubing	Description	Unanchored Tubing
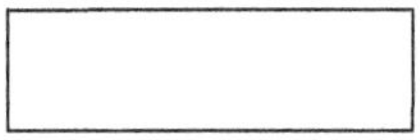	**Ideal card** Pump properly functioning completely filled with liquid.	
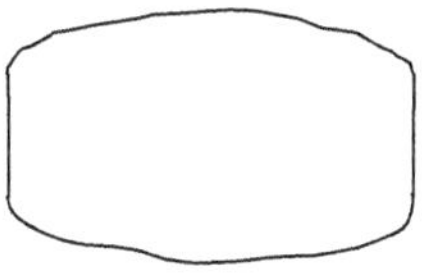	**Full pump with unaccounted friction** Extra friction along the rod string is not removed by the wave equation used to calculate the pump card.	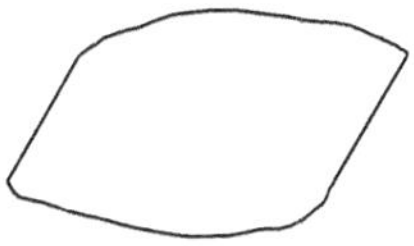
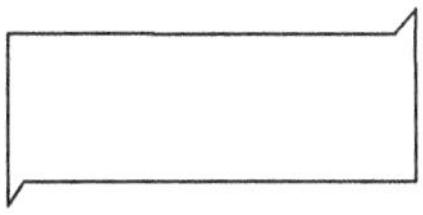	**Plunger tagging** Plunger hits up or down because of improper spacing of the pump.	
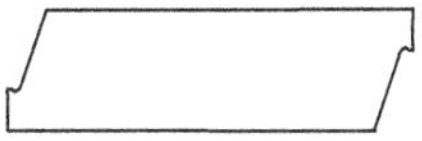	**Tubing anchor slipping** Malfunctioning tubing anchor allows tubing to stretch.	
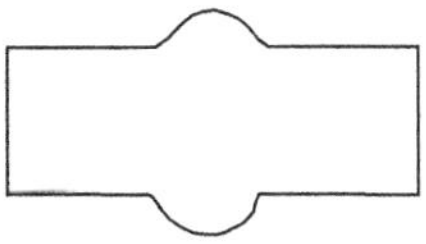	**Bent or sticking barrel** Load increases on upstroke, decreases on downstroke in defective section of barrel.	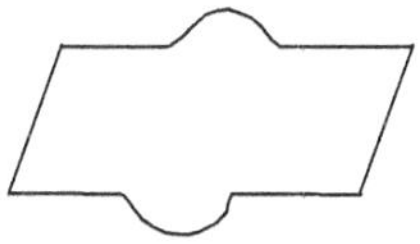
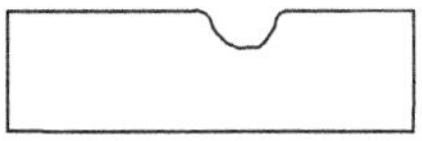	**Worn or split barrel** Rod load decreases in defective section of the barrel.	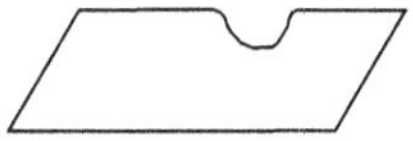
	**Sticking plunger** Load spike shows where plunger stopped; extra load is needed to overcome friction in the pump at this position.	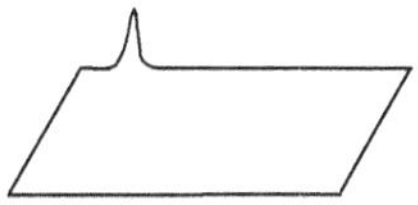
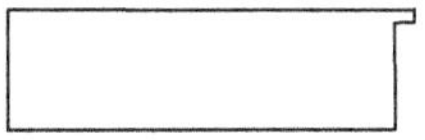	**Slight fluid pound** Fluid level falling to pump intake.	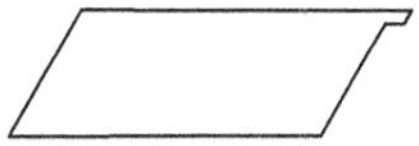

Continued

Table 6.1 Possible Synthetic Pump Card Shapes for Anchored and Unanchored Tubing Strings—cont'd

Anchored Tubing	Description	Unanchored Tubing
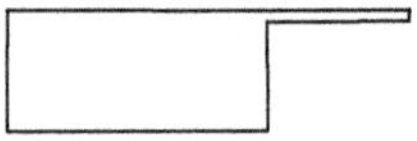	**Severe fluid pound** Barrel incompletely filling with liquid due to limited well inflow.	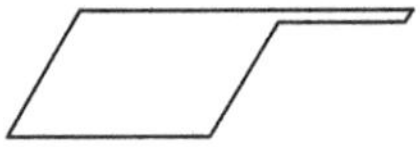
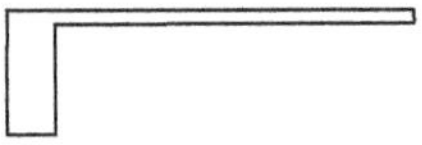	**Well pumped off** Pump displacement much greater than well inflow. PIP, pump fillage are low.	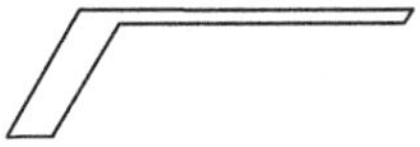
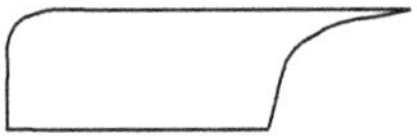	**Gas interference** Mixture of liquid/gas fills barrel. PIP is high, pump fillage is low. Unstable operation.	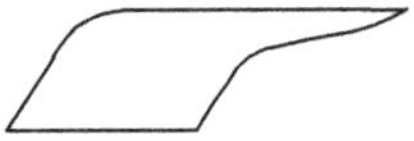
	**Gas-locked pump** Barrel filled with gas, valves remain closed, no liquid production. Low PIP.	
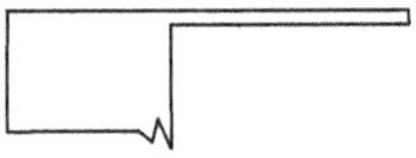	**Choked pump** Intake plugged, barrel incompletely fills during upstroke. PIP is high, pump fillage is low.	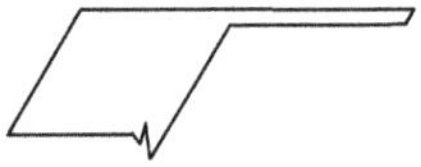
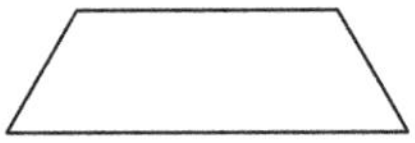	**Leaking TV or pump** TV leak or pump slippage causes delay in picking up and premature unloading of fluid load.	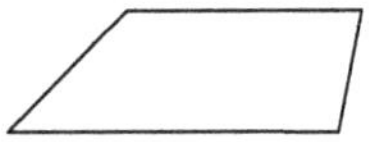
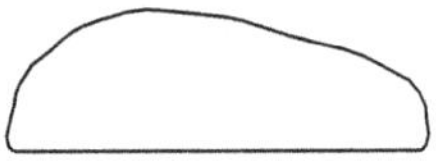	**Badly leaking TV or pump** TV or plunger/barrel completely worn out.	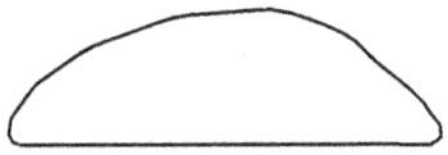
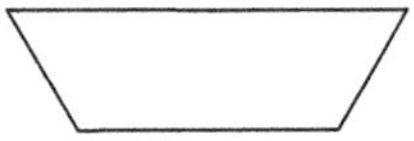	**Leaking SV** Premature loading at start of upstroke and delayed unloading at start of downstroke.	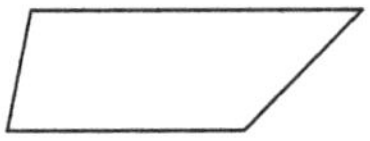
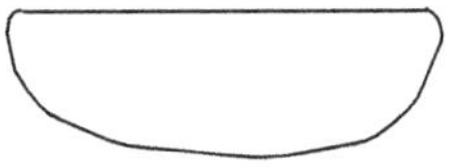	**Badly leaking SV** SV completely worn out.	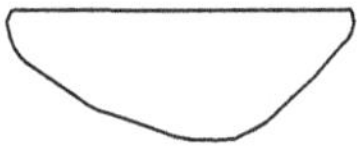
	Worn-out pump TV & SV valves and barrel/plunger completely worn out.	
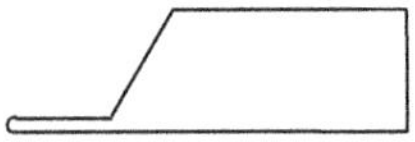	**Delayed closing of TV** TV ball does not seat as soon as upstroke starts.	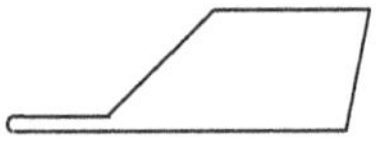
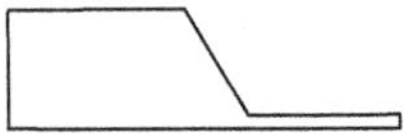	**Hole in barrel or plunger pulling out of barrel** Load drops as plunger reaches hole or pulls out.	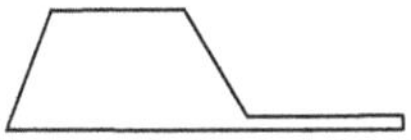

single problem only; cases with multiple problems have shapes combined from individual typical shapes. Once again, troubleshooting downhole pumping problems is much easier using pump cards than surface dynamometer cards because pump cards have more uniform and more easily recognizable shapes.

6.4.2.3.4 Cards in special cases

When diagnosing sucker-rod pumping installations, one often finds cases with **no fluid** production reaching the surface. The surface and pump dynamometer cards obtained on such wells have very specific **features** from which the actual downhole problems can be identified. The following discussions, after [60], describe the **six** different possible situations by presenting **typical** surface and pump cards along with their diagnosis. Proper identification of the existing downhole problems is facilitated by the use of the **reference lines** plotted on the surface and pump cards, as described in Section 6.4.2.2. It must be noted that dynamometer surveys for the six cases investigated have to be made using **horseshoe** transducers because PRTs always force the bottom of the pump card to lie on the **zero** load line. Additionally, the pump card must be plotted using **effective loads** because the use of true loads prevents a proper diagnosis in some of the investigated cases.

Deep Rod Part

In case the rod string breaks at or very close to the pump setting depth, or if the valve rod in the pump is broken, then the plunger and the rod string are **separated** and no pumping action is performed. The pump card lies on the zero load line and is very **flat** because the bottom of the string is not loaded (see Fig. 6.39). The polished rod loads as read from the surface dynamometer card are close to the buoyant **weight** of the rod string, W_{rf}, since there is no fluid load applied by the pump to the rod string. The surface card shape is similar to an **overtravel** card with the usual downward slope to the right.

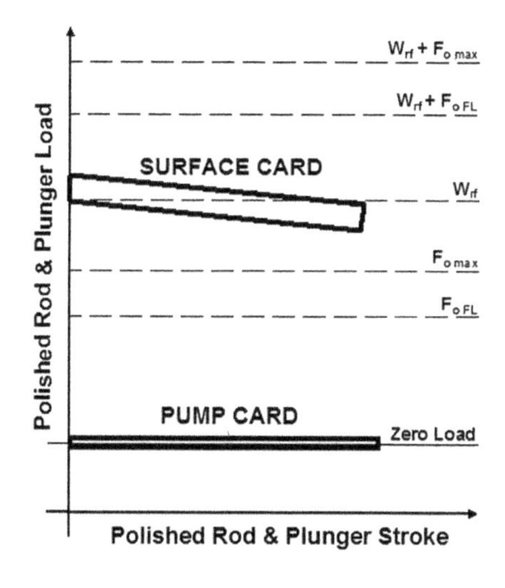

FIGURE 6.39

Surface and pump cards typical for a deep rod part.

Overtravel conditions occur because of the lack of rod string stretch due to the missing fluid load, and downhole stroke length becomes longer than the polished rod stroke. Note that a traveling valve stuck open during the pumping cycle produces exactly the same symptoms as described here because the rod string and the plunger are no longer attached to each other (see later).

Shallow Rod Part

The conditions with a rod break situated high above the pump are described in Fig. 6.40. Surface loads are less than the calculated buoyant rod string weight, W_{rf}, because the lower part of the string is missing. The very flat pump card is **shifted** below the zero load line by exactly the buoyed weight of the missing rod sections. This fact permits the determination of the **depth** of the rod break in the knowledge of the original taper lengths. To do so one has to repeat the calculation of the downhole card by **removing** different assumed lengths from the bottom of the rod string. The proper length of the broken string is found when the new pump card lies around the zero load line.

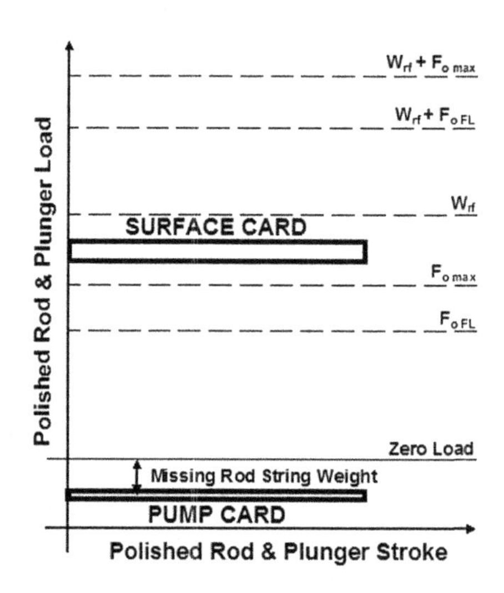

FIGURE 6.40

Surface and pump cards typical for a shallow rod part.

Standing Valve Stuck Open

The standing valve of the downhole pump may be kept open by some foreign material (scale, sand, trash, asphaltene, etc.) that causes its ball to **stick** in the valve cage; or keeps it off the valve seat. This condition prevents the **transfer** of the fluid load from the traveling valve to the standing valve during the downstroke. Polished rod loads, therefore, do not considerably change at the downstroke (see Fig. 6.41); the surface card may exhibit slight overtravel. Since the traveling valve **never** opens, polished rod loads are close to the sum of the buoyant weight of the string plus the fluid load, $W_{rf} + F_{o\,FL}$. The pump card is flat because the plunger carries the fluid load during both the up- and the downstroke; it lies on the load that can be calculated from the fluid level in the annulus, $F_{o\,FL}$.

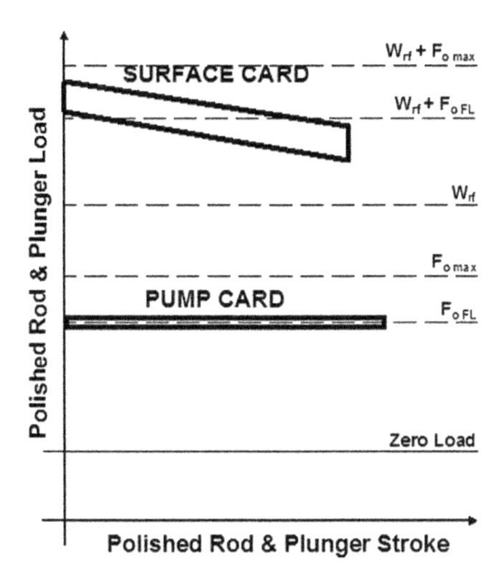

FIGURE 6.41

Surface and pump cards typical for a standing valve stuck open.

Traveling Valve Stuck Open

Just like in the standing valve, the ball of the traveling valve may **stick** in the cage or may be kept from closing on the valve seat by **foreign** material such as trash, sand, and asphaltene. Since there is no pumping action due to the permanently open traveling valve, pump loads are close to **zero** and the pump card lies around the zero load line (see Fig. 6.42). The loads on the surface dynamometer card

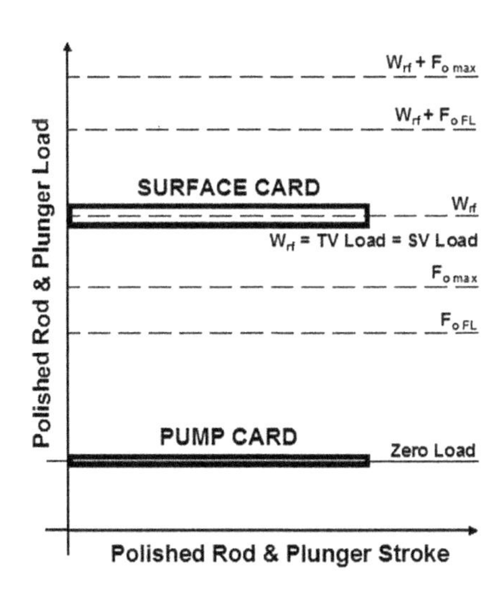

FIGURE 6.42

Surface and pump cards typical for a traveling valve stuck open.

are approximately equal to the buoyant rod string weight, W_{rf}. Under these conditions, testing of the pump's valves gives identical loads for the **TV test** and **SV test**; both are equal to the buoyant rod string weight, W_{rf}.

The symptoms of a stuck traveling valve are very similar to those observed in a deep rod part condition (see Fig. 6.39) because in both cases the connection between the rod string and the plunger is lost. A stuck traveling valve may be **cleared** by lightly **tagging** the plunger on the downstroke, whereas a parted string does not react to such an operation.

Blocked Pump Intake

The intake to the downhole pump may be partially or completely **choked off** by solids such as sand, scale, trash, and asphaltene and inflow to the pump barrel may become **restricted** or completely **blocked**. Blockage may also be caused by a sand screen **clogged** with fines or the pump intake may be covered by sand when the pump is set below the perforations. The common result of these problems is that the barrel does not fill with liquid during the upstroke or can even become empty. Due to the blocked intake the **pressure** inside the barrel during the upstroke goes down to zero and the plunger carries a load much higher than the load calculated from the measured fluid level, $F_{o\ FL}$, as shown in Fig. 6.43. The flat pump card, therefore, lies very close to the **maximum** possible fluid load, $F_{o\ max}$. The surface card, too, shows a relatively constant load equal to the sum of the buoyant rod string weight and the maximum possible fluid load, $W_{rf} + F_{o\ max}$.

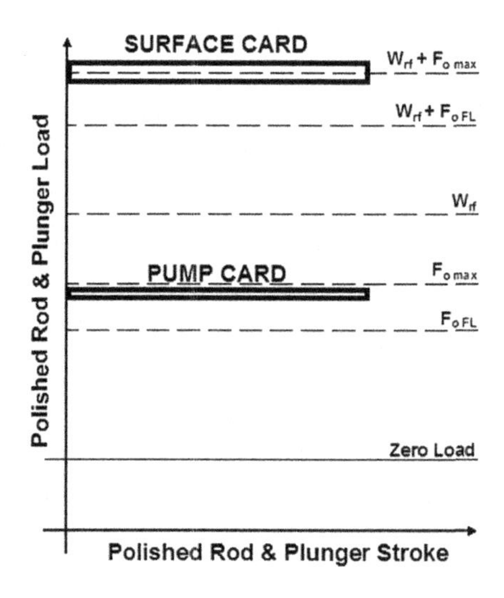

FIGURE 6.43

Surface and pump cards typical for a blocked pump intake.

Deep Hole in Tubing

If the tubing string has a hole at a relatively **deep** position, then the liquid produced by the pump is **circulated** through that hole back to the annulus. The pump, if in a good overall condition, does considerable work and the pump card is not as flat as in the other cases that do not produce fluid to the surface (see Fig. 6.44). If the tubing leak is big enough, the tubing above the depth of the hole is practically **dry**. Due to this fact, the measured **weight** of the rod string considerably **increases** because most of the string operates in air; the buoyancy force for the dry section is missing. The pump card, therefore, is **shifted** above the zero load line by the load equal to the **missing** buoyancy force. Fluid load on the plunger, as compared to previous cases, is considerably **greater** but much lower than normal because the pump lifts liquids up to the tubing leak only. The surface card, similarly to the pump card, is not flat and is situated above the buoyant rod string weight, W_{rf}.

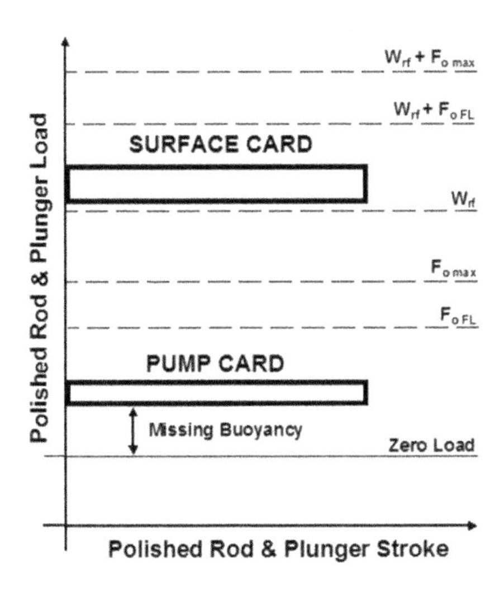

FIGURE 6.44

Surface and pump cards typical for a deep hole in the tubing.

6.4.2.4 Pump intake pressure calculations

The actual pressure valid at the downhole pump's suction level is called the **pump intake pressure** or *PIP*, which is an important paramcter for the evaluation of pumping conditions. One can, for example, use the *PIP* to detect the existence of any **free** gas at the pump suction by calculating the amount of gas in solution in the produced oil. The actual value of *PIP* also indicates the depth of the dynamic liquid **level** present in the well's annulus, which, in many cases, is not directly measured. Calculation of the pump intake pressure, therefore, is a standard procedure and the three different ways of its execution are detailed in the following [63,64]. As will be seen, all methods have different **accuracies** and reliabilities and require different input data. It must be noted that the analyst must not accept the result of a **single** calculation model but must use as many of them as possible so that, by comparing all results to any other available evidence, a **verified** *PIP* value can be selected.

6.4.2.4.1 Using acoustic survey data

As already discussed in Section 6.2.3.3, *PIP* (the fluid pressure at pump suction level) is made up of **three** components, as displayed in Fig. 6.45:

- the well's **casinghead** pressure,
- the pressure of the static **gas column** present in the annulus, and
- the pressure of the **gaseous liquid** (in stabilized wells, pure oil) column above the pump.

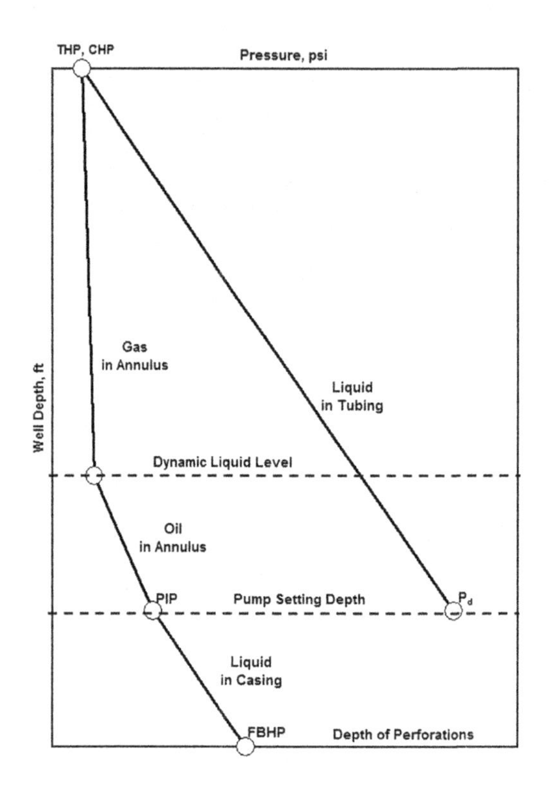

FIGURE 6.45

Pressure distributions in a pumping well's annulus and in the tubing string.

Of the components presented, the determination of the pressure of the gaseous oil column situated in the well's annulus is **critical** and the iterative procedure detailed in Section 6.2.3.3.1 is applied. Considering the calculation and measurement accuracy of the relevant terms, average *PIP* calculation **errors** (as compared to data received from downhole pressure recorders) of less than 5% can be expected according to **Rowlan et al.** [64]. This accuracy is valid for **stabilized** production conditions and consistent input data. Of the three possible procedures for finding *PIPs*, the use of acoustic survey data is the most **reliable** and most **accurate** one.

EXAMPLE 6.10: FIND THE PUMP INTAKE PRESSURE FOR THE FOLLOWING CASE WHEN THE WELL PRODUCES A NEGLIGIBLE AMOUNT OF GAS

Pump setting depth = 6,500 ft TVD	Oil $SpGr = 0.85$
Measured liquid level = 5,000 ft TVD	Gas $SpGr = 0.65$
Casinghead pressure = 300 psi	

Solution

The gas gradient in the annulus at the given casinghead pressure is found from Appendix A as 7.5 psi/1,000 ft; the gas pressure at the measured liquid level is found from this and the casinghead pressure:

$$p_g = 300 + 7.5 \times 5,000/1,000 = 37.5 \text{ psi.}$$

The hydrostatic pressure of the gasless oil column above the pump is found as:

$$p_l = (6,500 - 5,000) \times 0.433 \times 0.85 = 552 \text{ psi.}$$

Pump intake pressure is the sum of the calculated pressures:

$$PIP = 37.5 + 552 = 587.5 \text{ psi.}$$

6.4.2.4.2 Based on valve test results

As discussed previously, when testing the valves of a downhole pump with **perfectly** operating valves, the loads measured at the onset of the tests are as follows. The **TV test** gives an initial load equal to the sum of the buoyant **weight** of the rod string and the **fluid load**:

$$TV\ Load = W_{\text{rf}} + F_o \tag{6.47}$$

The **SV test**, on the other hand, furnishes the buoyant **weight** of the rod string only:

$$SV\ Load = W_{\text{rf}} \tag{6.48}$$

where:

$TV\ Load$ = load measured at the **TV test**, lb,
$SV\ Load$ = load measured at the **SV test**, lb,
W_{rf} = buoyant rod string weight, lb, and
F_o = fluid load on the plunger, lb.

From these two expressions the **fluid load** on the plunger can be easily determined, valid for conditions with negligible **leakage** in the pump's valves and no unaccounted **friction** in the pumping system:

$$F_{o\ \text{VT}} = TV\ Load - SV\ Load \tag{6.49}$$

where:

$F_{o\ \text{VT}}$ = fluid load calculated from valve test data, lb.

The fluid load on the plunger, F_o, however, can also be expressed based on the cross-sectional **area** of the plunger, A_p, and the **differential** pressure acting on the plunger. The pressure **above** the plunger (the pump's discharge pressure, designated p_d in Fig. 6.45) comes from the wellhead (tubing head) pressure and the hydrostatic pressure of the produced liquid, as shown in the same figure. Since the pressure **below** the plunger is equal to the pump intake pressure, *PIP*, one can write:

$$F_o = A_p \left(WHP + 0.433\ SpGr_t\ L_{\text{pump}} - PIP \right) \tag{6.50}$$

where:

F_o = fluid load on the plunger, lb,
A_p = cross-sectional area of the plunger, sq in,

WHP = wellhead (tubing head) pressure, psi,
$SpGr_t$ = specific gravity of the liquid in tubing, –,
L_{pump} = true vertical pump setting depth, ft, and
PIP = pump intake pressure, psi.

Expressing *PIP* from this equation and substituting $F_{o\ VT}$ from Eq. (6.49), we get the final formula for the calculation of the pump intake pressure utilizing valve test data:

$$PIP = WHP + 0.433\ SpGr_t\ L_{pump} - \frac{TV\ Load - SV\ Load}{A_p} \quad (6.51)$$

where:

PIP = pump intake pressure, psi,
WHP = wellhead (tubing head) pressure, psi,
$SpGr_t$ = specific gravity of the liquid in tubing, –,
L_{pump} = true vertical pump setting depth, ft,
TV Load = load measured at the **TV test**, lb,
SV Load = load measured at the **SV test**, lb, and
A_p = cross-sectional area of the plunger, sq in.

A comprehensive study [64] on the use of this method reported average calculation errors of ±10%, compared to data received from downhole pressure gauges. However, of the three models available for *PIP* calculations, this is the **least** reliable and accurate; the low reliability is caused by the many possible errors if valve test procedures are not followed correctly.

6.4.2.4.3 Using the pump card

The previous procedure used **static** loads on the plunger for the calculation of *PIP* values. If **dynamic** loads are considered then one can use the loads read from the downhole card received from the solution of the damped **wave** equation. Proper estimation of the average up- and downstroke loads on the plunger, as discussed in Section 6.4.2.2 permits the estimation of the pump intake pressure based on those F_{UP} and F_{DWN} loads. By substituting these loads in Eq. (6.51) we get:

$$PIP = WHP + 0.433\ \mathrm{SpGr}_t\ L_{set} - \frac{F_{UP} - F_{DWN}}{A_p} \quad (6.52)$$

where:

PIP = pump intake pressure, psi,
WHP = wellhead (tubing head) pressure, psi,
$SpGr_t$ = specific gravity of the liquid in tubing, –,
L_{set} = true vertical pump setting depth, ft,
F_{UP} = estimated average plunger load on the upstroke, lb, and
F_{DWN} = estimated average plunger load on the downstroke, lb.

The calculation model relies heavily on the accuracy of **solving** the damped wave equation because the loads F_{UP} and F_{DWN} are read from the calculated downhole card. Since the shape and scale of the pump card depend on the assumed **damping factor**, proper estimation of the damping factor is important. Another limitation is that accurate *PIP* data are received only in cases when there are negligible **unaccounted** friction forces during the pumping cycle. This procedure is considered to be more **reliable** than the one using valve test data but less accurate than the method based on acoustic surveys.

The merits of this procedure were investigated by **Pons et al.** [65] by comparing measured *PIP*s to values calculated with the help of the solution of the damped wave equation; for over 77% of nine cases the differences were below 30 psi.

6.4.3 MODERN INTERPRETATION METHODS

In analyzing surface or pump cards as well as in diagnosing pumping problems, the need for expert **knowledge** is essential. Only a person long experienced in rod-pumping analysis can be expected to properly and reliably interpret most dynamometer cards. Since expert knowledge is usually rare, the latest trend in dynamometer card analysis methods aims at reducing the **dependence** on human expertise by applying the methods of **artificial intelligence** (AI). The most commonly applied AI method in dynamometer card analysis is the use of **expert systems,** which are computer programs that simulate the problem-solving process of a human. They are based on detailed knowledge gained from rod pumping **experts** that is stored in computer memory and used in logical **decisions** that lead to the interpretation of pump cards. The diagnosis process starts with computer-generated questions to the user, then proceeds with the user's answers; finally, recommendations are given by the program based on the information input and the logic rules previously determined from a human expert's knowledge [66].

In analyzing dynamometer cards with the computer, one of the crucial problems is how the computer can be taught to **recognize** the shape of the card. To solve this, another field of AI is utilized: **pattern recognition**. One possible way to recognize different card shapes is by the use of **templates** previously set up. The templates are constructed according to the characteristic features of the different pumping problems to be detected. One such template to determine fluid pounding conditions [67] is reproduced in Fig. 6.46 with a dynamometer card on it. The card, as shown, is previously transformed into a dimensionless form, and the degree of **matching** is determined on the basis of the **number** of intersections between the card and the template. After all available templates have been used to match

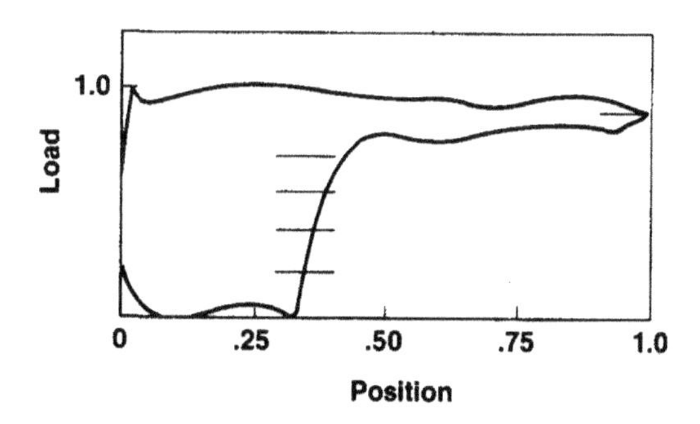

FIGURE 6.46

Use of a template in pattern recognition of pump cards, after **Foley and Svinos** [67].

the actual card, the template with the highest degree of match can be selected. Then, by applying logic rules built into the program, the computer gives a detailed **diagnosis** of the actual card.

The different pattern recognition methods used in dynamometer card analysis are compared by **Dickinson and Jennings** [68]. Another method involves a vector of load and position data derived from the pump card [69,70]. The card, after being transformed into a nondimensional form, is digitized with a sufficient number of points using a constant position increment. These data constitute the components of a **vector** that, after normalization, is claimed to fully **represent** the original card. The diagnosis is based on a comparison of the actual card's vector to data of cards with **known** problems contained in the program's library. The degree of **match** is defined as the dot product of the actual and the library vectors. After the library card with the highest level of match is found, the actual card is easily diagnosed.

The use of an artificial neural network (ANN) to pump card analysis has two modes: learning and retrieving. In the learning mode, pump cards with different problems are input; cards are first normalized, then 80 points and several slopes on the card are read. The many cards input into the system are classified into 11 problem conditions. The ANN program developed [71] could correctly detect the 11 problems and any of their combinations from test cards.

REFERENCES

[1] Dugan L, Howard L. Beyond pump-off control with downhole card well management. Proc. 49th annual southwestern petroleum short course. Lubbock. Texas. 2002. p. 55–61.

[2] Boyer L, Gibbs S, Nolen K, Cordoba A, Roberson A. Well production testing using a rod pump controller. Lufkin Automation; 2006. 8 p.

[3] Gibbs SG, Nolen KB. Inferred production rates of a rod pumped well from surface and pump card information. US Patent 7,212,923; 2007.

[4] Ahmed M, Sadek N. Automatic well testing and PIP calculations using Smart rod pump controllers. Paper SPE 150885 presented at the North Africa Technical Conference and Exhibition held in Cairo. Egypt. February 20–22, 2012.

[5] McCoy JN. Liquid level measurement. PE July 1975:62–6.

[6] ERCB Report 74-S Guide for calculating static bottom-hole pressures using fluid-level recording devices. Calgary (Canada): Energy Resources Conservation Board; November 1974.

[7] Thomas LK, Hankinson RW, Phillips KA. Determination of acoustic velocities for natural Gas. JPT July 1970:889–95.

[8] Godbey JK, Dimon CA. The automatic liquid level monitor for pumping wells. JPT August 1977:1019–24.

[9] Huddleston KL, Podio AL, McCoy JN. Application of portable microcomputers to field interpretation of acoustic well surveys. Paper presented at the 1985 SPE computer Technology Symposium. February 28–March 1, 1985. Lubbock (Texas).

[10] Podio AL, McCoy JN, Huddleston KL. Automatic pressure buildup data acquisition and interpretation using a microcomputer-based acoustic liquid level instrument. Paper SPE 16228 presented at the SPE production operations Symposium. Oklahoma City, Oklahoma. March 9–11, 1987.

[11] McCoy JN, Podio AL, Becker D. Improved acoustic liquid level surveys by digital data acquisition, processing and analysis. Proc. 38th annual Southwestern Petroleum short course. Lubbock, Texas. 1991. p. 165–183.

[12] McCoy JN, Huddleston KL, Podio AL. Data processing and display for Echo sounding data. US Patent 5,117,399; May 1992.

[13] Alexander LG. Determination of the "gas-free" liquid level and the annular gas flow rate for a pumping well. JCPT April–June 1976:66–70.
[14] Alexander LG. Testing and evaluating pumping wells. WO November 1980:74–8.
[15] Hasan AR, Kabir CS. Determining bottomhole pressures in pumping Wells. SPEJ December 1985:823–38.
[16] Walker CP. Method of determining fluid density, fluid pressure and the production capacity of oil wells. US Patent 2,161,733; 1939.
[17] Nind TEW. Principles of oil well production. 2nd ed. New York: McGraw-Hill Book Company; 1981.
[18] Agnew BG. Dynamometer stars in a new role. OGJ October 22, 1956:113–4.
[19] Agnew BG. The dynamometer as a production tool. API Drill Prod Pract 1957:161–8.
[20] Deax DW. A simple, low-cost method for determining the productivity index of high-GOR pumping wells. JPT December 1972:1417–22.
[21] McCoy JN, Podio AL, Huddleston KL, Drake B. Acoustic static bottomhole pressures. Paper SPE 13810 presented at the SPE production operations Symposium in Oklahoma City, Oklahoma. March 10–12, 1985.
[22] Gipson FW, Swaim HW. Designed beam pumping. Proc. 19th annual Southwestern Petroleum short course. Lubbock Texas. 1972. p. 81–107.
[23] Schmidt Z, Brill JP, Beggs HD. Prediction of annulus pressure gradients in pumping wells. Paper 80-Pet-53 presented at the ASME energy Technology Conference and Exhibition. New Orleans, Louisiana. February 3–7, 1980.
[24] Podio AL, Tarrillion MJ, Roberts ET. Laboratory work improves calculations. OGJ August 25, 1980. 137–146.
[25] Hasan AR, Kabir CS, Rahman R. Predicting liquid gradient in a pumping-well annulus. SPEPE February 1988:113–20.
[26] McCoy JN, Podio AL, Huddleston KL. Acoustic determination of producing bottomhole pressure. SPE FE September 1988:617–21.
[27] Vogel JV. Inflow performance relationships for Solution-Gas drive Wells. JPT January 1968:83–92.
[28] Eubanks JM, Franks BL, Lawrence DK, Maxwell TE, Merryman C. Pumping well problem analysis. Joe Chastain, Midland Texas. 1958.
[29] Slonegger JC. Dynagraph analysis of sucker rod pumping. Houston (Texas): Gulf Publishing Co.; 1961.
[30] Swaim HW, Hein Jr NW. Surface dynamometer card interpretation: a beam pumping problem-solving tool. Conoco Inc.; 1987.
[31] Dynamometer DYN 77. Adendorf (Germany): Fr. Leutert GmbH; June 2010.
[32] Garrett M, Rowlan L, Podio AL, Eguoto F, McCoy JN. Improved field measurements aid in sucker rod lift analysis. Proc. 43rd Southwestern Petroleum short course. Lubbock, Texas. 1996. p. 109–113.
[33] Garrett M, Rowlan L, Podio AL, Eguoto F, McCoy JN. Application of the Delta II dynamometer technique. JCPT April–June, 1966:66–74.
[34] Eickmeier JR. Diagnostic analysis of dynamometer cards. JPT January 1967:97–106.
[35] McCoy JN, Podio AL. Integrated well performance visualization and analysis. Paper presented at the SPE Workshop on microcomputer applications in artificial lift in Long Beach. California. October 16–17, 1989.
[36] Jennings JW, McCoy JN, Drake B. A portable system for acquiring and analyzing dynamometer data. Proc. 38th Southwestern Petroleum short course. Lubbock, Texas. 1991. p. 314–323.
[37] Electronic dynamometer DYC 77. Adendorf (Germany): Fr. Leutert GmbH; June 2010.
[38] McCoy JN, West JB, Podio AL. Method and apparatus for measuring pumping rod position and other aspects of a pumping system by use of an accelerometer. US Patent 5,406,482; 1995.
[39] Rowlan OL, McCoy JN, Podio JAL. Advances in dynamometer technology. Proc. 51st Southwestern Petroleum short course. Lubbock, Texas. 2004. p. 141–153.
[40] McCoy JN, Podio AL, Jennings JW. Method of using a polished rod transducer. US Patent 5,464,058; 1995.
[41] McCoy JN, Jennings JW, Podio AL. A polished rod transducer for quick and easy dynagraphs. Proc. 29th Southwestern Petroleum short course. Lubbock, Texas. 1992. p. 345–352.

[42] Boyer L, Dorado D, Cordoba A. New dynamometer technology allows quick setup and easy operation. Proc. 50th annual Southwestern Petroleum short course. Lubbock, Texas. 2003. p. 9–17.
[43] Gilbert WE. An oil-well dynagraph. API Drill Prod Pract 1936:94–115.
[44] Albert GD, Purcupile JC, Chacin JC. Downhole measurement on pumping wells. Paper SPE 17010 presented at the SPE production Technology Symposium at Lubbock, Texas. November 16–17, 1987.
[45] Waggoner JR. Insights from the downhole dynamometer database. Proc. 44th annual Southwestern Petroleum short course. Lubbock, Texas. 1997. p. 204–221.
[46] Waggoner JR, Mansure AJ. Development of the Downhole Dynamometer Database. SPE Prod Facil 2000; 15(No. 1):3–5.
[47] Fagg LW. Dynamometer charts and well weighing. Pet Trans AIME 1950;189:165–74.
[48] API Spec 11E Specification for pumping units. 18th ed. Washington DC: American Petroleum Institute; May 2009.
[49] Nolen KB, Gibbs SG. Quantitative determination of rod-pump leakage with dynamometer techniques. SPE Prod Eng August 1990:225–30.
[50] Kemler EN. Factors influencing pumping-well loads. OGJ March 8, 1954:92–6.
[51] Gipson FW, Swaim HW. The beam pumping design chain. Proc. 31st annual Southwestern Petroleum short course. Lubbock, Texas. 1984. p. 296–376.
[52] Kemler EN. What will dynamometer cards tell you. OGJ March 1, 1954:72–4.
[53] API bul 11L2 Catalog of analog computer dynamometer cards. 1st ed. Dallas (Texas): American Petroleum Institute; December 1969.
[54] Herbert WF. Sucker-rod pumps now analyzed with digital computer. OGJ February 21, 1966:81–5.
[55] Patton LD. A computer technique for analyzing pumping well performance. JPT March 1968:243–9.
[56] Price GS. The on-site computer-dynamometer after 1200 surveys. Proc. 19th annual Southwestern Petroleum short course. Lubbock, Texas. 1972. p. 137–143.
[57] Podio AL, McCoy JN, Rowlan OL, Becker D. Dynamometer analysis plots improve ability to troubleshoot and analyze problems. Proc. 50th annual Southwestern Petroleum short course. Lubbock, Texas. 2003. p. 160–176.
[58] McCoy JN, Rowlan OL, Podio AL. Pump card analysis simplified and improved. Proc. 52nd annual Southwestern Petroleum short course. Lubbock, Texas. 2005. p. 154–164.
[59] Rowlan OL, McCoy JN, Podio AL. Reference load lines aid in analysis of the downhole dynamometer pump card. Proc. 56th annual Southwestern Petroleum short course. Lubbock, Texas. 2009. p. 127–139.
[60] McCoy JN, Rowlan OL, Podio AL. The three causes of incomplete pump fillage and how to diagnose them correctly from dynamometer and fluid level surveys. Proc. 57th annual Southwestern Petroleum short course. Lubbock, Texas. 2010. p. 153–167.
[61] Cox JC, Lea J, Nickens H. Beam pump rod buckling and pump leakage considerations. Proc. 51st annual Southwestern Petroleum short course. Lubbock, Texas. 2004. p. 67–84.
[62] Yavuz F, Lea JF, Garg D, Oetama T, Cox J, Nickens H. Wave equation simulation of fluid pound and gas interference. Paper SPE 94326 presented at the Production and Operations Symposium held in Oklahoma City. April 17–19, 2005.
[63] Podio AL, McCoy JN, Rowlan OL. Pump intake pressure: comparison of values computed from acoustic fluid level, pump dynamometer and valve check measurement. Proc. 56th annual Southwestern Petroleum short course. Lubbock, Texas. 2009. p. 109–126.
[64] Rowlan OL, McCoy JN, Podio AL. Pump intake pressure determined from fluid levels, dynamometers, and valve-test measurements. J Can Pet Technol April 2011:59–66.
[65] Pons V, Rowlan OL, Skinner K. Fluid load line calculation and fluid level calculations: field data applications. Proc. 60th annual Southwestern Petroleum short course. Lubbock, Texas. 2013. p. 139–147.

[66] Derek HJ, Jennings JW, Morgan SM. Sucker rod pumping unit diagnostics using an expert system. Paper SPE 17318 presented at the SPE Permian Basin Oil and Gas Recovery Conference, Midland, Texas. March 10–11, 1988.
[67] Foley WL, Svinos JG. Expert adviser program for rod pumping. JPT April 1989:394–400.
[68] Dickinson RR, Jennings JW. Use of pattern-recognition techniques in analyzing downhole dynamometer cards. SPEPE May 1990:187–92.
[69] Keating JF, Laine RE, Jennings JW. Application of a pattern-matching expert system to sucker-rod dynamometer-card pattern recognition. Paper SPE 21666 presented at the SPE production operations Symposium. Oklahoma City, Oklahoma. April 7–8, 1991.
[70] Houang A, Keating JF, Jennings JW. Pattern recognition applied to dynamometer cards for sucker rod pumping diagnosis. Proc. 38th annual Southwestern Petroleum short course. Lubbock, Texas. 1991. p. 304–313.
[71] Ashenayi K, Nazi GA, Lea JF, Kemp F. Application of an artificial neural network to pump card diagnosis. SPE Comput Appl December 1994:9–14.

CHAPTER

LONG-STROKE SUCKER-ROD PUMPING

7

CHAPTER OUTLINE

7.1 INTRODUCTION

It is a generally accepted fact that sucker-rod pumping installations running at low pumping speeds are more energy efficient than those working at higher speeds. Of the reasons supporting this statement, the most important ones are: (1) better fluid **fillage** of the pump barrel at lower speeds, and (2) lower **dynamic** loads and lower energy losses along the rod string. Low pumping speeds, however, restrict

Sucker-Rod Pumping Handbook. http://dx.doi.org/10.1016/B978-0-12-417204-3.00007-8

the pumping capacity of the installation; this is the reason why higher pumping speeds must be used to lift higher liquid volumes. High pumping rates are, therefore, normally produced with high-capacity rod-pumping equipment running at higher pumping speeds. When increased rates are desired, of course, one tries to use the longest polished rod stroke length; pumping of extreme liquid volumes, however, usually requires high pumping speeds, causing an inevitable drop of the system's power efficiency.

Increasing the stroke length of traditional pumping units to facilitate the use of lower pumping speeds to produce the same liquid volumes has been done in the past, and pumping units with extremely long polished rod strokes are available today. Conventional units with 192 in, Mark II units with 216 in, and air-balanced units with 240 in stroke lengths are manufactured. Due to their mechanical design, however, these units need very high **torques** and the greatest API gearbox sizes of 1824 or even 2560 must be used. At the same time, upsizing of the usual pumping unit results in structures that become extremely big, **heavy**, and expensive [1].

The above problems of increasing the stroke length of sucker-rod pumping operations can only be solved if pumping units completely **different** from the traditional beam pump mechanisms are used. Such machines usually have polished rod stroke lengths greater than 24 ft and require significantly less torque than beam-pumping units. To produce high liquid volumes, they can be run at much **lower** speeds and can thus achieve greater overall system efficiencies: the final goal of high-rate sucker-rod pumping. The general **advantages** of long-stroke pumping over traditional pumping can be summed as: (1) greater liquid-producing **capacities** are achieved, (2) downhole pump problems are decreased, and (3) rod string **life** is substantially increased due to the reduced number of stress reversals.

This chapter provides complete coverage of present-day long-stroke rod-pumping methods and discusses the two main types of technologies available today: Rotaflex and DynaPump. After a short historical overview of long-stroke pumping, these two units are introduced and their technical and operational features are described in detail. At the end of the chapter the relative advantages and limitations of Rotaflex and DynaPump installations are summarized to facilitate their selection for artificial lift applications.

7.2 EARLY MODELS

The earliest long-stroke pumping systems introduced to the industry were hydraulically driven, had high maintenance costs, and were not too reliable. They almost completely disappeared from the oilfield by the early 1950s and mechanical driving mechanisms replaced their use. In the following sections some of the more important early mechanical systems are discussed.

Oilwell Model 3534

The unit's most striking feature was a more than 51-ft-tall steel structure (**tower**) inside which a hollow steel tank used to hold counterweight material was positioned; see Fig. 7.1 [2]. The heart of the system is a pair of **contoured** drums situated at the top of the structure whose common shaft is driven by a gear reducer, also located at the top. The drums are contoured so as to provide changing lever **arms** for the well and the counterbalance **loads** acting on the wirelines attached to them. The lines are fastened to the drums in an **opposite** spiraling way: one of them moves the polished rod via the carrier bar; the other is connected to the counterweight.

The prime mover is automatically switched off before the polished rod reaches any of its extreme positions, then **inertia** of the moving parts causes the system to stop. At the two extreme positions the

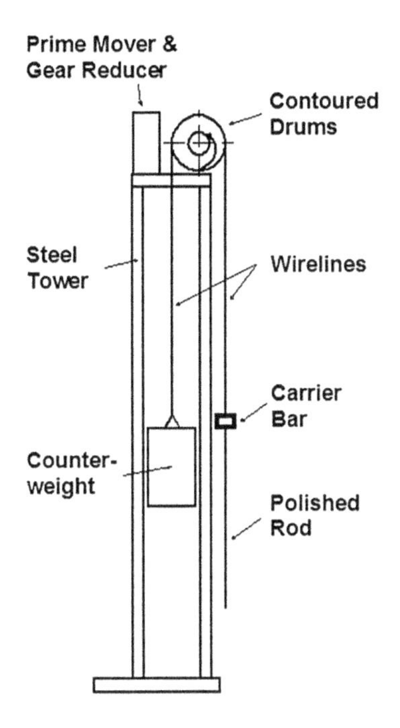

FIGURE 7.1

Schematic drawing of an Oilwell Model 3534 pumping unit.

lever arms of the well and the counterweight loads are such that the resulting torque starts the unit in the direction **opposite** to that before. Only after reversal of movement is completed is the motor **switched** back by the unit's controller, but it turns in the opposite direction.

Model 3534 units had a stroke length of 34 ft (408 in), a structural capacity of 35,000 lb, and a maximum pumping speed of 5 SPM. Field experience [3] indicated a rapid reduction of rod failures and reduced power requirements, as compared to traditional pumping.

Alpha I

Invented by **Gault** [4,5], the **Alpha I**, manufactured by Bethlehem Steel Corp., had a huge maximum stroke length of 40 ft (480 in) but did not use a tower; instead, the counterweight was placed **underground** in a cased hole. As shown in Fig. 7.2, operation of the unit is provided by a power **drum** driven by a standard gearbox. The power drum contains two contoured drums very similar to those in the **Model 3534** unit; these move the rod string and the counterweights through wireline **cables**. Two idler sheaves provide precise alignment with the well and the counterweight hole. As before, the contoured drums provide varying lever **arms** to the well load and the counterweights and ensure that the unit **stops** and **reverses** its movement at the two extremes of the stroke. The prime mover, under control of a microprocessor, turns in **opposite** directions during the up- and downstroke and is off around the stroke reversals.

No polished rod is used on the **Alpha I** unit because the rod string is attached to the end of the well cable. Sealing of the wellhead against well fluids is provided by a **traveling** stuffing box [6] situated

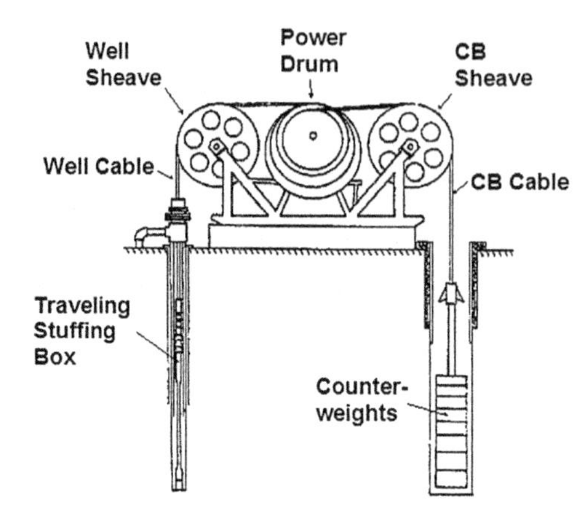

FIGURE 7.2

Schematic drawing of an Alpha I pumping unit.

below the surface. The unit had a structural capacity of 25,000 lb and the maximum recommended pumping speed was 3 SPM.

Operating experience showed a reduced power requirement and higher system efficiencies [7]. Other advantages were the low **profile** and the lower total weight than those of comparable beam pumping units.

Liftronic

Very similar to the **Alpha I**, this unit used heavy **chains** instead of wireline cables to move the rod string and the counterweights; the chains are **wrapped** and unwrapped on the contoured drums during the pumping cycle; see Fig. 7.3. On the upstroke the well chain **wraps** itself on its drum to **increase**

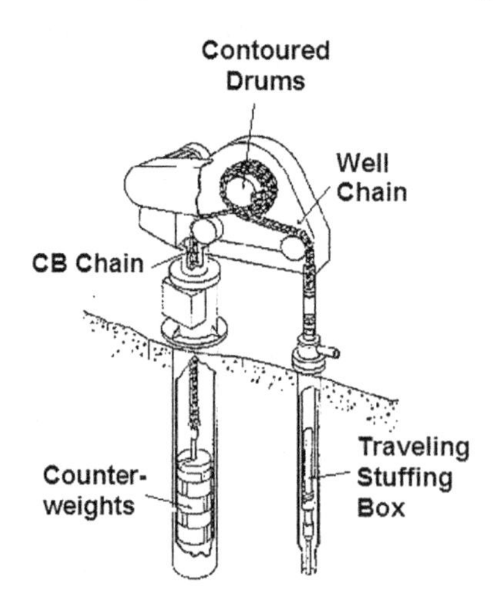

FIGURE 7.3

Schematic drawing of a Liftronic pumping system.

the lever arm of the well load while the counterbalance chain **unwraps** and the counterweight's lever arm decreases. Before the top of the stroke is reached, the motor is switched **off** by the microprocessor and the high torque resulting from the well load brings the unit to a **stop** and, because the counterbalance torque is diminishing, turns it in the **opposite** direction. The motor is switched back to continue the downstroke; reversal of stroke direction works similarly around the bottom of the stroke.

The Liftronic unit's polished rod stroke length was 360 in, its maximum structural capacity was 12,000 lb, and the pumping speed could be set between 1.5 and 3.5 SPM. Comparisons with conventional beam-pumping systems [8,9] proved that, on average, a reduction of power requirement of 15% was achieved; other benefits included extended service life of downhole equipment.

7.3 THE ROTAFLEX PUMPING UNIT

After several attempts in the late 1980s (see [10]) to develop a completely mechanical long-stroke pumping unit, it was **Lively**'s [11] patented version that became accepted in the oilfield. Its trade name is **Rotaflex** and by offering many advantages over traditional pumping units it is a strong competitor to beam-type pumping units and Electrical Submersible Pumping (ESP) equipment at the same time. As will be shown in the following sections all benefits are a direct consequence of the Rotaflex unit's revolutionary mechanical design. These units are getting more and more popular; the total number of Rotaflex installations was about 800 in 2002 [12] but increased to more than 7,000 a decade later.

Since Rotaflex units use standard API gear reducers, their **designation** follows the API nomenclature defined for beam pumping units [13]. The components of the API pumping unit code, separated by dashes, are the following:

- Rotaflex **geometry** is usually indicated by the letter **R** or **RF**,
- The next group of the code contains the torque **rating** of the gear reducer, in thousands of in-lb.
- The following numerals define the maximum polished rod **load** allowed on the pumping unit structure, in hundreds of pounds.
- The last group of the designation stands for the polished rod **stroke length** available in the given unit.

For example, the designation of an **M 1150** (manufacturer code) Rotaflex unit is R-320-500-366; it has a gearbox with a 320,000 in-lb maximum capacity, the peak polished rod load allowed on the unit is 50,000 lb, and the available stroke length equals 366 in.

7.3.1 CONSTRUCTION AND OPERATION

The construction and the basic operation of Rotaflex units are explained in Fig. 7.4, which shows a schematic drawing of the main parts and their functions. The unit is driven by a traditional pumping unit gearbox via the **drive sprocket** of a vertically arranged chain assembly with an idler sprocket situated vertically above the drive sprocket. The heavy-duty roller chain, as it is turned by the drive sprocket, drives a **weight box** (containing the counterweights of the unit) that is connected to one of the links of the chain; the weight box is allowed a vertical movement only. These components constitute the drive train of the unit.

On the well side, the polished rod is directly connected to an elastic **load belt** that runs on a drum situated higher up in such a way that polished rod loads generate only vertical forces in the belt.

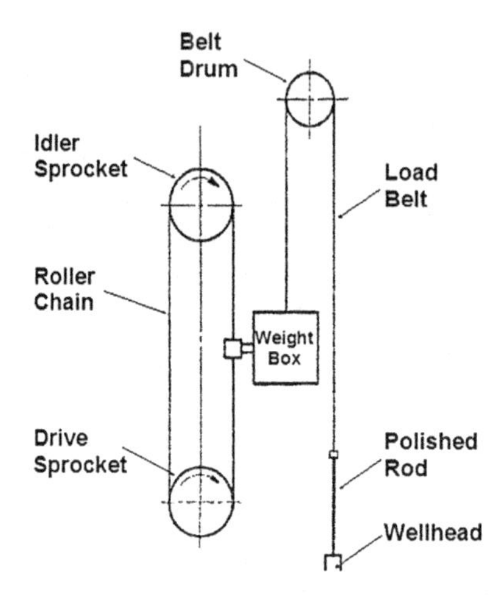

FIGURE 7.4

Operational principle of a Rotaflex pumping unit.

The other end of the belt hanging vertically from the belt drum is fixed to the weight box. The weight box, therefore, carries all operating loads: the pumping load on the polished rod as well as the counterbalance load. Since its movement is strictly vertical, it can always fully utilize the total weight of the counterweights in order to balance the load on the gearbox.

The Rotaflex mechanism just described is fully mechanical and it requires a **constant** direction of rotation from the prime mover; reversal of polished rod stroke direction happens automatically. This feature is a great **improvement** over previous long-stroke pumping units (see Section 7.2) that all necessitated stopping and reversing of the prime mover at both extremes of the polished rod stroke.

The structural components of a Rotaflex unit are shown in Fig. 7.5. The **derrick** or tower supports and contains most of the machinery and stands above the wellhead; it is erected from a folded-down position (the way it is shipped to location) using a pivot point built into the base **skid**. To facilitate workover operations the whole unit can be **rolled** back on rails from the wellhead.

The polished rod is connected to the unit in the traditional way: using a carrier bar, a polished rod clamp, and wireline hangers. The wireline hanger is fixed to a strong, flexible **load belt** hanging from the **belt drum** situated at the top of the derrick structure. The other end of the load belt is fixed to the counterweight assembly (**weight box**) that travels in a vertical direction up and down inside the derrick.

Proper positioning of the unit over the wellhead ensures that both the polished rod and the counterweight move **vertically** only, without any lateral movement. If the counterweight's travel is compared to that of a conventional pumping unit's **rotary** counterweights, one can conclude that this single feature alone can greatly reduce the energy requirements of the Rotaflex unit.

The Rotaflex pumping unit is driven by an electric **prime mover** that is connected to a standard API **gearbox** rotating in one direction only and operated continuously. The driving mechanism schematically depicted in Fig. 7.6 utilizes a **chain-and-sprocket** assembly with the **drive sprocket** directly connected to the gearbox. The drive sprocket drives a heavy-duty, endless **roller chain** led over an **idler sprocket** situated higher up in the derrick structure. Since the direction of the prime

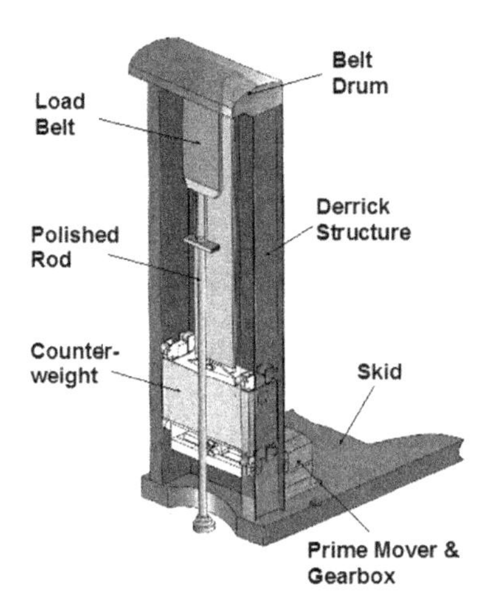

FIGURE 7.5

Structural components of a Rotaflex pumping unit.

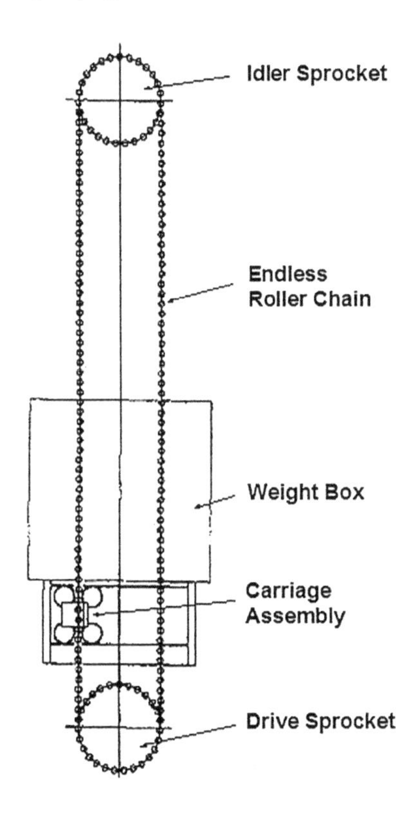

FIGURE 7.6

The driving mechanism of a Rotaflex pumping unit.

mover's rotation, in contrast to other long-stroke pumping units, is **constant** during the pumping cycle, a special solution was necessary to create the reciprocating motion required at the polished rod.

Reversing the movement of the polished rod (connected by the **load belt** to the **weight box** containing the counterweights) is facilitated by the introduction of the revolutionary **carriage assembly** fixed to the bottom of the weight box. This contains a rectangular frame structure, inside which the movement of a wheeled **carriage** is restricted so that the carriage can **slide** perpendicularly to the movement of the chain only. The carriage is fixed to the chain so that it moves along with it, raising or dropping the weight box, and consequently the polished rod, as the chain turns around the sprockets. The operation of the **carriage assembly** is very similar to that of the **Scotch yoke**; a mechanism used to convert linear motion into rotational motion or vice versa.

The construction of the **carriage** is detailed in Fig. 7.7, where it is shown as it negotiates the topmost position on the idler sprocket. A special link in the chain is connected to the carriage through a **swivel knuckle** so that the carriage is forced to travel vertically as the chain turns. At the same time, however, the carriage is free to move horizontally inside the carriage **frame**, its sideways movement being facilitated by the rotating **wheels** closely guided by the frame. This way the carriage along with the weight box can automatically change their direction of movement at the two extremes of the chain's path. Since the carrier bar is connected to the weight box through the **load belt**, the Rotaflex's unique driving mechanism provides the necessary alternating movement at the polished rod.

The movement of the carriage assembly is transmitted to the polished rod by the **load belt**, made of an elastic **composite** material with a safe working capacity of about 50,000 lb. The belt is composed of polyester fibers for strength encased in high-performance elastomers. The belt also acts as a **shock absorber** because its elasticity is much higher than that of the steel; shock loads cause the belt to stretch, thus reducing peak stresses in the sucker rods. If damaged, the belt can easily be replaced even in the field.

Because of the symmetric construction of the chain-and-sprocket assembly (see Fig. 7.6), the Rotaflex unit operates **identically** if run clockwise or counterclockwise. This means that its kinematic behavior does not change with the change of the prime mover's direction of rotation. Manufacturers, however, recommend changing the direction of rotation annually to prevent excessive wear on the sprockets and the chain.

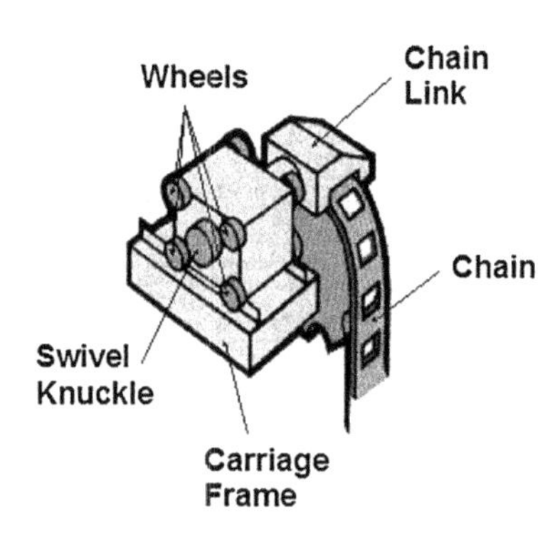

FIGURE 7.7

Details of the carriage assembly.

There are lots of **safety** features built into the Rotaflex unit. The only moving parts not enclosed are the polished rod and the load belt; all other components are **enclosed** in the derrick structure. All shafts of the prime mover and the gear reducer, as well as the drive belts, are protected with metal **guards**. If above-ground servicing is required, ladders with safety shrouds are available along with two guarded platforms on the derrick structure.

Even this elementary description of the main features of the Rotaflex pumping unit allows several important observations to be made:

- The **conversion** of the prime mover's output **torque** to the vertical reciprocating **force** at the polished rod is much more **efficient** than in traditional pumping units because only vertical loads are involved.
- Even great pumping loads act on relatively short torque **arms** (maximum the sprocket radius); torque requirements are thus much reduced.
- The counterweights move vertically only and their **full** weight is utilized over the complete pumping cycle. This is a radical **improvement** if compared to the operation of rotary counterweights used in traditional pumping units.
- The use of the patented carriage assembly [11] provides a **smooth** reversal of the polished rod movement's direction while the prime mover continuously turns in the same direction; thus, no need for **stopping** or reversing the rotation of the prime mover.
- The longitudinal axes of the chain-and-sprocket assembly, the counterweight assembly, and the load belt all lie in the same vertical plane. Therefore, all **forces** arising in these components must also fall in the same plane, with very limited sideways forces that could create energy losses in the system.

7.3.2 KINEMATIC BEHAVIOR

As discussed earlier, the Rotaflex pumping unit converts the prime mover's rotation into the reciprocating movement required for pumping with the help of the **chain-and-sprocket** and the **carriage** assemblies. Therefore, the unit's kinematic behavior can best be described through a thorough analysis of the movement of those components during the pumping cycle. In the following, we assume a **constant** rotational speed of the drive sprocket and will derive formulas for the description of the unit's kinematic parameters as a function of the sprocket's angle of rotation, θ. If a **rigid** load belt is assumed then the position of the carriage assembly defines the position of the **polished rod**; the stretch of the load belt that may affect the polished rod position will be accounted for in a later section.

The motion of the carriage assembly during a full pumping cycle is very **symmetric**, as seen from Fig. 7.6. At the start of the polished rod's upstroke the carriage is at the top of the idler sprocket, then it follows the perimeter of the sprocket, then it moves on a vertical line, then it turns on the drive sprocket, etc. Because the trajectories of the carriage's movement for the up- and downstroke are basically the same, the kinematic parameters of the carriage's movement do not depend on the **direction** of rotation of the prime mover; this fact simplifies the description and the operation of the unit.

In the following the kinematic parameters position, velocity, torque factor, and acceleration of the carriage are expressed as functions of the **sprocket angle**, θ. For simplicity, this is defined as the angle of rotation of the **idler sprocket** and is measured in the direction of the sprocket's rotation starting at $\theta = 0$, valid at the start of the upstroke. In order to cover the pumping cycle, the two sprockets make

several full turns; their total angular movement reaches the angle θ_{max} at the end of the downstroke. This angle is found from the length of the chain and the radius of the sprocket as follows:

$$\theta_{max} = \frac{L_{chain}}{R} \tag{7.1}$$

where:

θ_{max} = sprocket angle at the end of the downstroke, radians,
L_{chain} = length of the roller chain, in, and
R = sprocket radius, in.

7.3.2.1 Carriage position

The background for calculating the **position** of the carriage assembly is illustrated in Fig. 7.8. The left-hand part of the figure shows the chain-and-sprocket assembly, whereas the right-hand side indicates the **vertical** movement of the carriage assembly that defines the carriage position. The upstroke of the

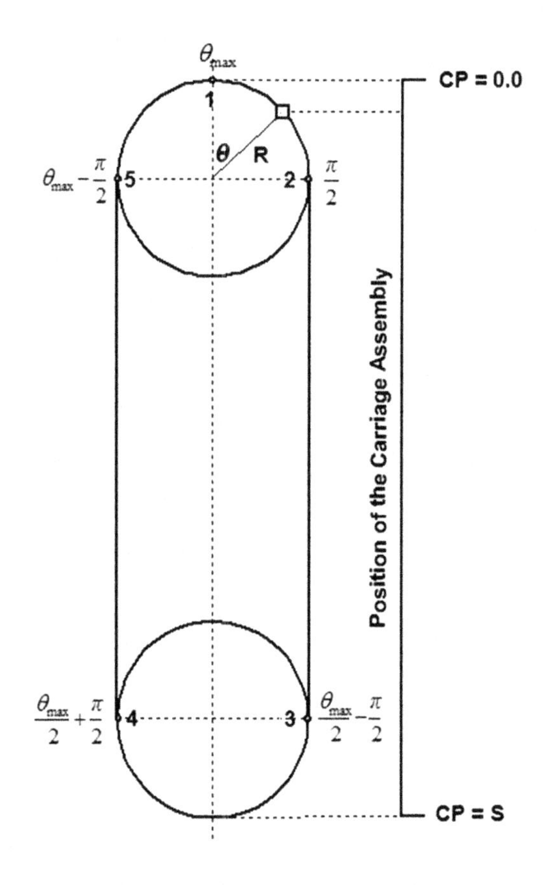

FIGURE 7.8

The ranges of the carriage assembly's motion during a complete pumping cycle.

polished rod starts when the carriage is at the top of the idler sprocket and ends when the carriage reaches the bottom-most position on the drive sprocket.

Carriage position, CP, is defined as zero at the start of the polished rod's upstroke ($CP = 0$) and reaches $CP = S$ at the top of the stroke, where S is the total stroke length of the carriage's travel. CP is calculated as the **vertical** component of the carriage assembly's motion at any sprocket angle, θ. The carriage assembly's movement exhibits two different behaviors during the pumping cycle: (1) it follows a **harmonic** motion while it rides on one of the sprockets, and (2) otherwise it moves **vertically** only. The description of carriage position, therefore, requires different treatments during the different ranges of the carriage's motion.

Figure 7.8 indicates the different ranges of the carriage assembly's motion during the pumping cycle. The sprocket is assumed to turn in the clockwise direction; the position of the carriage assembly at an arbitrary position is indicated by a small rectangle at a sprocket angle θ measured from the vertical. It is easy to understand that, due to the symmetry of the system, the **direction** of rotation does not affect the **position** of the carriage assembly or the position of the polished rod, as already indicated.

The small circles placed on the trail of the carriage indicate those positions where the movement changes from linear to harmonic and vice versa. A total of five distinct **ranges** of the sprocket angle θ can be distinguished and the following equations can be derived to find the carriage positions in those ranges.

For sprocket angles $0 \leq \theta \leq \pi/2$:

$$CP = R(1 - \cos\theta) \tag{7.2}$$

For sprocket angles $\pi/2 < \theta \leq \theta_{max}/2 - \pi/2$:

$$CP = R + R\left(\theta - \frac{\pi}{2}\right) \tag{7.3}$$

For sprocket angles $\theta_{max}/2 - \pi/2 < \theta \leq \theta_{max}/2 + \pi/2$:

$$CP = S - R + R\sin\left(\theta - \frac{\theta_{max}}{2} + \frac{\pi}{2}\right) \tag{7.4}$$

For sprocket angles $\theta_{max}/2 + \pi/2 < \theta \leq \theta_{max} - \pi/2$:

$$CP = S - R - R\left(\theta - \frac{\theta_{max}}{2} - \frac{\pi}{2}\right) \tag{7.5}$$

For sprocket angles $\theta_{max} - \pi/2 < \theta \leq \theta_{max}$:

$$CP = R(1 - \cos(\theta_{max} - \theta)) \tag{7.6}$$

where:

CP = carrier position, in
R = sprocket radius, in,
S = total stroke length of the carriage assembly, in, and
θ_{max} = sprocket angle belonging to the bottom of the downstroke, radians.

The variation of carriage position, i.e., polished rod position with time during a complete pumping cycle for a **Model 1151** Rotaflex unit, is plotted in Fig. 7.9. As seen, practically both the up- and

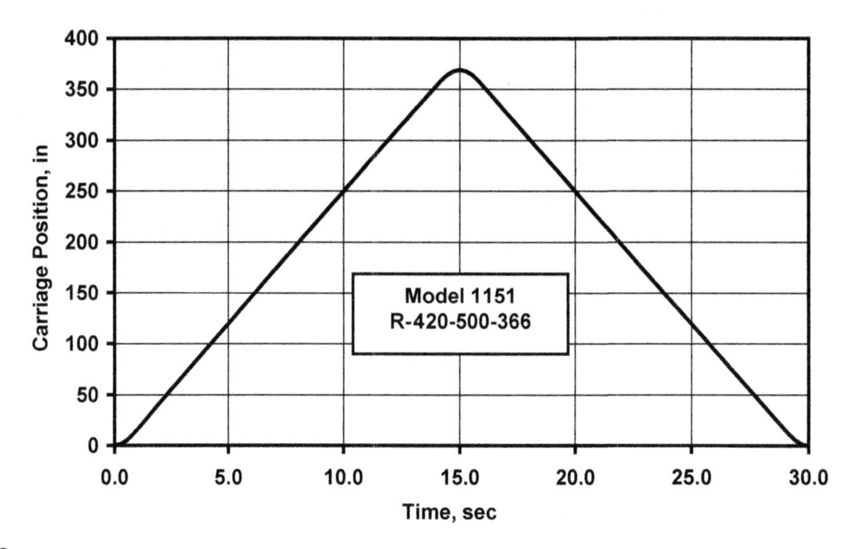

FIGURE 7.9

Carriage position vs time for a Model 1151 Rotaflex unit.

downstroke positions change **linearly** with time, i.e., with sprocket angle. The elapsed times, just like the sprocket angle ranges that belong to the up- and downstroke, are the same because the motion is symmetric. Changing the direction of rotation of the prime mover from clockwise to counterclockwise, or vice versa, results in the same kind of behavior; the Rotaflex unit, contrary to some beam-pumping units, is not sensitive to the direction of operation. It can also be observed that reversal of stroke direction is very **fast**, especially if compared to the reversals experienced on beam-pumping units.

7.3.2.2 Other kinematic parameters

The additional kinematic parameters of the Rotaflex pumping unit are carriage velocity and carriage acceleration, which are, for the reasons already discussed, identical to the velocity and acceleration of the polished rod. Carriage velocity, in principle, is found by **differentiating** carriage position, *CP*, with respect to time and must be evaluated differently in the different sprocket angle ranges indicated in Fig. 7.8. For this, knowledge of the variation of the sprocket angle with time is necessary. Since the gearbox of Rotaflex units usually turns at a constant rotational speed, the average sprocket angular velocity is found from the sprocket angle range, θ_{max}, and the cycle time of pumping, *T*. Sprocket angle, therefore, increases linearly with time, and using the angular frequency of the sprocket we get the following expression:

$$\theta = \frac{\theta_{max}}{T} t = \frac{N\ \theta_{max}}{60} t \tag{7.7}$$

where:

θ = sprocket angle, radians,
θ_{max} = sprocket angle at the end of the downstroke, radians,
N = pumping speed, 1/min,
t = time, s.

Using this formula when differentiating the carriage position function for the first two ranges of the sprocket angle indicated in Fig. 7.8, we find the carriage velocity as follows:

$$v = \frac{dCP}{dt} = R\frac{N\ \theta_{max}}{60}\sin\theta \quad \text{in the range 1–2} \tag{7.8}$$

$$v = \frac{dCP}{dt}R\frac{N\ \theta_{max}}{60} \quad \text{in the range 2–3} \tag{7.9}$$

where:

v = vertical velocity of the carriage, in/s,
θ_{max} = sprocket angle at the end of the downstroke, radians,
N = pumping speed, 1/min,
R = sprocket radius, in,
θ = sprocket angle, radians.

The **acceleration** of the sprocket assembly, which is identical to polished rod acceleration, is found by differentiating carriage velocity with respect to time. For the two ranges illustrated before, we get:

$$a = \frac{dv}{dt} = \left(R\frac{N\ \theta_{max}}{60}\right)^2\cos\theta \quad \text{in the range 1–2} \tag{7.10}$$

$$a = \frac{dv}{dt} = 0 \quad \text{in the range 2–3} \tag{7.11}$$

where:

a = vertical acceleration of the carriage, in/s^2,
θ_{max} = sprocket angle at the end of the downstroke, radians,
N = pumping speed, 1/min,
R = sprocket radius, in,
θ = sprocket angle, radians.

Torque factor, *TF*, values represent the torque arm to be used at any carriage position to find the gearbox torque valid at the given loading conditions; they facilitate the calculation of gearbox torques. Since the polished rod load is transferred by the load belt to the weight box, which, in turn, is connected to the carriage assembly, calculation of the Rotaflex unit's torque factors is based on the behavior of the chain-and-sprocket assembly. In analogy to traditional sucker-rod pumping units (see **Chapter 3**), torque factors are found by differentiating the carriage position with respect to sprocket angle. An alternative solution for torque factor calculations is presented by **McCoy et al.** [14]. For the two sample sprocket ranges used before, the following formulas are received:

$$TF = \frac{dCP}{d\theta} = R\sin\theta \quad \text{in the range 1–2} \tag{7.12}$$

$$TF = \frac{dCP}{d\theta} = R \quad \text{in the range 2–3} \tag{7.13}$$

where:

TF = torque factor, in,
R = sprocket radius, in,
θ = sprocket angle, radians.

The exact calculation formulas for carriage position, velocity, acceleration, and torque factor are contained in Table 7.1 for the five sprocket angle ranges indicated in Fig. 7.8.

Using the same **Model 1151** Rotaflex unit as before, velocity and acceleration values for a pumping speed of $N = 2$ SPM are given in Fig. 7.10 and Fig. 7.11. The **velocity** profile is characterized by a

Table 7.1 Kinematic Parameters of Rotaflex Pumping Units

For sprocket angles: $0 \le \Theta \le \pi/2$	
Carriage position, in	$R\,(1-\cos\theta)$
Carriage velocity, in/s	$R\,\frac{N\,\theta_{max}}{60}\sin\theta$
Torque factor, in	$R\,\sin\theta$
Carriage acceleration, in/s^2	$R\left(\frac{N\,\theta_{max}}{60}\right)^2\cos\theta$
For sprocket angles: $\pi/2 < \Theta \le \Theta_{max}/2 - \pi/2$	
Carriage position, in	$R + R\left(\theta - \frac{\pi}{2}\right)$
Carriage velocity, in/s	$R\,\frac{N\,\theta_{max}}{60}$
Torque factor, in	R
Carriage acceleration, in/s^2	0
For sprocket angles: $\Theta_{max}/2 - \pi/2 < \Theta \le \Theta_{max}/2 + \pi/2$	
Carriage position, in	$S - R + R\sin\left(\theta - \frac{\theta_{max}}{2} + \frac{\pi}{2}\right)$
Carriage velocity, in/s	$R\,\frac{N\,\theta_{max}}{60}\cos\left(\theta - \frac{\theta_{max}}{2} + \frac{\pi}{2}\right)$
Torque factor, in	$R\cos\left(\theta - \frac{\theta_{max}}{2} + \frac{\pi}{2}\right)$
Carriage acceleration, in/s^2	$-R\left(\frac{N\,\theta_{max}}{60}\right)^2\sin\left(\theta - \frac{\theta_{max}}{2} + \frac{\pi}{2}\right)$
For sprocket angles: $\Theta_{max}/2 + \pi/2 < \Theta \le \Theta_{max} - \pi/2$	
Carriage position, in	$S - R - R\left(\theta - \frac{\theta_{max}}{2} - \frac{\pi}{2}\right)$
Carriage velocity, in/s	$-R\,\frac{N\,\theta_{max}}{60}$
Torque factor, in	$-R$
Carriage acceleration, in/s^2	0
For sprocket angles: $\Theta_{max} - \pi/2 < \Theta \le \Theta_{max}$	
Carriage position, in	$R\,[1-\cos(\theta_{max}-\theta)]$
Carriage velocity, in/s	$-R\,\frac{N\,\theta_{max}}{60}\sin(\theta_{max} - \theta)$
Torque factor, in	$-R\,\sin(\theta_{max}-\theta)$
Carriage acceleration, in/s^2	$R\left(\frac{N\,\theta_{max}}{60}\right)^2\cos(\theta_{max} - \theta)$

Definition of variables: R = *sprocket radius, in;* S = *carriage stroke length, in;* N = *pumping speed, 1/min;* Θ = *sprocket angle, rad;* Θ_{max} = *max. sprocket angle, rad.*

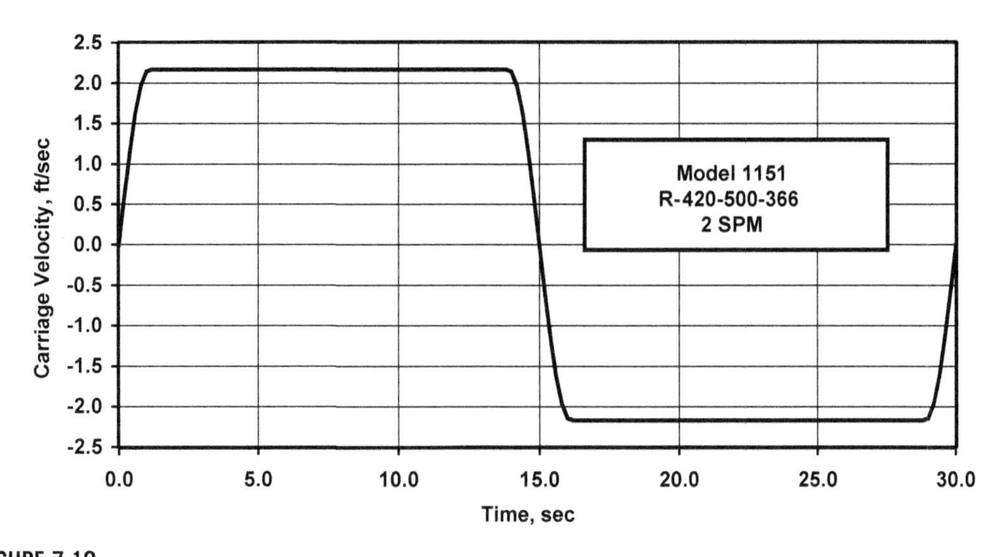

FIGURE 7.10

Carriage velocity vs time for a Model 1151 Rotaflex unit running at 2 SPM.

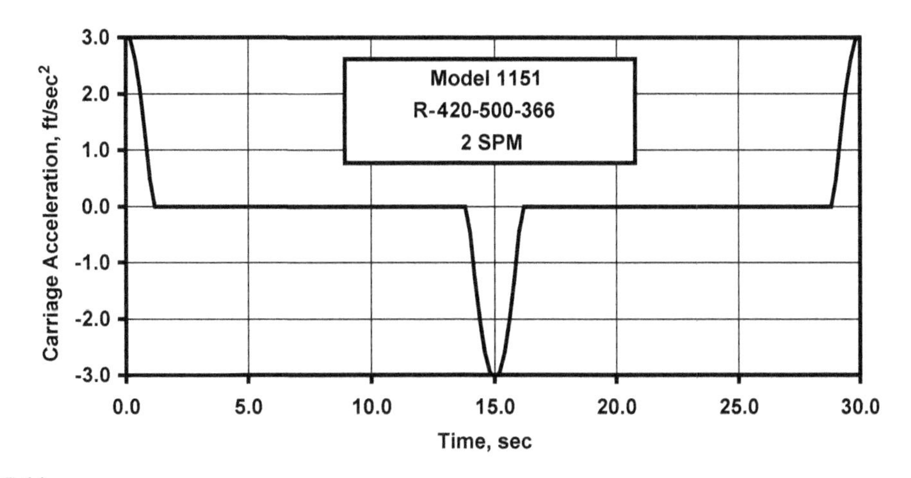

FIGURE 7.11

Carriage acceleration vs time for a Model 1151 Rotaflex unit running at 2 SPM.

constant speed during most of the up- and downstroke; only the direction of movement changes. Again, symmetrical operation is observed and the unit behaves **identically** if driven in opposite directions.

The long, quiet movement of the carriage results in zero **acceleration** for most of the pumping cycle. There are quite low acceleration levels around the two extremes of the stroke and they last for short periods only. Acceleration of the carriage assembly happens only for two 180° turns of the sprocket; this seems negligible if compared to the total sprocket rotation of 2,435° valid for the given unit.

In conclusion, the Rotaflex unit has an extremely **quiet** operation with very low acceleration and negligible dynamic effects and offers a kinematic behavior that differs very profoundly from that of any beam-pumping unit. The polished rod of traditional beam-pumping units continuously accelerates–decelerates during the up- and downstroke, causing varying velocities; those units, in contrast to Rotaflex units, experience high dynamic loads.

Torque factors for the same **Model 1151** Rotaflex unit as before are presented in Fig. 7.12 as a function of time. As seen, except for two short periods around the two extremes of the stroke, torque factors are **constant**; only their sign is different for the upstroke and the downstroke.

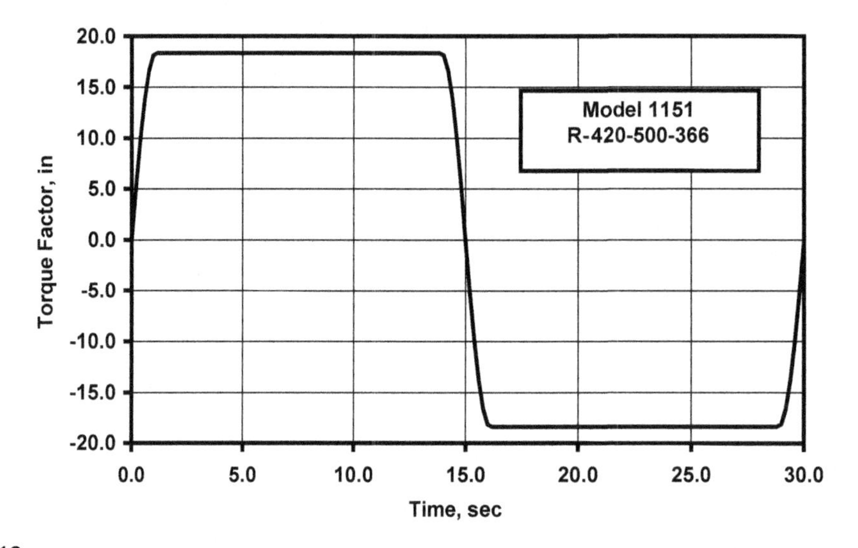

FIGURE 7.12

Torque factor values vs time for a Model 1151 Rotaflex unit.

EXAMPLE 7.1: FIND THE KINEMATIC PARAMETERS OF A ROTAFLEX UNIT AT THE SPROCKET ANGLE OF $\theta = 22.689$ RADIANS (1,300°). THE UNIT'S MAIN DATA ARE GIVEN AS:

Pumping speed = 2 SPM	Chain length = 780.4 in
Carriage stroke length = 369.2 in	Sprocket radius = 18.36 in

Solution

First find the sprocket angle belonging to the bottom of the downstroke from Eq. (7.1):

$$\theta_{max} = 780.4/18.36 = 42.5 \text{ rad } (2,435^{\circ}).$$

Since $42.5/2 - \pi/2 < 22.689 < 42.5/2 + \pi/2$, Eq. (7.4) is used:

$$CP = 369.2 - 18.36 + 18.36 \sin[22.689 - 42.5/2 + \pi/2] = 350.84 + 18.36 \sin(3.009) = 353.25 \text{ in.}$$

The velocity formula is found from Table 7.1 as follows:

$$v = \frac{RN\theta_{max}}{60 \cos[\theta - \theta_{max}/2 + \pi/2]} = \frac{18.36 \times 2 \times 42.5}{60 \cos[3.009]} = -25.8 \text{ in/s} = -2.15 \text{ ft/s}.$$

The acceleration is found from the formula given in Table 7.1:

$$a = -R(N\theta_{max}/60)^2 \sin[\theta - \theta_{max}/2 + 90] = -18.36\,(2 \times 42.5/60)^2 \sin[3.009]$$
$$= -4.84 \text{ in/s}^2 = -0.40 \text{ ft/s}^2.$$

Finally, the torque factor is calculated from the relevant formula in Table 7.1:

$$TF = 18.36 \cos(3.009) = -18.2 \text{ in.}$$

7.3.3 TORSIONAL ANALYSIS

7.3.3.1 Rigid load belt case

All calculations presented in this section will assume that the load belt transmitting the polished rod loads to the carriage assembly behaves as a **rigid** body, i.e., it does not **stretch** due to any loads acting on it. With this approach one assumes that the polished rod and the carriage assembly move in **unison**; this greatly simplifies the treatment of the torques acting on the drive sprocket and the gearbox. In reality, of course, the belt will stretch and absorb sudden polished rod load peaks and the movement of the carriage assembly, depending on the loads, will lag or lead the polished rod's movement. The effects of load belt stretch will be investigated in a later section.

7.3.3.1.1 Torque calculations

Let's investigate the loads acting on the carriage assembly during the pumping cycle and evaluate the torques arising on the drive sprocket. Figure 7.13 shows arbitrary positions of the weight box along with the loads acting on it during the upstroke and the downstroke while the drive sprocket turns in the clockwise direction. As described in Fig. 7.6, the weight box rides on the carriage assembly and has the load belt connected to it; this way both the polished rod **load**, F_c, and the **weight** of the counterweights, W_{CB}, act on the carriage. Since the torque factors are known for all possible positions of the carriage assembly, it is quite simple to find the torques to be developed by the gearbox.

The kinds of torques arising in the gearbox, similarly to those on a beam pumping unit [15], can be classified as follows:

- **Rod torque** represents the torque necessary to move the polished rod,
- **Counterbalance torque** is needed to keep the weight box and auxiliary weights in motion, and
- **Inertial torque** occurs due to the acceleration/deceleration of heavy moving parts.

In the following the calculation of the torque components defined here is detailed where gearbox torques are considered positive in the direction of rotation. It must be mentioned that because of the **symmetrical** operation of the chain-and-sprocket assembly discussed earlier, the direction of the drive sprocket's rotation does not affect the calculation of the torques.

Rod Torque

The dominant force causing rod torque, T_r, comes from the polished rod load acting on the weight box and the carriage assembly. An additional load is the weight of that part of the load belt hanging

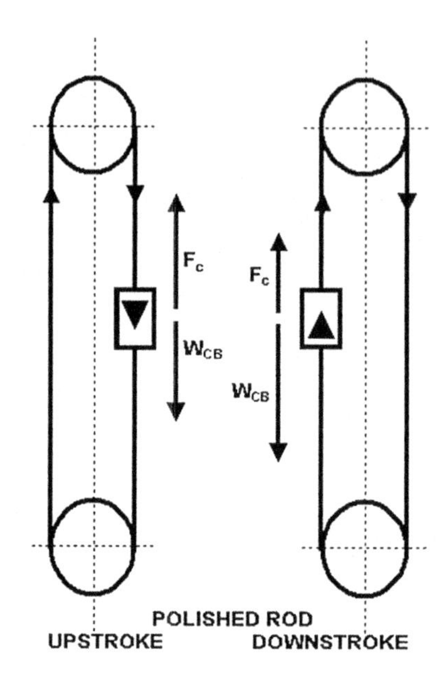

FIGURE 7.13

The loads acting on the weight box during the up- and downstroke.

between the load drum situated at the top of the derrick and the polished rod connection. If the **sum** of these loads is designated as F_c then the **torque** to be developed by the gearbox to overcome this force is found by using the torque factors of the unit:

$$T_r(\theta) = TF(\theta) \; F_c(\theta) \tag{7.14}$$

where:

$T_r(\theta)$ = rod torque, in lb,
$TF(\theta)$ = torque factor at sprocket angle θ, in, and
$F_c(\theta)$ = net load on the carriage assembly at sprocket angle θ, lb.

In the formula the term $F_c(\theta)$ is the sum of the polished rod **load** and the partial **weight** of the load belt, W_{bPR}, that varies with the actual position of the polished rod. At the start of the upstroke the belt weight is at its maximum, decreasing as the stroke continues and reaching its minimum at the start of the downstroke. During the downstroke it increases to its maximum value again; the process repeats with every stroke.

In conclusion, **net** load on the carriage assembly is found from:

$$F_c(\theta) = F_{PR}(\theta) + W_{bPR} \tag{7.15}$$

where:

$F_c(\theta)$ = net load on the carriage assembly at sprocket angle θ, lb,
$F_{PR}(\theta)$ = polished rod load at the given position, lb, and
W_{bPR} = variable weight of the load belt on the well side, lb.

Counterbalance Torque

Counterbalance torque is defined as the torque necessary to **move** the counterweights and is caused by the following components of the unit's counterbalance system:

- the weight box,
- auxiliary weights placed in the weight box, and
- the weight of the portion of the load belt measured from the load drum to the weight box.

It can be observed in Fig. 7.13 that the weight of the counterbalance **decreases** the torque needed to drive the polished rod during the upstroke while it must be lifted during the downstroke; counterbalance torque always **opposes** rod torque. The torque required to move the counterweights is calculated by using the torque factors, similarly to rod torque:

$$T_{CB}(\theta) = -TF(\theta)\ W_{CB} \tag{7.16}$$

where:

$T_{CB}(\theta)$ = counterbalance torque at sprocket angle θ, in lb,
$TF(\theta)$ = torque factor at sprocket angle θ, in,
W_{CB} = total weight of the counterbalance components, lb.

In this formula, W_{CB} includes the **portion** of the load belt's **weight**, W_{bCB}, which acts as a counterbalance component. This **varies** with the polished rod's travel in a similar fashion, as detailed in the section on rod torque calculations. The **total** weight of the counterbalance components, W_{CB}, can be calculated from:

$$W_{CB} = W_{WB} + W_{aux} + W_{bCB} \tag{7.17}$$

where:

W_{CB} = total weight of the counterbalance components, lb,
W_{WB} = weight of the weight box, lb,
W_{aux} = weight of auxiliary counterweights, lb, and
W_{bCB} = variable weight of the load belt on the counterbalance side, lb.

Inertial Torque

The torques previously calculated can be considered as **static** torques since they were calculated from the static loads on the pumping unit. All heavy components of the Rotaflex unit moving at speeds that change during the pumping cycle, however, cause **inertial** torques on the gearbox. The parts accelerating/decelerating while the polished rod completes its stroke are: the components of the counterweight system, the load belt, the sprockets, and the belt drum. Because of their great **deadweight** and relatively high **accelerations** (although for short periods in the pumping cycle), inertial forces and torques developed in the **counterweight** components and the load belt are the most significant. Neglecting articulating torques that develop in the sprockets and the belt drum, the inertial torque, T_i, is found from the inertial forces caused by the acceleration/deceleration of the total **mass** involved, as given in the following:

$$T_i(\theta) = -TF(\theta)\ W'_{CB}\,\frac{a(\theta)}{32.17} \tag{7.18}$$

where:

$T_i(\theta)$ = inertial torque at sprocket angle θ, in lb,
$TF(\theta)$ = torque factor at sprocket angle θ, in, and
W'_{CB} = total weight of the accelerating components, lb,
$a(\theta)$ = acceleration of the carriage assembly at sprocket angle θ, ft/s^2.

The total mass accelerating/decelerating includes the masses of the **counterbalance** system plus the whole load **belt**; the respective **total** weight, W'_{CB}, is calculated from:

$$W'_{CB} = W_{WB} + W_{aux} + W_b \tag{7.19}$$

where:

W'_{CB} = total weight of the accelerating components, lb,
W_{WB} = weight of the weight box, lb,
W_{aux} = weight of auxiliary counterweights, lb, and
W_b = total weight of the load belt, lb.

Net Torque

The torque to be developed by the drive sprocket, and consequently by the **gearbox**, is the **sum** of the torque components just described. This is called the net torque, T_{net}, and is calculated from the formula given in the following; the variation of this torque has a great impact on the power requirements of pumping as discussed later.

$$T_{net}(\theta) = TF(\theta)\left[F_c(\theta) - W_{CB} - W'_{CB}\frac{a(\theta)}{32.17}\right] \tag{7.20}$$

where:

$T_{net}(\theta)$ = net torque on the gearbox at sprocket angle θ, in lb,
$TF(\theta)$ = torque factor at sprocket angle θ, in,
$F_c(\theta)$ = net load (Eq. (7.15)) on the carriage assembly at sprocket angle θ, lb.
W_{CB} = total weight (Eq. (7.17)) of the counterbalance components, lb,
W'_{CB} = total accelerated weight (Eq. (7.19)) of counterbalance components, lb, and
$a(\theta)$ = acceleration of the carriage assembly at sprocket angle θ, ft/s^2.

7.3.3.1.2 Permissible loads

The concept of **permissible loads**, first introduced for beam-pumping units, can be applied to the analysis of Rotaflex units as well. Permissible loads are defined as the **maximum** polished rod loads at different sprocket angles θ that the gearbox can **handle** without being overloaded. If plotted on the dynamometer diagram, permissible load values provide an easy way to **check** the loading conditions of the gear reducer.

Permissible loads, F_p, vary with the sprocket **angle** and the actual amount of counterbalance on the unit; they produce a uniform loading on the gearbox equal to its **rated** torque capacity. They can be calculated by solving the net torque formula for the polished rod load:

$$F_p(\theta) = \frac{T_{al}}{TF(\theta)} + W_{CB} + W'_{CB}\frac{a(\theta)}{32.17} \tag{7.21}$$

where:

$F_p(\theta)$ = permissible load at sprocket angle θ, lb,
T_{al} = torque rating of the gearbox, in-lb,
$TF(\theta)$ = torque factor at sprocket angle θ, in,
W_{CB} = total weight (Eq. (7.17)) of the counterbalance components, lb, and
W'_{CB} = total accelerated weight (Eq. (7.19)) of the counterbalance components, lb.
$a(\theta)$ = acceleration of the carriage assembly at sprocket angle θ, ft/s^2.

7.3.3.1.3 Optimum counterbalancing

The objective of using counterweights on the Rotaflex unit, similarly to beam-pumping units, is to **smooth** the net torque load on the gear reducer during the pumping cycle. As seen before, the torque produced by the counterweights always **opposes** the rod torque and thus reduces the net torque load on the unit. Since adding or removing auxiliary weights to the weight box is easy, one surely can find **optimum** counterbalance conditions. Using the proper amount of counterweights may improve pumping operations in several ways:

- Compared to an unbalanced case, a gearbox with a smaller rating can be used.
- Prime mover size can also be reduced.
- Maintenance costs are reduced and gearbox life is extended because of the smoother overall operation.

There are several ways to ensure that the unit is optimally counterbalanced. They include setting the peak net torques or the peak motor currents **equal** on the up- and downstroke portion of the pumping cycle. Field practice follows these procedures but computerized solutions are more advanced and try to find the minimum of the mechanical **cyclic load factor** calculated from the variation of net torque, as given in the following formula:

$$CLF = \frac{\sqrt{\frac{\sum_{i=1}^{N} T_{net}^2}{N}}}{\frac{\sum_{i=1}^{N} T_{net}}{N}} \tag{7.22}$$

where:

CLF = cyclic load factor,
T_{net} = net torque values, in-lb, and
N = number of torque values.

According to its definition, *CLF* is the ratio of the root mean square and the average torques over the pumping cycle and is the best indicator of the **effectiveness** of counterbalancing. This is due to the fact that a theoretically **perfect** counterbalancing (never achievable because of the geometry of pumping units) with a constant torque load on the gearbox gives a cyclic load factor of unity, $CLF = 1$, while any other case results in values **greater** than unity. Trying to achieve the **minimum** of *CLF*, therefore, minimizes the fluctuations of gearbox torque, provides the **smoothest** possible operation of the pumping unit, and brings about all the advantages listed previously.

7.3.3.1.4 Practical considerations

The previous sections provided the theoretical **background** of torsional calculations for Rotaflex units, but practical applications require additional considerations.

The first problem is to find the polished rod loads and displacements required for **rod torque** calculations. The usual source of these data is the **dynamometer card** taken on the given well under the current conditions. Nowadays most dynamometer surveys use **electronic** dynamometers recording polished rod loads and displacements as a function of **time**. Measurements are done at constant time intervals and this feature enables one to find the sprocket **angle**, θ, values for each time increment by assuming a **constant** angular velocity of the drive sprocket. Since torque factors, *TF*, are a direct function of sprocket angle, **rod torque** is easily calculated; see Eq. (7.14).

The **actual** amount of counterbalance load is the sum of the weights of (1) the weight box and (2) any auxiliary counterweights. The first of these is provided by the manufacturer, so only the actual weight of auxiliary counterweights must be determined. Auxiliary weights are simple metal **blocks** and their weight is found from their volume and the density of the material used. Their total weight, therefore, is easily calculated, but a **direct** measurement with a dynamometer connected to the polished rod is also possible, as detailed in the following.

Measurement of the maximum load on the polished rod arising from the counterbalance torque, often called **counterbalance effect**, *CBE*, is based on the fact that loads indicated by a dynamometer are **identical** at the two ends of the load belt. Since the belt is connected to the weight box, the total weight of counterweights can be read directly on the dynamometer connected to the polished rod. The actual *CBE* measurement is done according to the following steps:

1. The Rotaflex unit is **stopped** at around the middle of the upstroke and the brake is applied. At this point the polished rod is much more loaded by the rod string's buoyant weight plus the fluid load on the plunger than the other end of the load belt by the total weight of the counterweights.
2. If the brake is slowly **released** the polished rod starts to fall, because of the actual imbalance of forces detailed previously. As soon as this happens, the brake is reset again to **stop** any movement.
3. At regular intervals, **releasing** and **setting** of the brake is repeated until no polished rod movement is experienced. During this phase, fluid load on the plunger gradually decreases due to fluid **slippage** in the downhole pump.
4. When releasing the brake does not cause the polished rod to move, **equilibrium** between the polished rod load just measured and the load on the other end of the load belt is reached; this load equals the total **weight** of the counterweights, i.e., the *CBE* value.

EXAMPLE 7.2: CALCULATE THE GEARBOX TORQUES AND THE PERMISSIBLE LOADS FOR ONE POINT IN THE POLISHED ROD STROKE WHERE THE MEASURED STROKE AND LOAD ARE 60.3 IN AND 32,912 LB, RESPECTIVELY. THESE VALUES WERE MEASURED AT 3.01 S FROM THE START OF THE POLISHED ROD UPSTROKE. ADDITIONAL DATA OF THE ROTAFLEX UNIT R-320-500-366 ARE GIVEN HERE:

Pumping speed = 2.11 SPM	Chain length = 770.6 in
Carriage stroke length = 366.1 in	Sprocket radius = 16.8 in
Weight of weight box = 9,800 lb	Weight of auxiliary weights = 12,000 lb
Weight of load belt = 2,010 lb	

Solution

First calculate the maximum sprocket angle at the bottom of the downstroke, as given in Eq. (7.1):

$$\theta_{max} = 770.6/16.8 = 45.87 \text{ rad } (2,628°).$$

The angular frequency of the sprocket is calculated from Eq. (7.8):

$$\frac{d\theta}{dt} = \frac{2.11 \times 45.87}{60} = 1.61 \text{ rad/s}.$$

The sprocket angle belonging to the measured time of 3.01 s can now be calculated:

$$\theta = \frac{d\theta}{dt} t = 1.61 \times 3.01 = 4.855 \text{ rad}.$$

The calculation of the kinematic parameters is started by finding the carriage position, *CP*, belonging to the sprocket angle just calculated; Eq. (7.3) is used:

$$CP = 16.8 + 16.8\,(4.855 - \pi/2) = 71.9 \text{ in}.$$

As seen from Table 7.1, acceleration is zero:

$$a = 0.$$

The torque factor is found from Table 7.1 as:

$$TF = R = 16.8 \text{ in}.$$

The gearbox torques can now be calculated, starting with the rod torque. When calculating the net load on the carriage assembly, F_c, the weight of the load belt on the well side was found as $W_{bPR} = 1{,}281$ lb; Eq. (7.15) gives:

$$F_c = F_{PR} + W_{bPR} = 32,912 + 1,281 = 34,193 \text{ lb}.$$

Rod torque from Eq. (7.14):

$$T_r(4.855) = TF \times F_c = 16.8 \times 34,193 = 574,442 \text{ in-lb}.$$

At the given position the weight of the load belt on the counterbalance side was found as $W_{bCB} = 526$ lb; the total weight of the counterbalance components is calculated from Eq. (7.17):

$$W_{CB} = W_{WB} + W_{aux} + W_{bCB} = 9,800 + 12,000 + 526 = 22,326 \text{ lb}.$$

The counterbalance torque can now be calculated from Eq. (7.16):

$$T_{CB}(4.855) = -TF \times W_{CB} = -16.8 \times 22,326 = -375,077 \text{ in-lb}.$$

Since sprocket acceleration at the given position is zero, inertial torque is also zero. The net torque acting on the gearbox is the sum of the rod and counterbalance torques:

$$T_{net}(4.855) = 574,442 - 375,077 = 199,365 \text{ in-lb}.$$

The permissible load can be found from Eq. (7.21) using the gearbox rating of $T_{al} = 320{,}000$ in-lb; since acceleration is zero, the formula is simplified:

$$F_p(4.855) = T_{al}/TF + W_{CB} = 320,000/16.8 + 22,326 = 41,374 \text{ lb}.$$

The calculated torques for the complete pumping cycle are displayed in Fig. 7.14. As can be seen, the gearbox is not overloaded, since all net torque values are below the rating of 320,000 in-lb. The balancing of the unit, however, is far from ideal because the upstroke peak torque is much higher than the downstroke peak. The mechanical cyclic load factor was calculated as $CLF = 1.261$; this also indicates that optimum counterbalancing was not attained.

The dynamometer card that served as the basis of this example is presented in Fig. 7.15, where the calculated permissible loads are also plotted. Since none of the permissible load curves cuts into the dynamometer diagram, the unit is not overloaded, as found earlier.

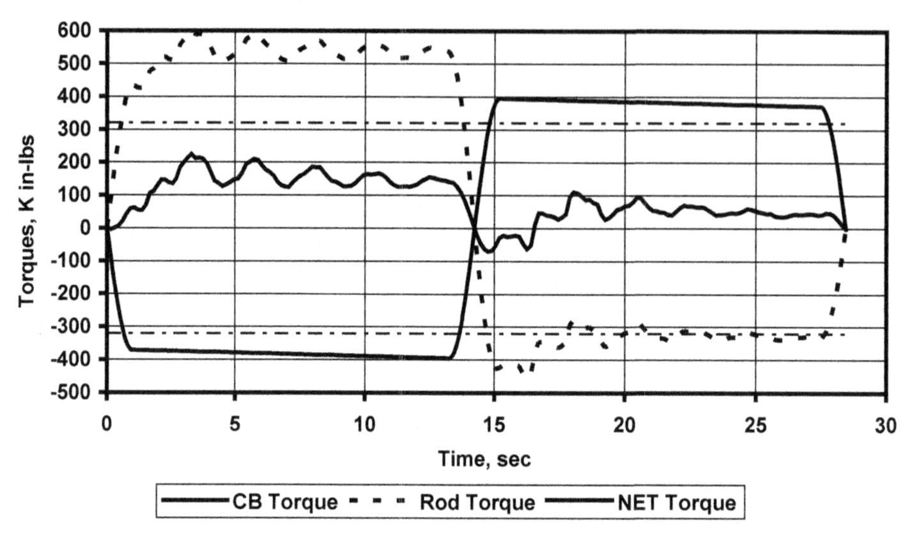

FIGURE 7.14

Variation of gearbox torque components vs time for Example 7.2.

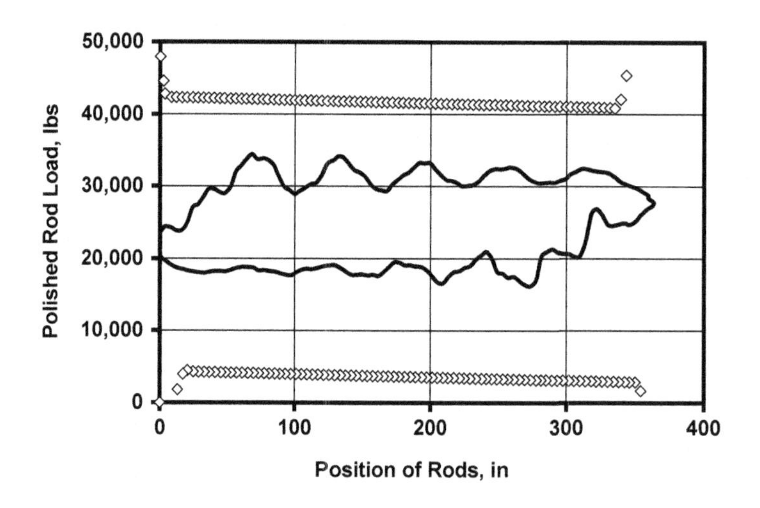

FIGURE 7.15

Dynamometer card with the permissible loads superimposed for Example 7.2.

In order to find the optimum counterbalance conditions, torque calculations assuming different amounts of auxiliary weights were performed; Fig. 7.16 shows how the calculated *CLF* values changed. The minimum of *CLFs* was found as $CLF = 1.121$ for an auxiliary counterweight of 15,000 lb. Using this optimum counterweight, the Rotaflex unit's torque conditions are depicted in Fig. 7.17. As seen, the variation of net gearbox torque during the pumping cycle is much more regular than before.

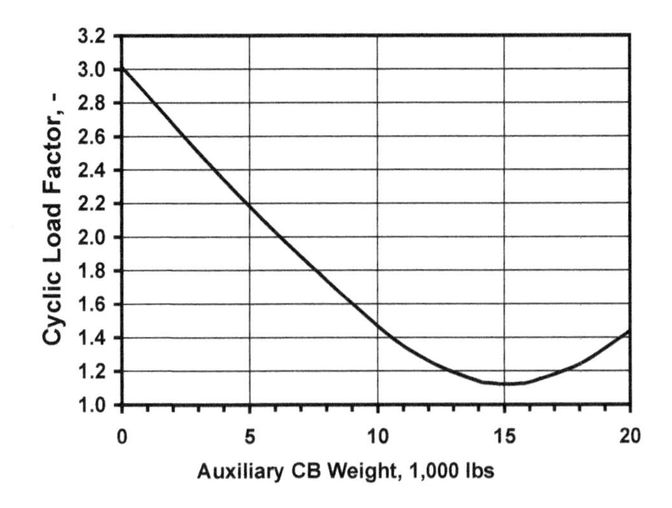

FIGURE 7.16

Change of CLFs for different auxiliary weights applied for Example 7.2.

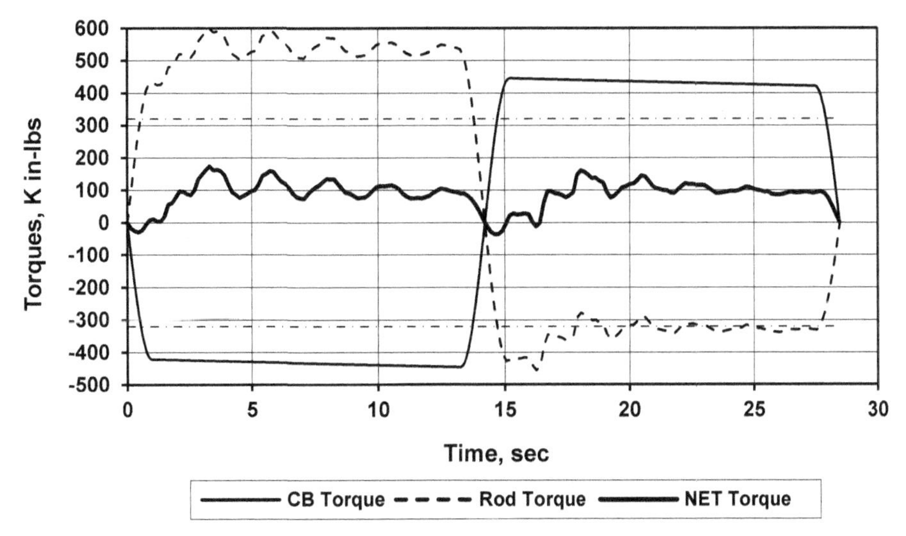

FIGURE 7.17

Variation of gearbox torque components vs time for optimum counterbalance conditions of Example 7.2.

7.3.3.2 The effect of load belt stretch

7.3.3.2.1 Calculation model

Up to this point we assumed that the load **belt** connecting the weight box to the polished rod was **rigid** and did not stretch. It follows from this assumption that the polished rod's position **exactly** defines the position of the carriage assembly, *CP*, which, in turn, defines the sprocket angle θ. Since all kinematic parameters are functions of the sprocket angle, gearbox torques can be determined if polished rod loads at different positions are available. When computerized dynamometers are used to analyze pumping conditions, then positions and loads are measured at constant time **intervals** and calculation of gearbox torques is a straightforward procedure.

For a **rigid** load belt case, sprocket angle is the **sole** function of time; calculated angles for each measured time step provide the basis for the calculation of the unit's **kinematic** parameters as well as for the torque **loading** on the gearbox. Sprocket angles can be found from the angular velocity of the sprocket as follows:

$$\theta_{\text{rigid}} = \frac{d\theta}{dt} t = \frac{N\,\theta_{\max}}{60} t \tag{7.23}$$

where:

θ_{rigid} = sprocket angles for a rigid load belt, radians,
$\theta_{\max}$ = sprocket angle at the end of the downstroke, radians,
N = pumping speed, 1/min, and
t = time, measured from the start of the upstroke, s.

In case the load belt's **elasticity** is also considered, then the belt **stretches** due to the polished rod load; belt stretch, Δl, is found from **Hooke's law**:

$$\Delta l = \frac{F_{PR}\,L_{\text{belt}}}{A_{\text{belt}}\,E} \tag{7.24}$$

where:

Δl = load belt stretch, in,
F_{PR} = measured polished rod load, lb,
L_{belt} = length of load belt, in,
A_{belt} = cross-sectional area of load belt, in^2, and
E = Young's modulus of the belt material, psi.

Now the **position** of the carriage assembly, *CP*, at any time, as compared to the case of a rigid belt, will **lead** the position valid for a rigid belt by exactly the calculated belt stretch, Δl. This position difference can be converted into sprocket angle **lead**, measured in radians, by the next formula [16].

$$\Delta\theta = \frac{\Delta l}{R} \tag{7.25}$$

where:

$\Delta\theta$ = sprocket angle lead, radians,
Δl = stretch of the load belt, in, and
R = sprocket radius, in.

The sprocket angles corresponding to the same polished rod positions for an **elastic** belt are **greater** than those for a rigid belt because of the belt **stretch** just described; they can be found from:

$$\theta_{\text{elastic}} = \theta_{\text{rigid}} + \Delta\theta \tag{7.26}$$

If sprocket angles are calculated from this formula for each measured time step, then the operation of the Rotaflex unit with an elastic load belt is properly described. The kinematic parameters calculated for these angles combined with the measured polished rod loads define the torque loading of the gearbox. In summary, the steps of gearbox torque calculations, including the effect of the load belt's stretch, can be listed as follows:

1. Based on dynamometer measurements, starting from the beginning of the polished rod's upstroke, sprocket angles, θ_{rigid}, for the measured times are calculated assuming a rigid load belt; see Eq. (7.23).
2. Using **Hooke's law** and the measured polished rod loads, the stretch of the load belt for each time step is found from Eq. (7.24).
3. Based on the calculated belt stretches, sprocket angles for the elastic belt case are determined from Eq. (7.26).
4. The determination of the necessary kinematic parameters, torque factors and polished rod accelerations, is done for each time step, based on the sprocket angles valid for the elastic belt case, $\theta_{elastic}$.
5. The different components of gearbox torque are calculated.

EXAMPLE 7.3: CALCULATE THE GEARBOX TORQUES AND THE PERMISSIBLE LOADS FOR OPTIMUM COUNTERBALANCE CONDITIONS GIVEN IN EXAMPLE 7.2 IF BELT STRETCH IS ALSO CONSIDERED. THE LENGTH OF THE BELT IS 466 IN, ITS CROSS-SECTIONAL AREA IS 21.87 IN^2. USE A YOUNG MODULUS OF 88,000 PSI FOR THE BELT MATERIAL

Solution

First calculate the stretch of the load belt from Eq. (7.24) based on the measured polished rod load:

$$\Delta l = 32,912 \times 466/(21.87 \times 88,000) = 7.97 \text{ in.}$$

The sprocket angle lead is found from Eq. (7.25):

$$\Delta\theta = 7.97/16.8 = 0.474 \text{ rad.}$$

The sprocket angle for the elastic case is, according to Eq. (7.26):

$$\theta_{elastic} = 4.855 + 0.474 = 5.329 \text{ rad.}$$

The calculation of the kinematic parameters follows. Carriage position, CP, belonging to the sprocket angle just calculated using Eq. (7.3):

$$CP = 16.8 + 16.8\,(5.329 - \pi/2) = 79.93 \text{ in.}$$

Acceleration is zero in this range; see Table 7.1; $a = 0$.
The torque factor from Table 7.1 is:

$$TF = R = 16.8 \text{ in.}$$

When calculating rod torque, only the weight of the load belt on the well side is different because of the different carriage position; it was found as $W_{bPR} = 1{,}247$ lb. Net weight on the load belt is found from Eq. (7.15) as:

$$F_c = 32,912 + 1,247 = 34,159 \text{ lb.}$$

Rod torque from Eq. (7.14):

$$T_r(5.329) = TF \times F_c = 16.8 \times 34,159 = 573,871 \text{ in-lb.}$$

The weight of the load belt on the counterbalance side was found as $W_{bCB} = 560$ lb; the total weight of the counterbalance components is calculated from Eq. (7.17):

$$W_{CB} = W_{WB} + W_{aux} + W_{bCB} = 9,800 + 15,000 + 560 = 25,360 \text{ lb.}$$

The counterbalance torque can now be calculated from Eq. (7.16):

$$T_{CB}(5.329) = -TF \times W_{CB} = -16.8 \times 25,360 = -426,048 \text{ in-lb.}$$

Since sprocket acceleration at the given position is zero, inertial torque is also zero. The net torque acting on the gearbox is the sum of the rod and counterbalance torques:

$$T_{net}(5.329) = 573,871 - 426,048 = 147,823 \text{ in-lb.}$$

The permissible load is found from Eq. (7.21) using the gearbox rating of $T_{al} = 320{,}000$ in-lb:

$$F_p(5.329) = T_{al}/TF + W_{CB} = 320,000/16.8 + 25,360 = 44,407 \text{ lb.}$$

The variation of net gearbox torque over the pumping cycle is shown in Fig. 7.18. The figure also contains the **difference** in net torques between the **rigid** and the **elastic** cases. As seen, for most of the polished rod's stroke length, there is a **negligible** amount of difference between the net torques of the two cases. Significant changes can only be expected around the two extremes of the stroke.

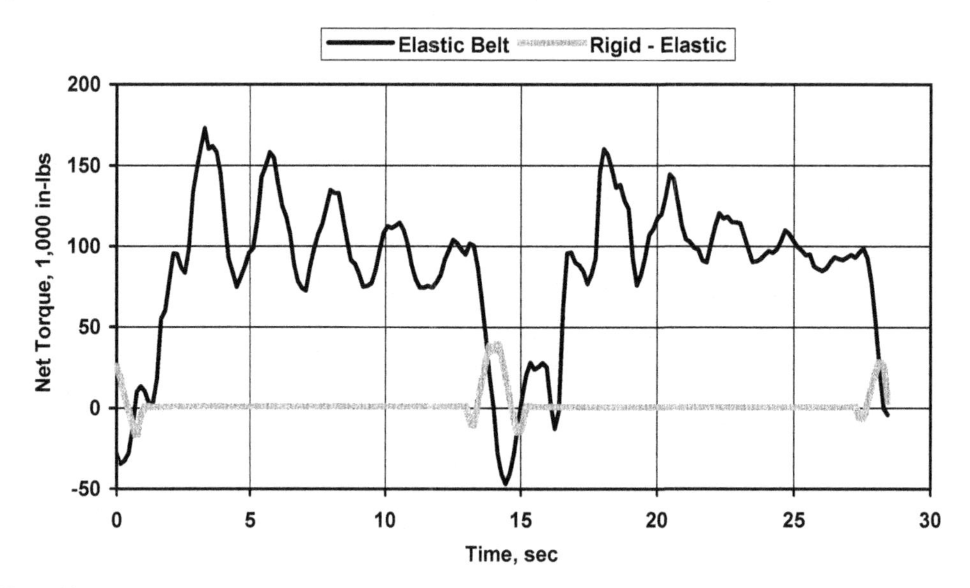

FIGURE 7.18

Variation of net gearbox torque, including the elasticity of the load belt and the difference between rigid and elastic cases vs time, for Example 7.3.

7.3.3.2.2 Conclusions

The investigation of the effect of load belt **stretch** on the net torque developed by a Rotaflex pumping unit resulted in the following conclusions:

- The effect of load belt stretch is negligible in those ranges where the carriage assembly moves on a vertical line, i.e., for the greatest part of the polished rod's stroke length.

- Belt stretch affects the net torque only in limited ranges of the sprocket angle around the two extremes of the polished rod stroke.
- Ignoring the elastic behavior of the load belt does not introduce great errors in torsional load calculations.

7.3.4 OTHER FEATURES

7.3.4.1 Available models, pumping capacities

Rotaflex units are presently manufactured in four sizes with a minimum stroke length of 288 in. Maximum stroke length is 366 in while maximum polished rod capacity is 50,000 lb. Other operational data are contained in Table 7.2.

Table 7.2 Main Technical Data of Available Rotaflex Pumping Units

Model	900	1100	1150	1151
Torque rating, 1,000 in-lb	320	320	320	420
Max. polished rod load, lb	36,000	50,000	50,000	50,000
Stroke length, in	288	306	366	366
Max. speed, SPM	4.50	4.30	3.64	3.75
Structural height, ft	40.5	44.5	49.5	49.5

Maximum possible liquid production capacities of the units are plotted in Fig. 7.19 according to **Beck** [17]. The values were calculated based on the use of oversize downhole pumps of up to 5¾ in

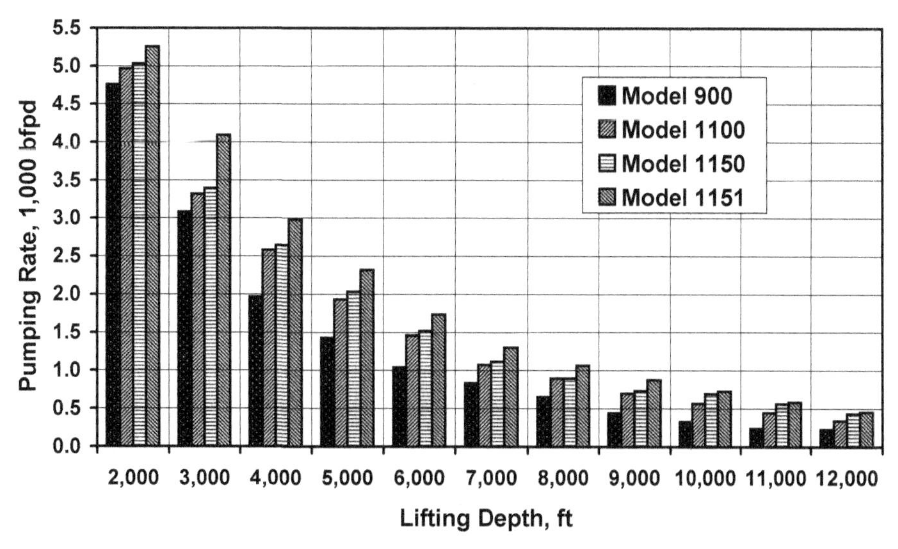

FIGURE 7.19

Production capacity of Rotaflex pumping units vs lifting depth according to **Beck** [17].

diameter and the units' maximum pumping speeds; the rates represent those that do not overload the rod string, the gearbox, or the pumping unit structure.

7.3.4.2 Case studies

From the time the first Rotaflex units were installed in the late 1980s, many papers had been published that compared this new technology to conventional rod-pumping installations. In the following a short review of the most important case studies is given.

Operational conditions of a conventional **C-640D-365-168** unit and a **Model 900** Rotaflex unit with a stroke length of 288 in using a 320 size gearbox were compared on the same well [18]. Production was about 1,000 blpd from 3,000 ft; the conventional unit ran at 10 SPM, while the Rotaflex ran at 3.3 SPM. Several electric motors were tried on the Rotaflex unit and the best solution was found with a **NEMA B** motor that gave a total system efficiency of 55%, as compared to 41% of the conventional installation.

Several studies investigated the use of Rotaflex units in the Permian Basin [19,20] and a detailed comparison of three beam pumping and two Rotaflex units was presented. Well depths were around 4,800 ft and production rates about 750 blpd. The air-balanced beam units had size 640 gearboxes while the Model 900 Rotaflex units had size 320 only. Average system efficiency for the beam units was 46%, but the Rotaflex units reached 60%.

Based on operational data of 13 wells on Rotaflex units [21] in a New Mexico oilfield, 50% electrical cost reduction was attained, as compared to the use of ESP equipment. The wells were about 4,400 ft deep and produced liquid rates between 400 and 600 blpd. A system efficiency increase of 4–5% was observed when NEMA C motors were used instead of NEMA D types.

Experience with Rotaflex units in Argentina [22] resulted in an average system efficiency of 70% for three wells; this extremely high effectiveness was credited to the use of variable speed drive units. The efficiencies of other artificial lift methods were 28.5% and 46.4% for ESP and beam-pumping installations, respectively.

7.4 THE DYNAPUMP UNIT

The **DynaPump** unit, a computer-controlled hydraulically driven long-stroke sucker-rod pumping unit, was invented by **A. Rosman** at the end of the 1980s [23,24]. Since its commercial introduction to the oilfield in 2001 the technology underwent several improvements that have led to the present-day models. Just like the Rotaflex units, DynaPump units can also compete with traditional rod-pumping and ESP installations by offering greater system efficiencies and lower production costs.

Due to their advantageous features, DynaPump units are getting popular; the worldwide number of installations reached 500 in the year 2008 [25].

7.4.1 CONSTRUCTION AND OPERATION

The two main components of a DynaPump installation are (1) the **pumping** unit and (2) the hydraulic **power** unit; the first converts the hydraulic power to lift the well load, and the second provides the power fluid of a controlled volumetric rate to drive the pumping unit. A schematic drawing is given in Fig. 7.20 that depicts the main parts of the system. The power and the pumping units are connected by high-pressure hoses not shown in the figure.

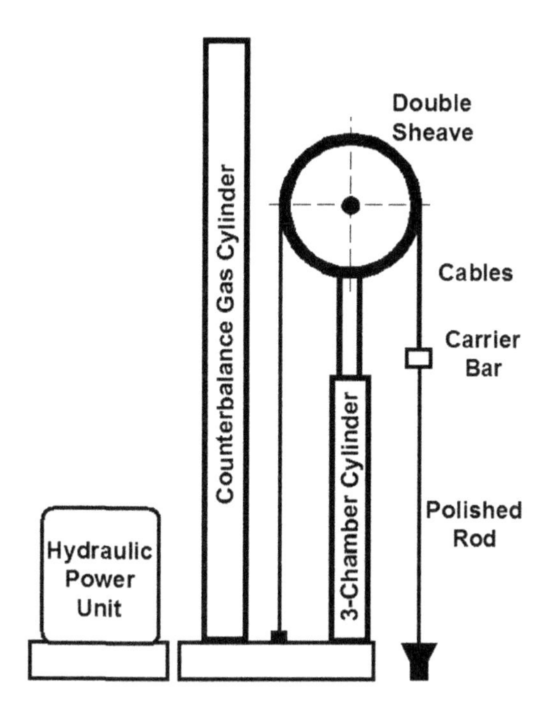

FIGURE 7.20

The main components of a DynaPump system.

7.4.1.1 The pumping unit

The pumping unit of the DynaPump system drives a traditional downhole pump attached to the sucker-rod string. The polished rod is lifted by the usual carrier bar attached to two wireline cables. The other ends of the cables are fixed to the unit base and the cables run on two **sheaves** to form a 2:1 **pulley** system. The sheaves are situated at the top of a plunger that protrudes from a special hydraulic **three-chamber cylinder**, vertical position of the sheaves being the function of the hydraulic fluid rate received from the power unit. To counterbalance the variation of the well load during the pumping cycle, **gas** pressure is utilized that also acts in the three-chamber cylinder, gas pressure being provided by a high-volume gas (usually nitrogen) storage cylinder. Because of the pulley system, the **displacement** of the polished rod is exactly **twice** as much as the vertical movement of the sheaves caused by the three-chamber cylinder. For the same reason, the **load** on the cylinder at any time equals **twice** the polished rod load.

As seen from this elementary description, the **three-chamber cylinder** is the heart of the Dyna-Pump's mechanical operation; the construction and operation of this ingenious device is described in the following.

The three-chamber cylinder (see Fig. 7.21) contains three **concentric** cylinders, two of which are stationary; the moveable one, designated as "**plunger**," provides the vertical movement of the unit's sheaves and thereby lifts the polished rod [23]. An annular **piston** is fixed to the plunger that has seals both on the inside of the outer cylinder and on the outside of the inner cylinder. This arrangement of the

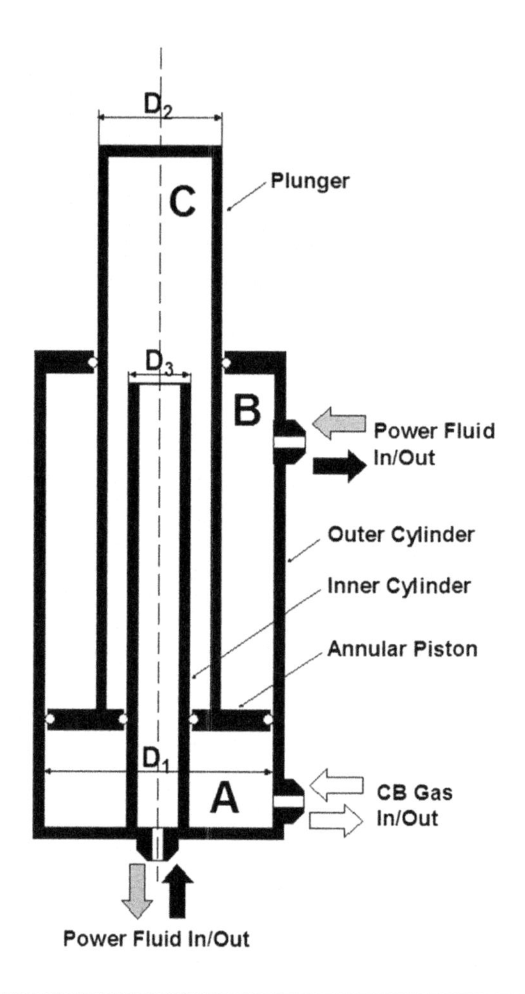

FIGURE 7.21

Construction details of the three-chamber cylinder in a DynaPump unit.

cylinders creates the three sealed **chambers A**, **B,** and **C**; the power fluid driving the pumping unit and the gas pressure providing the counterbalance are connected to those. As shown in the figure, gas pressure is connected below the annular piston in chamber **A** and power fluid provided by the power unit is connected to chambers **B** and **C**.

During the **upstroke** of the pumping unit, power fluid is pumped in chamber **C** (see the black arrow) and the plunger moves upward, lifting the polished rod. However, **part** of the well load is supported by the counterbalance **gas pressure** acting on the annular piston's area; this effect decreases the **force** needed for pumping. As the plunger lifts during the upstroke, the volume of chamber **B** decreases and power fluid is forced out of this chamber (see the black arrow). When the downstroke starts, the flow direction of the power fluid is **reversed** (see the gray arrows); fluid is pumped in chamber **B** and fluid is displaced from chamber **C**. Gas pressure, as before, counterbalances the forces acting on the plunger.

It can be seen from the previous discussion that the **reciprocating** motion of the unit's plunger and that of the polished rod is controlled by two parameters of the power fluid: (1) flow **direction** and (2) flow **rate**. The change of flow direction **changes** the direction of the **movement** of the unit's plunger from the upstroke to the downstroke and vice versa, as indicated by the black and gray arrows in Fig. 7.21. The flow **rate** of the power fluid, on the other hand, determines the **speed** of the plunger; the speed of the polished rod will be **twice** that value because of the pulley system used. Since the flow direction and volumetric rate of the power fluid are controlled by a **computer** situated in the DynaPump power unit, the **kinematic** behavior of the driving mechanism can be **changed** in a broad range, as will be discussed later.

During normal operation, chamber **A** of the three-chamber cylinder is connected to the **counter-balance** gas cylinder (see Fig. 7.20) that contains nitrogen gas at a constant pressure. To compensate for environmental temperature variations, the gas storage cylinder has ample volume, usually 10 times the maximum volume of chamber **A**. Gas pressure regulated by the DynaPump's control unit acts on the area of the annular piston so that the force thus created shall equal the weight of the sucker-rod string plus part of the fluid load on the downhole pump. This force is transmitted to the plunger of the three-chamber cylinder during both the up- and the downstroke and it greatly reduces the forces to be developed by the power fluid during a complete pumping cycle. It must be noted that the active areas available for the power fluid in both chambers **B** and **C** are relatively smaller than the area of the annular piston, so the required volume of power fluid to achieve a given stroke length of the plunger is relatively small. This is a greatly beneficial feature of the three-chamber cylinder that results in low hydraulic power requirements and low power costs; the high power efficiency of the DynaPump unit is a direct consequence of this ingenious operational principle.

The DynaPump uses a **closed** hydraulic system to supply power to the pumping unit so the fluid volume **pumped** into chamber **C** must be **equal** to the volume simultaneously **displaced** from chamber **B**. Assuming a Δx vertical displacement of the plunger, the increase of volume in chamber **C** is found as:

$$\Delta V = \Delta x \frac{D_3^2 \pi}{4} \tag{7.27}$$

At the same time, the fluid volume displaced from chamber **B** equals:

$$\Delta V = \Delta x \frac{\left(D_1^2 - D_2^2\right)\pi}{4} \tag{7.28}$$

Since the two volumes must be equal, the **relationship** between the diameters of the three cylinders is described by the following formula:

$$D_3 = \sqrt{D_1^2 - D_2^2} \tag{7.29}$$

where:

D_3 = outside diameter of the inner cylinder, in,
D_1 = inside diameter of the outer cylinder, in, and
D_2 = outside diameter of the plunger, in.

The formula allows one to select the proper **combination** of cylinder sizes. Normally the plunger size (D_2) is fixed for a given pumping unit size and Eq. (7.29) provides a way to choose the proper diameters of the other cylinders. In practice, however, the cross-sectional area of chamber **C** is usually selected to be smaller than that of chamber **B** by 5–30%. This modification provides a higher upward speed of the three-chamber cylinder for the same flow rate and also helps to filter and cool the oil in the closed loop circuit.

7.4.1.2 The power unit

The DynaPump power unit is normally driven by electric motors (there are also gas-driven units using the well's casinghead gas), and the high-pressure pumps the motors drive provide the hydraulic power required to operate the pumping unit. The main task of the power unit is to create the necessary volume and direction of the power fluid as well as the following:

- to control the operation of the hydraulic pumps,
- to provide the required speed, acceleration, and deceleration of the polished rod,
- to cool and filter the power fluid during normal operation,
- to monitor the performance of the system,
- to communicate with external devices for remote operation, and
- to provide pump-off control capabilities.

The power unit is a self-contained, complex device, the most significant feature of which is that it **incorporates** its own variable frequency drive (VFD). The main components of the power unit of a usual DynaPump installation are the following:

- A programmable logic controller (PLC) that controls the operation of the motor driving the main pump and performs monitoring and safety functions as well.
- A VFD that provides electrical frequencies that are changed during the pumping cycle and are fed to the motor driving the hydraulic pump.
- An induction-type electric motor connected to the main pump of the unit; its speed changes with the electrical frequency driving the motor.
- Power fluid is provided by a high-pressure pump, the output rate of which is proportional to motor speed but independent of the discharge pressure because the pump used is of the positive displacement type. Pump discharge pressure is created by the load on the three-chamber cylinder acting on the area of chambers **B** or **C** and varies during the pumping cycle in sync with actual polished rod loads.
- The necessary hydraulic piping and valves as well as electrical wiring.

Alternatively to the configuration described here, DynaPumps using **swash-plate** hydraulic pumps driven by constant-speed electric motors or by gas engines using casinghead gas are also available. In these cases the necessary variable power fluid rate, direction, accelerations, and decelerations are created by controlling the swash-plate position at the constant rotational speed provided by the motor.

7.4.1.3 Available models, pumping capacities

DynaPump units are classified according to the **size** (in inches) of the three-chamber cylinder's **plunger**; the models are named accordingly. Main technical data of available models are contained in Table 7.3. The units have two primary **ratings** related to their lift capacity:

- Maximum polished rod load refers to the maximum **structural** load that can be sustained for a design life beyond 30 years. The biggest units can handle polished rod loads up to 80,000 lb that are never attainable by beam-type pumping units.

Table 7.3 Main Technical Data of Available DynaPump Units

Model	5	7	9	11	13
Unit plunger size, in	5.0	7.0	9.0	11.0	13.0
Max. hydraulic pressure, psi	1,800	1,800	1,800	1,800	1,800
Max. counterbalance gas pressure, psi	1,000	1,000	1,000	1,000	1,000
Max. polished rod load, lb	15,000	25,000	40,000	60,000	80,000
Max. stroke length, in	168	240	288	336	360
Max. speed, SPM	6.8	4.8	4.0	3.4	3.0
Structural height, ft	23	28	34	39	41

- Maximum **hydraulic** load (analogous to the torque rating of a beam-pumping unit) depends on two parameters: (1) the maximum hydraulic pressure in the three-chamber cylinder, and (2) the maximum counterbalance gas pressure. Those two pressures acting on their respective areas (see Fig. 7.21) give the maximum lifting force achievable on the three-chamber cylinder's plunger; polished rod loads half of this value will overload the unit.

As with beam-type pumping units, under different well conditions either the maximum structural or the maximum hydraulic load is reached first when selection of a bigger-capacity unit is needed. Maximum allowable pumping **speeds** (SPMs) are based on the maximum allowed speed of the three-chamber cylinder's plunger, limited to 200 ft/min. This converts to a polished rod speed of 400 ft/min or 6.7 ft/s, which is comparable to the polished rod velocities of bigger air-balanced sucker-rod pumping units.

Power units come in seven different **power** capacities in the range of 15–150 HP. Different powers mean different power fluid **rates** available for pumping, which permit different polished rod velocities and accelerations. Each power unit can be used with several pumping units, but bigger models usually require more power; proper selection of the combination of power and pumping units is important.

Approximate maximum liquid production capacities of **DynaPump** units are shown in Fig. 7.22 according to **Rosman** [26]. The figure gives pumping rates for the largest (Model 13) unit using oversize (maximum 5¾ in) pumps and normal pumping speeds.

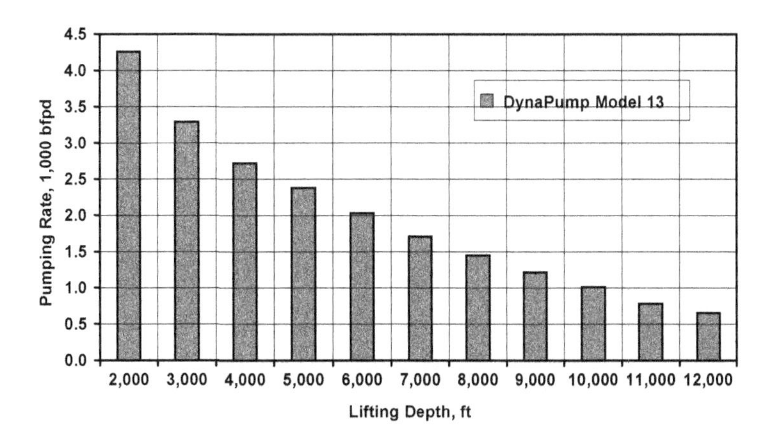

FIGURE 7.22

Approximate maximum liquid production capacities of DynaPump units vs lifting depth according to **Rosman** [26].

7.4.2 KINEMATIC BEHAVIOR

In order to properly describe the kinematic behavior of the DynaPump unit, one has to understand how the polished rod's movement varies with the electrical frequency provided by the unit's variable frequency drive. This can be found if the operating conditions of the drive train, consisting of (1) the VFD unit, (2) the electric motor, (3) the hydraulic pump, (4) the three-chamber cylinder, and (5) the polished rod, are identified. During the pumping cycle the VFD unit alters the electrical frequency according to a schedule programmed into the PLC that controls the operation of the DynaPump unit. Since induction motors are used, motor speed changes proportionally with the frequency provided by the VFD unit. The pump is usually a positive displacement pump whose output fluid rate directly changes with the speed of the motor driving it. Thus, the volumetric rate of the power fluid received in the three-chamber cylinder will follow the variation of the frequency provided by the VFD unit.

As discussed in conjunction with Fig. 7.21, the three-chamber cylinder's plunger ascends/descends as power fluid is alternatively pumped in its chambers **C** or **B**. Plunger movement is proportional to the fluid volume pumped in the given chamber, so a constant pumping rate results in a constant velocity of the plunger. This implies that the velocity of the three-chamber cylinder linearly changes with the electrical frequency driving the unit's main prime mover. Further, as discussed earlier (see Fig. 7.20), the velocity of the polished rod is twice the velocity of the plunger because of the pulley system utilized. The final conclusion is that polished rod velocity is a direct function of the frequency output of the VFD unit. Therefore, in the following discussions the polished rod's kinematic parameters (position, velocity, and acceleration) will be used to describe the kinematics of the DynaPump.

In contrast to beam-pumping units, DynaPump units allow the operator to select the variation of most of the kinematic parameters during the pumping cycle; these can be input to the unit's controller and are discussed in the following [27,28].

Up- and downstroke velocities. The DynaPump unit provides **constant** polished rod velocities for long **portions** of both the up- and downstroke periods; this feature greatly decreases dynamic forces and reduces energy requirements for most of the stroke. The operator can choose from **eight** different motor speeds in the range of 500–2,200 RPM that provide the same number of constant polished rod velocities. Upstroke and downstroke speeds are selected **independently** of each other. Normally the upstroke speed chosen is higher than the downstroke speed; thus the period of the polished rod's upstroke is shorter than that of the downstroke. This feature of the unit has several advantages, to be detailed later.

Acceleration and deceleration periods. Both the upstroke and the downstroke of the polished rod's movement includes periods of (1) acceleration, (2) constant speed, and (4) deceleration. The lengths of the acceleration and deceleration periods (four in total) can be **independently** selected on the DynaPump, allowing an excellent control of the unit's dynamics at the stroke reversals.

Starts of decelerations. The starting point of deceleration is **selectable** for both the up- and the downstroke. The so-called "switches" are set based on the position of the three-chamber cylinder's plunger, which is continuously monitored and easily translated to polished rod position, as used in the following.

After the parameters just described are set, the polished rod stroke length and the pumping speed are automatically established. Polished rod stroke **length** is determined by the "switch" settings, i.e., the positions in the upstroke and the downstroke where deceleration starts, and the selected up- and downstroke velocities. Acceleration and deceleration periods have a minor effect only. Polished rod stroke length and pumping speed are **interrelated**; pumping **speed** in SPM units is found from the calculated length of the pumping cycle.

It is important to understand the basic **difference** of the DynaPump's and the beam-type pumping unit's kinematic behavior. As discussed previously, the polished rod's velocity and acceleration pattern for a DynaPump can be **freely** and independently chosen; this is made possible by the very low **inertia** of the system. The low inertia is the result of the relatively small total **mass** of the DynaPump unit's moving components and the **weightless** counterbalance (see Fig. 7.20); that is why the unit can stop or start its movement instantly. Beam-type pumping units, in contract, have **great** inertias because of the heavy machinery and cannot easily change their velocities. In addition to the effects of the inertia, the kinematic parameters (velocity, acceleration) of beam-type units are directly **pre-determined** by the pumping unit's geometrical dimensions, as detailed in Section 3.7.5 of this book. For a given beam unit, therefore, the velocity and acceleration pattern of the polished rod's movement cannot be controlled, while it is easily done for a DynaPump. The same difference is found between the Rotaflex and the DynaPump unit due to the great inertia of the Rotaflex equipment. This advantageous feature of the DynaPump greatly reduces the **dynamic** loads during the pumping cycle, as well as decreases the unit's power requirements.

For illustrating the kinematic behavior of DynaPump installations a **Model 9** unit will be used with the following settings:

Upstroke velocity, ft/s	5.00	Acceleration time upstroke, s	2.0
Downstroke velocity, ft/s	3.33	Acceleration time downstroke, s	1.0
Top switch, in	228	Deceleration time upstroke, s	2.0
Bottom switch, in	20	Deceleration time downstroke, s	1.0

Under these conditions the polished rod stroke length is calculated as 288 in, and the pumping speed is found as 4 SPM. The unit's velocity and acceleration pattern during the pumping cycle is depicted in Fig. 7.23. As shown, big portions of both the up- and downstroke are performed at **constant** polished rod velocities and the accelerations at the top and bottom stroke reversals are fully controlled by the operator for an optimum dynamic performance.

The acceleration and velocity profiles of the polished rod are completely **defined** by the **input** variables listed earlier; polished rod position, therefore, is found by **integration** of the velocity with respect to time. For the given case, calculated positions during the pumping cycle are given in Fig. 7.24; the figure reveals the different phases of the movement of the unit. From point **1** to **2** during the upstroke acceleration period the polished rod reaches the set upstroke speed, then movement with a constant velocity follows from point **2** until the top switch position is reached at point **3**. Now the deceleration period comes until point **4,** where the polished rod's travel reaches its maximum. Downstroke is starting next, with the downstroke acceleration period from point **4** to **5**. From here, downward travel continues with a constant velocity; note that this velocity is less than that during the upstroke. Finally, from point **6,** belonging to the bottom switch position, the unit slows down to reach the bottom of the downstroke at point **7**.

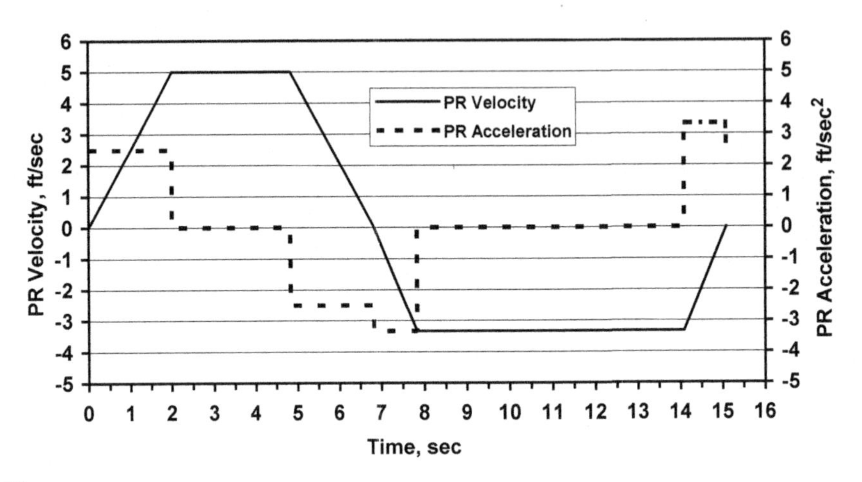

FIGURE 7.23

Example polished rod velocity and acceleration profiles of a Model 9 DynaPump unit.

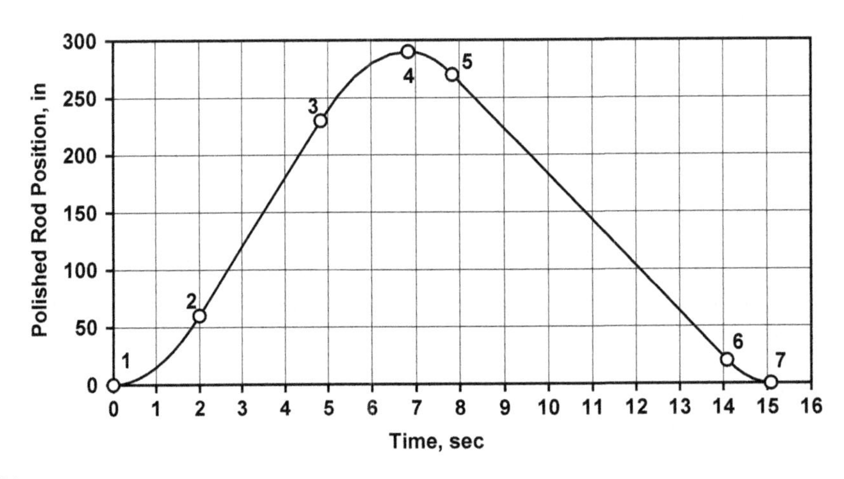

FIGURE 7.24

Example polished rod position vs time for a Model 9 DynaPump unit.

Investigation of the shape of the polished rod position versus time curve reveals the following characteristic features of the DynaPump unit:

- The period for the upstroke is **shorter** than that of the downstroke. This condition is beneficial in reducing the volume of fluid lost due to pump **leakage** in worn downhole pumps, especially in deep wells. Since leakage rate through the clearance between the pump barrel and plunger is constant at given conditions, reduction of the time spent on the upstroke definitely decreases the volume of fluid lost and thereby increases the production rate of the pump.

- Polished rod velocity for the downstroke is usually selected to be **lower** than for the upstroke. This feature has two advantageous consequences:
  - The **buckling** tendency of the sucker-rod string is greatly reduced. Buckling is caused by **compressive** forces generated in the rods during the downstroke and those forces greatly increase at higher rod velocities. The slower downstroke velocities of the DynaPump, therefore, can almost completely eliminate rod buckling because compressive forces in the rod string are much reduced.
  - Due to the lower downward velocities of the rods, viscous **friction** (damping) forces along the string are reduced.
- The polished rod stroke length (within physical limits) is infinitely **selectable**, i.e., it can be set at any value in its possible range. This kind of operation is made possible by the proper selection of the input parameters: (1) up- and downstroke velocities, (2) acceleration and deceleration periods, and (3) "switch" points. Thus the DynaPump's lifting **capacity** can easily be modified to follow the changes in the well's inflow performance.

The interrelationship of polished rod stroke length and pumping speed is illustrated in Fig. 7.25, which contains calculated polished rod positions for the **Model 9** unit investigated before. The three curves presented were found for different combinations of up- and downstroke velocities while acceleration/deceleration periods as well as the starts of decelerations (switches) were held constant. As shown, lower velocity settings **increase** the up- and downstroke periods (see the elapsed times between points **2** and **3** as well as **5** and **6** in Fig. 7.24) and hence the total pumping

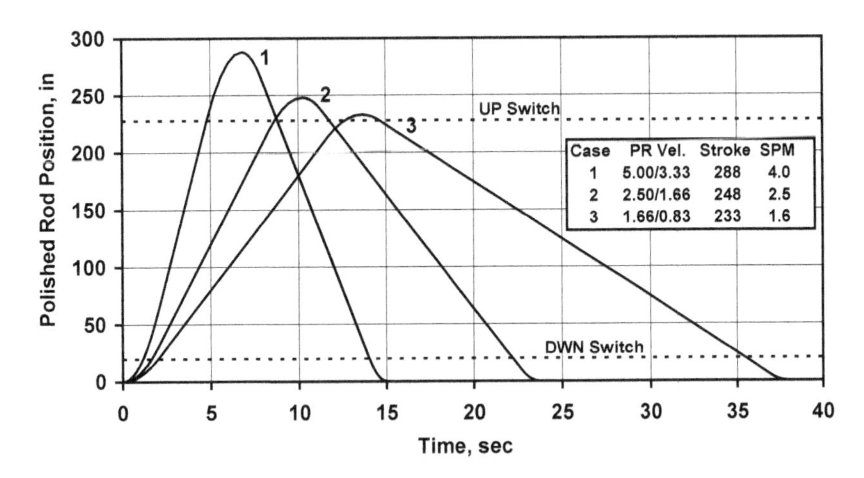

FIGURE 7.25

Interrelationship of polished rod stroke length and pumping speed for a Model 9 DynaPump unit. PR Vel = polished rod velocity.

cycle. Pumping speed (being the number of pumping cycles per minute) will, of course, decrease for longer cycles. Polished rod stroke length also changes for the different velocities selected, but its decrease is not as significant as that of the pumping speed; a desired stroke length could be maintained by increasing the value of the upstroke switch. The conclusion of this discussion is that the lifting **capacity** of DynaPump units (which is a function of the stroke length and the pumping speed) can be very **easily** changed by simply changing the input parameters to the control unit. This enables the operator to easily follow any changes in well inflow parameters that may happen with time.

7.4.3 OPERATIONAL FEATURES

7.4.3.1 Counterbalancing

As described in conjunction with Fig. 7.21, the DynaPump unit is counterbalanced by using high-pressure nitrogen **gas** contained in the counterbalance cylinders and acting on the annular piston's area in chamber **A** of the three-chamber cylinder. The load resulting from the gas pressure acting on the piston **helps** lift the polished rod during the upstroke and thus reduces the energy required to operate the system. During the downstroke, when the rod string falls and the pumping unit becomes the driven component, the force arising on the annular piston due to gas pressure again **opposes** the high loads and helps the operation of the unit. The **approximate** value of the required counterbalance effect, just like in beam-type rod pumping, equals the buoyant rod string weight plus half of the fluid load on the downhole pump's plunger. The gas pressure that corresponds to this value must be properly maintained in the gas cylinders during the complete pumping cycle.

The proper way to find **optimum** counterbalance conditions is based on the measurement of the electrical **current** taken by the DynaPump unit's motor, and **peak** currents during the up- and downstroke must be approximately equal. Monitoring of counterbalance gas **pressures** also provides the possibility to check the unit's balance conditions: up- and downstroke peak pressures should be in a $\pm 10\%$ range. Since **ambient** temperature affects the gas pressure in the counterbalance cylinders, pressure readings will be different under different **climatic** conditions. To minimize such effects, the storage volume of the counterbalance cylinders is at least 10 times greater than the volume of chamber **A** of the three-chamber cylinder.

The main **advantage** of using nitrogen gas for counterbalancing the DynaPump unit is the **easy** adjustment of counterbalance conditions; one only needs to add or release nitrogen to/from the counterbalance cylinders. This is a great improvement compared to the tedious procedure to be followed on beam-type pumping units.

7.4.3.2 Pump-off control

The DynaPump's computer-controlled operation permits an easy application of pump-off control on wells with low or decreasing inflow rates. A **pump-off** condition, as discussed in Section 5.3.2 of this book, occurs when the downhole pump's capacity **exceeds** the **inflow** from the well. Detection of this condition in DynaPump installations is facilitated by checking the **dynamometer cards** automatically recorded by the unit on every stroke. Using a load set point on the downstroke that can be adjusted as required, pump-off is indicated every time the measured polished rod load is more than the set value.

As soon as pump-off is identified, the power unit's computer (the PLC) sends a signal to the VFD unit to **decrease** its output **frequencies** for the downstroke. In response to the lower frequencies, the electric motor slows down, causing the hydraulic pump's fluid **rate** and the polished rod's downstroke **velocity** to decrease. All other parameters (acceleration, deceleration periods, the starts of decelerations) being unaltered, the kinematic behavior of the DynaPump unit will change just as was described in conjunction with Fig. 7.25. The unit's pumping speed **slows** down, resulting in a reduction of the system's liquid production capacity; the whole process is repeated until pump-off is eliminated. The computer will check at every stroke if pumping could continue at the original speeds without the well being pumped off. This kind of automatic pump-off control has several **advantages** over conventional solutions:

- No extra equipment is needed; operation is controlled by the DynaPump unit itself.
- In contrast to many POC controllers, the pumping unit is not shut down when pump-off is detected.
- Because of the continuous operation of the installation allowed by the pump-off control mechanism, daily liquid production from the well increases. This is the result of a deeper average dynamic liquid level in the annulus and a consequently lower average flowing bottomhole pressure.

7.4.3.3 Case studies

One of the first papers on the DynaPumps [29] reports on a four-year experience with smaller units and shows that the computer-controlled operation of the unit greatly reduces production costs. A very detailed evaluation of DynaPump operational conditions on a single well [30] gave a system efficiency of 46%, comparable to beam-type sucker-rod pumping installations. The energy requirement in kWh/bbl units was close to that of Rotaflex units. One of the important conclusions presented in the paper showed that the DynaPump could be adjusted to stroke only in the **undamaged** part of the downhole pump. Since the barrel was sticking close to the top of its stroke, use of the DynaPump's special features eliminated pulling of the otherwise good pump.

A major operator in the Permian Basin started to use DynaPumps in 2002 and conducted experimental measurements on wells placed on different artificial lift methods [31]. Based on power measurements on four DynaPump, six beam pump, and six ESP installations operating under similar conditions, the results shown in Fig. 7.26 were obtained, where power consumption as a function of pumping rate is plotted. Power requirement per barrel of fluid produced was found as:

DynaPump	0.70 kWh/bbl
Beam pump	1.36 kWh/bbl
ESP	4.15 kWh/bbl

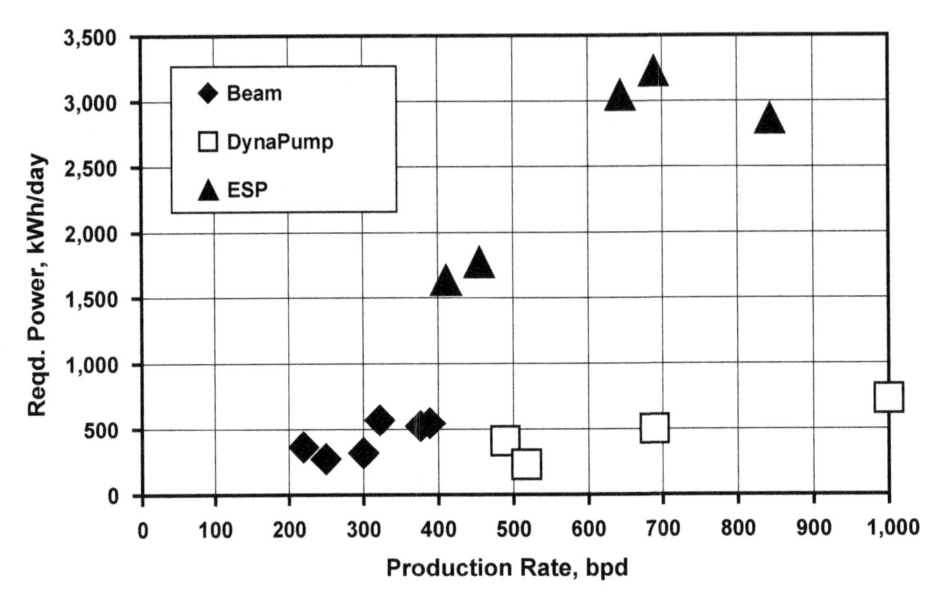

FIGURE 7.26

Power consumption vs pumping rate for different artificial lift methods according to **Tovar** [31].

As seen, the DynaPump unit consumed about **half** the power of a beam-pumping and about one-sixth of the power of an ESP installation under similar conditions. Field experience with several units also indicated their benefits in **deviated** wells; the number of downhole failures was **drastically** reduced in wells converted from both beam-pumping and ESP operations. Similar results were reported in [32].

7.5 CONCLUSIONS

After the detailed discussion of the current types of long-stroke sucker-rod pumping systems, this section **compares** their advantages and limitations to conventional rod-pumping installations and to each other. The most basic advantage of **any** type of long-stroke pumping is the increased liquid **production** capacity made possible by the long polished rod stroke length. Typical stroke lengths make the effect of rod stretch on plunger displacement negligible, especially in deep wells where in beam-pumping systems rod stretch can be a significant portion of the plunger's stroke length, reducing the lifting capacity of the system.

Another important advantage of long-stroke pumping is that severely **reduced** pumping speeds can be used because of the extremely long polished rod stroke lengths available. As previously shown, pumping speeds can be as low as a few SPM for both the Rotaflex and the DynaPump units. Such low speeds bring about the following inherent **advantages**:

- Reduced pumping speeds very significantly increase the **life** of the rod string and reduce the number of rod **failures** because sucker rods experience a much reduced number of fatigue cycles.
- Peak polished rod loads are much lower because **dynamic** loads diminish at low pumping speeds. Since maximum rod loads decrease along the rod string, rod-tubing **wear** is also decreasing due to the reduced side loads in deviated well sections.
- Minimum polished rod loads and rod loads during the **downstroke** increase. Although dynamic loads tend to decrease rod loads on the downstroke they are much less significant at lower pumping speeds so minimum rod loads increase. The higher minimum rod loads, especially in lower rod tapers, decrease the **buckling** tendency of the rods and increase the life of the rod string.

The power requirement of both the Rotaflex and the DynaPump units is significantly **less** than that of beam-pumping units or ESP installations producing the same liquid volume from the same depth. This is easy to explain if the driving **mechanisms** of the units are compared: in traditional beam-pumping operations the different components of the pumping unit as well as the counter-weights have accelerations of various magnitudes and directions during the pumping cycle. Rotaflex and DynaPump units, in contrast, have all loads, including counterbalance forces, acting along the **same** line and do not waste much energy on accelerating/decelerating different parts of the unit.

The production capacities of the biggest Rotaflex and DynaPump units are compared in Fig. 7.27; as shown, they can produce very **similar** liquid rates from the same pumping depths. Some other important operational features of the two different units are compared in the following.

Adjustment of Pumping Capacity:

- Rotaflex units permit changing of the pumping **speed** only, and because of the fixed polished rod stroke length the possible range of pumping rates is limited.
- Pumping capacity of DynaPump units can be changed in very **broad** ranges, made possible by an almost unlimited possibility of changing polished rod stroke lengths and pumping speeds at the same time.

Counterbalance Adjustment:

- Rotaflex units need to be stopped to change the unit's counterbalance conditions.
- DynaPump units must not be stopped to adjust their counterbalancing; change of counterbalance effect is simple by changing the gas **pressure** in the counterbalance cylinders.

Pumping Speed Adjustment:

- Speed changes on Rotaflex units require changing of V-belt **sheaves**; pumping speed is the **same** for the up- and the downstroke.

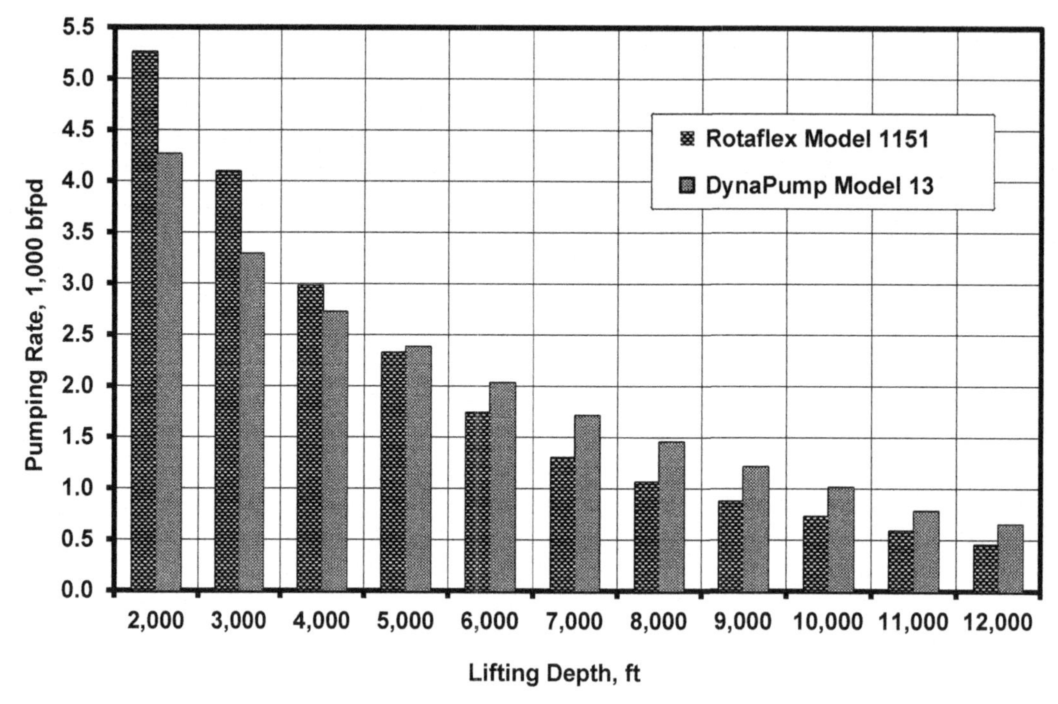

FIGURE 7.27

Comparison of the production capacities of the biggest Rotaflex and DynaPump units.

- Pumping speed of DynaPump units is **easily** changed on their controller, **different** up- and downstroke speeds can be set. Higher upstroke and lower downstroke speeds, as normally set, decrease leakage in the downhole pump and reduce buckling tendency of sucker rods.

  Pump-Off Control:

- Rotaflex units necessitate the use of **separate** pump-off controllers; most controllers shut down the unit for intermittent operation, thus reducing daily liquid rates.
- Pump-off operation is controlled by the DynaPump unit **itself**; no extra equipment is needed. The built-in variable frequency controller slows the unit down, reducing the system's liquid production capacity to match the inflow from the well. Since the unit must not be shut down, daily liquid production from the well increases.

Finally, Table 7.4 presents an interesting aspect of comparing **long-stroke** and **conventional** pumping units. The table contains main operational parameters like maximum polished rod load and stroke length of beam, DynaPump, and Rotaflex units of comparable capacities along with their approximate **weights**. Since the weight of machines made of steel is a good indicator of their **price**, investment costs of the three units are easily compared. According to the data contained in the table, beam and Rotaflex units are in the same weight range, while DynaPump units of comparable capacities have weights of about 50% less.

Table 7.4 Comparison of Main Features of Long-Stroke and Conventional Pumping Units

Conv. Beam Pumping Units				DynaPump Units				Rotaflex Units			
	Maximum Polished Rod				Maximum Polished Rod				Maximum Polished Rod		
Size	**Load (lb)**	**Stroke (in)**	**Unit Weight (lb)**	**Size**	**Load (lb)**	**Stroke (in)**	**Unit Weight (lb)**	**Size**	**Load (lb)**	**Stroke (in)**	**Unit Weight (lb)**
160	20,000	74	17,300	**5**	15,000	168	3,900				
228	21,300	120	26,800	**7**	25,000	240	7,400				
320	25,600	144	36,500								
456	36,500	120	40,000	**9**	40,000	288	18,600	**900**	36,000	288	42,900
640	36,500	120	42,500								
912	36,500	192	53,900								
1280	36,500	192	61,900					**1100**	50,000	306	52,300
1824	36,500	192	65,000					**1150**	50,000	366	53,900
				11	60,000	336	24,900				
				13	80,000	360	28,000				

REFERENCES

[1] Clegg JD. High rate sucker-rod pumping and its economics. Proc. 43rd southwestern petroleum short course. Lubbock, Texas; 1996. p. 53–67.

[2] Kuhns JP, Rizzone ML. Well-pumping apparatus. US Patent 3,285,081; 1966.

[3] Metters EW. A new concept in pumping unit technology. Paper SPE 3193 presented at the Hobbs petroleum technology symposium held in Hobbs, New Mexico, October 29–30, 1970.

[4] Gault RH. Longstroke pumping apparatus for oil wells. US Patent 4,076,218; 1978.

[5] Gault RH. The 40-foot stroke, Winch type pumping Unit. Proc. 24th Southwestern Petroleum short course. Lubbock, Texas; 1977. p. 153–7.

[6] Gault RH. Oil well pumping unit having traveling stuffing box. US Patent 3,640,342; 1972.

[7] Brinlee L. Operating experience with the Alpha I. Pet Eng Int August 1978:71–6.

[8] Tait HC. A rod pumping system for California lift requirements. Paper SPE 11747 presented at the California regional meeting held in Ventura, California, March 23–25, 1983.

[9] Tait HC, Hamilton RM. A rod pumping system to reduce lifting costs. J Pet Technol November 1984: 1971–8.

[10] Lang JM, Lively GR. Well pumping unit with counterweight. US Patent 4,719,811; 1988.

[11] Lively GR. Long stroke well pumping unit with carriage. US Patent 4,916,959; 1990.

[12] Rotaflex long stroke pumping units. Weatherford Artificial Lift Systems Brochure; 2002.

[13] Rotaflex 900-1100-1150-1151 pumping units. Installation, operation, maintenance. Weatherford International; 2007.

[14] McCoy JN, Rowlan OL, Podio AL, Becker D. Rotaflex efficiency and balancing. Paper SPE 67275 presented at the production and operations symposium held in Oklahoma city, March 24–27, 2001.

[15] Takacs G. Modern sucker-rod pumping. Tulsa, Oklahoma: PennWell Books; 1993.

[16] Takacs G, Chokshi R. Calculation of gearbox torques of rotaflex pumping units considering the elasticity of the load Belt. Paper SPE 152229 presented at the Latin American and Caribbean petroleum engineering conference held in Mexico city, April 16–18, 2012.

[17] Beck S. Technical consultation. Production Systems, Weatherford International; November 2013.

[18] Hicks A, Jackson A. Improved design for slow long stroke pumping units. Proc. 38th Southwestern petroleum short course. Lubbock, Texas; 1991. p. 289–308.

[19] Wiltse DJ. Long stroke pumping units. Paper 95-28 presented at the 46th annual technical meeting of the petroleum society of CIM in Banff, Canada. May 14–17, 1995.

[20] Adair RL, Dillingham D. Ultra long stroke pumping system lowers lifting costs. Petroleum Engineer International; July 1995. 42–7.

[21] Dollens DL, Becker RL, Adair, RL. Artificial lift optimizations: a case study of ultra long stroke pumping systems performance, enhanced with modified NEMA C motors. Paper SPE 37503 presented at the production operations symposium held in Oklahoma city, March 9–11, 1997.

[22] Antonioli M, Stocco A. Long stroke pumping system improves the energy efficiency of the production. Paper SPE 108122 presented at the Latin American and Caribbean petroleum engineering conference held in Buenos Aires, April 15–18, 2007.

[23] Rosman AH. Hydraulically operated lift mechanism. US Patent 4,801,126; 1989.

[24] Rosman AH. Oil-well pumping system or the like. US Patent 4,848,085; 1989.

[25] Huard W. Intelligent long stroke hydraulic pumping system reduces well intervention costs while maximizing energy efficiency. Paper presented at the 4th annual sucker-rod pumping workshop. Houston Texas, September 9–12, 2008.

[26] Rosman AH. Personal communication. November 2012.

[27] Rev. F DynaPump Operator's Manual. The Woodlands, Texas: DynaPump, Inc.; November 2011.
[28] Ver. 1.30 DynaPilot operations Manual. The Woodlands, Texas: DynaPump, Inc.; July 2006.
[29] Rosman AH, Nofal M. Computer controlled pump unit cuts power, increases output. 53–6. World Oil; November 1996.
[30] Pena AR, Rowlan OL. DynaPump field evaluation. Proc. 50th Southwestern petroleum short course. Lubbock, Texas; 2003. p. 148–59.
[31] Tovar S. DynaPump project update. Proc. 51st Southwestern petroleum short course. Lubbock, Texas; 2004. p. 173–80.
[32] Huard W. Optimizing production and overall efficiency with Intelligent long stroke hydraulic pumping system. Proc. 56th Southwestern petroleum short course. Lubbock, Texas; 2009. p. 89–95.

APPENDICES

STATIC GAS PRESSURE GRADIENT CHART

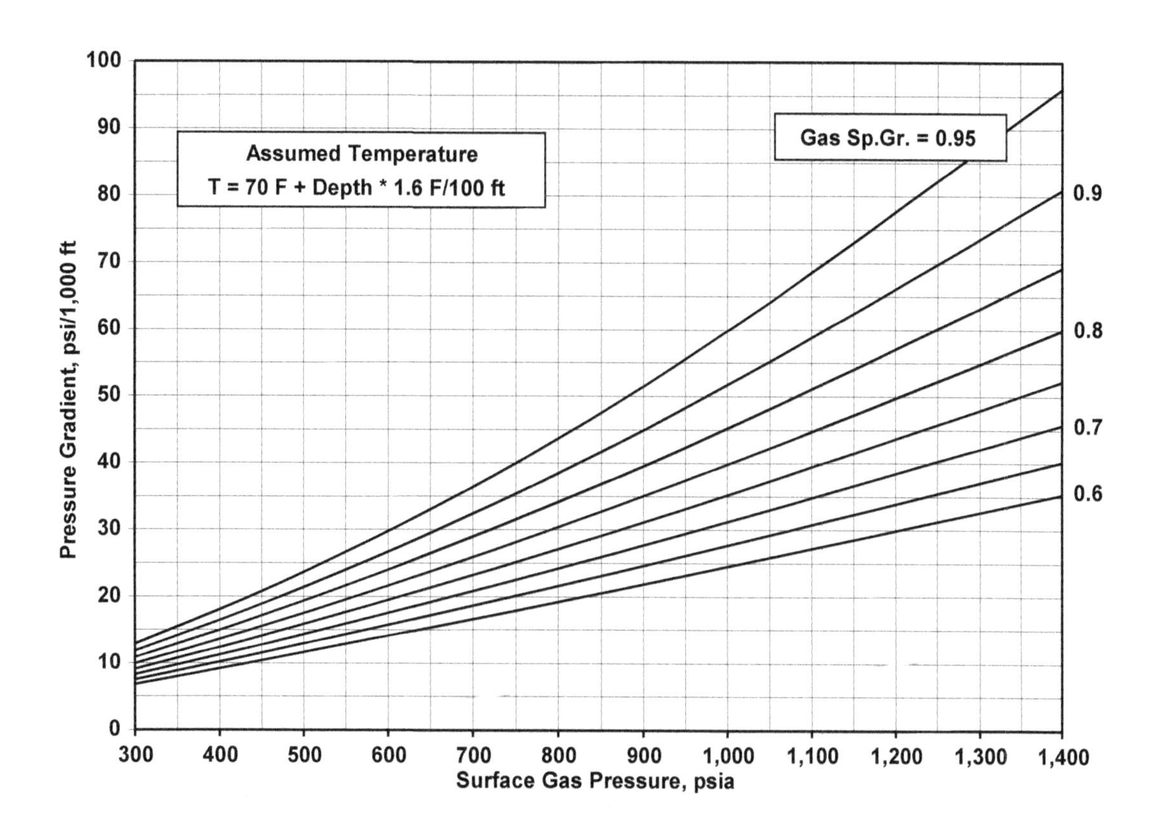

APPENDIX B

API GEOMETRY DIMENSIONS OF CONVENTIONAL PUMPING UNITS

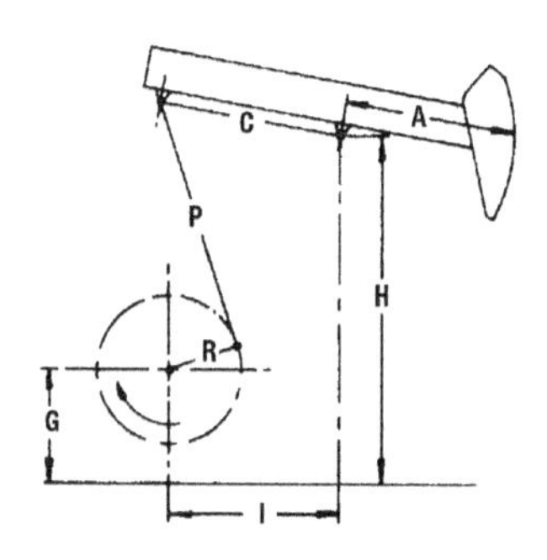

API Designation	A (in)	C (in)	I (in)	P (in)	H (in)	G (in)	R1 (in)	R2 (in)	R3 (in)	R4 (in)	SU (lbs)
C-1824D-365-192	210	120.03	120	172.50	286.00	111.00	53	46	39	32	−1,985
C-1280D-365-192	210	120.03	120	172.50	286.00	111.00	53	46	39	32	−1,985
C-912D-365-192	210	120.03	120	172.50	286.00	111.00	53	46	39	32	−1,985
C-912D-305-192	210	120.03	120	172.50	286.00	111.00	53	46	39	32	−1,985
C-912D-365-168	210	120.03	120	148.50	262.00	111.00	47	41	35		−1,500
C-912D-305-168	210	120.03	120	148.50	262.00	111.00	47	41	35		−1,500
C-640D-365-168	210	120.03	120	148.50	262.00	111.00	47	41	35		−1,500
C-640D-305-168	210	120.03	120	148.50	262.00	111.00	47	41	35		−1,500
C-456D-305-168	210	120.03	120	148.50	262.00	111.00	47	41	35		−1,500
C-912D-427-144	180	120.03	120	148.50	262.00	111.00	47	41	35		−650
C-912D-365-144	180	120.03	120	148.50	262.00	111.00	47	41	35		−650
C-640D-365-144	180	120.03	120	148.50	262.00	111.00	47	41	35		−650
C-640D-305-144	180	120.03	120	144.50	260.00	111.00	47	41	35		−520
C-456D-305-144	180	120.03	120	144.50	260.00	111.00	47	41	35		−520
C-640D-256-144	180	120.03	120	144.50	260.00	111.00	47	41	35		−400
C-456D-256-144	180	120.03	120	144.50	260.00	111.00	47	41	35		−400
C-320D-256-144	180	120.03	120	144.50	260.00	111.00	47	41	35		−400
C-640D-365-120	152	120.03	120	148.50	262.00	109.00	47	41	35		570
C-456D-365-120	152	120.03	120	148.50	262.00	109.00	47	41	35		570
C-640D-305-120	155	111.09	111	133.50	234.00	96.00	42	36	30		−120
C-456D-305-120	155	111.09	111	133.50	234.00	96.00	42	36	30		−120
C-456D-256-120	155	111.07	111	132.00	232.00	96.00	42	36	30		55
C-320D-256-120	155	111.07	111	132.00	232.00	96.00	42	36	30		55

Continued

—Cont'd

API Designation	A (in)	C (in)	I (in)	P (in)	H (in)	G (in)	R1 (in)	R2 (in)	R3 (in)	R4 (in)	SU (lbs)
C-456D-213-120	155	111.07	111	132.00	232.00	96.00	42	36	30		0
C-320D-213-120	155	111.07	111	132.00	232.00	96.00	42	36	30		0
C-228D-213-120	155	111.07	111	132.00	232.00	96.00	42	36	30		0
C-456D-256-100	129	111.07	111	132.00	232.00	96.00	42	36	30		550
C-320D-256-100	129	111.07	111	132.00	232.00	96.00	42	36	30		550
C-320D-305-100	129	111.07	111	132.00	232.00	96.00	42	36	30		550
C-228D-213-100	129	96.08	96	113.00	196.13	79.13	37	32	27		0
C-228D-173-100	129	96.05	96	114.00	196.13	79.13	37	32	27		0
C-160D-173-100	129	96.05	96	114.00	195.88	78.88	37	32	27		0
C-320D-246-86	111	111.04	111	133.00	232.00	96.00	42	36	30		800
C-228D-246-86	111	111.04	111	133.00	232.00	96.00	42	36	30		800
C-320D-213-86	111	96.05	96	114.00	196.13	79.13	37	32	27		450
C-228D-213-86	111	96.05	96	114.00	196.13	79.13	37	32	27		450
C-160D-173-86	111	96.05	96	114.00	195.88	78.88	37	32	27		450
C-114D-119-86	111	84.05	84	93.75	166.00	69.25	32	27	22		115
C-320D-246-74	96	96.05	96	114.00	196.13	79.13	37	32	27		800
C-228D-200-74	96	96.05	96	114.00	196.13	79.13	37	32	27		800
C-160D-200-74	96	96.05	96	114.00	195.88	78.88	37	32	27		800
C-228D-173-74	96	84.05	84	96.00	168.25	69.25	32	27	22		450
C-160D-173-74	96	84.05	84	96.00	168.25	69.25	32	27	22		450
C-160D-143-74	96	84.05	84	93.75	166.00	69.25	32	27	22		300
C-114D-143-74	96	84.05	84	93.75	166.00	69.25	32	27	22		300
C-160D-173-64	84	84.05	84	93.75	166.00	69.25	32	27	22		550
C-114D-173-64	84	84.05	84	93.75	166.00	69.25	32	27	22		550
C-160D-143-64	84	72.06	72	84.00	146.50	59.50	27	22	17		360
C-114D-143-64	84	72.06	72	84.00	146.50	59.50	27	22	17		360
C-80D-119-64	84	64.00	64	74.50	126.13	51.13	24	20	16		0
C-114D-173-54	72	72.06	72	84.00	146.50	59.50	27	22	17		500
C-114D-133-54	72	64.00	64	74.50	126.13	51.13	24	20	16		330
C-80D-133-54	72	64.00	64	74.50	126.13	51.13	24	20	16		330
C-80D-119-54	72	64.00	64	74.50	126.13	51.13	24	20	16		330
C-57D-76-54	72	64.00	64	74.50	115.13	47.13	24	20	16		330
C-80D-133-48	64	64.00	64	74.50	126.13	51.13	24	20	16		440
C-80D-109-48	64	56.05	56	65.63	115.13	46.13	21	16	11		320
C-57D-109-48	64	56.05	56	65.63	115.13	46.13	21	16	11		320
C-57D-95-48	64	56.05	56	65.63	115.13	46.13	21	16	11		320
C-40D-76-48	64	48.17	48	57.50	106.63	45.13	18	14	10		0
C-57D-89-42	56	48.17	48	57.50	106.63	45.13	18	14	10		150
C-57D-76-42	56	48.17	48	57.50	106.63	45.13	18	14	10		150
C-40D-89-42	56	48.17	48	57.50	106.63	45.13	18	14	10		150
C-40D-76-42	56	48.17	48	57.50	106.63	45.13	18	14	10		150
C-40D-89-36	48	48.17	48	57.50	106.30	45.13	18	14	10		275
C-25D-67-36	48	48.17	48	57.50	106.30	45.13	18	14	10		275
C-25D-56-36	48	48.17	48	57.50	106.30	45.13	18	14	10		275
C-25D-67-30	45	36.22	36	49.50	90.50	37.00	12	8			150
C-25D-53-30	45	36.22	36	49.50	90.50	37.00	12	8			150

API GEOMETRY DIMENSIONS OF AIR BALANCED PUMPING UNITS

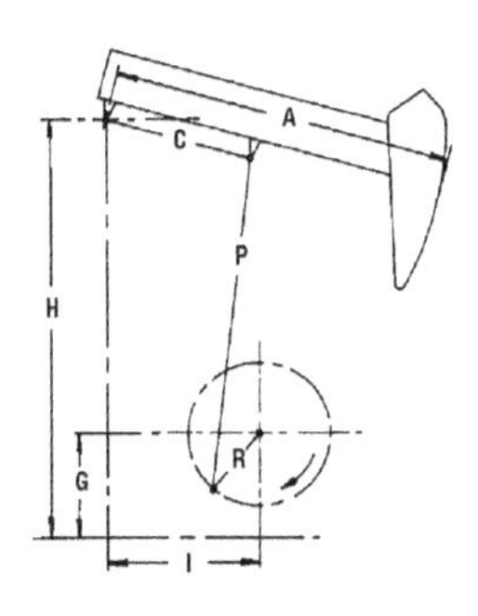

API Designation	A (in)	C (in)	I (in)	P (in)	H (in)	G (in)	R1 (in)	R2 (in)	R3 (in)
A-2560D-470-240	336	134.5	130.0	261.5	303.5	42	47.0	39.4	
A-1824D-470-240	336	134.5	130.0	267.5	303.5	36	47.0	39.4	
A-1280D-470-240	336	134.5	130.0	267.5	303.5	36	47.0	39.4	
A-912D-470-240	336	134.5	130.0	267.5	303.5	36	47.0	39.4	
A-1824D-427-216	308	121.5	114.5	246.0	282.0	36	41.8	36.6	32.0
A-1280D-427-216	308	121.5	114.5	246.0	282.0	36	41.8	36.6	32.0
A-912D-427-216	308	121.5	117.5	252.0	282.0	30	41.8	36.6	32.0
A-1824D-427-192	276	121.5	114.5	216.0	252.0	36	41.8	36.6	32.0
A-1280D-427-192	276	121.5	114.5	216.0	252.0	36	41.8	36.6	32.0
A-912D-427-192	276	121.5	117.5	222.0	252.0	30	41.8	36.6	32.0
A-1280D-305-168	231	88.0	84.0	208.0	244.0	36	31.3	26.2	22.0
A-912D-305-168	231	88.0	84.0	214.0	244.0	30	31.3	26.2	22.0
A-640D-305-168	231	88.0	85.0	216.0	244.0	28	31.3	26.2	22.0
A-912D-427-144	200	88.0	84.0	184.0	214.0	30	31.3	26.2	22.0
A-640D-427-144	200	88.0	85.0	186.0	214.0	28	31.3	26.2	22.0
A-640D-305-144	208	77.0	74.5	186.0	214.0	28	26.2	22.0	18.9
A-456D-305-144	208	77.0	74.5	186.0	214.0	28	26.2	22.0	18.9
A-640D-365-120	175	77.0	74.5	159.0	187.0	28	26.2	22.0	18.9

Continued

—Cont'd

API Designation	A (in)	C (in)	I (in)	P (in)	H (in)	G (in)	R1 (in)	R2 (in)	R3 (in)
A-456D-365-120	175	77.0	74.5	159.0	187.0	28	26.2	22.0	18.9
A-456D-256-120	184	69.0	66.0	159.0	187.0	28	22.0	18.9	16.3
A-320D-256-120	184	70.0	68.0	159.0	187.0	28	22.4	19.4	16.8
A-320D-305-100	155	70.0	68.0	132.0	160.0	28	22.4	19.4	16.8
A-228D-246-86	151	56.0	54.0	122.0	149.0	27	18.3	15.8	13.8
A-2280-246-86	131	56.0	54.0	122.0	149.0	27	18.3	15.8	13.8
A-160D-200-74	120	50.0	48.0	114.0	141.0	27	15.3	13.3	11.3
A-114D-173-64	115	48.0	46.5	114.0	132.0	18	13.3	11.3	

API GEOMETRY DIMENSIONS OF MARK II PUMPING UNITS

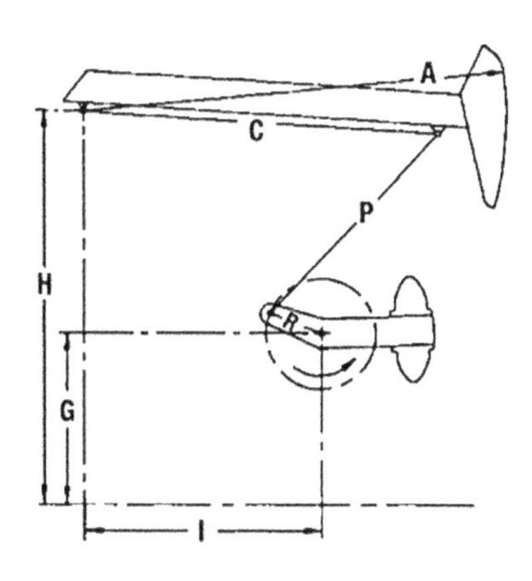

API Designation	A (in)	C (in)	I (in)	P (in)	H (in)	G (in)	R1 (in)	R2 (in)	R3 (in)	τ (deg)	SU (lbs)
M-1824D-427-216	384	306	228.0	234.38	346.00	132.00	80.1	71.1	62.1	22.0	−7,450
M-1280D-427-216	384	306	228.0	234.38	346.00	132.00	80.1	71.1	62.1	22.0	−7,450
M-912D-305-216	384	306	228.0	234.38	346.00	132.00	80.1	71.1	62.1	22.0	−7,450
M-912DS-365-216	384	306	228.0	234.38	346.00	132.00	80.1	71.1	62.1	22.0	−7,450
M-1280D-427-192	384	306	228.0	228.06	346.00	132.00	71.7	62.7	53.7	19.5	−7,160
M-912DS-427-192	384	306	228.0	228.06	346.00	132.00	71.7	62.7	53.7	19.5	−7,160
M-912D-305-192	384	306	228.0	228.06	346.00	132.00	71.7	62.7	53.7	19.5	−7,160
M-640D-305-192	384	306	228.0	228.06	346.00	132.00	71.7	62.7	53.7	19.5	−7,160
M-456D-305-192	384	306	228.0	228.06	346.00	132.00	71.7	62.7	53.7	19.5	−7,160
M-912DS-427-168	334	270	203.0	196.50	295.13	112.13	63.6	56.6	49.6	19.0	−6,820
M-912D-365-168	334	270	202.6	193.50	295.13	112.13	63.6	56.6	49.6	19.0	−5,385
M-912D-305-168	334	270	202.6	193.50	295.13	112.13	63.6	56.6	49.6	19.0	−4,860
M-640D-365-168	334	270	202.6	193.50	295.13	112.13	63.6	56.6	49.6	19.0	−4,860
M-640D-305-168	334	270	202.6	193.50	295.13	112.13	63.6	56.6	49.6	19.0	−4,860
M-456D-305-168	334	270	202.6	193.50	295.13	112.13	63.6	56.6	49.6	19.0	−4,860
M-912D-365-144	312	258	186.0	182.38	271.13	112.13	53.8	47.8	41.8	23.0	−4,860

Continued

—Cont'd

API Designation	A (in)	C (in)	I (in)	P (in)	H (in)	G (in)	R1 (in)	R2 (in)	R3 (in)	τ (deg)	SU (lbs)
M-640D-365-144	312	258	186.0	182.38	271.13	112.13	53.8	47.8	41.8	23.0	−4,680
M-456D-365-144	312	258	186.0	182.38	271.13	112.13	53.8	47.8	41.8	23.0	−4,680
M-912D-305-144	312	258	186.0	182.38	271.13	112.13	53.8	47.8	41.8	23.0	−4,300
M-640D-305-144	312	258	186.0	182.38	271.13	112.13	53.8	47.8	41.8	23.0	−4,300
M-456D-305-144	312	258	186.0	182.38	271.13	112.13	53.8	47.8	41.8	23.0	−4,300
M-640D-256-144	312	258	186.0	182.38	271.13	112.13	53.8	47.8	41.8	23.0	−4,010
M-456D-256-144	312	258	186.0	182.38	271.13	112.13	53.8	47.8	41.8	23.0	−4,010
M-320D-256-144	312	258	186.0	182.38	276.88	117.88	53.8	47.8	41.8	23.0	−4,010
M-456D-365-120	312	258	186.0	173.75	271.88	112.13	45.1	39.1	33.1	24.0	−4,510
M-640D-305-120	312	258	186.0	173.75	271.13	112.13	45.1	39.1	33.1	24.0	−4,130
M-456D-305-120	312	258	186.0	173.75	271.13	112.13	45.1	39.1	33.1	24.0	−4,130
M-320D-305-120	312	258	186.0	173.75	271.13	118.13	45.1	39.1	33.1	24.0	−4,130
M-456D-256-120	312	258	186.0	173.75	271.13	112.13	45.1	39.1	33.1	24.0	−3,840
M-320D-256-120	312	258	186.0	173.75	276.88	117.88	45.1	39.1	33.1	24.0	−3,620
M-228D-256-120	312	258	186.0	173.75	276.88	117.88	45.1	39.1	33.1	24.0	−3,435
M-320D-213-120	312	258	186.0	173.75	276.88	117.88	45.1	39.1	33.1	24.0	−3,560
M-228D-213-120	312	258	186.0	173.75	276.88	117.88	45.1	39.1	33.1	24.0	−3,235
M-320D-305-100	312	258	186.0	173.75	277.13	118.13	37.6	31.6	25.6	24.0	−3,700
M-320D-256-100	312	258	186.0	173.75	276.88	117.88	37.6	31.6	25.6	24.0	−3,470
M-228D-256-100	312	258	186.0	173.75	276.88	117.88	37.6	31.6	25.6	24.0	−3,285
M-228D-173-100	312	258	186.0	173.75	276.88	117.88	37.6	31.6	25.6	24.0	−3,175
M-228D-246-86	222	186	124.0	135.75	212.50	97.13	31.5	26.5	21.5	24.5	−2,140
M-228D-213-86	222	186	124.0	135.75	209.50	94.13	31.5	26.5	21.5	24.5	−2,040
M-160D-213-86	222	186	124.0	135.75	209.50	94.13	31.5	26.5	21.5	24.5	−2,040
M-160D-173-86	222	186	124.0	135.75	209.50	94.13	31.5	26.5	21.5	24.5	−1,930
M-114D-143-86	189	162	111.0	112.19	163.63	68.13	32.3	27.8	23.3	27.0	−1,535
M-228D-200-74	222	186	124.0	130.50	209.50	94.13	27.3	22.3	17.3	24.5	−1,960
M-160D-200-74	222	186	124.0	130.50	209.50	94.13	27.3	22.3	17.3	24.5	−1,890
M-228D-173-74	222	186	124.0	130.50	163.63	68.13	27.3	22.3	17.3	24.5	−1,860
M-160D-173-74	222	186	124.0	130.50	163.63	68.13	27.3	22.3	17.3	24.5	−1,860
M-114D-173-74	222	186	124.0	130.50	163.63	68.13	27.3	22.3	17.3	24.5	−1,820
M-114D-143-74	189	162	111.0	107.94	163.63	68.13	27.9	23.4	18.9	27.0	−1,440
M-114D-173-64	189	162	111.0	107.94	163.63	68.13	24.2	19.7	15.2	28.0	−1,420
M-114D-143-64	189	162	111.0	107.94	163.63	68.13	24.2	19.7	15.2	28.0	−1,420

API GEOMETRY DIMENSIONS OF REVERSE MARK PUMPING UNITS

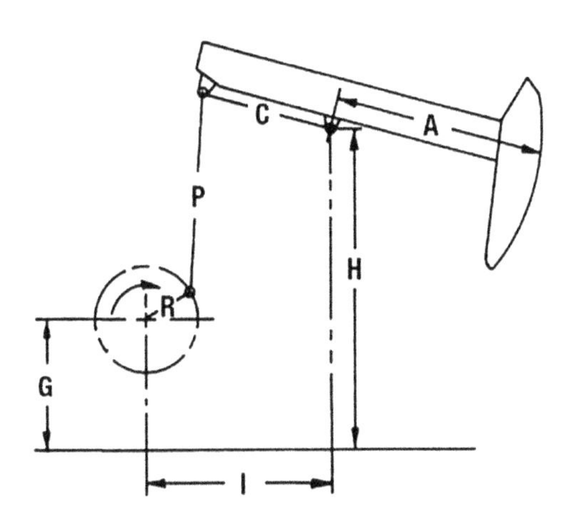

API Designation	A (in)	C (in)	I (in)	P (in)	H (in)	G (in)	R1 (in)	R2 (in)	R3 (in)	τ (deg)	SU (lbs)
RM-1824D-427-192	229	118	171	163	274.00	111	44.7	38	31	−9.0	−1,650
RM-1824D-365-192	229	118	171	163	274.00	111	44.7	38	31	−9.0	−1,650
RM-1824D-427-168	200	118	171	163	274.00	111	44.7	38	31	−9.0	−515
RM-1824D-365-168	200	118	171	163	274.00	111	44.7	38	31	−9.0	−515
RM-1280D-427-192	229	118	171	163	274.00	111	44.7	38	31	−9.0	−1,650
RM-1280D-365-192	229	118	171	163	274.00	111	44.7	38	31	−9.0	−1,650
RM-1280D-427-168	200	118	171	163	274.00	111	44.7	38	31	−9.0	−515
RM-1280D-365-168	200	118	171	163	274.00	111	44.7	38	31	−9.0	−515
RM-912D-427-192	229	118	171	163	274.00	111	44.7	38	31	−12.0	−1,650
RM-912D-365-192	229	118	171	163	274.00	111	44.7	38	31	−12.0	−1,650
RM-912D-305-192	229	118	171	163	274.00	111	44.7	38	31	−9.0	−1,600
RM-912D-427-168	200	118	171	163	274.00	111	44.7	38	31	−12.0	−515
RM-912D-365-168	200	118	171	163	274.00	111	44.7	38	31	−9.0	−515
RM-912D-305-168	200	118	171	163	274.00	111	44.7	38	31	−9.0	−1,070
RM-912D-427-144	172	118	171	163	274.00	111	44.7	38	31	−9.0	630
RM-912D-365-144	172	118	171	163	274.00	111	44.7	38	31	−9.0	50

Continued

—Cont'd

API Designation	A (in)	C (in)	I (in)	P (in)	H (in)	G (in)	R1 (in)	R2 (in)	R3 (in)	τ (deg)	SU (lbs)
RM-912D-305-144	172	118	171	163	274.00	111	44.7	38	31	−9.0	50
RM-640D-305-192	229	118	171	163	274.00	111	44.7	38	31	−12.0	−1,600
RM-640D-365-168	200	118	171	163	274.00	111	44.7	38	31	−12.0	−515
RM-640D-305-168	200	118	171	163	274.00	111	44.7	38	31	−9.0	−1,070
RM-640D-427-144	172	118	171	163	274.00	111	44.7	38	31	−12.0	630
RM-640D-365-144	172	118	171	163	274.00	111	44.7	38	31	−9.0	50
RM-640D-305-144	172	118	171	163	274.00	111	44.7	38	31	−9.0	50
RM-640D-256-144	172	118	171	163	274.00	111	44.7	38	31	−9.0	−140
RM-640D-305-120	145	103	149	142	237.88	96	39.0	33	27	−12.5	335
RM-640D-256-120	145	103	149	142	237.88	96	39.0	33	27	−12.5	235
RM-456D-305-168	202	103	149	142	237.88	96	39.0	33	27	−14.0	−1,580
RM-456D-365-144	173	103	149	142	237.88	96	39.0	33	27	−14.0	−765
RM-456D-305-144	173	103	149	142	237.88	96	39.0	33	27	−12.5	−700
RM-456D-256-144	173	103	149	142	237.88	96	39.0	33	27	−12.5	−870
RM-456D-365-120	145	103	149	142	237.88	96	39.0	33	27	−12.5	680
RM-456D-305-120	145	103	149	142	237.88	96	39.0	33	27	−12.5	335
RM-456D-256-120	145	103	149	142	237.88	96	39.0	33	27	−12.5	235
RM-320D-256-144	173	103	149	142	237.88	96	39.0	33	27	−14.0	−870
RM-320D-305-120	145	103	149	142	237.88	96	39.0	33	27	−14.0	335
RM-320D-256-120	145	103	149	142	237.88	96	39.0	33	27	−12.5	235
RM-320D-213-120	145	103	149	142	237.88	96	39.0	33	27	−12.5	40
RM-320D-305-100	121	103	149	142	237.88	96	39.0	33	27	−12.5	600
RM-320D-256-100	121	103	149	142	237.88	96	39.0	33	27	−12.5	600
RM-320D-246-86	104	92	134	127	206.00	79	34.3	30	25	−14.0	720
RM-228D-213-120	145	103	149	142	237.88	96	39.0	33	27	−14.0	40
RM-228D-256-100	121	103	149	142	237.88	96	39.0	33	27	−14.0	600
RM-228D-213-100	120	92	134	127	206.00	79	34.3	30	25	−14.0	90
RM-228D-173-100	120	92	134	127	206.00	79	34.3	30	25	−14.0	90
RM-228D-246-86	104	92	134	127	206.00	79	34.3	30	25	−14.0	720
RM-228D-213-86	104	92	134	127	206.00	79	34.3	30	25	−14.0	340
RM-228D-200-74	89	92	134	127	206.00	79	34.3	30	25	−14.0	680
RM-228D-173-74	89	92	134	127	206.00	79	34.3	30	25	−14.0	680

Index

Note: Page numbers followed by "f" and "t" indicate figures and tables respectively.

B

C

E

F

G

H

I

N

O

P

Q

R

S

Printed in the United States
By Bookmasters